W9-AFK-167

AN INTRODUCTION TO

COMPUTER SCIENCE

An Algorithmic Approach

McGRAW-HILL SERIES IN ARTIFICIAL INTELLIGENCE

Consulting Editor
Edward A. Feigenbaum, *Stanford University*

Allen **ANATOMY OF LISP**
Feigenbaum and Feldman **COMPUTERS AND THOUGHT**
Nilsson **PROBLEM-SOLVING METHODS IN ARTIFICIAL INTELLIGENCE**
Winston **THE PSYCHOLOGY OF COMPUTER VISION**

AN INTRODUCTION TO

COMPUTER SCIENCE

An Algorithmic Approach

Jean-Paul Tremblay

Richard B. Bunt

Department of Computational Science
University of Saskatchewan, Saskatoon
Canada

McGraw-Hill Book Company

New York St. Louis San Francisco Auckland Bogotá Düsseldorf
Johannesburg London Madrid Mexico Montreal New Delhi
Panama Paris São Paulo Singapore Sydney Tokyo Toronto

AN INTRODUCTION TO COMPUTER SCIENCE: An Algorithmic Approach

Copyright ©1979 by McGraw-Hill, Inc. All rights reserved. Printed in the United States of America. No part of this publication may be reproduced, stored in a retrieval system, or transmitted, in any form or by any means, electronic, mechanical, photocopying, recording, or otherwise, without the prior written permission of the publisher.

123456789 DODO 7832109

This book was set in Electra by Monotype Composition Company, Inc.
The editors were Charles E. Stewart, Michael Gardner, and James S. Amar;
the designer was Charles A. Carson;
the production supervisor was Dominick Petrellese.
The drawings were done by Fine Line Illustrations, Inc.
R. R. Donnelley & Sons Company was printer and binder.

Library of Congress Cataloging in Publication Data

Tremblay, Jean-Paul, date
    An Introduction to computer science.

    (McGraw-Hill computer science series)
    Includes index.
    1.   Electronic digital computers—Programming.
2.   Algorithms.   3.   Data structures (Computer science)
I.   Bunt, Richard B., date   joint author.
II.   Title.
QA76.6.T73   519.7   78-21653
ISBN 0-07-065163-9

To Deanna Tremblay
To Gail and Andrea Bunt

Contents

Preface

The first course in a computer science curriculum is certainly one of the most important. For most students this constitutes their initial exposure to fundamental notions such as the algorithm, and to the description of solutions in a manner sufficiently precise for computer interpretation. It is important that these notions be properly taught for, as the ancient Roman poet Horace observed, "A new cask will long preserve the tincture of the liquor with which it was first impregnated."

In the mid 1960s, students in the first course became well-versed in the syntax of the instructor's favorite programming language (usually FORTRAN), and probably were given a limited view of the organization of a computer. It was assumed, however, that the students brought with them an ability to solve problems. Assistance in organizing their solutions was provided in the form of the flowchart. The emphasis was largely on the solution of numerical problems to which the FORTRAN language was well-suited. Since the class was populated (and taught) largely by engineers, usually late in their academic programs, it seemed to fulfill their short-term requirements and there was little motivation for change.

Today the needs of such a class have changed. The emergence of computer science as an undergraduate discipline has created a need for a different orientation. The class is now generally offered in the first year of study, meaning that today's students seldom have the mathematical or problem-solving maturity of their predecessors. The increased use of computers in nonnumeric applications such as text processing requires the presentation of new concepts in the handling of nonnumeric data. Many curricula now require that students take courses in areas such as data structures and discrete mathematical structures early in their programs [1]. It is important that the first course provide proper motivation for these areas by providing some knowledge of nonnumeric computation and applications.

We feel that considerations such as these dictate strong requirements for a textbook for the first course. It must do substantially more than simply train the student in rules of syntax for a particular programming language. Our focus is on problem-solving—in a rigorous, systematic fashion—through the use of algorithms. We concentrate on the design of algorithms for the solution of nonnumeric as well as numeric problems. Considerable attention is devoted to the data structures appropriate to the application. In order that the student not become mired in programming language details, we introduce a carefully considered algorithmic language that is well-suited to the preparation of computer solutions. The constructs of the algorithmic language are not introduced all at once, but rather are presented throughout the book to meet new requirements as they arise. Detailed examples and applications serve the dual purpose of reinforcing in the student the ideas presented, and providing motivation for upcoming concepts.

A departure from many contemporary texts is the heavily decreased emphasis on flowcharts. We find flowcharts to be awkward and clumsy, and incomplete in the representation of several key concepts. It has been our experience that most of our students created their flowcharts posthumously to fulfill a lab requirement rather than as an aid to solving the problem. In this book we use flowcharts only to illustrate the action of important constructs in the algorithmic language, such as the "looping" construct and the "selection of alternatives" construct (with their variations), and not for complete applications.

Clearly the use of a programming language is an important part of the first course. It must not, however, obscure other important concepts. Our book is based on the premise that a student should learn to program *into* a programming language rather than *in* one. The translation of the algorithms developed in this book into most programming languages ought to be a straightforward exercise. The final choice of language to be used in any class is left to the discretion of the instructor. We have prepared a series of supplementary integrated programming guides (including PL/I and FORTRAN) to provide the needed support.

Much has been said and written in the past few years about the merits of an approach to programming loosely termed "structured programming." Studies of the programming task itself have shown that adherence to certain basic principles can result in the production of better quality programs. Our approach is based on many of these principles, and we have designed our algorithmic language to be well-structured. Chapter 7, on programming style, examines the process of programming itself in more depth.

Despite the fact that programming per se is not considered in this book, the student is given ample opportunity to hone his/her problem-solving skills. Each of the chapters, with the exception of the first, contains carefully worked-out examples in which the material introduced in that chapter is applied to the solution of representative problems. The choice of applications reflects our concern with the nonnumeric aspects of computing. The same concern is carried over into the exercises as well. Exercises are found at the end of most sections and at the end of most chapters.

SUMMARY BY CHAPTERS

The book begins with a brief historical overview of the development of computers and programming. Although this material is not required by subsequent chapters, we have found that it provides the student with an important perspective on the field.

Chapter 2 can be viewed as the real introduction to the material of the book. It lays an important foundation by presenting a number of basic computing concepts as well as the first constructs of the algorithmic language. Some simple applications are described.

The notion of "flow of control" is introduced in Chapter 3, along with two fundamental control structures: the selection from alternative actions and the loop. Algorithmic notation for these constructs is presented and important variations are discussed. Solutions to several fairly elaborate applications are developed.

The concept of the array is the topic of Chapter 4. Processing of single-dimensional arrays, or vectors, is discussed first. Following this discussion the chapter moves to a consideration of arrays of higher dimension. Some typical applications of vectors and arrays are discussed. Among these are the important applications of searching and sorting, which are discussed for the first time.

String processing is the topic of Chapter 5. The representation of strings in a computer and some basic operations on strings are described. A number of simple applications involving string processing are developed. More advanced topics are deferred to Chapter 9.

Chapter 6 deals with functions and procedures. Topics discussed include the correspondence of arguments and parameters, and the way in which functions and procedures are invoked and values are returned. Three applications involving the use of functions and procedures are considered.

Programming style is the topic of Chapter 7. We feel this to be an important topic, but one which many computer science textbooks ignore. The programming process involves more than the simple recall of language rules. Topics discussed include considerations of program quality, defensive programming, managing complexity, writing readable programs, programming by abstraction and refinement, and programming as a human activity. The concept of "top-down" development of a solution is illustrated with the aid of a major example.

Chapter 8 deals with the subject of numerical computation. The chapter begins with a discussion of the nature of errors in numerical computations and their causes. From there the discussion moves to a treatment of a number of important numerical applications. These include root finding, numerical integration, the solution of simultaneous linear equations, and curve fitting. For some of the material in this chapter, familiarity with elementary calculus would be an asset.

Chapter 9 returns to the topic of string processing, with the presentation dealing with more advanced applications such as KWIC indexing and text editing.

Chapter 10 offers an introduction to the study of linear data structures. Simple structures such as linear lists, stacks, and queues are discussed, as are basic operations on these structures. A number of important applications are described. These include the implementation of recursion, the translation of expressions, and the symbolic manipulation of polynomials. Also discussed in this chapter are hash-table techniques.

The book concludes with a look at the most important of the nonlinear data structures—the tree. Topics include the binary tree, operations on binary trees, and storage representation of trees. Some mention is made of general trees, although the presentation is slanted toward binary trees. Techniques for converting general trees to binary trees are discussed. Finally, the application of trees to problems such as the symbolic manipulation of expressions, searching, and sorting is discussed.

HOW TO USE THE BOOK

This book is intended for a two-semester course with an organization similar to that advocated for courses CS1 and CS2 in the revised curriculum proposals of the Association for Computing Machinery [2]. In particular, Chapters 1 through 8 align with CS1; and Chapters 9 through 11, with CS2. The material is sufficiently modularizable, however, that it can also be used for a single-semester course. This can be accomplished in a number of ways. Certainly any course should include the material in Chapters 2 through 6. Chapters 1, 7, 8, and 9 are self-contained, and can be included or not at the instructor's discretion. The material in Chapter 11 draws on material from Chapter 10; however, an instructor can elect to go only as far as he/she wishes. For example, an instructor wishing to introduce only linear data structures need go only as far as Chapter 10.

Material can be tailored to a specific course or set of interests in yet another way. We have made an effort in our choice of applications and exercises to cover a wide range of interests, including scientific computing, business data processing, engineering applications, societal issues, and topics of general interest. By selecting appropriate applications and problems, it is possible to realize a number of different course orientations. We have also made an effort to keep to a minimum the mathematics required to understand the material in the book. Our own course attracts many students from fields of study other than the physical sciences. We do expect, however, that most students will have had the equivalent of high school mathematics.

We advocate a laboratory environment of some sort for the parallel presentation of issues relating to the actual programming component of the course. The instructor may wish, on occasion, to deal with particularly difficult notions in class, but too much of this detracts from continuity of presentation. Students should not view the class as a class in programming. The laboratory also provides the student with the opportunity to work on programs, with assistance available readily when needed.

For convenience of presentation we have made several assumptions as to the nature of available computing facilities. We have assumed throughout a card reader/line printer environment; we recognize that this may not be the case for many students. The dependency on such matters is minor. Should an alternative environment exist, a simple comment from the instructor should suffice to overcome any possible problems of comprehension.

ACKNOWLEDGMENTS

We owe a debt of thanks to the many people who assisted us in the preparation of this manuscript. Grant Cheston devoted a great deal of time to the reading of our notes and contributed many valuable suggestions. Lorna Stewart assisted in the production of Chapters 1 and 8. The comments and suggestions of Ken Kozar played a large part in the development of Chapter 7. Ram Manohar contributed to both the material and the form of Chapter 8. Other contributions were made by Mike Williams and C. C. Gottlieb who supplied important information for Chapter 1, Tom Austin, who provided valuable criticism of early drafts of Chapters 2 and 3, and Doug Bulbeck, who assisted in Section 6-5.2. We owe a large debt of thanks to our proofreaders, in particular Judy Richardson and Lyle Opseth, who read the entire final manuscript and helped to pick out many errors. Some proofreading was also contributed by Dave Hrenewich, Murray Mazer, and Gail Bunt. Murray Mazer and Lyle Opseth solved all the exercises in the book en route to the production of an instructor's manual. The excellent typing of Janet Morck, with some assistance from Arlene Looman and Gail Walker, under the combined pressures of deadlines and illegible scrawls is most gratefully acknowledged. We are also grateful for the support and comments of our colleagues and students in the Department of Computational Science at the University of Saskatchewan, who have class-tested preliminary versions of the book over the past 3 years. Their experiences, as well as our own, have done much to shape its final form. Finally, we acknowledge the financial assistance provided by the University of Saskatchewan, without which the production of the book would have been most difficult.

Jean-Paul Tremblay
Richard B. Bunt

REFERENCES

1. **Sloan,** M. E., "Survey of Electrical Engineering and Computer Science Departments in the U.S.," *Computer*, Vol. 8, No. 12, Dec. 1975, pp. 35–42.
2. **Austing,** R. H., Barnes, B. H., Bonnette, D. T., Engel, G. L., and Stokes, G., "Curriculum Recommendations for the Undergraduate Program in Computer Science: A Working Report of the ACM Curriculum Committee on Computer Science," *SIGCSE Bulletin*, Vol. 9, No. 2, June 1977, pp. 1–16.

1

A Computer History

Many aspects of modern society that we have come to accept as commonplace would not be possible if there were no computers. Computers today are used extensively in many areas of business, industry, science, and education. What is a computer? To many, the computer epitomizes future shock. In films and on television, we see large, complex machines, flashing lights, and spinning tape reels. Too often, the computer has a personality: cold and impersonal, sinister and scheming. The operator of the computer is a brilliant, usually eccentric, technician, who spends night and day with his/her computer, and as a result teeters on the brink of madness. The entire operation is cloaked in mystery as the computer, with or without the aid of the operator, plots to control mankind.

This scenario is very theatrical and makes entertaining fiction. In actual fact, the computer was not the product of a mad scientist's warped ambition. It has evolved in parallel with mankind's growing need for fast, accurate calculations, and can trace its ancestry back more than 3,000 years. In this chapter we will look at some of the branches of the computer's family tree. We will consider the role played by some important people along the way, people whose contributions were often not fully appreciated by their contemporaries. In addition to the history of the computers themselves, we will trace some of the efforts made in systems for easier programming of the computers. This will lead us ultimately to the point where we can begin learning to use a computer ourselves.

1-1 EARLY COMPUTATION

The human need for computation dates back thousands of years. We have all heard stories of primitive peoples counting their sheep by moving sticks or stones. Our base 10 number system undoubtedly grew from the use of 10 fingers as counting objects.

Early computational abilities were surprisingly well developed. Clay tablets containing mathematical calculations have been unearthed by archeologists in the Middle East. Tablets containing multiplication tables and tables of reciprocals found near Babylon are believed to have been written about 1700 B.C. The Babylonians worked in a sexagesimal (that is, base 60) number system, from which came our present time units of hours, minutes, and seconds. There is evidence that the Babylonians solved many types of algebraic equations. Formulas were represented by step-by-step lists of rules for evaluation along with actual numerical examples. The rules given were sufficiently general to allow for substitution of different numbers, thus the solution of the equation with different arguments. In this way the rules seem to resemble what we will be calling an algorithm. As evidence of their computational sophistication, it is believed that the Babylonians were able to predict eclipses accurately as early as 500 B.C.

From the earliest times, man has recognized his limitations with regard to mental calculations and has devised a seemingly endless string of aids, ranging from the very simple to the grandiose. For many hundreds of years visitors to the Salisbury Plain in southern England have been fascinated by a mysterious cluster of giant stones known as Stonehenge. The subject of story, legend, and speculation as to its origins and purpose since as early as the sixth century, Stronehenge has been viewed variously as a marketplace, a temple, and even a site of human sacrifice.

In 1965, in a book entitled *Stonehenge Decoded*, Gerald S. Hawkins presented the results of a careful consideration of the purpose of Stonehenge. His conclusion was that Stonehenge was built sometime in the period 1900 to 1600 B.C. and served as a combination observatory/computer. In his research (conducted, in fact, with the aid of a computer), Hawkins discovered an astonishing number of alignments of various combinations of the stones with important positions of the sun and moon (some are shown in Fig. 1-1), far too many such alignments, in fact, to be coincidental. He theorized that the operators of the Stonehenge computer used it not only to chart the coming and passing of the seasons, but also as an extremely accurate predictor of eclipses. The operation of the Stonehenge computer, although quite simple, reveals an elaborate and careful design.

The very site of Stonehenge is particularly well chosen. There is only one latitude in the northern hemisphere where, at their extreme positions, the sun and moon azimuths are 90 degrees apart, and Stonehenge is within a few miles of it. If the site were much farther south or north, the important figure joining four key stones (see Fig. 1-1) would be a skewed parallelogram instead of a rectangle, considerably complicating the geometry. At Stonehenge, this figure differs from a perfect rectangle by only $\frac{1}{5}$ of a degree. Was this accurate placement of Stonehenge accident or design? It seems likely that if the builders of Stonehenge were sophisticated enough to construct a "computer," they could well have chosen the site for its astronomic geometry.

Structures appearing to perform similar functions to Stonehenge have

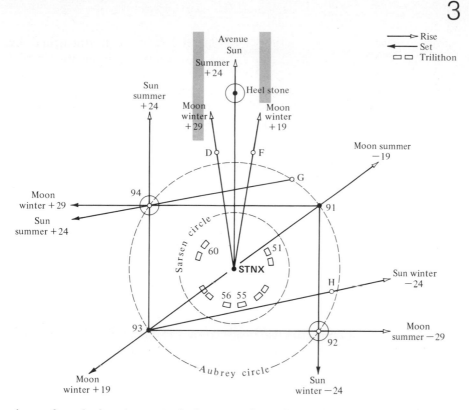

FIGURE 1-1
Sample of Stonehenge
alignments. *(From G. S.
Hawkins, "Stonehenge
Decoded," Nature, Vol.
200, October 1963.
Reproduced by
permission of the
publisher.)*

been found elsewhere, including North and South America. A series of "medicine wheels" in the western regions of the United States and Canada, some believed to be as old as Stonehenge, show remarkably similar characteristics.

Unfortunately, some of the skills displayed in the design of these structures disappeared through the ages, or we might have had the advantages of computer technology much earlier. The need for computation has remained nonetheless, and human beings themselves seem ill equipped for the precision and tedium required. In the next section we will trace the development of tools to assist us in the performance of computation, culminating in the twentieth century with the electronic digital computer.

1-2 THE DEVELOPMENT OF AUTOMATIC CALCULATING DEVICES

One of the earliest mechanical computational aids was the abacus, early versions of which were used in the Middle East as early as 2500 B.C. The familiar Chinese abacus (dating from approximately 1200 A.D.) is composed of a frame and a number of wires. Along each wire slide seven beads, two above a crossbar, and five below. The wires correspond to positions of digits in a decimal number—units, tens, hundreds, and so on—and the beads

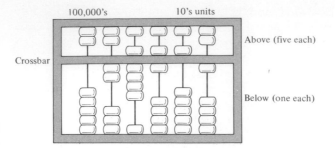

100,000's   10's units

Above (five each)

Crossbar

Below (one each)

FIGURE 1-2
The abacus.

represent digits—beads above the crossbar represent 5, and those below, 1. Numbers are represented by the beads close to the crossbar. Figure 1-2 shows the number 29,651 on an abacus capable of representing any number up to 999,999. (The standard abacus has several variants. The Japanese abacus, for example, has a slightly different arrangement, with four beads below the bar and one above.)

Addition of two numbers on the abacus can be performed by representing the first number, and then, without resetting it, the second. On any wire showing 10 or more, the two beads above the crossbar are moved back, and an extra 1 (the carry) is added to the wire on the left. This process can be easily generalized to addition and subtraction of more than two numbers. The abacus is still popular in many circles. In fact, a skilled abacus operator can add columns faster than many operators of electronic calculators.

Napier's bones were developed as an aid to multiplication by a Scottish nobleman named John Napier, who is also noteworthy as the originator of logarithms. Although devices similar to the bones had been used since late in the sixteenth century, they first appeared in print in 1614. A complete set of Napier's bones consists of nine rods, one for each of the digits 1

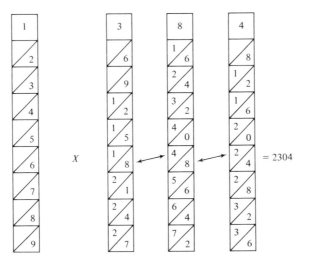

FIGURE 1-3
Napier's bones.

through 9. A rod is essentially one column of a multiplication table. Four
of the rods of a set are shown in Fig. 1-3, arranged for the multiplication
of 384 by 6. To obtain the product, the digits in each diagonal are added,
right to left, taking care to handle the carries.

In 1633, an English clergyman, William Oughtred, invented a computing device based on Napier's logarithms, which he called *circles of
proportion*. This later became the familiar slide rule, shown in Fig. 1-4. The
markings on the slide rule represent logarithms of numbers; consequently,
products and quotients are obtained by adding and subtracting lengths. This
is done by sliding a movable center piece (the cursor) back and forth along
the rule. Other scales permit calculations involving exponents, trigonometric
functions, and other mathematical functions. Fast, portable, and inexpen-
sive, the slide rule remained popular among scientists and engineers until
only very recently, when its place was usurped by the electronic pocket
calculator.

In 1642, at the age of 19, the French philosopher and mathematician
Blaise Pascal developed a rotating wheel calculator, the predecessor of the
later popular desk calculator. Built largely to assist his father, who was a tax
collector in the town of Rouen, Pascal's calculator (shown in Fig. 1-5) had
one wheel corresponding to each power of 10; each wheel had 10 positions,
one for each of the digits 0 through 9. The operation was similar to that of
an automobile odometer. Although Pascal's calculator could only add and
subtract, it could be used indirectly for multiplication (by successive
additions) and division (successive subtractions) as well. Pascal had hoped
to market his calculator, but although he made over 50 different versions,
none of them worked reliably, and in the end, it brought him little money.

Charles Babbage (1792–1871), a British mathematician and engineer,
is considered by many to be the real father of today's computer. Only
recently has the import of much of his work been fully appreciated. Babbage
was disturbed by the errors in mathematical tables of his day, and by 1822,
had built a working model of a machine to calculate tables—the *difference
engine*. The difference engine (shown in Fig. 1-6), also based on the rotating
wheel principle, was operated by means of a single crank. In July 1823, the
British government agreed to finance the building of an expanded version
of the difference engine. Unfortunately, the toolmaking industry at the time
was not sufficiently sophisticated to build parts for his machine. As a result,

FIGURE 1-4
The slide rule.

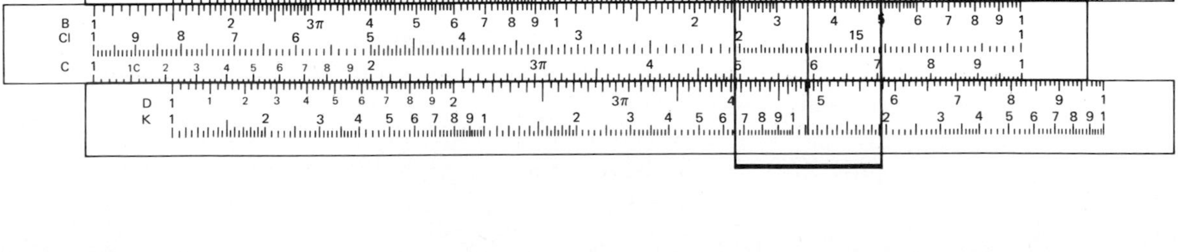

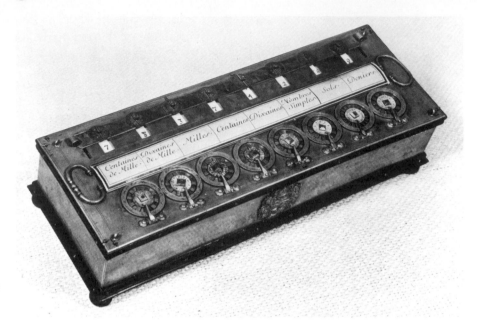

FIGURE 1-5
Pascal's wheel calculator.
(Courtesy of IBM.)

Babbage was forced to spend much of his time designing parts and tools for making them. This severely retarded progress. The project continually exceeded its budget, and several times production came to a halt through lack of funds. During one of these breaks, Babbage's chief mechanic walked out on him, taking all the tools that had so far been built. It was about this time, in 1833 in fact, that Babbage conceived the idea of a significantly improved difference engine, which he dubbed the *analytical engine*. More general than the difference engine, the analytical engine could be "programmed" to evaluate a wide range of different functions. Progress on the difference engine came nearly to a halt, and when it was not completed by 1842, the government withdrew its support. Babbage then devoted his time to the design of his analytical engine. Although the design was eventually completed, the machine was never built, largely because the technology of the day was not far enough advanced. It was a full century later before similar ideas were actually put into practice.

In the 1880s, a statistician named Herman Hollerith was commissioned by the U.S. Census Bureau to develop a technique for speeding up the processing of census data. A census is taken every 10 years in the United States. Since the data from the 1880 census had taken almost 8 years to analyze, it was feared that the 1890 census data would not be processed before the 1900 census—a critical situation indeed. Hollerith proposed that the census data be punched onto cards and automatically tabulated with specially designed machines (see Fig. 1-7). Punched cards had been used by Joseph Marie Jacquard for setting the pattern on a weaving machine in the early 1880s, but cards were not yet part of the data-processing environment.

FIGURE 1-6
Babbage's difference
engine. *(Courtesy of IBM.)*

With this new procedure, the data of the 1890 census were processed in less than 3 years. Influenced by the success of the American effort, the governments of Canada, Austria, and Russia also used Hollerith machines for census processing in the 1890s. Many large organizations, such as insurance companies and other agencies of government, began to apply Hollerith machines to their own data-handling problems and the demand

FIGURE 1-7
Hollerith's tabulating
machine. *(Courtesy of
IBM.)*

grew. In the 1890s, Hollerith left the Census Bureau and began the Tabulating Machine Company, which later became part of the International Business Machines Corporation (IBM). Hollerith's name remains associated with the punched card, now almost universally the symbol of automated data processing.

The late 1930s and early 1940s witnessed a flurry of activity in the development of computers. This period, called the "effervescent years" by Tropp, was influenced largely by the onset of World War II. The war effort intensified the need for scientific calculations, such as the production of ballistics tables, for example, and funding of several key projects was made possible. In addition, the needed technology was now available. This mixture of technology with incentive produced the necessary climate. Because of the intensity of the activity in this period, with many projects under way simultaneously, there remains to this day considerable confusion over the actual order of developments or the influence various projects had on one another.

Among the projects in this period was the construction of a series of five large-scale computers under the direction of George Stibitz of Bell Telephone Laboratories and at the behest of the U.S. Army. Called the *Bell relay computers*, because they used electromechanical relays as their basic operational component, they represented a significant advance on calculators of the day. In terms of operating speed, they outstripped even the most advanced rotating-wheel calculators. Although they were designed for specific calculations, the Bell relay computers provided important evidence that relay calculators could perform calculations 24 hours a day, 7 days a week, with few errors and little time lost due to malfunction.

In a basement room of the physics building at Iowa State College in Ames, Iowa, in the late 1930s, John Vincent Atanasoff began formulating the principles of the first automatic *electronic* calculator. With the assistance of a graduate student, Clifford Berry, he began the construction of a prototype of the Atanasoff-Berry computer in the fall of 1939. Upon its completion this prototype was capable of solving, with a high degree of accuracy, simultaneous equations in as many as 29 unknowns. Among the important principles incorporated in this machine were the use of the base 2, or binary, rather than the standard base 10, or decimal, number system and the incorporation of a machine-regenerative memory. Atanasoff and his work have remained relatively unknown until only recently. In October 1973, U.S. District Judge Earl R. Larson of Minneapolis, ruling on a $200 million patent-infringement lawsuit involving two present-day computing companies, was among the first to draw public attention to Atanasoff's pioneering efforts.

During the same period, a German by the name of Konrad Zuse was extensively involved in the design and construction of computers. His first effort, eventually called the Z1, was actually begun in the living room of his parents' apartment in Berlin. Although much of his work was destroyed during the war, he was able to remain active, going on to design improved models and eventually forming a successful computer manufacturing company. In his early machines Zuse pioneered some fundamental ideas of automatic computing despite having no knowledge of work going on elsewhere.

During the war British Intelligence was involved in the construction of a series of electronic computers known as *Colossi* in a highly classified project at Bletchley Park. Developed as a special-purpose computer to be used in code breaking, the first Colossus became operational in December 1943. Much of the information on the Colossus computers and their use is still officially classified.

A machine known as the *Automatic Sequence-Controlled Calculator*, later called the MARK I, was built at Harvard University during the period 1937–1944 by Howard Aiken with the support of IBM and the U.S. Navy (see Fig. 1-8). Like the Bell computers, the MARK I was a relay computer and was able to perform arbitrary sequences of arithmetic operations controlled by a coded sequence of instructions. Unlike some of his contemporaries, Aiken was familiar with the work of Babbage a century earlier and acknowledged his influence in his writings. In fact, the MARK I was later called "Babbage's dream come true." Aiken went on to develop MARKs II through IV.

A project funded by the Ballistic Research Laboratory of the Aberdeen Proving Grounds in Maryland led to the construction at the Moore School of Electronic Engineering at the University of Pennsylvania of the first large-scale, fully electronic computer (see Fig. 1-9). The ENIAC (*Electronic Numerical Integrator And Calculator*) was built during 1943 to 1946 under the direction of John W. Mauchly and J. Presper Eckert. Instead of the electromechanical relays used, for example, in the MARK I, the ENIAC used electronic vacuum tubes, which made it capable of calculating at 1,000 times the speed of its mechanical predecessors. Mauchly was familiar with the work of Atanasoff at Iowa State College, but the extent of the influence of Atanasoff's work is not clear.

This was the most ambitious and the most important of the pioneering efforts in electronic computation. Among its many components, the ENIAC contained some 18,000 vacuum tubes, 70,000 resistors, and 10,000 capacitors, and used over 150 kilowatts of power. It occupied more than 15,000 square feet of floor space and weighed 30 tons. Many predicted that the ENIAC would never work because the reliability of vacuum tubes had not yet been

FIGURE 1-8
The Automatic Sequence Controlled Calculator (Harvard Mark I). *(Courtesy of IBM.)*

FIGURE 1-9
The Electronic Numerical
Integrator And Calculator
(ENIAC). *(Courtesy of
Sperry UNIVAC.)*

established. However, it was successfully completed, and during its lifetime
(almost 10 years) it processed 80,223 hours of work. Although data input
and output were on punched cards, programs were constructed by the
wiring of connections between components. The ENIAC could perform
5,000 additions or subtractions, or 300 multiplications, per second, which
made it much faster than any of its contemporaries. Its greatest shortcoming
was its limited capacity for information storage. Outshone by new machines,
the ENIAC finally retired from active service in 1955.

Programming these early computers was a formidable task, requiring complete familiarity with the details of the machine's operation, considerable ingenuity, and a great deal of patience. Programs for ENIAC, for example, were installed and changed by changing the wiring among its various components—possibly a 1- or 2-day job. This situation soon became intolerable. In a computer capable of solving problems in minutes, it was unreasonable to spend days in setup.

John von Neumann, a consultant to the ENIAC project, along with the ENIAC staff, was first to propose the concept of a *stored* program. He proposed that the instructions be stored in the computer along with the data. This idea, now known as the *von Neumann concept*, increased the flexibility and the versatility of the computer in two ways. First, the instructions could be changed without manually rewiring the connections (and therefore more quickly), and second, since the instructions would be stored as numbers, the computer could process instructions as if they were data, thus making possible the automatic modification of instructions and the alteration of their sequence. This concept opened the door for a series of new projects. In 1946, von Neumann, the ENIAC group, and H. H. Goldstine began the construction of a stored program computer, the EDVAC (*Electronic Discrete Variable Automatic Computer*), but its completion was delayed until 1952. The EDSAC (*Electronic Delay Storage Automatic Calculator*), built at Cambridge University under the direction of M. V. Wilkes, was, in fact, the first stored program digital computer completed. It went into operation in May 1949. In 1952, a stored program computer, the IAS computer, was completed at the Institute for Advanced Study at Princeton University under von Neumann's supervision. During this time (late 1940s and early 1950s), other stored program computers were being developed, with names that convey the excitement of the period: the ILLIAC at the University of Illinois, the JOHNIAC at the RAND Corporation, the MANIAC at Los Alamos, and the WHIRLWIND at the Massachussetts Institute of Technology.

The developers of the ENIAC machine, Mauchly and Eckert, went from the ENIAC project to the construction of a machine called the *Universal Automatic Computer*, or UNIVAC, the first production-line digital computer (see Fig. 1-10). The UNIVAC I was able to achieve greater speed by the use of crystal diodes instead of vacuum tubes, foreshadowing the solid-state era. Other new features included the ability to read information into the computer, perform computations, and write information out, all at the same time. This was accomplished through the use of independently operating, comparatively high-speed peripheral devices called the UNITYPER and UNIPRINTER. The UNIVAC I also had a sophisticated magnetic tape system. The first installation of a UNIVAC I was at the U.S. Census Bureau in 1951; its first commercial installation shortly afterward (the first *commercial* installation anywhere) was at a new General Electric plant in Louisville, Kentucky.

FIGURE 1-10
The UNIVAC I: the first computer commercially available. *(Courtesy of Sperry UNIVAC.)*

IBM did not enter the commercial large-scale computer field until relatively late. Its incredible success story is largely the result of one man's efforts and some timely opportunism. In 1911, the Computing Tabulating Recording Company was formed by the amalgamation of four small companies. Three years later, Thomas J. Watson was hired as manager. In 1924, Watson renamed the company the International Business Machines Corporation. Before 1950, the only automatic computing machines produced by IBM were designed to complement its punched card equipment. IBM's first large-scale computer, later called the IBM 701, was completed in 1953, just in time to meet the rising need for computation caused by the Korean War. Eighteen 701s were installed over the next 3 years. By 1955, the 701 had been replaced by the 702; however, it was thought by many to be inferior to the earlier UNIVAC. As a result, the 705 was soon announced and the 702 withdrawn from the market. Instead of the cathode-ray-tube memory of its predecessors, the 705 had faster and more reliable magnetic core memory. By 1959, the 705 was firmly established in the data-processing field. The 704, completed in 1956, was a successful large-scale scientific computer.

Input and output were slow on the early computers. IBM attempted to remedy the situation in 1958, with the announcement of the 709. By permitting input/output operations to proceed simultaneously with calculations, it wasted less precious computation time. The 709 was a vacuum-tube computer, however, and soon became obsolete with the development of transistorized circuitry. In 1959, IBM released the 7090 (see Fig. 1-11), a transistorized computer compatible with the 709. Hundreds of 7090s were sold at an average price of $3 million. In 1962 and 1963, IBM produced the 7040 and 7044 computers. These were less powerful than the 7090 but considerably less expensive.

FIGURE 1-11
The IBM 7090. *(Courtesy of IBM.)*

On a smaller scale, in 1953 IBM announced a medium-sized computer, the 650. The demand for the 650s was greatly underestimated—instead of building 50, as originally intended, IBM produced and sold over a thousand 650 computers. Their successors, in the early 1960s, were transistorized computers—the small scientific 1620 and the 1400 series computers, which were widely used for data processing.

The design of the IBM System/360 family of computers began in 1961, with the objective of standardizing IBM equipment. They were more powerful and less expensive than earlier computers and supported a wide range of peripheral devices. The 360 (shown in Fig. 1-12) was actually a spectrum of computers with increased capability as you moved through the product line. It had a tremendous impact on the whole computer industry. Thousands of 360s have been installed throughout the world and IBM's dominant position in the industry has been firmly established.

Since the mid-1960s there has been a flurry of activity in the area of minicomputers and, more recently, microcomputers. Prominent in the manufacture of minicomputers has been the Digital Equipment Corporation with its PDP line. The first PDP-1 was installed at MIT in 1961. While the price of these computers has been dropping steadily, their capabilities have been growing rapidly, and more and more small businesses are finding it possible to own their own computer rather than simply purchasing time on

FIGURE 1-12
The IBM System/360.
(Courtesy of IBM.)

an outside installation. Today as many small computers are being purchased as large ones. Microcomputers, too, are rapidly finding their way into our daily lives, from pocket calculators to recreational devices. The "computer hobbyist" is a rapidly growing phenomenon as more and more parts and kits become available at low cost. Indications are that within the next two decades computers in some form may well become household articles as common as the television set is today.

1-3   THE DEVELOPMENT OF PROGRAMMING AND PROGRAMMING AIDS

The introduction of the stored program computers gave birth to a new profession, the computer programmer. Since then, significant advances have been made in the field of computer programming, specifically in the development of techniques to make it less difficult for the human programmer.

Many acknowledge that the distinction of being the first programmer should go to a very colorful lady who died almost a century before the first stored program computers appeared—Ada Augusta, the Countess of Lovelace. Lady Lovelace lived an eventful life. She was born in 1815, one of the many offspring of the prolific English poet, Lord Byron. A few months after her birth, her parents separated and she never saw her father again.

Around the time of her marriage to the Earl of Lovelace in 1835, Ada formed an acquaintance with Charles Babbage, who was at the time launching his analytical engine project. Possessing aptitudes for mathematics and things mechanical, she offered to work with Babbage on his project, and in 1842 she translated an early Italian description of the engine into English, adding many notes of her own. She referred to "cycles of operations" and the repeated use of cards in subroutinclike structures. She also spoke of nonnumerical computation and symbolic manipulation. She observed that the analytical engine could not "originate anything" but could only do "what we know how to order it to perform." One of her notes was a step-by-step description for computing Bernoulli's numbers with the analytical engine, cited by many as the first "program." Ada and Charles later collaborated in a scheme to apply the analytical engine to the problem of beating the odds at horse races. For the rest of her life, Ada gambled recklessly and lost a considerable part of the Lovelace fortune. She died of cancer in 1852.

Just as natural language serves as the vehicle of communication between human beings, so too are there languages to effect communications between human beings and computers. These languages permit the expression of the programs, or sets of instructions, that the human being wishes to have the computer process. Computer *languages* take different forms. Programs for the early computers, such as the ENIAC and the EDSAC, were composed in the actual language of the machines themselves. In *machine language*, instructions are expressed simply as a string of *binary digits*, or *bits*. The difficulty of programming early machines in this manner severely limited their usability and provided great incentive toward the development of higher-level programming languages more oriented toward the expression of solutions in the notation of the problems themselves. A specially designed program would then be called upon to perform a translation into the actual language used by the machine.

The first programming languages were known as *assembly languages*, an example of which was TRANSCODE, developed for the University of Toronto's FERUT computer by Pat Hume and Beatrice Worsley. In an assembly language, a special code (called a *mnemonic*) is defined for each of the operations of the machine, and a notation is introduced to specify the data on which the operations are to be performed. A special program, called an *assembler*, translates the symbolic assembly language instructions to the machine language instructions required for execution. *Assembly languages* are still popular today for certain applications. Although a significant improvement over machine language programming, this was still not sufficient to meet the needs of all would-be programmers.

The mid-1950s saw the introduction of the first general-purpose problem-oriented programming languages. One that soon revolutionized the field of programming was known as FORTRAN (FORmula TRANslating system), first published in 1954. The leader of the FORTRAN project was John Backus, who worked for IBM and was later responsible for the development of a formal method for defining the syntax of programming languages—Backus-Naur Form or BNF. FORTRAN was implemented in 1957, with new versions released in 1958, 1960, and 1962. The latest version became known as FORTRAN IV. FORTRAN is a language oriented to numerical scientific problems. It is easy to learn, read, and write. With FORTRAN, the user was able for the first time to write a program knowing little about the physical characteristics of the machine on which the program was to be run. Indeed, the language is independent of the machine, and in theory, although sometimes difficult in practice, FORTRAN programs written for one machine ought to be easily transferred to other machines. This is not the case with programs written in assembly language or machine code. The FORTRAN language had a major impact on the computer industry because it allowed people to program their own problems without the aid of a professional programmer.

At first, FORTRAN was not readily accepted because of fears of formidable translation costs. Instead of an assembler, higher-level problem-oriented languages, because of their generality, require a more sophisticated form of translator known as a *compiler*—more complex, and therefore more expensive to run. Usage increased, nonetheless. Over the years the cost of compilation has been significantly reduced, and FORTRAN has become the most widely used programming language in the world—and a very significant factor in the increased use of computers. The introduction of fast, student-oriented compilers such as PUFFT, developed at Purdue University, and WATFOR and WATFIV, developed at the University of Waterloo, has made practical the teaching of FORTRAN programming, and through this, computing itself, to large numbers of students.

Other programming languages followed quickly on FORTRAN's heels. The ALGOrithmic Language, ALGOL, was designed by an international committee in 1958 and revised in 1960. ALGOL is a very effective language for solving a wide class of problems in numerical mathematics, but it is inadequate (as is FORTRAN) in its ability to handle nonnumeric data. Today, ALGOL is much more popular in Europe than it is in North America.

Both FORTRAN and ALGOL are oriented primarily to scientific computations. In May 1959, a meeting was called by the U.S. Department of Defense to discuss the problem of developing a common language for business applications. In attendance were about 40 representatives from users, government installations, computer manufacturers, and other interested parties. An initial version of COBOL (COmmon Business Oriented Language) was released in December 1959.

The objectives of COBOL included a natural (that is, English-like) expression of programs, making the language easy to learn and largely self-documenting, and machine independence, thus allowing the transfer of COBOL programs from one installation to another. Although the specifications of COBOL have been revised and extended several times since its first release, the language itself has remained essentially the same. It is widely used in business data-processing environments.

BASIC (Beginner's All-purpose Symbolic Instruction Code), a scientific programming language that was designed with the objective of being as easy as possible to learn and to use, was developed at Dartmouth College in 1965 by John Kemeny and Tom Kurtz. The BASIC system was the first system used on a network or distributed basis, and also the first available in timesharing or conversational mode. Each command submitted by a user from a BASIC terminal causes an immediate response from the computer. This enables the user to have closer control over the processing of his/her program. BASIC remains very popular today, and timesharing has become a very common mode of operation, with many more languages and facilities available to the user.

In September 1963, a committee comprising IBM personnel and customers was formed "to provide a language which would encompass more users while still remaining as a useful tool to the engineer." At the time it was thought that the committee would simply extend FORTRAN appropriately, but after studying FORTRAN, ALGOL, and COBOL, and talking to a variety of users, the committee decided to develop a new language. The committee presented a report on the proposed new language on March 1, 1964. (Originally called NPL for New Programming Language, the name was later changed at the request of the National Physical Laboratory.) The language was revised in June and in December, and was eventually named PL/I. The first official manual was issued early in 1965, and the first PL/I compiler was implemented on the IBM System/360 in August 1966.

Because PL/I is a very general language, it has an extremely wide application area. Its usage is increasing, and some people think it may eventually replace its parents—FORTRAN, ALGOL, and COBOL. Recently, a number of student-oriented compilers have been produced. Such systems as Cornell University's PL/C and the University of Toronto's SP/k ought to increase the acceptance of PL/I as a teaching language.

1-4 CLOSING REMARKS

Although this brings us to the end of our history, the development of new technology in both computer hardware and programming languages continues at a feverish pace. (In 1972, the number of languages in use in the United States alone was estimated by Sammet at 170.) Increasingly, we are moving from an age of general-purpose systems and languages to a time of systems and languages specially tailored to individual applications. With respect to programming languages, this permits the use of notation more closely oriented to the specific application area, and also simplifies the task of designing good compilers for these languages. Keeping this trend in mind, we have chosen not to base this book on any specific computer configuration or any particular programming language. We have chosen rather to present principles of computer programming that transcend languages and installations, in the belief that once you learn how to prepare computer solutions for problems, training in particular systems is easily acquired.

BIBLIOGRAPHY

Eames, C., and Eames, R.: *A Computer Perspective*, Harvard University Press, Cambridge, Mass., 1973.

Gleiser, M.: "Men and Machines before Babbage," *Datamation*, vol. 24, October 1978, p. 125.

Hawkins, G. S.: *Stonehenge Decoded*, Doubleday, New York, 1965.

Huskey, H. D., and Huskey, V. R.: "Chronology of Computing Devices," *IEEE Transactions on Computers*, vol. C-25, December 1976, p. 1190.

Kean, D. W.: "The Computer and the Countess," *Datamation*, vol. 19, May 1973, p. 60.

Knuth, D. E.: "Ancient Babylonian Algorithms," *Communications of the ACM*, vol. 15, July 1972, p. 671.

Mollenhoff, G. G.: "John V. Atanasoff, DP Pioneer," *Computerworld*, March 13, 1974; March 20, 1974; March 27, 1974.

Randell, B. (ed.): *The Origins of Digital Computers: Selected Papers*, Springer-Verlag, New York, 1973.

Renfrew, C.: *Before Civilization*, Jonathan Cape, London, 1973.

Rosen, S.: "Electronic Computers: A Historical Survey," *Computing Surveys*, vol. 1, March 1969, p. 7.

Sammet, J.: *Programming Languages: History and Fundamentals*, Prentice-Hall, Englewood Cliffs, N.J., 1969.

————: "Programming Languages: History and Future," *Communications of the ACM*, vol. 15, July 1972, p. 601.

Tropp, H.: "The Effervescent Years: A Retrospective," *IEEE Spectrum*, February 1974, p. 70.

Wexelblat, R. L. (ed.): Preprints from the History of Programming Languages Conference Sponsored by ACM SIG PLAN, *SIG PLAN Notices*, vol. 13, August 1978.

Williams, M. R.: "The Difference Engines," *The Computer Journal*, vol. 19, February 1976, p. 82.

2

Computers and Solving Problems

I'm a private shamus, so I'm used to seeing peculiar sights and hearing peculiar pitches. But when a curvy girl scientist dressed in nothing but a mink wrap showed me a long, coffin-shaped box and asked me to find the million bucks worth of machinery that used to be inside, I had to smile. "Baby, that may be a computer to you," I said, "but it's just a burying-box to me."

Raymond Banks, The Computer Kill *(Toronto: Popular Library, 1961)*

On your first exposure to computers, it is very easy to be overpowered by technological details. The technology is fascinating, but although it may be of interest to some of you, it is not necessary for this discussion and can obscure other important concepts. For this reason we will adopt a very simplified view of a computer, initially. As your use of computers becomes more extensive, they will seem much less formidable, and the picture will become more complete.

In this chapter we first present a very brief overview of the makeup of a typical computer. We then consider the use of a computer. Finally, we turn to the main point of this text—solving problems with the use of a computer. Some key concepts necessary to solve problems on a computer are introduced, and some simple problems are solved. The first problems may seem quite simple, but once mastered, the techniques extend easily, and will be a great aid as the complexity of the problems grows.

2-1   COMPUTER SYSTEMS: A BRIEF INTRODUCTION

A modern computer system is a collection of hardware and software components designed to provide a productive and, to some extent, friendly environment for computing. The terms hardware and software may not be familiar. By *hardware* we mean the physical computing equipment itself; by *software* we mean the collection of support programs written to provide various services to the users of the system. We will provide examples of both hardware and software in this section.

It is difficult to talk about a computing environment without dealing with both the hardware and the software aspects of that environment. For this reason we often use the term *computer system* to describe what the user interacts with, rather than the term *computer*, which has definite hardware overtones. For most users of computer systems, in fact, software components are much more visible. Users may not even be aware of the hardware that is involved as their programs are processed.

In the early days of computing, users of the system interacted much more closely with the actual hardware than is common today. Now the trend is away from direct user interaction with hardware, largely because human reaction times tend to be too slow by comparison with machine speeds and unacceptable idle periods are introduced. Many of the functions that once were performed by the users themselves are now handled by software known as the *operating system*. The operating system is essentially responsible for the processing of programs by the computer system; it ensures, for example, that the appropriate hardware resources of the system are made available as required, that software resources are supplied as needed, and, in most systems, that the users are charged for their use of the system's resources in proportion to the amount they use. In summary, the operating system creates an environment in which users can prepare programs and run them without being overly concerned with hardware details.

To meet the dramatic growth in the demand for computing, it has become common to move to a form of operation, known as *multiprogramming*, in which several users use the system simultaneously. It is the function of the operating system to coordinate the *sharing* of the system's resources (both hardware and software) in such a way as to provide the type of service that the users require, and to minimize the amount of interference that takes place among their executing programs. The overall objective of the computer system guides all decisions made by the operating system.

Broadly, we can divide computer systems into four categories based on the type of service they provide. As a point of fact, many actual systems provide several types of service, but for the sake of simplicity we will deal with them individually.

The two most common types of computer system are batch processing systems and interactive timesharing systems. Both the method of access and the user requirements are different. In a *batch processing system*, users normally submit their programs (or *jobs*) as decks of punched cards and then return at some later point in time to pick up their (usually printed) output. As jobs arrive, the operating system collects them into batches, which are kept in external storage, chooses jobs for execution as the needed system resources become available, and replaces completed jobs from those awaiting execution. The primary objective is high throughput, that is, a large amount of computation per unit time, with, hopefully, fast turnaround of the user jobs.

In an *interactive timesharing system*, users submit work to the system from individual terminals. These terminals are similar to typewriters in appearance, and may or may not have cathode-ray-tube (CRT) displays. The operation is interactive in the sense that, during his/her session at the terminal, the user is continually engaged in a dialogue with the operating system as the machine and the user cooperate in the processing of the job. Through what seems to the user to be an immediate response to his/her requests, an illusion is presented that the user has the complete and undivided attention of the computer system. In actual fact, there may be many other users simultaneously operating in a similar mode, as shown in Fig. 2-1. The operating system creates this illusion by giving to each user in turn a small amount of attention in the form of usable CPU time. The amount is "small" in user terms, but "large" enough in computer system terms to execute many instructions. For example, in a half a second, most computer systems could execute many thousands of instructions. Fast response to user requests is the dominant objective of these operating systems. Studies have shown that although more total computing time may be used, interactive users often are able to solve their problems in a shorter period of time than batch users. Terminals connected to an interactive timesharing system can be far removed geographically from the actual computer site, transmitting data by means of communications paths such as standard telephone lines. Microwave or satellite communications may be used when very long distances are involved.

Two other forms of computer system have gained wide acceptance only relatively recently. *On-line transaction systems* have a strong surface similarity to interactive timesharing systems in that user access is by means of terminals. However, the objectives are quite different. Whereas users of timesharing systems are programmers, users of on-line transaction systems are not; instead, they are simply entering data to be processed by a program that runs continuously. Familiar examples are found in the areas of airline reservations and banking. In the latter, the users are tellers who enter such information as customer name, date, and amount of deposit, and receive back a printed statement of the transaction, showing the customer's account updated. In the former, the users are reservation clerks providing services to travellers in the form of flight information, seat reservations, and so forth. On-line transaction systems are becoming increasingly popular as a business tool to provide information and services rapidly. The design of such systems is considered in a later chapter.

A fourth type of computer system worthy of mention is the *computer network*. A computer network consists of a number of interconnected

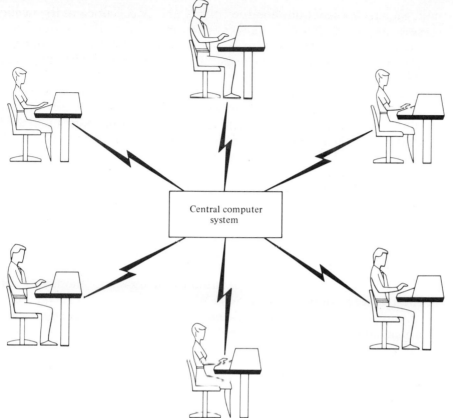

FIGURE 2-1
User view of a timesharing
system.

computer systems, which again may be quite far apart geographically. Users at any one of the systems in the network may have work processed at any other system in the network and may thereby take advantage of some special features available at other installations. An example of a computer network is the ARPANET network connecting a large number of computer centers throughout the United States. Although still largely in the experimental stage, computer networks are expected by many to play a large role in computing in the years to come.

Although different computer systems may perform radically different functions, and may also be radically different in appearance, their basic hardware organizations are, in fact, quite similar. We now turn to a consideration of the basic hardware components of a typical computer system.

2-1.1   COMPONENTS OF A TYPICAL COMPUTER

All computers, from the very smallest microsystems to the largest complexes, consist of four basic components. These are storage, the arithmetic/logic unit, the input/output unit, and the control unit. We shall consider each of these in more detail.

Let us look first at the *storage* component. Every computer comes with a certain amount of *internal storage*, sometimes referred to as its *main memory*, along with *external storage* such as magnetic disk and magnetic tape devices. The two forms of storage differ in characteristics and in purpose. Programs currently in execution, along with some of the data required for execution, must reside in main memory. Main memory is expensive and, in most systems, a scarce resource. Information not immediately needed is normally relegated to external storage, where the capacity can be almost unlimited, although the time to retrieve the information is considerably longer. Also, the cost of external storage is comparatively low, so information can be stored for long periods of time.

A computer's main memory is divided into small, equal-sized units called *words*, each of which has a unique *address*. Each memory word is capable of holding essentially one piece of information, such as the result of a numeric computation. The size of a word is an important design parameter of the computer, and determines, for example, the largest and smallest number that can be stored. We might imagine, for example, a machine designed with a word size capable of storing up to 8 decimal digits, or a machine with a word size capable of storing 16 binary digits, and so forth. The consequences of this decision will be considered in more detail later in the book.

As stated, each word in main memory has a unique address. Figure 2-2 illustrates how this might be represented in an arrangement similar to a set of post office boxes. The memory address serves to identify the particular word uniquely, so that the desired information can be read from it or written into it. In this way, the address serves the same function as the box number on the post office box. For purposes of illustration, the addresses in this portion of storage are shown above each word, although in practice they would not need to be physically present. Also, we have adopted a two-

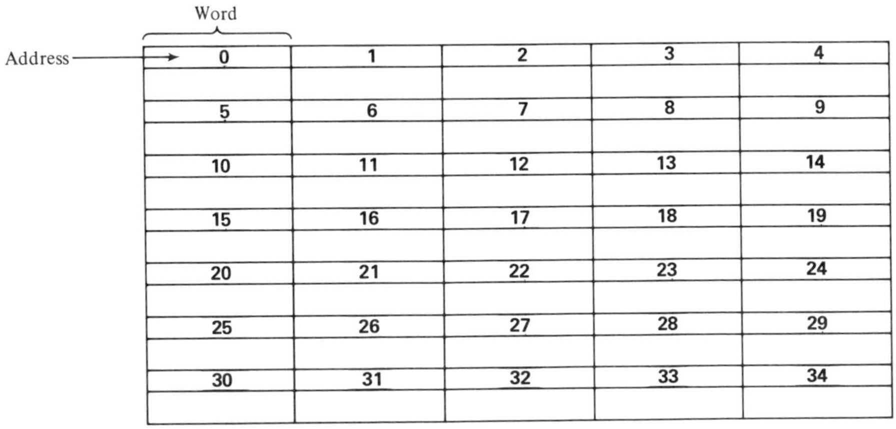

Word				
0	1	2	3	4
5	6	7	8	9
10	11	12	13	14
15	16	17	18	19
20	21	22	23	24
25	26	27	28	29
30	31	32	33	34

Address ⟶

FIGURE 2-2
Memory representation of
a computer.

dimensional "row, column" presentation, when, in fact, memories are normally viewed as a linear string of words.

Computer memories can be built in many ways. In early machines, for example, vacuum-tube "flip-flops" were used to represent the stored information. Later, the technology of magnetized "cores" prevailed. Since the late 1960s semiconductor memories have become increasingly popular. These technological advances have improved the reliability of memories and reduced the time required to access information (thus increasing the overall speed of the computer), while at the same time vastly reducing their physical size. To illustrate how information is actually stored, let us imagine that memory is composed of a series of two-position switches. A switch in the "on" position, by agreement, represents the binary digit 1; a switch in the "off" position represents the binary digit 0. In our illustrations we denote a switch in the "on" position by a blackened rectangle; a white rectangle denotes a switch in the "off" position.

At this point it is necessary to digress for a short introduction to the binary (or base 2) number system. The binary system, like our familiar decimal (or base 10) system, is a *positional* number system. This means that the value of a digit is determined by its position in the number. In the decimal system, the number 2,562 has the following interpretation:

$$
\begin{array}{rcl}
2 + 1{,}000\ (10^3) & = & 2{,}000 \\
+5 \times\ \ 100\ (10^2) & = & 500 \\
+6 \times\ \ \ \ 10\ (10^1) & = & 60 \\
+2 \times\ \ \ \ \ \ 1\ (10^0) & = & \underline{\ \ \ \ 2} \\
 & & 2{,}562
\end{array}
$$

Note that the digit 2 has a different interpretation on its two occurrences in the number. The first (leftmost, or *high-order*) occurrence indicates a contribution of 2,000 to the value of the number; the second occurrence (in the rightmost, or *low-order*, position) indicates a contribution of 2.

In general, the interpretation of the number *abcde* in the base *n* number system is

$$a \times n^4 + b \times n^3 + c \times n^2 + d \times n^1 + e \times n^0$$

The extension to higher-order digits for larger numbers proceeds in the obvious way. Thus, in a positional number system, the position of a digit in the number specifies the power by which the base of the number system (also called the *radix*) is raised to give the contribution of that digit in the final value.

In the binary number system, the radix is 2. Thus, the value of the number 110101 is

$$
\begin{aligned}
1 \times 2^5 + 1 \times 2^4 &+ 0 \times 2^3 + 1 \times 2^2 + 0 \times 2^1 + 1 \times 2^0 \\
&= 32 + 16 + 0 + 4 + 0 + 1 \\
&= 53
\end{aligned}
$$

Although straightforward methods exist for converting from decimal to binary, and vice versa, we will not go into them. To follow the examples used in this section, it will suffice to be able to perform the calculation described.

Returning to our previous discussion, Fig. 2-3 shows a portion of a memory with a "word" composed of eight consecutive switches. The settings yield the decimal values 91 for word 0, 110 for word 1, and 1 for word 2. In memory terminology, any device capable of representing a binary digit (that means that the device must have two readily identifiable states) is said to define a "bit" of information. Word sizes are normally quoted in bits; thus, in our example, we are dealing with a word size of 8 bits. A small amount of reasoning will convince you that the largest decimal value that can be represented with n bits is $2^n - 1$.

The representation of information is somewhat more complicated than we have let on. The complication arises from the fact that we wish to represent information of a variety of types. For instance, we require nonnumeric information as well as numeric information. The representation of nonnumeric (or character) information is done through a coding scheme, whereby certain patterns of bits are by agreement taken to stand for certain characters. Two coding schemes are most popular in the computing industry: the 8-bit EBCDIC code (*Extended Binary Coded Decimal Interchange Code*) and the 7-bit ASCII code (*American Standard Code for Information Interchange*). These are described in more detail in Chap. 5.

With respect to numeric information, several issues remain to be considered, for example, the question of negative numbers and the distinction between integers and real numbers. The matter of sign is normally handled by designating one bit of the word (usually the leftmost bit) as the *sign bit*, with 0 signifying that the number is positive and 1 signifying that it is negative. If the word size is m bits, this leaves $m - 1$ bits for the magnitude of the number. The representation of real numbers is considerably more complicated and will not be dealt with here.

The second basic component of any computer system is the *arithmetic/logic unit* (ALU). All calculations are carried out here. These may involve arithmetic operations such as addition, subtraction, multiplication, or division, or logical operations such as the comparison of two values to see

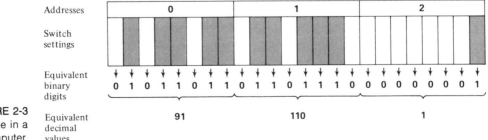

FIGURE 2-3
Information storage in a computer.

which of them is larger. As a complex computation is performed, it is often necessary for information to be moved to and from main memory, during the course of the computation.

Any computing system that is meant to interface with the outside world must do so by means of *input and output devices*. Input devices provide the means by which data (and, as we shall see, instructions) are transmitted to the computer. Output devices are necessary for the display of results. There exists a wide variety of both input and output devices. Familiar input devices include punched-card readers, teletypewriters, and cathode-ray-tube display devices (CRTs) with keyboards. The most common output devices are line printers, card punches, and, once again, teletypewriters and cathode-ray-tube display devices, which can serve as both input and output devices.

The final basic component of a computer system is to some extent the most important, since it controls the actions of all the other components. This is known as the *control unit*. Operating under the control of instructions from the programmer that reside in main memory, the control unit causes data to be read from the input device, passes the appropriate values from storage to the arithmetic/logic unit for the required calculations, stores and retrieves data and intermediate results from main memory and passes results to the output device for display. The combination of the control unit and the arithmetic/logic unit is commonly referred to as the *central processing unit* (CPU). The components of a typical computer system are summarized in Fig. 2-4.

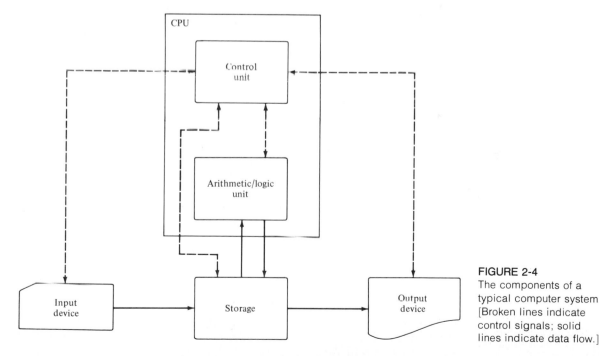

FIGURE 2-4
The components of a typical computer system. [Broken lines indicate control signals; solid lines indicate data flow.]

2-1.2 MACHINE LANGUAGE AND PROGRAMS

From the foregoing discussion, you now have some understanding of just what a computer program is. A *program* is simply a sequence of instructions that directs the CPU in the performance of some computation. Ultimately, this program must be expressed in a form that can be understood by the CPU. Although this is seldom of direct concern to the programmer, a brief discussion of machine language and machine-language programs can provide insight into the operation of a CPU.

A CPU can only understand instructions that are expressed in terms of its *machine language*. To illustrate the concept of a machine-language program and the action of the CPU in executing that program, let us consider a simple CPU with a very primitive machine language. Any language is nothing more than a set of rules to specify how things must be expressed to be understood. For the language of this particular machine, we will assume that each instruction is 8 bits in length and consists of two parts: a 3-bit *operation code* that specifies the operation to be performed, followed by a 5-bit *operand* that gives the address of the data on which the operation is to be performed. We will also assume the existence of an *accumulator* in the arithmetic/logic unit, in which arithmetic operations are performed. Table 2-1 defines the *instruction set* of this simple CPU; this means that this particular CPU has hardware circuitry that directly executes these particular instructions.

Let us assume that the following short program has been loaded in the first 10 words of memory. To aid in the understanding of this example the operation code and operand for each instruction are shown separately, although in memory the entire instruction is one 8-bit string (as it is, for example, in Fig. 2-5).

Address of Instruction (in Binary)	Operation Code	Operand
00000	001	01010
00001	010	01100
00010	001	01110
00011	011	01011
00100	010	01110
00101	001	01100
00110	100	01101
00111	010	01100
01000	110	00010
01001	111	00000

In addition to the words indicated, other words of memory have been given values as well. The complete memory layout as execution begins is shown in Fig. 2-5. We will assume that execution begins at word 00000 with the instruction 00101010, and, unless otherwise stated, instructions will be

TABLE 2-1 Sample Instruction Set

Operation Code	Meaning
001	*Load:* copy the value of the word addressed into the accumulator
010	*Store:* copy the value of the accumulator into the word addressed
011	*Add:* replace the present value of the accumulator with the sum of its present value and the value of the word addressed
100	*Subtract:* replace the present value of the accumulator with the result obtained by subtracting from its present value the value of the word addressed
101	*Branch:* jump to the instruction at the word addressed
110	*Branch if not zero:* jump to the instruction at the word addressed only if the present contents of the accumulator are other than 0
111	*Halt:* terminate execution

taken in the order they appear. From Table 2-1 we see that the first instruction is a *load* instruction (operation code 001) that is asking the CPU to copy the value of word 01010 into the accumulator. Therefore, following this instruction, the value 00000011 (or 3 in base 10) is in the accumulator. That this is a "copy" operation means that the value of word 01010 remains unaffected; however, the previous contents of the accumulator have been overwritten. Execution proceeds with the next instruction in the sequence, that is, the instruction at word 00001.

The second instruction is a *store* instruction (operation code 010) that is directing the CPU to copy the value of the accumulator into word 01100, replacing what was there previously with the value 00000011.

The third instruction is another *load*. The present contents of word 01110 are copied into the accumulator. The value copied is 00000000.

Word

Address (in binary)			
00000	00001	00010	00011
00101010	01001100	00101110	01101011
00100	00101	00110	00111
01001110	00101100	10001101	01001100
01000	01001	01010	01011
11000010	11100000	00000011	00000100
01100	01101	01110	01111
00000000	00000001	00000000	00000000

FIGURE 2-5
Memory contents of a sample program and its associated data.

The fourth instruction is an *add* (operation code 011). The contents of memory location 01011 (presently 00000100) are added to the current contents of the accumulator (00000000). The addition is performed in binary in the accumulator, yielding the result 00000100, or 4 in base 10. This value replaces the previous accumulator contents.

The fifth instruction is a *store* that causes the result of the addition just performed to be copied into location 01110, replacing what was there previously.

The sixth instruction is another *load*, causing the value of the word at location 01100 to be copied into the accumulator. This turns out to be the value that was stored by the second instruction, which is 00000011.

The seventh instruction is a *subtract* instruction (operation code 100). The value in location 01101 (00000001) is to be subtracted from the contents of the accumulator (00000011). As was the case with the previous arithmetic operation, the addition in the fourth instruction, the subtraction is performed in binary in the accumulator, and yields the result 00000010.

The eighth instruction is another *store*, which causes the value just obtained to be stored in location 01100.

At this point we pause and consider what has happened up to now. The present state of the memory and the accumulator is illustrated in Fig. 2-6. Note that some changes have taken place from the original memory of Fig. 2-5. The words affected have been words 01100 and 01110. In fact, as the computation proceeds, we should keep a close watch on these two words.

We turn now to the ninth instruction, which is 11000010. Table 2-1 tells us that this instruction is *branch if not* 0 (operation code 110). This is a different type of instruction from those executed up to now. It tells us to do two things. First, examine the present contents of the accumulator. If the value contained in the accumulator is other than 0, which in our case it is (being 00000010), then jump (or branch) immediately to the instruction whose address is given as the operand (00010). This will then be the next instruction executed. If the accumulator contents did happen to be 0, then

Memory:

00000	00001	00010	00011
00101010	01001100	00101110	01101011
00100	00101	00110	00111
01001110	00101100	10001101	01001100
01000	01001	01010	01011
11000010	11100000	00000011	00000100
01100	01101	01110	01111
00000010	00000001	00000100	00000000

Accumulator: 00000010

FIGURE 2-6
Memory contents after executing eight instructions.

instead of jumping to the instruction at 00010, we would proceed directly to the next instruction in the sequence, that is, the instruction at location 01001.

This conditional type of branch instruction is important, and we will consider similar ideas in Chap. 3.

Since we have returned to what was originally our third instruction, it is apparent that instructions three through eight will now be executed again. We will leave the execution of these as an exercise, but remember that you are now beginning from the situation shown in Fig. 2-6. Following your second pass through this group of instructions, you should arrive at the situation shown in Fig. 2-7. The contents of word 01100 are 00000001, the contents of word 01110 are 00001000 (8 in base 10), and the contents of the accumulator are 00000001.

At this point the conditional branch instruction is encountered for the second time. As in the previous instance, the contents of the accumulator are nonzero, so the branch is taken again. As a result, instructions three through eight are executed a third time. Again, we leave this as an exercise. Following the third execution of instruction eight, the contents of memory and the accumulator are as shown in Fig. 2-8. In particular, the contents of word 01100 are 00000000, the contents of word 01110 are 00001100 (12 in base 10), and the contents of the accumulator are 00000000.

This time, the conditional branch instruction is not taken, since the contents of the accumulator are 0. Consequently, we move to the instruction at location 01001, which is the next in sequence. This is a *halt* instruction (operation code 111), so execution of this program terminates. From Fig. 2-8 we take our final values. In particular, the value of the word at location 01110 is 12 in base 10. Can you reflect back on what has taken place, paying particular attention to the initial values given in Fig. 2-5, and figure out the purpose of this particular program? (*Hint:* Look in particular at words 01010 and 01011.)

The foregoing discussion described how a typical computer operates.

Memory:

00000	00001	00010	00011
00101010	01001100	00101110	01101011
00100	00101	00110	00111
01001110	00101100	10001101	01001100
01000	01001	01010	01011
11000010	11100000	00000011	00000100
01100	01101	01110	01111
00000001	00000001	00001000	00000000

Accumulator: 00000001

FIGURE 2-7
Memory contents after second pass through loop.

Memory:

00000	00001	00010	00011
00101010	01001100	00101110	01101011
00100	**00101**	**00110**	**00111**
01001110	00101100	10001101	01001100
01000	**01001**	**01010**	**01011**
11000010	11100000	00000011	00000100
01100	**01101**	**01110**	**01111**
00000000	00000001	00001100	00000000

Accumulator: 00000000

FIGURE 2-8
Memory contents at
program termination.

The particular machine language used in the example was designed solely for the purpose of illustration, yet actual machine languages have very similar properties. The role of a *program* was identified as providing the particular instructions under whose control the CPU is to operate to perform some required computation. The task of a programmer is to produce the appropriate sequence of instructions to solve the problem at hand.

2-1.3 USING A COMPUTER

Clearly, to produce a program in raw machine language such as that used in the previous example could be a difficult task. Fortunately, software is provided in most computer systems to make this task considerably easier. Programs called *compilers* are available to translate users' programs (which are written in a form more suited to human comprehension) into the machine language required for execution of the programs. Compilers for the most popular programming languages are normally supplied by the computer manufacturer, although they may come from other sources as well. Compiler writing is an important field of computer science and is considered in more detail later in this book.

Figure 2-9 shows the steps involved in using most computer systems. The scenario begins with the programmer writing his/her program, perhaps using specially prepared coding forms. In order for the program to be processed by the computer, it must be submitted on a form that can be read by the computer. Quite commonly, this requires the preparation of a deck of punched cards, using a keypunch. This deck of cards is then submitted to the computer through the card reader, as shown. Alternatively, the program may be submitted through one of the teletypewriter terminals described previously, eliminating the need for punched cards. After the program has been processed, its results will be displayed as output, perhaps on a line printer, as shown, or a terminal.

Let us now trace the history of a program as it is processed in the cycle described above. Along the way we will introduce some important termi-

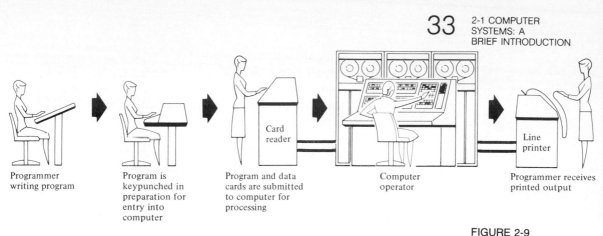

Programmer
writing program

Program is
keypunched in
preparation for
entry into
computer

Program and data
cards are submitted
to computer for
processing

Computer
operator

Programmer receives
printed output

FIGURE 2-9
Using a computer.

nology. The program as it is prepared by the programmer is referred to as the *source program*. This is usually written on one of the programming languages mentioned earlier, such as PL/I, FORTRAN, ALGOL, or COBOL. Before the steps expressed can be processed by the computer to return the required results, this source program must be translated into the machine language for the particular computer by means of a *compiler* for the programming language being used. During the translation process, the compiler checks that the rules of the language used have not been violated by the programmer in his/her source program. Appropriate error messages, or *diagnostics*, are issued if necessary. The result of the compilation is referred to as the *object program*. This object program is now ready for execution. Figure 2-10 shows these stages.

The total time frame involved in running a program can be divided into two separate and distinct phases. *Compilation* (or *compile*) *time* relates to the translation of the source program into its equivalent object program; *run time*, to the execution of the object program. Certain errors are detected at run time, often by the operating system. Run time is also the time at which the particular data that the programmer wishes to supply for the computation are made available to the program. We will illustrate this point later in the chapter.

2-1.4   SUMMARY
The material in this section has been presented to give you a better perspective on computers and computing. We have tried to give you some

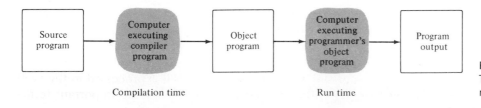

FIGURE 2-10
The stages of a computer run.

idea of how a typical computer is organized and how it operates, and also how a modern computer system is used. Clearly, we have merely scratched the surface. We could explore any of these areas in much more detail, but to do so would detract from the main purpose of this book.

The primary objective of this book is to teach you to solve problems using a computer. A computer programmer is first and foremost a solver of problems. To become proficient as a programmer, one must first learn to deal with problems in a rigorous, systematic fashion. Much of this book is concerned with a methodology for solving problems that is appropriate to the development of good computer programs. Central to this methodology is the notion of an *algorithm*, a notion we introduce in the next section.

2-2 ALGORITHMS

Computers, unfortunately, do what we tell them to do, not necessarily what we want them to do. There must be no ambiguity in the instructions that we give to a computer in our programs, no possibility of alternative interpretations. The computer will always take some course of action; great care must be taken to ensure that there is only one possible course of action so that the results we get are those we intended. Although a statement such as "compute the average grade on this test" seems to specify the computation we wish to have performed, it is, in fact, much too imprecise. Too much detail is left unspecified: for example, where are the grades, how many are there, are the absentees to be included, and so on. Herein lies the essence of computer programming.

Most interesting problems appear to be very complex from a programming standpoint. For some problems (such as difficult mathematical problems), this complexity may be inherent in the problem itself. In many cases, however, it can be due to other factors that may be within our control: for example, incomplete or unclear specification of the problem. In the development of computer programs, as we shall see, complexity need not always be a problem if it can be properly controlled.

Computer programming can be difficult. If it were not, good programmers could not command the high salaries that they receive. It is difficult largely because it itself is a complex activity, combining many mental processes. We can do a great deal, however, to make it easier. For instance, the programming task can be made much more manageable by systematically breaking it up into a number of less complex subtasks (the divide-and-conquer approach). We will use this approach of subdividing, which has seen considerable success in practice, over and over again in this book.

First, it is important that we separate the *problem-solving phase* of the task from what we will term the *implementation phase*, as shown in Fig. 2-11. In the problem-solving phase we concentrate on the design of an *algorithm* to solve the stated problem. Only after we are satisfied that we have formulated a suitable algorithm do we turn to the details of the

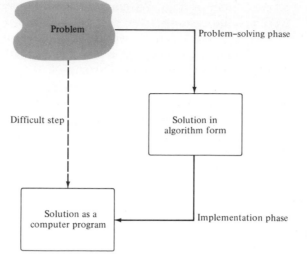

FIGURE 2-11
Problem solving and
programming.

implementation of this algorithm in some programming language. Given an
algorithm that is sufficiently precise, the translation to a computer program
is actually quite straightforward. We will devote considerable space in this
book to the design of algorithms. We will be presenting many problems to
which computers can be (and are being) applied. Algorithms for the solution
of these problems will be carefully formulated.

An *algorithm* can be defined as an unambiguous, ordered sequence of
steps that leads to the solution of a given problem. Although the term itself
may be new, the concept of an algorithm should be very familiar. Directions
given to a particular street constitute an algorithm for finding the street. A
recipe is a very familiar form of algorithm. A blueprint serves the same
purpose in a construction project. At Christmas time, many parents spend
exasperating hours following algorithms for assembling their children's new
toys.

We will require that our algorithms have several important properties.
First, the steps in an algorithm must be simple and unambiguous and must
be followed in a carefully prescribed order. Further, we will insist that
algorithms be *effective*; that is, they must always solve the problem in a

*(By permission of Johnny
Hart and Field Enterprises,
Inc.)*

finite number of steps. We could ill afford to pay the computing costs if this
were not the case.

As an illustration, we will design an algorithm to carry out a familiar
process, the replacement of a burned-out light bulb. The basic operation
can be expressed in two simple steps:

1. Remove burned-out bulb
2. Insert new bulb

This appears to solve the problem, but suppose that we are trying to instruct
a new household robot to change this light bulb. In its present form, this
particular algorithm would not be suitable for our purposes; it presupposes
too many things. The individual steps are not sufficiently simple for our
robot to understand. There are too many essential details unspecified. With
our ultimate objective in mind, we now proceed to specify the details of
each of these steps.

First, what is required in removing the old light bulb? Let us assume it
is in a ceiling fixture. This requires a ladder. Once we have positioned the
ladder and climbed it, how, specifically, is the old bulb removed? It is done
by turning it in a counterclockwise direction until it comes free. (One is
advised to turn the bulb rather than the ladder—fewer people are required
that way.) Thus, the single step "Remove burned-out bulb" has been
expanded into the following:

Position ladder under burned-out bulb

Climb ladder until bulb can be reached

Turn bulb in counterclockwise direction until it comes free

Inserting a new bulb requires the selection of a bulb with the same
wattage (which should really be done prior to climbing the ladder), the
placement of the bulb in the vacated socket, the turning in a clockwise
direction until secure, and, finally, the descent of the ladder. Thus, the step
"Insert new bulb" has been expanded into the following:

Select new bulb of same wattage as that of burned-out bulb

Place new bulb in socket

Turn bulb in clockwise direction until secure

Descend ladder

Our algorithm for replacing a burned-out light bulb now consists of
seven steps.

1. Position ladder under burned-out bulb

2. Select new bulb of same wattage as that of burned-out bulb
3. Climb ladder until bulb can be reached
4. Turn bulb in counterclockwise direction until it comes free
5. Place new bulb in socket
6. Turn bulb in clockwise direction until secure
7. Descend ladder

We are not yet at a sufficiently precise specification. Several of the steps in this algorithm imply more elaborate operations that must be stated explicitly. The selection process in step 2 is a case in point. Assume that we have a box containing new bulbs. We must examine these bulbs in turn until we find one having the same wattage as the old bulb. This requires the following operations:

Select a candidate for replacement

If wattage is not the same as wattage of old bulb, then repeat until a
    match is found
        Discard the bulb selected
        Select a new one

We have introduced two important concepts: the concept of decision ("if") and the concept of repetition ("repeat"). The decision capability allows us to defer the actual selection of a course of action until run time. In this case we cannot determine what to do until the wattage of an actual bulb has been examined. It is also possible to specify that a basic operation be repeated over and over again; in this case, selecting and discarding are repeated until we find a bulb of the appropriate wattage. A similar process is involved in step 3 (climbing steps of the ladder until the bulb can be reached), step 4 (turning bulb counterclockwise until it comes free of its socket), and step 6 (turning bulb clockwise until it is secure in the socket). More will be said about these operations in the next chapter. With these additions, our algorithm for replacing the burned-out light bulb is now the following:

1. Position ladder under burned-out bulb
2. Select candidate for replacement
    If wattage is not the same as wattage of old bulb, then repeat until
    a match is found
        Discard the bulb selected
        Select a new one
3. Repeat until bulb can be reached
    Advance one step up the ladder
4. Repeat until bulb comes free of socket
    Turn bulb in counterclockwise direction
5. Place new bulb in socket

6. Repeat until bulb is secure in socket
   Turn bulb in clockwise direction
7. Descend ladder

As you can see, this process of increasing the detail in an algorithm can continue almost indefinitely. Our present algorithm for changing a light bulb, however, serves to illustrate how algorithms can be expressed. The operations indicated are simple and unambiguous, the decisions required are straightforward, and the order in which the steps are to be carried out is clearly stated. The method by which this algorithm was developed is important. We began with a very general statement of the solution of the problem, and proceeded to the final algorithm by systematically increasing the level of detail.

How do we know when we have a sufficient level of detail in the algorithm? This is a function of the agent who will eventually carry out the algorithm. If we are communicating a recipe to the head chef in a large hotel, we can assume familiarity with many aspects of cooking that may need to be explained to the person boiling his/her first pot of water. The recipe must be described in terms that can be understood by the person who will do the actual cooking. The same is true with algorithms intended for computer processing. Computers have a very limited set of instructions and the algorithm must ultimately be expressed in terms of these.

The expression of an algorithm is very important, as it assists not only in the translation to a computer program, but even in the development of the algorithm itself. Because precision is so essential, we will be introducing a specific notation for algorithms that resembles the intuitive English-like notation used here. Before we do this, however, we must establish a sound foundation by introducing a number of key computing concepts. This is done in the next two sections.

EXERCISES 2-2

Using the method described in this section, develop algorithms to solve the following problems.

1. Develop an algorithm for changing a flat tire. Assume that a good spare and jack are available.

2. Develop an algorithm for making popcorn in a pot on the stove using butter, salt, and popping corn.

3. Develop an algorithm to replace a broken pane of glass in a window. Describe any materials needed in a separate list.

4. Develop an algorithm to get yourself out of the house in the morning. Begin from the state "asleep in bed" and include all your regular morning activities.

5. (*a*) Develop an algorithm for making a telephone call.
   (*b*) Expand part (*a*) to an algorithm for making a long-distance telephone call. Include the possibilities for a station-to-station call, a collect call, and a person-to-person call.

2-3 DATA, DATA TYPES, AND PRIMITIVE OPERATIONS

The primary purpose of any computer is the manipulation of information or data. These data can be weekly sales figures in a department store's inventory system, names and addresses in a mailing list, final grades in a computer science class, measurements from a scientific experiment, and so on. Most computers are capable of dealing with several different types (or modes) of data. When most people think of computers, they think of numeric data. Computers are very adept at performing computations on numeric data, such as the measurements from the scientific experiment, but as we will see, their capabilities extend beyond this type of data. Computer people are increasingly finding applications involving nonnumeric data, and many of these will be discussed. For example, computer processing of textual material has become quite common. These data are largely nonnumeric, involving strings of alphabetic characters.

The different types of data are represented in different ways in a computer and, in fact, the computer even has different instructions to deal with the different types. At this time it is not important to go into the details, but certain facts are important. Numeric data can be represented in two distinctly different ways: as *integer numbers* or as *real numbers*. It is important to remember that these are different representations and are handled differently by the computer.

Integers (sometimes referred to as *fixed-point numbers*, for reasons we will not go into) correspond to whole numbers. They have no fractional or decimal component and may be either positive or negative. The following are examples of integers:

```
  13      7
  -6    208
7,830    16
-295     25
```

Many applications require numbers that go beyond the realm of integers into that of *real numbers*. In our computer representation, real numbers always contain a decimal point. Fractions in a computer must be stored as their decimal equivalent, since there is no mechanism for storing numerators and denominators. Thus, the decimal portion of a real number is used to represent any fractional part. Unlike integers, real numbers can theoretically take on any value on the real number line. Again, the numbers can be either positive or negative. The following are examples of real numbers.

$$\begin{array}{ll} 23.8 & 3{,}738.72 \\ 3.6752 & -56.321 \\ -8.910 & -7.7 \end{array}$$

In some scientific applications, we require a special representation to handle very large numbers, such as the mass of the sun, or very small numbers, such as the thickness of a very thin film. A computer representation of any number, however, be it integer or real, has only a certain fixed number of digits. The actual number may vary from machine to machine, but eight significant digits is typical. Thus, we face a problem when we must represent a number such as 3,863,213,632 or 0.00000002857, for example. Neither of these numbers can be represented in only eight digits. (Note that the first is integer while the second is real.) Science has traditionally coped with this problem with a special notation in which a certain number of *digits of accuracy* (or precision) are expressed, followed by the magnitude of the number expressed as a power of 10. For example, we might write 3,863,213,632 to four digits of accuracy as 3.863×10^9. Although these values are clearly not the same, the second version is shorter to write and is probably sufficiently accurate for most applications. Most measuring devices have associated with them a certain limit of accuracy anyway. Try to measure the width of your thumb to the nearest 1/1,000 of an inch. For most applications, three or four digits of accuracy is enough. In the previous example we have given four. The power of 10 used (here, 9) converts the number to the correct magnitude. The conversion rule is:

1. Decide how many digits of accuracy you wish to retain (if necessary, the last digit is rounded)
2. Add 1 to the power of 10 used as the multiplier for each place that we moved the decimal point to the left (the decimal point may be shown or else only implied)

Suppose that we wished to retain seven digits of accuracy. The representation is derived as follows:

$$\underbrace{3\ 8\ 6\ 3\ 2\ 1\ 3}_{\substack{\text{digits to be}\\\text{retained}}}\underbrace{\ 6\ 3\ 2}_{\substack{\text{digits to}\\\text{be ignored}}}\quad \times \quad \underline{10^9}$$

$$\text{must be rounded} \qquad\qquad \text{current magnitude}$$

The decimal point moves three positions to the left.

$$3\ 8\ 6\ 3\ 2\ 1\ 3\ 6\ 3\ 2$$

giving us the representation

$$3\ 8\ 6\ 3\ 2\ 1\ 4 \times 10^3$$

This is the general principle for the representation of very large numbers. In actual practice, we usually carry it one step further. We always try to write the number in the same way. We will assume that in our *normalized representation*, the decimal point is always placed immediately before the first significant digit. In the example above, this requires moving the decimal point an additional seven places to the left. Thus, our final representation, to seven digits of accuracy, is

$$.\ 3\ 8\ 6\ 3\ 2\ 1\ 4 \qquad \times \qquad 10^{10}$$

digits of accuracy   magnitude
(precision)

We will refer to a number represented in this fashion as a *floating-point number*. This is nothing more than a special representation for real numbers. Incidentally, readers familiar with the use of logarithms will note a certain similarity in representation.

For very small numbers, we apply the same principles to get their floating-point representation. In this case, however, we shift the decimal point to the right to get the normalized representation, and as we do, we subtract one from the power of 10 used as the multiplier, for each place moved. The rest of our conversion rule applies as stated.

For example, to represent the number .00000002857 (or .00000002857 $\times 10^0$) to three digits of accuracy, we move the decimal point seven places to the right and round the 5 up to a 6 to get

$$.2\ 8\ 6 \times 10^{-7}$$

To summarize the conversion of very large or very small numbers to normalized floating-point form:

1. Decide how many digits of accuracy you wish to retain in the final representation. Round the last digit if necessary.
2. Adjust the magnitude portion of the number accordingly. For each place the decimal point is moved to the left, add one to the power of 10 used as multiplier; for each place moved to the right, subtract one from the power of 10.

These procedures may seem very tiresome at this point. In fact, they are quite important. The word size of the computer places very real limitations on the length of number that can be stored, as described earlier. In order to live within these limitations, yet retain the ability to perform calculations with a wide range of numbers, special representations are needed. The representation described is very close to the internal representation of floating-point numbers inside the computer itself. The representation has two components, called the *fraction* portion and the *exponent* portion. Each of these has an associated sign. To illustrate these terms on our earlier examples:

$$.3\ 8\ 6\ 3\ 2\ 1\ 4 \times 10^{10} \qquad \text{fraction:} \quad +.3863214$$
$$\text{exponent:} +10$$
$$.2\ 8\ 6 \times 10^{-7} \qquad \text{fraction:} \quad +.286$$
$$\text{exponent:} -7$$

The decimal point itself is not actually stored. Since it is assumed to be in the same position always (an advantage of normalized representation), we need not waste storage on it. It is inserted automatically whenever the number is read or written.

In addition to numbers, we are going to become extensively involved in the storage and manipulation of nonnumeric items, one example of which is something we will call the *character string*. A character string is as you might expect, a string of characters. These characters can be any of some large number of possible characters that are known to the computer: the familiar alphabetic characters (A, B, . . . , X, Y, Z, a, b, . . . , x, y, z), the digits (0, 1, 2, . . . , 8, 9), and a variety of special characters (blank, $, #, _, etc.). A character string is represented in programs as a collection of these characters beginning and ending with the special character quote ('), so we know where the string begins and ends. We refer to the quotes as string *delimiters*. A character string can be any length in theory. (In practice, there may be a limit, such as no string can exceed 255 characters.) Examples of character strings are:

'RICK BUNT'
'PAUL TREMBLAY'
'THE UNIVERSITY OF SASKATCHEWAN, SASKATOON'
'COMPUTING SCIENCE IS, OF COURSE, MY FAVORITE CLASS'
'333 + 444 GIVES 777'

Internally, the computer records two important pieces of information about a character string: the characters that make it up (excluding the delimiting quotes, which are not part of the string itself) and its length.

Occasionally, we may require a quote mark itself as one of the characters of a string. Suppose, for example, that we wished to process the string O'DELL. To distinguish such a quote from one of the string delimiters, we represent it by two consecutive quotes, as

'O''DELL'

Only one of these is actually stored as part of the string.

Other types of data also exist, although we will not be involved with them just yet. One of these is the *logical datum*. A logical datum has only two possible values, which we will call "true" and "false." The ability to perform logical operations is basic to computing; however, we will defer discussion of these operations until later in the book, when we will also make use of a type of datum known as a *pointer*. In a sense a pointer is an abstraction of the concept of address described in Sec. 2-1.

In summary, Fig. 2-12 shows a complete taxonomy of data types.

The representation of data is only one side of the coin; computations require that we be able to manipulate these data elements as well. We are all familiar with the four basic operations on numeric data: addition, subtraction, multiplication, and division. For nonnumeric data, the operations are not so familiar. First, let us consider the numeric operations briefly. Nonnumeric operations are dealt with in later chapters.

Addition and subtraction are represented in the usual way. Multiplication, however, has a small notational problem. To avoid possible confusion with the letter x, the multiplication operation is denoted by *, as in **36.8 *2.59**. Division also has a notational problem. Normally, we are required to keep the entire operation confined to a single line, a constraint dictated by limitations of traditional input media such as punched cards. This means that something like $\frac{8}{2}$ cannot be written. Instead, we use the slash to denote the division operation, as in **8 / 2**.

We have been very careful to distinguish between the two types of numeric data. We write them differently, and the computer stores them differently. As long as we continue to preserve this distinction in our mathematical operations, things remain fairly straightforward. The general rule is that the result of any operation has the same type as its two operands. For example, if we add two real numbers, we get a real result. If we multiply two integer numbers, we get an integer result. If we divide two real numbers, we get a real result. But a problem can arise if we divide two integer

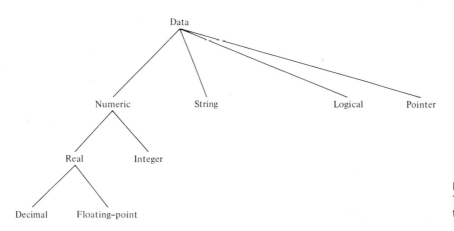

FIGURE 2-12
Taxonomy of primitive data types.

numbers. Our rule tells us that we get an integer result. Thus, the result of
8 / 2 is 4, an integer. What happens, however, if this were 8 / 5? We know
that the answer is 1.6, but this is not an integer. There are two possibilities,
in fact, for an integer result: either 1 or 2. If the resulting answer is 1, we
say that it has been *truncated*; if it is 2, we say it has been *rounded*.
Normally, the results are truncated. In any event, a degree of uncertainty
exists—an undesirable situation.

The situation becomes even more undesirable if allowed to go un-
checked. Consider the expression

$$\frac{1}{10} \times 10$$

The answer is obviously 1, right? Wrong! First, it must be expressed as
1/10 * 10. Now, since all numbers involved in the expression are integers,
all results (including intermediate results) must be integer. Let us suppose
we carry out these operations in the normal order, from left to right, that
is, dividing before multiplying. The first operation is 1/10. We know that the
real answer is .1, but for an integer result it is truncated to 0. We refer to
this as an *intermediate* result, since we have not yet completed the
calculation. The next operation is 0 * 10, which is clearly 0. Thus, 1/10 * 10
gives the answer 0. Observe that this obviously undesirable situation is a
direct consequence of the hazards of integer division. With real numbers,
this problem would not have arisen.

Fortunately, it is usually possible to mix types in numeric operations,
although this was not so in the early days of programming. That is, one
operand can be integer and the other, real. In such cases, the result is
always expressed as a real number. Thus, division need not pose a problem
as long as at least one of the operands is real. The result of the operation
1.0 / 10 is .1 as expected. Since the first operand (1.0) is real, the result will
be real.

The exponentiation operation is a fifth numeric operation provided in
most programming languages. This operation, like the previous four, is a
binary operation (that is, involving two operands), but it is expressed in a
different fashion in traditional mathematics. For example, 2^5 says to raise 2
(the base) to the power (or exponent) 5. This can be calculated as 2 * 2 * 2 * 2 * 2
(a total of five terms), which gives the result 32. As another example, 10^3 is
calculated as 10 * 10 * 10, which gives the result 1000. The two operands of
the exponentiation operation (the base and the exponent) can be either
integer or real, with the same rules as before governing the type of the
result. Thus, 3^4 is 81 (type integer), 3.0^4 is 81.0 (type real), and $2^{3.5}$ is 11.31
(type real). In some cases, such as $2^{3.5}$, exponentiation will, in fact, be
calculated using logarithms rather than repeated multiplications.

In traditional mathematical notation there is no explicit operator
denoting exponentiation. It is indicated by the presence of an exponent in

a raised position relative to the base. As in the case of division, however, we require the operation to be confined to a single line in our computer notation. For this reason, we will introduce a special symbol, ↑, to denote exponentiation. Thus, 8^2 is written for our purposes as **8 ↑ 2**. Thus, in programming notation, its appearance is more like that of the other four operations.

We now have some appreciation of the concept of data and manipulations that can be performed on numeric data. We have seen some of the basic operations that can be performed on numeric data. In later chapters we will see manipulations of nonnumeric data. Familiarity with data of all types is fundamental to success in programming.

EXERCISES 2-3

1. Give the type of each of the following constants:
   (a) 613
   (b) 613.0
   (c) −613
   (d) '613'
   (e) $−3.012 \times 10^{15}$
   (f) 17×10^{12}
   (g) $−28.3 \times 10^{-33}$
   (h) 'END OF QUESTION'

2. Give the result and its type for each of the following expressions:
   (a) 5 ↑ 20 + 3
   (b) 6 + 19 − 0.3
   (c) 3.0 ↑ 5.0 + 1
   (d) 1 / 4 + 2
   (e) 29.0 / 7 + 4
   (f) 3 / 6.0 − 7

3. Express each of the following as a real number in floating-point form with six significant figures (consult appropriate sources for the values required):
   (a) π
   (b) e
   (c) Avogadro's number
   (d) Mass of an electron (in kg)
   (e) Diameter of an atom (in cm)
   (f) Value of a parsec (in mi)

2-4 VARIABLES AND EXPRESSIONS

In the previous section some fundamental operations on numeric data were presented. We saw that it is possible to manipulate these data in a variety of ways and, by combining these fundamental operations, to form more elaborate expressions. In computer programs these expressions make up a major part of the specification of a computation. In this section we introduce an important concept that ties much of this together, the concept of a variable.

In a computer program, a *variable* is an entity that possesses a value and is known to the program by a name (sometimes called an *identifier*). In mathematics, the concept is very familiar. For example, if we denote the sides of a right-angled triangle by *a*, *b*, and *c*, as shown in Fig. 2-13, the

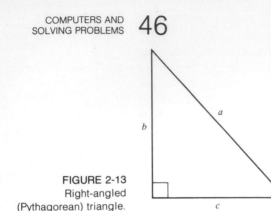

FIGURE 2-13
Right-angled
(Pythagorean) triangle.

Pythagorean theorem gives us the following relationship between the lengths of the three sides:

$$a^2 = b^2 + c^2$$

A relationship expressed using variables in this way defines a general formula that can then be applied to specific computations. For example, if side a is 5 in, and side b is 4 in, our formula tells us that side c is 3 in

$$(5)^2 = (4)^2 + c^2$$
$$c = \sqrt{(5)^2 - (4)^2}$$
$$= \sqrt{25 - 16}$$
$$= \sqrt{9}$$
$$= 3$$

A second triangle might have the following measurements: b is 5 in, c is 12 in. Although this is a different triangle from the first one, the general Pythagorean formula still applies and can be used to calculate the length of side a.

$$a^2 = (5)^2 + (12)^2$$
$$a = \sqrt{(5)^2 + (12)^2}$$
$$= \sqrt{25 + 144}$$
$$= \sqrt{169}$$
$$= 13$$

In computer programs too, the use of variables permits the specification of a general computational formula. As in mathematical formulas, variables in programs have names and take on values. In the examples above, we saw two different sets of values, as shown in Table 2-2. A variable can take on many different values, but at any particular time, it can have only one such value. Likewise, in Western society, a person may have many wives/

TABLE 2-2

Case	a	b	c
1	5	4	3
2	13	5	12

husbands, as long as he/she has only one at any point in time. The names
of the variables used in the Pythagorean formula were simply a, b, and c.
In computer programs we prefer more descriptive names, such as possibly
LEFTLEG, RIGHTLEG, and HYPOTENUSE in this case. This makes their
purpose more apparent to the reader of the program.

There are usually some simple rules concerning the naming of a
variable. These rules may vary slightly, depending on the particular programming language or computer system you are using. For our purposes,
we will require that a variable name begin with a letter. After this the
characters may be chosen from a set of possible characters containing
letters, numeric digits, and some special characters. Blanks will not be
allowed within a variable name. The following are examples of variable
names that we will consider to be valid:

TOTAL   A123
SIDE2   BLACK_BOX
XSQUARED   A_LONG_NAME

Note the use of the special character _ in the last two examples. Since
blanks are not allowed, we may use this character (called the *break character*)
to separate words in a multiword name; A_LONG_NAME is much easier to
read than ALONGNAME. The following are examples of variable names that
we will consider to be invalid:

3FILE   Does not begin with a letter

X+Y   "+" is not one of the special characters permitted
               in a variable name; otherwise, how would we dif-
               ferentiate this from the symbol for addition?

TWO WORDS   Blanks are not allowed

A variable stands for something, specifically some piece of data, in an
expression. In the Pythagorean formula, a, b, and c stood for the lengths
of the sides of the right-angled triangle. In the expression

LENGTH * WIDTH

LENGTH and WIDTH might be variables standing for the sides of a rectangle.
The names chosen tell us something about how we intend these variables
to be used. If we give them specific values, such as **20** and **15**, the value of

the expression (**300**) gives us the area of the rectangle. If we supply different values, the expression has a different result. The expression

BASE * HEIGHT / 2.0

gives the area of a triangle if we supply the appropriate values to the variables **BASE** and **HEIGHT**, as shown in Fig. 2-14. This particular expression contains a total of three data items: the variables **BASE** and **HEIGHT**, and a constant, **2.0**. **BASE** and **HEIGHT** can take on different values at different times, but the value of the constant **2.0** never changes.

In the previous section we introduced five important data types: integer, real, string, logical, and pointer. The types apply to variables as well as constants. For constants it is easy to recognize the types. For example, an integer constant is a number with no decimal point (for example, **2, 41, −6**), a real constant has a decimal point (for example, **16.5, −2.789, 0.000467**), and a string constant is a sequence of characters enclosed in single quotes (for example, **'THIS IS A TYPICAL STRING'**). All the expressions in the previous section, were, in fact, composed solely of constants of various types. When variables are used in expressions, they, too, must be of a certain type. An integer variable can take on only integer values, a real variable can take on only real values, a string variable can take on only string values, and a logical variable can take on only the values "true" or "false."

Now that we have some idea of how variables are used, let us see how we give them their values. There are two ways of doing this. One way is to read a value into the variable by means of an input operation. We will look at the input operation later in this section along with its companion, the output operation. The other way is to "assign" the variable a value by means of the assignment operation.

2-4.1 THE ASSIGNMENT OPERATION

The *assignment operation* is a very natural way to specify that a variable be given a value. We simply assign that value to the variable. The assignment

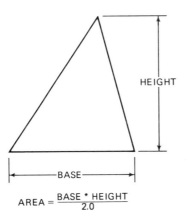

FIGURE 2-14
Computing the area of a triangle.

$$AREA = \frac{BASE * HEIGHT}{2.0}$$

operation will be denoted by the symbol ←. For example, the statement

    A ← 3

means "variable **A** is assigned the value **3**." If you like, you can think of **A** as representing a word in the memory of the computer. After this assignment statement has been executed, this word contains the number **3** (in integer format). Notice that since a memory word can contain only one value at a time, the number **3** replaces whatever value, if any, had been previously stored in that word. We say that assignment is a *destructive* operation, since whatever value the variable possessed prior to the processing of the assignment operation has been lost; it has been replaced by the new value. Thus, if the sequence of operations

    A ← 16
    A ← −27
    A ← 1

were executed, the value of the variable **A** after the three operations would be 1. The values **16** and **−27** have been destroyed—overwritten by subsequent assignments to the same variable. Note that when a sequence of assignments is given, the individual operations are always executed in the order given. We will deal at greater length with the sequence in which statements are executed in later sections.

2-4.2 TYPE CONVERSIONS

Earlier we said that variables must be of a certain type and can only take on values of that type. In the examples above we were operating under the implicit assumption that **A** is an integer variable since it is being given integer values. Suppose that **B** were a real variable. **B** can then take on real values as shown in the following assignment statements:

    B ← −295.3
    B ← 0.0006

Suppose that **C** were a variable of type string. We could then assign string values to **C**, as in

    C ← 'A STRING CONSTANT'
    C ← 'CHICAGO, ILLINOIS, 60611'

    Special action must be taken, however, if we try to assign a value to a variable that is not of matching type. An error will result if we try to assign a string value to a numeric variable, or a numeric value to a string variable. These are simply not compatible. We can, however, assign an integer value to a real variable, or a real value to an integer variable. The value that the variable actually possesses will always match the type of the variable, although the value being assigned may have to be converted to this type to conform. For example, suppose that the variables **VI** and **VR** are of type integer and real, respectively. Suppose further that the following assignment operations are specified:

VR ← −17
VI ← 283.682

In the first case, the integer constant, −17, must be converted to real to conform with the type of the variable VR. This is done automatically by the computer, and the value is stored as −17.0, a real value equal to the original constant. This is, in fact, a good way of converting integer values to real for use, say, in a division operation (see earlier remarks on division). In the second case, the real constant, 283.682, must be converted to integer to conform with the type of the variable VI. This is done (again, automatically by the computer) by truncation, that is, by chopping the number off at the decimal point. Thus, the value actually stored is 283, an integer value. Note, however, that this is not equivalent in value to the original constant. It is therefore important to remember that conversion of real numbers to integers may alter the value. This can lead to problems for the unwary programmer. It is best to avoid such conversions unless they are required for a special purpose.

2-4.3 EXPRESSIONS

The form of assignment operation that we have used to this point, that of assigning a constant to a variable, is only a special case of a more general form. The right-hand side of the assignment can, in fact, be any expression, where an expression is a combination of variables, constants, and operators. The result of the evaluation of that expression is the value that is then assigned to the indicated variable. The comments on type compatibility remain in effect. The general form of the assignment statement is then

variable ← expression

As an example of an expression used in an assignment statement, consider the following. Assume INT to be an integer variable.

INT ← 3 + 16 + 8

We begin by evaluating the expression on the right-hand side of the assignment arrow. This yields a value of 27, which we then assign to the variable INT. The effect of this particular assignment, then, is that the variable INT receives the calculated value 27.

The expressions on the right-hand side in an assignment statement may also contain variables, the values of which have been previously assigned. Consider the following sequence of assignment statements, where TERM1, TERM2, and RESULT are all real variables.

TERM1 ← 13.6 + 7.4
TERM2 ← 0.7 * 28.6
RESULT ← TERM1 / TERM2

We begin by evaluating the expression 13.6 + 7.4 to get a real value 21.0. By assignment, this becomes the value of the real variable TERM1. In the second statement, the expression 0.7 ∗ 28.6 yields the real value 20.02, which is assigned to real variable TERM2. In the third statement, we divide the value currently possessed by TERM1 (21.0) by the value currently possessed by TERM2 (20.02) to get 1.048951. This is then assigned to the real variable RESULT.

After execution of this sequence of statements, the values of the three variables are as shown in Table 2-3. It is important to observe that the variables TERM1 and TERM2 have retained the values they were last assigned. The statement "RESULT ← TERM1 / TERM2" does not alter the value of any variable on the right-hand side of the assignment symbol. The values of TERM1 and TERM2 can be used in subsequent operations.

A point worth emphasizing is that any variables used in any expression (such as the expression on the right-hand side of an assignment symbol) must have values at the time this expression is evaluated. For example, if the order of the statements in the last example were altered to read

```
TERM1 ← 13.6 + 7.4
RESULT ← TERM1 / TERM2
TERM2 ← 0.7 ∗ 28.6
```

an error would result even though the same three statements are present. The statements are carried out in the indicated order, and at the time the expression "TERM1 / TERM2" is to be evaluated, the variable TERM2 has not yet received its value. Thus, it is not possible to complete this operation. It is the programmer's responsibility to ensure that all variables appearing in an expression have values at the time the expression is to be evaluated.

2-4.4 MODIFYING A STORED VALUE

Suppose that we have a value stored in a variable X (let us assume that X is an integer variable for this example). We know that if we assign a new value to that variable it replaces the value that it had prior to the assignment. Suppose that we encounter the following sequence of statements, where A is also an integer variable.

```
X ← 0
A ← 0
X ← A + 1
```

TABLE 2-3

Variable	Value
TERM1	21.0
TERM2	20.02
RESULT	1.048951

As a result of the first two statements, X and A both have the value 0. The third statement says "take the current value of the variable A (0), add 1 to it, and assign the result to the variable X." Thus, after execution of the third statement, X has a new value, 1, while A retains its value, 0.

Suppose that, instead, we had written

    X ← 0
    X ← X + 1

How would this second statement be interpreted? It would be processed in precisely the same way as the third statement in the previous example: "Take the current value of the variable X (0), add 1 to it, and assign the result to the variable X." Thus, after execution of the second statement, X has a new value, 1. We have modified the value of the variable X by adding 1 to it.

This is a fundamental concept in computer programming. The whole point of variables is that they may hold different values at different times. The ability to modify these values, relative to what values the variables have, is very important indeed. Do not be perplexed by the appearance of the same variable on both sides of the assignment arrow. Recognize that these appearances mean different things in the processing of the statement. On the right-hand side the appearance of a variable indicates that its value is to be used; on the left-hand side it indicates that a value is to be assigned to it.

Many programming languages (including FORTRAN, PL/I, and BASIC) use the equal sign (=) as the symbol for assignment. Taken to mean equality, the statement "X = X + 1" is nonsense mathematically. As an assignment statement, it means, simply, increase the value of X by 1. This is confusing for many beginning programmers. To alleviate this problem somewhat, we have chosen to use the symbol ← for assignment in the algorithmic notation, reserving the equal sign for the equality relation.

2-4.5   PRECEDENCE OF OPERATORS

Earlier in the book we spoke of the dangers of ambiguity. Consider the following assignment statement:

    VAR ← 3 + 6 * 13

What value does VAR receive? This depends on the order in which the two mathematical operators (* and +) are processed. If we process them strictly left to right, that is, performing the addition before the multiplication, we get the result 117. This evaluation process can be represented diagramatically as in Fig. 2-15.

Suppose, instead, that we performed the multiplication before the addition, as shown in Fig. 2-16. In this case our result is 81. Which result is correct?

To get around such difficulties, traditional mathematics has defined some additional rules on the evaluation of expressions, specifically on the

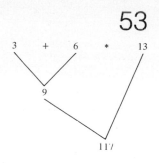

FIGURE 2-15
Evaluation of a simple
expression.

order in which the individual operations are performed. With each operator is associated a *precedence* or priority. Operators with highest precedence are processed first, again proceeding from left to right. For example, multiplication is given a higher precedence than addition. Thus, in the expression **3 + 6 * 13**, the term **6 * 13** is processed first, giving **78**; then the addition operator is processed, giving the final result **81**, as shown in Fig. 2-16. Notice that with this rule, the same result is obtained if the expression is written as **13 * 6 + 3** (see Fig. 2-17).

This evaluation procedure is easily applied to more complex examples. Consider the expression

8 + 7 * 3 + 4 * 5

We see two multiplications and two additions. The multiplications, being of higher precedence, are processed first (from left to right), yielding the intermediate results **21** and **20**. The additions are then processed to give the final result, **49**. This evaluation is shown in Fig. 2-18.

Before we give the precedences of the other mathematical operators, let us introduce one additional wrinkle, the use of parentheses. Parentheses can be used by a programmer to group terms as an explicit indication of the order in which he/she wishes the operations in an expression to be processed. Operations in parentheses are evaluated first. For example, suppose that in the last case we had wanted the additions to be processed before the multiplications. For this we would write the expression as

(8 + 7) * (3 + 4) * 5

This gives the result **525**, as shown in Fig. 2-19.

Parentheses can be nested, as in

(3 * (6 + 2)) * 8

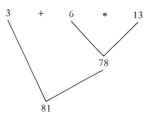

FIGURE 2-16
Alternative evaluation of
the same expression.

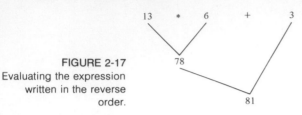

FIGURE 2-17
Evaluating the expression
written in the reverse
order.

In this case the inner parentheses are processed first. Thus, the result of this expression is **192**, as shown in Fig. 2-20.

Now, let us look at the complete picture. The standard mathematical operators can be grouped by precedence into four classes. The highest precedence is given to the exponentiation operator, ↑. Next come the sign operators, − for negative and + for positive (which is rarely present, as an unsigned value is assumed to be positive). These are sometimes called *unary minus* and *unary plus,* to distinguish them from the binary operators for subtraction and addition, which use the same symbols. Then we have the operators ∗ and / for multiplication and division. Finally, the lowest precedence goes to + and −, used for addition and subtraction. These are summarized in Table 2-4.

TABLE 2-4 Standard Operator Precedences

Class	Operators*	Meaning
1	↑	Exponentiation (applied right to left)
2	−, +	Sign operators (unary minus and plus, also applied right to left)
3	∗, /	Multiplication, division (applied left to right)
4	+, −	Addition, subtraction (applied left to right)

* Operators grouped together have the same precedence.

You will notice that the exponentiation operator is applied somewhat differently from the four more common operators. Exponentiation operations are processed from *right to left*, whereas multiplication, division, addition, and subtraction are all processed from *left to right*. Thus, the value of **2 ↑ 3 ↑ 2** (written mathematically as $2^{3^{2}}$) is **512** rather than **64**. The

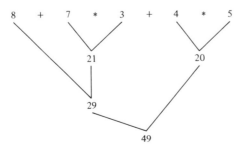

FIGURE 2-18
Evaluating a more
complex expression
according to the rules of
operator precedence.

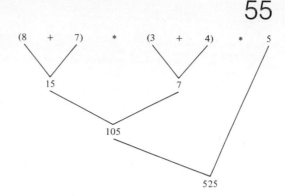

FIGURE 2-19
Evaluating a complex
expression containing
parentheses.

sign operators + and − (not addition and subtraction) are, in fact, also
processed from right to left, although the need to apply this rule occurs very
infrequently.

2-4.6 EVALUATION OF EXPRESSIONS

Once we have precedences associated with the operators, the evaluation of
an expression is straightforward. We simply scan the expression from left
to right a separate time for each class of operators represented in the
expression. Consider the following expression as an example:

$$-3 * 7 + 2 \uparrow 3 / 4 - 6$$

Each of the four classes of operators is represented in this expression. This
means a total of four scans are required, or four passes over the expression.
On the first pass we process all exponentiation operations, applying them
from right to left; in this case, there is only one: **2 ↑ 3**, giving **8** as an
intermediate result. Our expression now reads

$$-3 * 7 + 8 / 4 - 6$$

       ↑
       intermediate result from first pass

On our second pass we process all sign operators. Once again, there is
only one: **−3**. Although this does not alter the form of the expression for
the next pass, it tells us to interpret the first term as the constant **−3**. Note

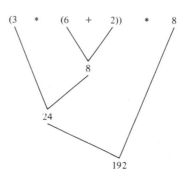

FIGURE 2-20
Evaluating an expression
with nested parentheses.

that the remaining + and − are not sign operators, but rather operators indicating addition and subtraction. This is clear from their context.

On the third pass over the expression we process the multiplications and divisions. There is one of each—they are processed from left to right. The term −3 * 7 yields the intermediate result −21; the term 8 / 4 yields the intermediate result 2. The expression now reads

−21 + 2 − 6
↑   ↑
intermediate results

On the fourth pass we process the addition and subtraction operations. Again, there is one of each, and we take them in turn from left to right. This gives us our final result for this expression, which is −25. Diagrammatically, the complete evaluation of this expression is shown in Fig. 2-21.

We mentioned earlier the use of parentheses to group terms in an expression. Since parenthesized terms are evaluated before any others, a programmer can use parentheses to overcome the effects of the standard operator precedences. For example, suppose that you were asked to represent the following fraction:

$$\frac{12 - 2}{2 + 3}$$

Recall that limitations of input and output devices force us to write all expressions in programs on a single line. If we try to put this on one line exactly as it appears, we get

12 − 2 / 2 + 3

Because of precedence rules, however, this does not give the required value. Since the division is of higher precedence than the subtraction and addition,

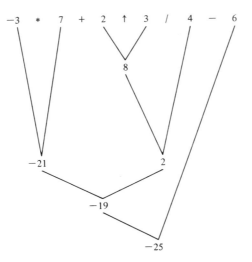

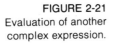

FIGURE 2-21
Evaluation of another
complex expression.

the calculated result would be 14 instead of the required 2. To get the correct interpretation we must use parentheses to indicate that the terms of the numerator and denominator are to be grouped. The correct expression is

$$(12 - 2) / (2 + 3)$$

Need a programmer ever bother to learn the rules of precedence? Through parenthesization, any order of evaluation can be imposed on an expression. Excessive use of parentheses, however, gives a strange appearance to programs and can make them hard to read. In most cases the standard precedence rules lead to a natural interpretation. You will not find them difficult to remember as you become used to them.

It is worth commenting at this point that this tiresome procedure is not, in fact, actually followed by the computer itself as it evaluates expressions. If it were, it would take unwarranted amounts of time to process very simple expressions. The computer uses faster, more sophisticated methods, some of which are described in later chapters. They give the same results, however, as the "brute-force" methods that we have described here.

2-4.7   BUILT-IN FUNCTIONS

Despite the best intentions of language designers, it is not possible to specify all operations required in programs in the form of conventional operators such as +, −, *, and /. Quite often the conventional set is supplemented by a very useful set of special operators that we will call *built-in functions*. For example, we will use **SQRT** to denote the square-root operation rather than introducing another special symbol. Thus, if we know the lengths of the legs of a right-angled triangle (say, **LEFTLEG** and **RIGHTLEG**), the following expression gives the length of the hypotenuse:

HYPOTENUSE ← SQRT(LEFTLEG ↑ 2 + RIGHTLEG ↑ 2)

Built-in functions are prewritten routines supplied by the designers of the programming language to assist the programmer in performing computations that require more than the conventional set of operators. Table 2-5 lists some of the common built-in functions that are provided in most programming languages, along with instructions for their use. We will introduce additional functions later in the book. Appearing in expressions, built-in functions have very high precedence—higher, in fact, than that of the exponentiation operator.

2-4.8   INPUT AND OUTPUT

Computer calculations are of little value to us unless we can, first, supply the data on which we wish the calculations to be performed, and second, see the results of these calculations. Since input and output operations are very much concerned with interacting with programmers, their form is highly dependent on the specific programming language used, and occasionally, even on the computer system itself. For this book we will define special forms that will be useful in later applications.

TABLE 2-5   Some Useful Built-In Functions

Function Name	Argument and Type	Meaning and Type of Result
ABS(ϵ)	ϵ: a real or integer expression	Absolute value: result is $|\epsilon|$, same type as ϵ
SQRT(ϵ)	ϵ: a real or integer expression (must be ≥ 0)	Square root: result is $\sqrt{\epsilon}$, type real
TRUNC(ϵ)	ϵ: a real expression	Truncate: result is the largest integer small than or equal to ϵ
ROUND(ϵ)	ϵ: a real expression	Round: result is the largest integer smaller than or equal to ($\epsilon + 0.5$)
LOG(ϵ)	ϵ: a real expression	Logarithm base e: result is the natural (Naperian) logarithm of ϵ, type real
LOG10(ϵ)	ϵ: a real expression	Logarithm base 10: result is the base 10 logarithm of ϵ, type real
EXP(ϵ)	ϵ: a real expression	Exponent: result is e^ϵ, type real
SIN(ϵ)	ϵ: a real expression	Sine: result is the sine of ϵ (radians), type real
COS(ϵ)	ϵ: a real expression	Cosine: result is the cosine of ϵ (radians), type real
TAN(ϵ)	ϵ: a real expression	Tangent: result is the tangent of ϵ (radians), type real

We define two new types of statement to handle input and output. The *read* statement allows us to read given values into indicated variables; the *write* statement allows us to display results. Input may come from punched cards by way of a card reader, or from some type of terminal device. Output may appear on a display screen or be printed on paper. We will not concern ourselves with the details of the devices.

The form of the read statement that we will use is the following:

Read(*input list*)

The *input list* gives the names of the variables to which values are to be given as they are encountered in the input stream. If, for example, the statement is

Read (A, B, C)

the next three values encountered would be given to the variables A, B, and C—the first to A, the second to B, and the third to C.

Many of you will be using punched cards for the input of both programs and the data used by these programs. Figure 2-22 shows a standard 80-*column punched card* (sometimes referred to as a *Hollerith card* in honor of the man who made them popular). It measures $7\frac{3}{8}$ in by $3\frac{1}{4}$ in and consists

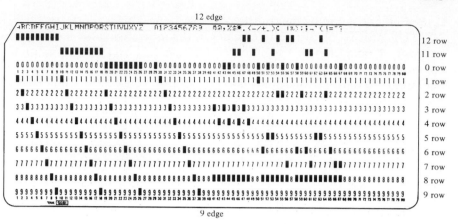

FIGURE 2-22
The format of a standard 80-column punched card.

of 12 rows. Ten of the rows are numbered 0 through 9 on the card; the top two have no numbers but are sometimes referred to as *rows 12* and *11*. The upper edge of the card is known as the *12 edge*, and the bottom as the *9 edge*.

Each column of the card is capable of representing a single character. Each character is represented by a specific pattern of punched holes. A standard coding scheme associates certain patterns with certain characters. As the card passes through the card reader, special sensing units detect the presence of holes and determine the particular character according to the pattern found. A special machine known as a *keypunch* is used to prepare punched cards. It operates much like a typewriter with special facilities for the movement and positioning of cards.

Different programming languages require the information on program cards and data cards to be formatted in different ways. We will not deal with the format of program cards here. With respect to data cards, we will adopt a very general format. We will assume that the individual data items are simply punched in a series on the data cards, with either a comma or a blank separating successive items. Character strings must be enclosed in delimiting quotes.

Those of you who will be programming on interactive terminals can replace all occurrences of "card" with "input line." Everything else is basically the same.

Let us return now to our read statement. Suppose that values were being read from punched cards. Suppose further that the values punched on the cards were −16, 3, 7, 21, 16, 0, 4, 8, 1. These could be on one card, or spread over several. Suppose that we had the following three read statements:

```
Read(A, B, C)
Read(D, E, F, G)
Read(X, Y)
```

The values from the input stream would be given to the variables A, B, C,

D, E, F, G, X, Y one by one in the order written. This is illustrated in Fig. 2-23. The process of giving these values to the indicated variables is like the assignment operation in two important respects. First, it is a destructive operation in the same sense that the assignment operation was a destructive operation. Whatever values, if any, that the variables had previously possessed are overwritten. Also, the comments on type compatability hold here as well. If the value in the input stream is of a different type from that of the variable in the input list, a conversion is required.

The output statement is like the input statement in format, but a bit more general in terms of what can be specified in the *output list*. It is possible to display the contents of any variable, the result of any expression, or the value of any constant. The general form of the write statement that we will use is

Write(*output list*)

We will assume that, unless otherwise stated, the values from a single output list appear on the same line. To simplify matters we will allow ourselves to take liberties with the format of printed output. We will assume whatever output format is convenient to the presentation. We will ignore questions of spacing and questions of the number of decimal places shown for real numbers. Details such as these vary from language to language, and are dealt with in the companion manuals.

The following sequence of statements illustrates the action of a write statement. Assume that all variables are real.

MARK1 ← 73.0
MARK2 ← 65.0
MARK3 ← 94.0
MARK4 ← 87.0
AVERAGE ← (MARK1 + MARK2 + MARK3 + MARK4) / 4.0
Write(AVERAGE)

The first four statements assign to the indicated variables a student's grades on four tests. The fifth statement computes the average grade. The write statement then displays this calculated result:

79.75

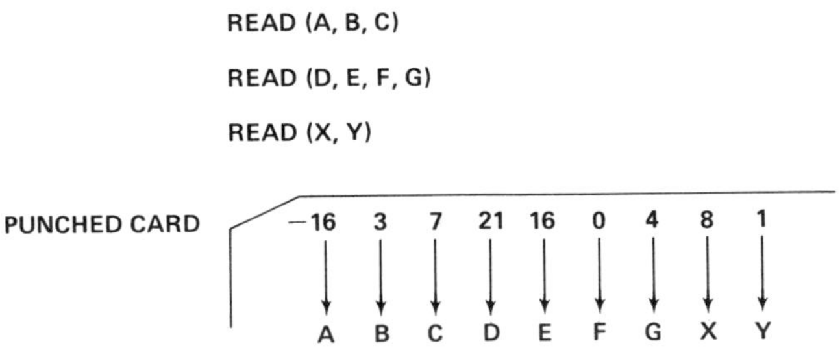

READ (A, B, C)

READ (D, E, F, G)

READ (X, Y)

PUNCHED CARD

−16   3   7   21   16   0   4   8   1

A   B   C   D   E   F   G   X   Y

FIGURE 2-23
Reading data from a
punched card.

Perhaps the instructor wishes to see, as well as the calculated average, the individual grades used to compute it. Replacing this write statement with

    Write(MARK1, MARK2, MARK3, MARK4, AVERAGE)

would do the trick. The resulting output line would now be

    73.0 65.0 94.0 87.0 79.75

Note that all numbers are written as real numbers. This is because the variables themselves are all real variables. Now, however, we have simply five numbers displayed. Looking at this output, you can see that there is nothing to tell someone reading it what the numbers signify. This could easily lead to confusion and/or misinterpretation. Suppose, instead, that we gave the following output statements:

    Write('INDIVIDUAL GRADES ARE', MARK1, MARK2, MARK3, MARK4)
    Write('FINAL AVERAGE IS', AVERAGE)

In this case the output would be displayed as

    INDIVIDUAL GRADES ARE 73.0 65.0 94.0 87.0
    FINAL AVERAGE IS 79.75

(Note that delimiting quotes do not appear in displayed output.) In terms of information content, the last form is definitely superior. Not only is all the information presented, but it is also clearly labeled. Those of you who receive itemized bills with particulars such as the cryptic "**BX STY PNS**" ought to appreciate the value of clear, identifiable labels. This is also important, as we will see, in the choice of variable names. Do not force your readers into word games they may not want to play.

This last example shows the use of two different kinds of items in an output list. We have used variables, in this case numeric variables such as **MARK1** and **AVERAGE**. We have also used constants, in this case string constants such as '**FINAL AVERAGE IS**'. We can also use numeric constants or string variables. For example, suppose that **GIVEN_NAME** and **SURNAME** are string variables in the following:

    GIVEN_NAME ← 'DONALD'
    SURNAME ← 'DUCK'
    Write(SURNAME, ',', GIVEN_NAME, 3)

The output is the current value of the string variable **SURNAME**, followed by the string constant ',', followed by the current value of the string variable **GIVEN_NAME**, each displayed without delimiting quotes. The final item in the output list is the numeric constant **3**. The resulting output line is

    DUCK, DONALD 3

In many programming languages it is also possible to include an expression as part of an output list. The expression is first evaluated, then the result is printed. For instance, in our final version of the averaging example, the same output would result from the following:

```
Write('INDIVIDUAL GRADES ARE', MARK1, MARK2, MARK3, MARK4)
Write('FINAL AVERAGE IS', (MARK1+MARK2+MARK3+MARK4) / 4.0)
```

There can be many details and special features associated with input and output in a specific programming language, for example, formatting details. What we have given here is a very general approach. More details will be added later as warranted by particular applications.

EXERCISES 2-4

1. Give the value of the variable **RESULT** after execution of the following sequences of operations (assume all variables to be real):
   (a) RESULT ← 3.0 * 6
   (b) X ← 2.0
       Y ← 3.0
       RESULT ← X ↑ Y − X
   (c) RESULT ← 4
       X ← 2
       RESULT ← RESULT * X

2. Give the value of each of the following integer expressions:
   (a) 16 * 6 − 3 * 2
   (b) −2 ↑ 3
   (c) (28 + 3 * 4) /4
   (d) 3 + 2 * (18 − 4 ↑ 2)
   (e) 2 ↑ 2 * 3
   (f) 8 − 30 / 6

3. Assume that **A**, **B**, and **C** are real variables and that **I**, **J**, and **K** are integer variables. Given **A = 4.0**, **B = 6.0**, and **I = 3**, what is the final value requested in each of the following?
   (a) C ← A * B − I   C = _____
   (b) K ← I / 4 * 6   K = _____
   (c) C ← B / A + 1.5   C = _____
   (d) K ← TRUNC(B / A + 4.7)   K = _____
   (e) J ← ROUND(A / (5 / I))   J = _____
   (f) K ← ABS(A − B) * 2 + I   K = _____

4. Write the following mathematical expressions as computer expressions:

   (a) $\dfrac{a}{b} + 1$

   (b) $\dfrac{a + b}{c - d}$

   (c) $\dfrac{a + \dfrac{b}{c}}{d - \dfrac{e}{f}}$

   (d) $a + \dfrac{b}{c - d}$

   (e) $(a + b)\dfrac{c}{d}$

   (f) $[(a + b)^c]^d$

   (g) $\dfrac{\sin a + \cos a}{\tan a}$

   (h) $\dfrac{-b + \sqrt{b^2 - 4ac}}{2a}$

2-5 DESCRIPTION OF ALGORITHMS

The development of a suitable algorithm is the first stage in the preparation of a computer program for a given application. In Sec. 2-2 we saw that an algorithm is a sequence of steps to be followed in a carefully prescribed order. The importance of precision was emphasized. The steps of an algorithm must be clearly and unambiguously expressed.

The description of an algorithm in a clear, easy-to-follow manner aids in its development, its subsequent transformation to a running program, and the documentation of the program after it has been completed. The latter aspect is a particularly important stage of the programming process about which more can be said after you have had more experience writing programs. In this section we will investigate several approaches to the description of algorithms.

2-5.1 NARRATIVE DESCRIPTION

A straightforward method of expressing an algorithm is simply to outline its steps verbally. Recipes, for example, are normally expressed in this fashion. Since apparently very little in the way of specialized training is required for narrative description, it seems easy to do on the surface. However, this is not always the case. Natural language is wordy and imprecise and often quite unreliable as a vehicle for transferring information. It is not impossible to be precise with natural language; that it is difficult, however, ought to be clear to anyone who has ever struggled with the fine print on a contract or legal document. Comprehensibility is sacrificed for precision. The difficulty of transferring information with natural language is, in fact, the basis of a popular game that you probably played as a child. The players are seated in a circle and the game begins with one player whispering a story to his/her neighbor. He/she, in turn, passes the story on to his/her neighbor, and so on around the circle. The last player then repeats the story aloud. The final version of the story usually bears shockingly little similarity to the original.

Because of the imprecision of natural language, the danger of misinterpretation or information loss looms large. Subtle nuances and shades of meaning may stimulate the reader's imagination in a novel, but in technical writing, unclear details can lead quickly to errors.

For these reasons, natural language is unsuitable as the sole vehicle for the expression of algorithms. Despite its drawbacks, however, prose does have a place, possibly in presenting clarifying remarks or outlining special cases. For this reason we will not discard it entirely in our search for better methods of expression.

2-5.2 THE FLOWCHART

It is often easier to convey ideas pictorially than verbally. Maps provide a convenient representation of the topology of a city, and may be of more value when you are lost than verbal directions. Assembly directions for a newly acquired piece of equipment are usually much more comprehensible if diagrams are included. A very early attempt at providing a pictorial format for the description of algorithms, which dates back to the days of von Neumann, involved the use of the flowchart. A flowchart shows the logic of an algorithm, emphasizing the individual steps and their interconnections. Over the years a relatively standard symbolism has emerged (see Fig. 2-24). A computation is represented by a rectangular shape, a decision by a diamond shape, and input and output operations by shapes symbolizing the media used. The boxes are connected by directed lines indicating the order in which the operations are to be carried out. A simple flowchart, with annotations, is shown in Fig. 2-25.

Over the past few years the attitude of the computing profession toward the once popular flowchart has cooled considerably. In his book of essays entitled *The Mythical Man-Month*, Frederick P. Brooks offers that "the flowchart is a most thoroughly oversold piece of documentation." A recent study led by Shneiderman at Indiana University raises serious questions as to the utility of detailed flowcharts. Thoughts on the flowchart vary. By some, it is considered simply as an unnecessary appendage to a program; by others, it is viewed as an instigator of bad programming habits. A new generation of programmers is being schooled in new methods of program development—methods to which the flowchart in its familiar form adds little. Very often, while showing the *logic* of an algorithm, the flowchart obscures its *structure*. We feel that the structure is at least as important (and quite possibly more). As a result, we will make very little use of flowcharts in this book. The reader who wishes more information on the subject will find it readily available.

2-5.3   AN ALGORITHMIC LANGUAGE

Our approach in this text is to extract the best features of the previous two approaches and combine these in a special language for the expression of algorithms. From the narrative approach we borrow the descriptiveness of prose. To this we add the conciseness of expression of the flowchart. The result is a kind of "pidgin programming language," similar in flavor to several programming languages. We have avoided tying the algorithmic language too closely to any particular programming language, although there are obvious influences. It is sufficiently general that the translation from algorithm to program ought to be quite straightforward regardless of the programming language to be used. The final decision on which programming language to use depends on many factors, ranging from the nature of the

FIGURE 2-24
Common flowcharting
symbols.

Compute   Decision   Input   Output

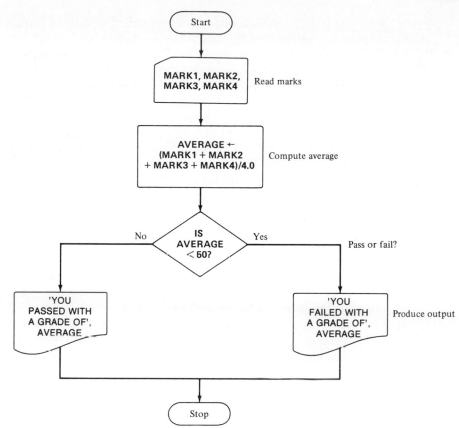

FIGURE 2-25
Simple flowchart.

particular application and whether the language has features suited to it, to the availability of languages on the host computer system.

The algorithmic language that we will use is perhaps best described through example. Consider the following algorithm, for a problem described earlier.

Algorithm PYTHAGORAS. Given the lengths of the legs of a right-angled triangle (in variables **LEFT_LEG** and **RIGHT_LEG**), this algorithm determines and prints the length of the third side (**HYPOTENUSE**). All variables are assumed to be integer.

1. [Input known lengths]
   Read(LEFT_LEG, RIGHT_LEG)
2. [Compute squares of known lengths]
   SQ1 ← LEFT_LEG ↑ 2
   SQ2 ← RIGHT_LEG ↑ 2
   (the purpose of this step is to reduce the complexity of the expression in step 3)

3. [Calculate length of third side]
    HYPOTENUSE ← SQRT(SQ1 + SQ2)
    (SQRT indicates the square-root operation)
4. [Output]
    Write('SIDES ARE', LEFT_LEG, RIGHT_LEG, HYPOTENUSE)
5. [Finished]
    Exit   □

Study this example carefully. The algorithm is given an identifying name, in this instance **PYTHAGORAS**. A brief narrative describing the function of the algorithm follows. The algorithm itself is then expressed as a sequence of numbered steps.

Each step of the algorithm begins with a short remark enclosed in square brackets. This phrase gives a brief description of that particular step and may, in fact, be one of the steps from a more general description of the algorithm. Following the brief description is a statement or series of statements that describes the actions required by that step. The steps are executed in sequence, beginning at step 1; the statements in each step are executed in the order given. The special verb "Exit" causes execution of the algorithm to terminate at the point where it is encountered. We will see later that algorithms may have several *logical* ends, depending on conditions within the algorithm, but can have only one *physical* end. "Exit" marks a logical end of the algorithm; its physical end is denoted by the symbol □.

Let us examine the steps of this particular algorithm more closely. Step 1 causes the input data to be read into the variables **LEFT_LEG** and **RIGHT_LEG**. This is indicated by a read statement. Step 2 contains two statements, so closely related that they are expressed as part of the same step. This step is further explained by a comment enclosed in parentheses. Comments are purely descriptive, intending to help the reader better understand the step, and do not indicate any action to be carried out. The comment in step 2 explains that the purpose of this particular step is to simplify the expression required in the following step.

In step 3 the value of **HYPOTENUSE** is computed. As the attached comment tells us, the square-root operation is indicated by the built-in function **SQRT**, with the quantity involved given as the argument of the function in parentheses. Step 4 causes the lengths of the three sides to be displayed along with an explanatory message. Step 5 causes the algorithm to terminate.

In Chap. 3 we will be describing some additional features of this algorithmic language. It is important that you understand the basic features described here first. Reread the previous example to be sure.

2-5.4 TRACING AN ALGORITHM

Once we have developed an algorithm, how can we be sure that it does what it is supposed to do? One way is to execute the algorithm by hand, on

TABLE 2-6   Trace of Algorithm **PYTHAGORAS**

Step	LEFT_LEG	RIGHT_LEG	SQ1	SQ2	HYPOTENUSE	Output
1	3	4	?	?	?	
2	3	4	9	16	?	
3	3	4	9	16	5	
4	3	4	9	16	5	SIDES ARE 3 4 5

some representative data, using a pen and paper to record the values taken by the variables at each step of the algorithm. This is known as a *trace*. Let us do this with our algorithm **PYTHAGORAS** and see if we get the correct result.

Assume we know that the two legs of a right-angled triangle measure 3 in and 4 in. Thus, our input data are the pair of numbers **3, 4**. We begin executing the algorithm at step 1. This causes the data values to be assigned to the variables **LEFT_LEG** and **RIGHT_LEG**. Because **LEFT_LEG** appears first in the read statement, it gets the first value in the input stream. Thus, **LEFT_LEG** and **RIGHT_LEG** get values **3** and **4**, respectively. Step 2 introduces two new variables: **SQ1** and **SQ2**. Executing the two given assignment statements results in **SQ1** and **SQ2** getting the values **9** and **16**, respectively.

Step 3 calculates the length of the hypotenuse of the triangle using the intermediate values calculated in step 2. Substituting these values, the expression becomes

   SQRT(9 + 16)

which evaluates to the integer value **5**. This is then assigned to the integer variable **HYPOTENUSE**. Step 4 causes the following output:

   SIDES ARE   3   4   5

Step 5 causes the algorithm to terminate. Table 2-6 shows the complete trace. The symbol **?** is used to show that a variable has no value at this point.

On the basis of this particular trace, involving one set of data values, we are firmer in our conviction that the algorithm **PYTHAGORAS** does what it purports to do. Our confidence in this conclusion would be increased by additional tests, which we will leave as an exercise.

As a second example of the use of the algorithmic language, we present the following algorithm, **GRADES**. Study this example and test it with some appropriate data values.

Algorithm GRADES. This algorithm reads in four marks (denoted by **MARK1**, **MARK2**, **MARK3**, and **MARK4**) and computes the average grade, placing it in **AVERAGE**. The final result is then printed along with an explanatory message. All variables are assumed to be real.

1. [Input individual marks]
   Read(MARK1, MARK2, MARK3, MARK4)

2. [Compute average grade]
   AVERAGE ← (MARK1 + MARK2 + MARK3 + MARK4) / 4.0
3. [Output result]
   Write('FINAL GRADE IS', AVERAGE)
4. [Finished]
   Exit ☐

2-5.5 SUMMARY OF SOLUTION METHODOLOGY

Confronted with a problem for computer solution, we advocate that you follow these steps:

1. Make sure that you understand the problem specifications completely. A useful practice is to indicate the output that you expect from several sample inputs.
2. Formulate a rough general algorithm for the solution of the problem, paying little attention to specific details. This will be the first attempt at expressing the strategy of your solution. Assure yourself that the strategy is correct by tracing through the various steps with some sample inputs. If it is not, revise it.
3. Identify and list any variables that you feel might be necessary. Many of these will be suggested directly by the general algorithm developed in the previous step and by the problem statement itself. Include with the name of the variable its type and an indication of its purpose. Recognize that this list is only a first approximation to the final set of variables you will use; the list may be lengthened or shortened as necessary.
4. Return to the individual steps of the algorithm and proceed to fill in the details. Each time a step is broken down into a number of more detailed steps, be sure to check that the collection of more detailed steps performs the function expressed in the original step.
5. When you feel that you have produced an appropriately detailed algorithm, again carefully trace through its steps with some sample inputs to satisfy yourself that your solution satisfies the stated requirements. The tracing technique described earlier is appropriate at this stage. Testing an algorithm thoroughly is never easy. It can, in fact, be as difficult as producing it in the first place, but it is a step in the process that cannot be ignored. Some comments on conducting a good test are given in the next section.
6. Only now is it appropriate to consider implementing the algorithm in a particular programming language.

For simple problems this approach may seem unnecessarily tedious. However, it is important that you master the technique. As an analogy, consider the problem of long division. For simple division problems, most people arrive at the answer by inspection, without requiring the methodology of long division. Unless the scope of problems you can hope to solve is to remain extremely limited, however, you must learn through practice the methodology of long division. The same is true of our approach. Simple algorithms (and programs) can perhaps be produced "by inspection," but more difficult problems require a much more orderly approach. The methodology that we propose extends naturally as the problems become more difficult.

We end this section with a short philosophical note on the existence of algorithms. We have been saying that a key step in the solution of any problem is the development of a suitable algorithm. We have evaded the question, "Will there always be a suitable algorithm?" Certainly this is an important issue. We do not want to waste many hours trying to construct something that cannot be constructed.

It turns out that there are many problems for which it is impossible to develop suitable algorithms, that is, algorithms that guarantee a result after the execution of a finite number of steps. We can specify an algorithm to get from point A to point B in a strange city. If the instructions are correct and precise, so that, for example, we do not circle the same block indefinitely, we will get to point B eventually. Thus, a suitable algorithm does exist for this problem. Suppose, however, that someone were to ask you, "How can I become a concert pianist?" A little reflection ought to convince you that there can be no suitable algorithm for this problem; the person may lack the innate talent to play the piano well, and no algorithm will ever turn him/her into a concert pianist.

As you study more computer science, you will learn that there are certain classes of problems that are unsolvable, or certain questions that are undecidable. The question "Does a suitable algorithm exist for this particular problem?" is a case in point. It turns out, in fact, that the very question "Is this question undecidable?" is itself undecidable. Unfortunately, just about the only way of determining if a suitable algorithm exists for a given problem is to see if one can be constructed. If so, you have shown that at least one algorithm does exist—the one that you have constructed. Suppose, however, that you are unable to construct a suitable algorithm after some measure of effort. This does not mean that one does not exist; it means simply that you have not been able to construct one. Not a satisfying state of affairs.

We will not say any more about this problem in this book, but it is something on which you might care to reflect. The theoretical study of algorithms is an exciting field of computer science. Rest assured that all problems and exercises in this book do have suitable algorithms, even though late some night you may think otherwise.

EXERCISES 2-5

1. In which of the following pairs is the order of the statements important? In other words, when will changing the order of the statements change the final results? (Assume that $X \neq Y \neq Z$.)

(a) $X \leftarrow Y$
      $Y \leftarrow Z$

(b) $X \leftarrow Y$
      $Z \leftarrow X$

(c) $X \leftarrow Z$
      $X \leftarrow Y$

(d) $Z \leftarrow Y$
      $X \leftarrow Y$

2. The Canadian weather office has recently undergone a conversion to the metric system. Design algorithms to perform the following conversions:

  (a) Read a temperature given on the Celsius scale and print its Fahrenheit equivalent (conversion formula: $°F = \frac{9}{5}°C + 32$).

  (b) Read a rainfall amount given in inches and print its equivalent in millimeters (25.4 mm = 1 in).

3. Formulate an algorithm to read a person's name in the form "given name" followed by "surname," and print the name in the form "surname" followed by "given name." Example:

  Input: 'URIAH', 'HEEP'

  Output: HEEP, URIAH

4. Devise an algorithm to compute the statistical standard deviation, σ, of five numbers. The formula required is

$$\sigma = \sqrt{\frac{1}{4.0} \sum_{i=1}^{5} (x_i - \bar{x})^2}$$

where $x_1, x_2, \ldots, x_5$ are five values to be read in; $\bar{x}$ denotes their mean; and $\sum_{i=1}^{5}$ means the summation of the five indicated terms.

5. The following is the menu at a local hamburger emporium. Design an algorithm that reads in the number of each item purchased and computes the total bill.

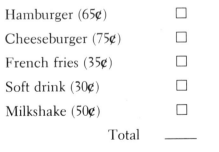

  Bertha's Burgers

  Hamburger (65¢)   ☐

  Cheeseburger (75¢)   ☐

  French fries (35¢)   ☐

  Soft drink (30¢)   ☐

  Milkshake (50¢)   ☐

                Total   _____

6. The roots of a quadratic equation of the form

$$ax^2 + bx + c = 0$$

are real if and only if the discriminant given by

$$b^2 - 4ac$$

is nonnegative. Design an algorithm to read the values of the coefficients a, b, and c and print the value of the discriminant.

7. The cost to the consumer of a new car is the sum of the wholesale cost of the car, the dealer's percentage markup, and the provincial or state sales tax (applied to the "mark-up" price). Assuming a dealer's mark-up of 12% on all units, and a sales tax of 6%, design an algorithm to read the wholesale cost of a car and print the consumer's cost.

8. Design an algorithm to read the lengths of the three sides of a triangle (S1, S2, and S3), and compute the area of the triangle according to the formula

$$\text{area} = \sqrt{T(T - S1)(T - S2)(T - S3)}$$

where

$$T = \frac{S1 + S2 + S3}{2}$$

2-6 APPLICATIONS

In this section we present several problems to which the techniques introduced to date can be applied. We will discuss each of the applications in turn, develop an algorithm for the solution of the problem, and give sample input and output. Some comments are made on testing the algorithms.

The first application involves the preparation of reports on student class performance, where a final grade is determined from a weighted combination of several factors. The second application is very topical. It concerns comparing the cost of items purchased to their cost a month ago. The result is a crude measure of the rate of inflation. The third and final application is the calculation of parimutuel payoffs at a racetrack.

2-6.1 REPORTING STUDENT GRADES

In a particular computer science class, a student's final grade is derived from his/her performance in three aspects of the year's work. There is a midterm examination which counts 30% toward the final grade. Laboratory work is given a mark that counts 20% toward the final grade. The final examination makes up the remaining 50%.

The problem is to compute the final grade from the supplied marks in each area. The input data consist of four separate items: the student's name, the mark received for laboratory work, the mark received for the midterm examination, and the mark received on the final examination. Each mark is given out of a possible 100. The output should be the student's name, the three supplied marks, and the calculated final grade. So that the individual items in the output can be readily identified, we will supply appropriate headings.

The solution to this problem consists of the following steps:

1. Read in student's name and his/her marks on the individual items
2. Compute and print final grade along with name of student

As this is a reasonably straightforward restatement of the original problem, we will not dwell on its verification. Based on the information at hand we will require the following variables. The names have been chosen to reflect their purpose. The types are shown in parentheses.

NAME (string)	Student name
LAB_WORK (integer)	Mark on lab work
MIDTERM_EXAM (integer)	Mark on midterm exam
FINAL_EXAM (integer)	Mark on final exam
GRADE (real)	Final grade

The variable NAME is clearly a string variable. If we assume that the supplied marks are whole numbers, they can be integer quantities. We will, however, allow a fractional portion in the final grade.

Step 1 of our original algorithm is a straightforward read statement. Since we have introduced the needed variables, we can write this statement directly as

Read(NAME, LAB_WORK, MIDTERM_EXAM, FINAL_EXAM)

It is worthwhile to pause at this point to ensure that the items have been given in the order that matches the actual input data.

Having verified that this is the case, we can proceed to step 2 of our original algorithm. Two basic functions are included in this step: the computation of the final grade, and the printing of the required information with appropriate headings. Thus, we could consider expressing this step as two more detailed steps

1. Compute final grade
2. Display results

Clearly these two together perform the function of the original second step. Neither of these two steps, however, is sufficiently detailed that we can stop at this point. The calculation of the final grade involves forming an expression in which the terms are a weighted sum of the three supplied marks. Using the defined variables, we arrive at

GRADE ← 0.20 * LAB_WORK + 0.30 * MIDTERM_EXAM
+ 0.50 * FINAL_EXAM

The formulation of the appropriate output statements is straightforward. The result can be seen in the final algorithm.

Putting all of this together according to the format described in the previous section, we obtain our final algorithm, which we will call **REPORT**.

Algorithm REPORT. Given individual marks in three aspects of a year's work in a course (**LAB_WORK, MIDTERM_EXAM, FINAL_EXAM**) for a student named **NAME**, this algorithm computes the final grade with weightings of 20%, 30%, and 50%, respectively. The printed output gives the student's name, his/her individual marks, and final grade. The variable **GRADE** is real.

1. [Get input data]
    Read(NAME, LAB_WORK, MIDTERM_EXAM, FINAL_EXAM)
2. [Compute final grade]
    GRADE ← 0.20 * LAB_WORK + 0.30 * MIDTERM_EXAM
    + 0.50 * FINAL_EXAM
3. [Display results]
    Write('STUDENT NAME:', NAME)
    Write('LABORATORY MARK:', LAB_WORK)
    Write('MIDTERM EXAMINATION:', MIDTERM_EXAM)
    Write('FINAL EXAMINATION:', FINAL_EXAM)
    Write('FINAL GRADE:', GRADE)
4. [Finished]
    Exit   □

Let us take some hypothetical input data and check the output of the algorithm. Suppose that the following values were supplied as input:

   'ARTHUR FONZARELLI', 72, 68, 65

To this input we apply the appropriate weighting factors to determine the contribution of each of the individual items (72, 68, and 65) toward the final grade. We take 20% of the lab mark (0.20 * 72 gives 14.4), 30% of the mark on the midterm exam (0.30 * 68 gives 20.4), and 50% of the mark on the final exam (0.50 * 65 gives 32.5). We sum these to get the value assigned to GRADE (67.3). Thus, the following is printed:

STUDENT NAME:	ARTHUR FONZARELLI
LABORATORY MARK:	72
MIDTERM EXAMINATION:	68
FINAL EXAMINATION:	65
FINAL GRADE:	67.3

We remind you that we are taking liberties with the format of the printed output. We have not said how an alignment such as this would be produced.

Such details are rightly part of specific programming languages, and thus are not dealt with here. We will continue to assume whatever output format is convenient to the presentation. Notice once again, as well, that delimiting quote marks do not appear in the output of string values.

We know now that the algorithm does produce output, but are we confident that it produces the correct output? To answer this question we proceed to the testing of this algorithm. Suppose that instead of an algorithm, we were testing the accuracy of the speedometer on a car. To test a speedometer you might drive a certain number of independently measured miles at a fixed speed and compare the time you took to the time you know you should have taken. For example, if you drive 5 measured miles with the speedometer registering 60 mph, you know it should take exactly 5 minutes. If, in fact, it takes you more than 5 minutes, you can conclude that the speedometer reads high. You can adjust this and try again.

Likewise, it is a good practice to test an algorithm on data that will give a known result. In this example we know that a student who receives identical marks on all three items should also receive this mark as a final grade. To verify this let us assume that a particular student received marks of 75 on each item. For this student the lab component of his final grade is 0.20 * 75 or 15.0, the midterm exam component is 0.30 * 75 or 22.5, and the final exam component is 0.50 * 75 or 37.5. Summing these, we get 75.0, as expected.

In testing an algorithm, it is also important to test extreme values of the input data. Errors often result from a failure to account for the extreme values. For example, suppose that our student received a mark of 100 on all three items. His/her final grade in the course should then be 100. Does the algorithm give this result? A little checking shows that it does. Also, suppose some unfortunate student received a mark of 0 on all items. Checking through the algorithm shows a final grade of 0 is computed, as expected.

Testing can never show that an algorithm has no errors; at best, it shows that you have not found any errors. Either there are no errors, or they have remained hidden throughout the testing. We hope that it is the first case, but we can never be sure. Proper choice of test cases can help to increase our confidence. Errors often result from circumstances that were not anticipated. A good tester tries to anticipate possible problems in his/her programs and force errors into revealing themselves. A sabotage mentality helps here. Errors that are forced to reveal themselves in the testing phase will not return to plague the programmer after he/she has delivered the program to the client with his/her stamp of approval.

2-6.2 GAUGING INFLATION

To get an indication of the current rate of inflation we would like to compare the present price of an article with the price we paid the last time it was purchased. For the sake of uniformity, let us assume that we shop regularly once a month. The following algorithm reads three pieces of input data: a short description of the item purchased (a character string), the price paid (in dollars and cents), and the price paid last month (again, in dollars and cents). For example,

'LAUNDRY SOAP', 2.49, 2.25

shows that we paid **$2.49** for a box of laundry soap this month, whereas we paid **$2.25** for the same product last month. (Note that the dollar signs are not part of the input stream but part of the interpretation of the values found there.)

For our inflation indicator we will compute two statistics. The first of these is the straight algebraic difference between the two prices. In this case the algebraic difference is $2.49 − $2.25 = $0.24. As well, we compute the percentage difference over the month. For this particular product, the percentage difference is calculated as

$$\frac{2.49 - 2.25}{2.25} \times 100\% = 10.67\%$$

Our first solution attempt has the following four steps:

1. Read item description, current price, and last month's price
2. Compute algebraic difference in price
3. Compute percentage difference
4. Print results

From this brief description of the solution, and the problem statement, the following variables are suggested. Again, names are chosen to describe their purpose and types are given in parentheses.

ITEM (string)	Description of item
CURRENT_PRICE (real)	Price paid this month
OLD_PRICE (real)	Price paid last month
ALG_DIFF (real)	Algebraic difference
PC_DIFF (real)	Percentage difference

The refinement of the steps in our general solution proceeds in the same way as in the previous section. Since the difficulty is roughly the same in the two cases, we will not go into the details, but simply present the final algorithm, GAUGE.

Algorithm GAUGE. This algorithm reads in a description of an article (ITEM), the current price of the article (CURRENT_PRICE), and its price one month ago (OLD_PRICE). ITEM is a string variable; CURRENT_PRICE and

OLD_PRICE are real variables. The algorithm computes and prints the algebraic difference (ALG_DIFF) between the two prices and the percentage difference (PC_DIFF). These values may be positive or negative.

1. [Input]
      Read(ITEM, CURRENT_PRICE, OLD_PRICE)
2. [Compute the algebraic difference in price]
      ALG_DIFF ← CURRENT_PRICE − OLD_PRICE
3. [Compute the percentage difference]
      PC_DIFF ← ALG_DIFF / OLD_PRICE * 100
4. [Output]
      Write('ITEM PURCHASED:', ITEM)
      Write('PRICE THIS MONTH:', '$', CURRENT_PRICE)
      Write('PRICE LAST MONTH:', '$', OLD_PRICE)
      Write('ALGEBRAIC DIFFERENCE:', '$' ALG_DIFF)
      Write('PERCENTAGE DIFFERENCE:', PC_DIFF, '%')
5. [Finished]
      Exit   ☐

For demonstration purposes, let us try three different sets of data and see what output we get. We take the following three sets of data:

'COOKING OIL', 4.79, 4.38
'POUND OF CHEESE', 1.57, 1.54
'LIMA BEANS', .38, .39

Applying the algorithm to the first set of input data results in the string variable ITEM getting the value 'COOKING OIL', and the real variables CURRENT_PRICE and OLD_PRICE being set to 4.79 and 4.38, respectively. In step 2 OLD_PRICE is subtracted from CURRENT_PRICE, resulting in the value 0.41, which is assigned to the variable ALG_DIFF. Step 3 converts this value to a percentage of the previous month's price. The calculation is

$$\frac{0.41}{4.38} \times 100$$

which gives the value 9.36 (we will express it here to only two decimal places, although, in reality, more would probably be given). This is assigned to the variable PC_DIFF. Output is produced in step 4. For this set of data values the following output results. (Notice again that we have assumed a convenient output format. Notice also the use of the special symbols $ and %.)

ITEM PURCHASED:	COOKING OIL
PRICE THIS MONTH:	$4.79
PRICE LAST MONTH:	$4.38

ALGEBRAIC DIFFERENCE: $0.41

PERCENTAGE DIFFERENCE: 9.36%

Try the algorithm yourself on the other two sets of data values. You should get the following outputs:

ITEM PURCHASED: POUND OF CHEESE

PRICE THIS MONTH: $1.57

PRICE LAST MONTH: $1.54

ALGEBRAIC DIFFERENCE: $0.03

PERCENTAGE DIFFERENCE: 1.95%

and

ITEM PURCHASED: LIMA BEANS

PRICE THIS MONTH: $0.38

PRICE LAST MONTH: $0.39

ALGEBRAIC DIFFERENCE: $−0.01

PERCENTAGE DIFFERENCE: −2.56%

Note in the third example the effect of the price reduction in lima beans. The resulting algebraic and percentage differences are negative. In the case of the algebraic difference, this does not adapt well to our output format. How might you adjust for this?

As with any algorithm, this one should now be thoroughly tested. From the example of the previous section, see if you can come up with an appropriate set of test cases.

2-6.3   PARIMUTUEL PAYOFFS

Horse racing is one of the most popular spectator sports in North America. Wagering at most racetracks is on the parimutuel system, a system in which the winning bettors share the total amount wagered, less a certain percentage to the operators of the track to cover track expenses and purses for the race, and also a certain percentage to the government.

At our local racetrack, the various levels of government take a combined total of 10.6% of all money wagered. The track operators take an additional 12%. From the amount remaining in the "pool" of money wagered, payoffs are made to all those who bet on the winning horse. (We are assuming "win" bets only at this track; the calculation of payoffs for "place" and "show" bets is more complicated.) The calculation of the payoff is based on the premise that money remaining in the pool is to be divided among the winners in proportion to the amount they wagered. The payoff is normally posted for the standard $2 bet.

We require an algorithm that takes as input the total amount wagered to win (the "win pool") for a given race, the name of the winning horse, and the amount wagered on the winning horse. The output is to be the name of the winning horse and the payoff on a $2 bet.

As our first algorithm, we come up with the following:

1. Read total amount wagered, name of winning horse, and amount bet on winning horse
2. Compute amount taken by governments and track operators from total wager pool
3. Compute payoff to bettors, expressed per $2 bet
4. Print out the name of winning horse and amount of payoff

From this, we generate the following list of variables.

WIN_POOL (real)	Total amount bet to win on all horses
WINNER (string)	Name of winning horse
BET (real)	Amount bet (to win) on winning horse
GOVT_SHARE (real)	Government taxes
TRACK_SHARE (real)	Amount taken by track
PAYOFF (real)	Payoff on $2 bet

We now proceed to refine each of these steps. Step 1 is straightforward, so we can go directly to the appropriate read statement.

    Read(WIN_POOL, WINNER, BET)

Step 2 requires a statement to generate each of the amounts taken from the win pool. Some thought results in the following assignment statements:

    GOVT_SHARE ← 0.106 * WIN_POOL
    TRACK_SHARE ← 0.12 * WIN_POOL

Step 3 is more involved still. The money remaining in the win pool must be divided among the successful bettors in proportion to the amount that they bet. First, we must determine how much remains in the pool, then the correct proportion. Two steps are suggested:

1. Compute amount of win pool remaining
2. Compute payoff ratio

Referring back to our original step 3 reminds us that the payoff is to be expressed per $2 bet. Therefore, our refinement needs a third step to adjust the computed ratio (which is expressed per $1 bet) to a payoff on a $2 bet. This gives us:

1. Compute amount of pool remaining
2. Compute payoff ratio (per $1 bet)
3. Multiply payoff ratio by 2

To handle steps 2 and 3, we will introduce an additional variable.

RATIO (real)   Amount paid per dollar bet

These steps can be expressed as the following assignment statements:

WIN_POOL ← WIN_POOL − (GOVT_SHARE + TRACK_SHARE)
RATIO ← WIN_POOL / BET
PAYOFF ← RATIO ∗ 2

The refinement of step 4 of our general algorithm is straightforward. Combining all these steps and expressing them in our algorithmic notation, we get the following algorithm PARIMUTUEL.

Algorithm PARIMUTUEL. This algorithm computes the payoff for a standard $2 bet, given the size of the win pool (WIN_POOL), the name of the winning horse (WINNER), and the total amount bet on the winning horse (BET). BET and WIN_POOL are real variables; WINNER is a string variable. Variables GOVT_SHARE, TRACK_SHARE, RATIO, and PAYOFF are real.

1. [Input]
    Read(WIN_POOL, WINNER, BET)
2. [Determine government's and track's share of pool]
    GOVT_SHARE ← 0.106 ∗ WIN_POOL
    TRACK_SHARE ← 0.12 ∗ WIN_POOL
3. [Reduce pool accordingly]
    WIN_POOL ← WIN_POOL − (GOVT_SHARE + TRACK_SHARE)
4. [Compute payoff ratio for each dollar wagered]
    RATIO ← WIN_POOL / BET
5. [Compute posted payoff]
    PAYOFF ← RATIO ∗ 2
6. [Output]
    Write(WINNER, 'WINS and PAYS $', PAYOFF)
7. [Finished]
    Exit   □

For demonstration purposes, we will try the algorithm with the following sets of data values.

10000, 'COMPUTER DELIGHT', 550
24003, 'SASKATOON STAR', 18335

In the first case, WIN_POOL, WINNER, and BET get the values 10000, 'COMPUTER DELIGHT', and 550, respectively, as a result of the READ statement in step 1. Deductions are made in steps 2 and 3, reducing the size of the win pool from $10,000 to a final total of $7740. The return on a $1 investment is computed in step 4 to be $14.07 (expressed here to two decimal figures). This represents the original $1 plus a profit of $13.07. In step 5 this is converted to a payoff on the standard $2 bet for posting purposes. The resulting output line is

COMPUTER DELIGHT WINS AND PAYS $28.14

In case 2, SASKATOON STAR was a heavy favorite. As a result, much of the money bet was bet on him. This will mean a small payoff. The government and the track take their share first. This reduces the size of the win pool from its original $24,003 to a final amount of $18,578.32 (again, expressed to two decimal places). The calculated value of RATIO in step 4 is small, 1.01 (expressed to two decimal places), meaning a return of 1¢ (or 1%) on each dollar invested. Thus, the final output line is

SASKATOON STAR WINS AND PAYS $2.02

Once again, we leave the detailed testing of this algorithm as an exercise.

In many regions, by law, the track is obliged to guarantee to its patrons a certain minimum percentage return on investment. In most cases this is 5%. Thus, the payoff to the bettors on SASKATOON STAR in the example above must be no smaller than $2.10. The track is forced to make up the difference (in this case 8¢ on each $2 bet) from its own funds. Fortunately for the operators of a race track, this does not happen very often. How might you modify the algorithm PARIMUTUEL to incorporate this minimum return on investment?

CHAPTER EXERCISES

1. You have been commissioned by the organizers of an international track and field meet to provide "simultaneous translation" services for results of events that are reported in metric units. Develop algorithms to handle the following events.
   (a) Convert high-jump results reported in meters to feet and inches (1 m = 39.37 in).
   (b) Given a sprinter's time for the 100-m dash, compute his/her time over 100 yards assuming that he/she runs at constant speed for the entire distance.

2. A 1977 listing of foreign exchange rates gave the following table of equivalents:

100 French francs = 21.55 Canadian dollars
1 British pound = 1.84 Canadian dollars
100 Greek drachmas = 2.95 Canadian dollars
100 Dutch guilders = 43.20 Canadian dollars
100 Swedish kronors = 24.25 Canadian dollars
1 U.S. dollar = 1.06 Canadian dollars

Develop algorithms to make the following conversions:
(a) Read an amount in French francs and print the equivalent in Canadian dollars.
(b) Read an amount in Greek drachmas and print the equivalent in British pounds.
(c) Read an amount in Canadian dollars and print the equivalent in both U.S. dollars and Dutch guilders.
(d) Read an amount in U.S. dollars and print the equivalent in both Swedish kronors and French francs.

3. Honest John's Used Car Company pays its sales staff a salary of $250 per month plus a commission of $15 for each car they sell plus 5% of the value of the sale. Each month, Honest John's bookkeeper prepares a single punched card for each salesperson, containing his/her name, the number of cars sold, and the value of cars sold. Design an algorithm to compute and display a salesperson's salary for a given month. Test the algorithm thoroughly on some appropriate sample data.

4. Three masses, m_1, m_2, and m_3, are separated by distances r_{12}, r_{13}, and r_{23} as shown in Fig. 2-26. If G is the universal gravitational constant, the binding energy holding the mass particles together is given by the formula

$$E = G\left(\frac{m_1 m_2}{r_{12}} + \frac{m_1 m_3}{r_{13}} + \frac{m_2 m_3}{r_{23}}\right)$$

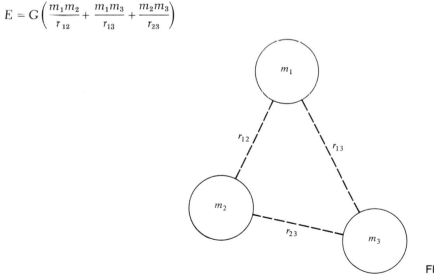

FIGURE 2-26

Design an algorithm to read values of m_1, m_2, m_3, r_{12}, r_{13}, and r_{23}; then compute and print the binding energy along with the initial data values. For mass in kilograms and distance in meters, $G = 6.67 \times 10^{-11}$ newton-meter^2/kg^2. The values of m_1, m_2, and m_3 are punched on the first data card and the values of r_{12}, r_{13}, and r_{23} are punched on the second. Assume that all data are punched as real values.

5. A system of linear equations of the form

$$ax + by = c$$
$$dx + ey = f$$

can be solved by using the following formulas:

$$x = \frac{ce - bf}{ae - bd} \quad \text{and} \quad y = \frac{af - cd}{ae - bd}$$

Design an algorithm to read in the two sets of coefficients (a, b, and c, and d, e, and f) and print out the solution values for x and y. Are there any cases for which this algorithm will not work?

6. The cost of hail insurance in a typical farming community is 3.5% of the desired amount of coverage per acre, multiplied by the number of acres seeded. Assuming that the crop possibilities are limited to wheat, oats, and barley, design an algorithm that reads the desired coverage and number of acres planted for each of the three crops and computes the total cost of hail insurance for this customer.

7. An experiment was recently conducted to determine the acceleration due to gravity in Saskatoon. A ball was dropped from rest and allowed to fall freely to the ground, from the top of a number of buildings. The time to reach the ground was recorded in each case. A total of five trials was made, with the following results:

Height of Building, ft	Time to Reach Ground, s
227	3.74
375	4.84
710	6.64
423	5.13
158	3.11

Design an algorithm to compute the gravitational acceleration g from these data using the formula

$$y = \tfrac{1}{2}gt^2$$

where y denotes the distance fallen and t the time taken. Each of the five cases shown will yield a value for g; the "best guess" for g for Saskatoon based on the results of this experiment is to be the average of the five separate trials. Your algorithm must read the data supplied and print this "best guess."

8. Although the speed of light remains constant regardless of the relative speeds of the light source and the observer, the measured frequency and wavelength do change—an effect first predicted by Johann Doppler and so dubbed the "Doppler effect." The wavelength λ emitted by a source moving toward an observer with velocity v appears to be compressed by an amount $\Delta\lambda$, which is given by the formula

$$\Delta\lambda = \frac{v\lambda}{c}$$

where c is the speed of light. Suppose that an airplane is traveling toward a radio station at a constant velocity of 360 km/h (10^4 cm/s). If the radio station is broadcasting at a wavelength of 30 m, the change in wavelength due to the Doppler effect is

$$\Delta\lambda = \frac{v\lambda}{c} = \frac{(10^4 \text{ cm/s}) \times (3 \times 10^3 \text{cm})}{3 \times 10^{10} \text{ cm/s}}$$
$$= 10^{-3} \text{ cm}$$

Thus, the pilot of this airplane must adjust his/her receiving set to a wavelength of 3,000 cm *minus* 10^{-3} cm, or 2,999.999 cm, until he/she reaches the station, and then to a wavelength of 3,000.001 cm as he/she moves away from the station. Design an algorithm to read the broadcast wavelength of a radio station and the speed of an approaching plane, and then print out the actual setting at which the pilot will receive the signal.

BIBLIOGRAPHY

Brooks, F. P. Jr.: *The Mythical Man-Month: Essays on Software Engineering*, Addison-Wesley, Reading, Mass., 1975.

Shneiderman, B., Mayer, R., McKay, D., and Heller, P.: "Experimental Investigations of the Utility of Detailed Flowcharts in Programming," *Communications of the ACM*, vol. 20, June 1977, p. 373.

3

Decision
Structures

In Chap. 2 we saw how to develop algorithms to solve a class of problems on a computer. The algorithms followed more or less a standard pattern: the input of some data values, a series of calculations, and the output of some results. Much of the real power of a computer, however, comes from its ability to make decisions and determine a course of action at run time, based on the value of some piece of data read or on the result of some calculation.

In this chapter we introduce two powerful programming concepts that require the power of decision: the ability to select one action from a set of specified alternatives, and the ability to repeat a series of actions. Algorithmic notation for these programming concepts is introduced. Examples and applications using these new features are discussed.

3-1 MOTIVATION

There is a significant amount of manual activity required to get any computer program running. For example, after keypunching, the data card or cards must be appended to the program deck, and the entire deck then read in through the card reader. For very simple applications, many people feel that the effort is not justified. They would rather use pen and paper or a pocket calculator. The algorithm REPORT of Chap. 2, for example, is designed to compute and print the final grade for a single student. In a large class of, say, 250 students, the calculation would have to be repeated 250 times for a complete class report. As the algorithm stands now, this would require rerunning the corresponding program 250 times, each time with a new data card. Clearly, this would be a tedious activity.

We introduce now a very important concept, the concept of *flow of control* through an algorithm. This refers to the order in which the individual steps of the algorithm are executed. To this point we have relied almost exclusively on *linear* flow of control; that is, the steps are executed sequentially, from step 1 through to the last step of the algorithm. In this chapter we investigate two methods by which we can depart from normal linear flow through the use of *control structures*. Control structures are program or algorithm constructs that directly affect the flow of control in a program. One control structure we will introduce will allow us to repeat a group of steps automatically. Another will allow us to select a course of action from a pair of specified alternatives, based on an evaluation of certain conditions.

The introduction of these new control structures will allow the computer to do automatically some of what we have had to do manually up to now. In this way it will make some applications much less tedious. Also, it permits the solution of applications which were up to now impossible to solve with only linear flow of control. We will discuss applications of each type.

3-2   THE SELECTION FROM ALTERNATIVE ACTIONS

In Sec. 2-1 we presented the arithmetic/logic unit as one of the major components of a typical computer. The purpose of the arithmetic portion has been amply illustrated. The logic portion gives the computer the ability to make decisions. This will form the basis of the *if–then–else* algorithmic construct, which will allow us to select from alternative actions.

(By permission of Johnny Hart and Field Enterprises, Inc.)

A decision is specified in a logical expression much as a calculation is specified in a numerical expression. Suppose we have two numeric variables, call them A and B, that have unequal values, and suppose we want to print the value of the larger. If A is greater than B, we want to print the value of A. On the other hand, if B is greater than A, we want to print the value of B. Clearly, we have two alternative actions—"print the value of A" and "print the value of B." The choice of which action to take rests on a decision of whether or not A is greater than B. This structure can be illustrated by the simple flowchart shown in Fig. 3-1.

The expression $A > B$ is a logical expression describing a condition that we wish to test. If $A > B$ is true, that is, if A is in fact greater than B with the current values of A and B (note that this is a *run-time* test), we will take the action on the left in Fig. 3-1 and print the value of A. If $A > B$ is false, that is, if A is not greater than B (it may be less than B), we will take the action on the right and print the value of B. The result of the entire operation is that the larger of the two values is printed.

To check this, assume that A has the value 5 and B the value 3. Since 5 is greater than 3, the condition $A > B$ is true. As a result, we take the action on the left and print the value of A, or 5. Suppose these values were reversed so that A had the value 3 and B the value 5. Now $A > B$ is false, since 3 is smaller than 5. Because the condition is false, according to the flowchart, we take the action on the right and print the value of B, or 5. In either case the larger value is printed, as required.

Although this example is simple, its structure is very important. The selection of one action from the two specified alternatives is directed by a condition. If the condition happens to be true at the time it is tested, one course of action is taken; otherwise, the other is taken. This structure forms the basis of our new algorithmic construct, the if–then–else construct.

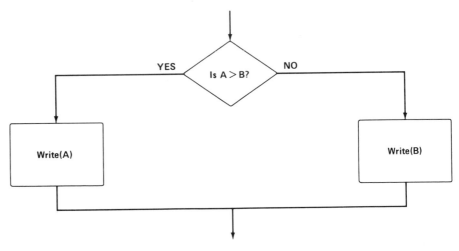

FIGURE 3-1
Flowchart of a typical decision.

3-2.1 THE IF–THEN–ELSE CONSTRUCT

The format of the if–then–else construct is as follows:

```
If "condition"
then "true alternative"
else "false alternative"
```

The construct begins with the special word "If" followed by some condition to be tested. The alternative to be taken if the condition happens to be true (the "true alternative") is preceded by the special word "then." The other alternative, to be taken if the condition happens to be false (the "false alternative") is preceded by the special word "else." The example in Fig. 3-1 would be written

```
If A > B
then Write(A)
else Write(B)
```

Each of the components of this control structure can be considerably more complex. For instance, either or both of the alternatives can be more than a single statement. We will be using appropriate indenting of statements to make the interpretation clear. This will be illustrated shortly.

A special set of operators is required to express conditions. In the example above, $>$ is known as a relational operator since it describes a possible relation between two values. The complete set of possible relational operators is shown in Table 3-1. Simple conditions, such as $A > B$, can be joined by logical connectives or operators, such as "and," "or," and "not" to form more complex conditions. This will be discussed in Sec. 3-4.

Relational operators can be applied to either numeric or nonnumeric data items. Their meaning in a numeric context is clear. What do they mean in connection with strings? It turns out that there is a definite sequence posed on strings by their representation in a computer referred to as the *collating sequence* (see Chap. 5). This will allow us, for example, to sort a list of names into alphabetical order. We will defer this material until Chap. 5. In the meantime we will restrict our tests of strings to equality and inequality, whose interpretations are obvious.

TABLE 3-1 Relational Operators

Operator	Meaning
$>$	Greater than
$<$	Less than
$=$	Equal to
$\geq$	Greater than or equal to
$\leq$	Less than or equal to
$\neq$	Not equal to

Let us examine more carefully the flow of control through the if–then–else construct. The control structure itself is to be viewed as a complete entity that is entered at the top and exited at the bottom. The idea of a single entry point and a single exit point is an important one. Figure 3-2 shows this in diagrammatic form. The control structure is entered at the point where the condition is to be evaluated. Following the evaluation of the condition, either the true alternative or the false alternative is executed. After the completion of the selected alternative, control passes to the next indicated action. In some cases there may be no special action if the condition is false (that is, there is no "else" part). This type of structure is shown in Fig. 3-3.

Consider the following algorithm, in which a condition is used.

Algorithm DISPLAY This algorithm reads two values, determines the largest, and prints the value with an identifying message. All variables are assumed to be integer.

1. [Input the respective values]
   Read(VAL1, VAL2)
2. [Determine the largest]
   If VAL1 > VAL2
   then MAX ← VAL1
       Write('FIRST VALUE IS THE LARGEST')
   else MAX ← VAL2
       Write('SECOND VALUE IS THE LARGEST')
3. [Output largest value]
   Write('VALUE IS', MAX)
4. [Finished]
   Exit

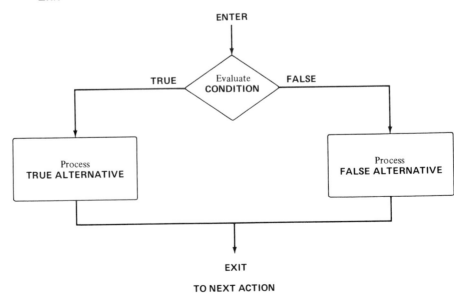

FIGURE 3-2
General flowchart of the selection construct.

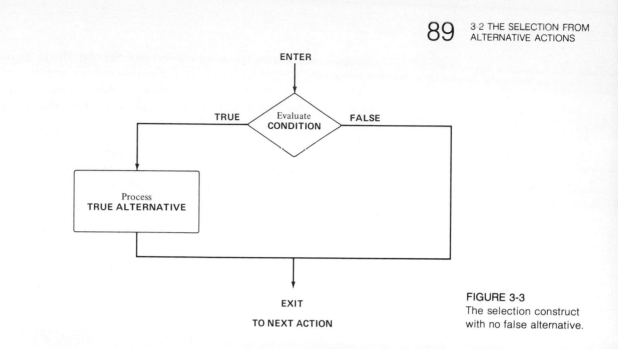

ENTER

TRUE   Evaluate
CONDITION   FALSE

Process
TRUE ALTERNATIVE

EXIT

TO NEXT ACTION

FIGURE 3-3
The selection construct
with no false alternative.

This is one of several ways to solve the problem of finding the largest
(or smallest) of a group of numbers; we will see others. This particular
method is used here to illustrate a number of important points. The execution of the algorithm begins at step 1 with the reading of two values, say, −2
and 10, into VAL1 and VAL2, respectively. Then comes step 2, the if–then–
else control structure. First, the condition is evaluated. With this particular
pair of data values, it turns out to be false. Thus, control passes to the false,
or "else," alternative. The variable MAX is assigned the value of VAL2 (10),
and the message "SECOND VALUE IS THE LARGEST" is printed. This is the
end of the execution of step 2. We then proceed to step 3, and on through
the algorithm.

Suppose the data values were interchanged so that VAL1 received the
value 10 and VAL2, − 2. Consequently, the true, or "then," alternative would
be selected. The variable MAX would be assigned the value of VAL1 (10) and
the message "FIRST VALUE IS THE LARGEST" would be printed. This is the
end of this step, so we proceed directly to step 3, *skipping over the statements
in the "else" alternative*.

The key point in this example is that you execute only one of the alternatives in an if–then–else control structure on each occasion, never both.
This is shown clearly in Fig. 3-2.

Note carefully the indentation scheme used in step 2 of the algorithm
DISPLAY. Each of the three major parts of the if structure (the "if" part, the
"then" part, and the "else" part) begins on a new line, in the same position in
fact. Additional statements in the "then" or "else" parts are lined up directly
under one another. With this scheme the major components of the structure
are clearly visible, and the statement or statements defining each alternative

are also clearly seen. We will be adopting this convention for all the algorithms in this book.

As a second example of the use of this particular control structure, we turn to an application relating to a political nominations convention. A number of ballotings are held. A candidate is declared nominated on any given balloting only if his/her vote total exceeds the sum of the vote totals of all other candidates currently on the ballot. If no candidate can be nominated on a particular balloting, the candidate with the lowest vote total is dropped from consideration and a new balloting is conducted.

Suppose that on this particular balloting, four candidates remain in contention. We require an algorithm to determine the outcome of the balloting. First we will determine if the candidate with the largest vote total has a total that is higher than the sum of the other vote totals; if so, this candidate is declared nominated. If not, we must drop the candidate with the lowest vote total from the next ballot. This is a clear application of the "if" statement. For the sake of simplicity, the following algorithm, NOMINATIONS, assumes that the totals have already been sorted such that the first total read is the highest and the last is the smallest. We will assume that two data cards are punched: one containing the totals in this order, and a second containing the names of the four candidates in the same order.

Algorithm NOMINATIONS. This algorithm determines the outcome of a particular balloting as just described. Integer variables COUNT1, COUNT2, COUNT3, and COUNT4 are used for the four vote totals described, and character variables NAME1, NAME2, NAME3, and NAME4 are used for the names of the respective candidates.

1. [Input]
    Read(COUNT1, COUNT2, COUNT3, COUNT4)
    Read(NAME1, NAME2, NAME3, NAME4)
2. [Post results of ballot]
    If COUNT1 > (COUNT2 + COUNT3 + COUNT4)
    then Write(NAME1, 'IS NOMINATED.')
    else Write('NO NOMINATION.', NAME4, 'IS DROPPED FROM BALLOT')
3. [Finished]
    Exit   □

Let us assume that Table 3-2 gives the results of the current balloting. Two data cards are then prepared as follows:

| 764 | 419 | 307 | 175 | (card 1) |
| 'PROBERT' | 'CHESTON' | 'COOKE' | 'KOZAR' | (card 2) |

Table 3-3 shows the trace of the algorithm NOMINATIONS with these data. Note the inclusion of a separate column for the results of the condition evaluation in step 2.

TABLE 3-2

Name	Vote Count
Probert	764
Cheston	419
Cooke	307
Kozar	175

3-2.2   NESTED IFs

In some applications, one of the alternatives in an if–then–else structure may involve a further decision. For example, before you take action on a certain grade on a test, you may want to discount the possibility of a data entry error. This requires an additional check to see if the data item read is a valid item. This suggests the use of an additional if–then–else structure embedded within each of the original alternatives. Such a process is an example of *nesting*.

Consider the problem of determining the largest of three input values (say A, B, and C) rather than two. These must be tested in pairs. We first test to see if A is greater than B. If so, we test to see if it is also greater than C, and so forth. The following algorithm, MAX_3, solves this problem. To simplify matters, we will assume the three values to be distinct.

Algorithm MAX_3. This algorithm reads three numbers A, B, and C and prints the value of the largest. Assume distinct values.

1. [Input data values]
   Read(A, B, C)
2. [Determine largest value by pairwise comparisons]
   If A > B
   then If A > C
           then MAX ← A   (A > B, A > C)
           else MAX ← C   (C > A > B)
   else If B > C
           then MAX ← B   (B > A, B > C)
           else MAX ← C   (C > B > A)
3. [Print largest value]
   Write('LARGEST VALUE IS', MAX)
4. [Finished]
   Exit   □

Study this example carefully. Notice that there is a nested if–then–else within each of the original alternatives. If you are not careful, you can get quite confused with such structures. It is very important, for example, that the "then" and "else" portions are matched with the proper "if" test. A lack of care can lead to ambiguities. Consider the following example, where c_1 and c_2 denote conditions and s_1 and s_2 denote statements:

```
If c₁
then If c₂
then s₁
else s₂
```

Under what condition is s_2 executed? Is it when c_1 is false, or when c_2 is false? It is unclear whether the "else" refers to the first or second "if"—an ambiguity that cannot be allowed. Such a possibility (called the "dangling else") exists in any language that has a similar construct (PL/I and ALGOL 60 are examples) and must be resolved through additional language rules. Rather than impose more rules in the algorithmic notation, we will try to handle this problem through our indentation convention. We will always line up an "else" directly under its corresponding "if" and "then" as we have in the earlier examples. The following forms show the interpretation of the indentations.

```
If c₁                    If c₁
then If c₂               then If c₂
    then s₁                  then s₁
    else s₂              else s₂
```

In the case on the left, the statement "else s_2" is part of the inner "if," and thus statement s_2 is executed if c_2 is false and c_1 is true; in the case on the right, the statement "else s_2" is part of the outer "if," and thus s_2 is executed if c_1 is false (in which case c_2 is never evaluated). In addition to specifying an unambiguous interpretation, such a convention can aid considerably in the overall readability of the algorithms.

As a second example of the use of nesting, we present the following problem. Its solution illustrates the use of nesting to test a series of alternatives rather than just a pair.

The Genco Pura Oil Company has decided to give its deserving employees a Christmas bonus. The bonus is to be based on two criteria: the number of overtime hours worked, and the number of hours the employee was absent from work. A card is punched for each of the company's employees, containing the employee's name, number of overtime hours worked, and number of hours absent.

TABLE 3-3   Trace of Algorithm NOMINATIONS

Step	COUNT1	COUNT2	COUNT3	COUNT4	NAME1
1	764	419	307	175	PROBERT
2	764	419	307	175	PROBERT

The company has decided to use the following formula to calculate the bonus: subtract two-thirds of the employee's absent hours from his/her overtime hours, and distribute the bonus according to Table 3-4.

We require an algorithm to read in the employee's information and calculate the amount of bonus he/she is to get. Clearly, we have a set of mutually exclusive alternatives; no employee can fall into more than one category. To determine the appropriate category, we employ nesting in a special way.

Algorithm BONUS. This algorithm reads an employee name (NAME), overtime hours worked (OVERTIME), and hours absent (ABSENT), and determines a Christmas bonus (PAYMENT). Variables representing hours are integer and the variable representing a payment is real. NAME is a character string.

1. [Input employee information]
   Read(NAME, OVERTIME, ABSENT)
2. [Complete bonus]
   If OVERTIME − (2 / 3.0) ∗ ABSENT > 40
   then PAYMENT ← 50.0
   else If OVERTIME − (2 / 3.0) ∗ ABSENT > 30
        then PAYMENT ← 40.0
        else If OVERTIME − (2 / 3.0) ∗ ABSENT > 20
             then PAYMENT ← 30.0
             else If OVERTIME − (2 / 3.0) ∗ ABSENT > 10
                  then PAYMENT ← 20.0
                  else PAYMENT ← 10.0
3. [Print results]
   Write('BONUS FOR', NAME, 'IS $', PAYMENT)
4. [Finished]
   Exit ☐

Notice carefully the "cascading" test in step 2. This could also be written as a series of unnested "if" statements. The nested version is sometimes better, for reasons we shall now discuss. Notice also the use of 2/3.0 to cause real rather than integer division.

Suppose that an employee has worked 50 hours of overtime and has been absent only 3. Subtracting $\frac{2}{3}$ of his/her hours absent from his/her overtime

TABLE 3-3 Trace of Algorithm NOMINATIONS (Continued)

NAME2	NAME3	NAME4	Condition	Output
CHESTON	COOKE	KOZAR		
CHESTON	COOKE	KOZAR	False	NO NOMINATION. KOZAR IS DROPPED FROM BALLOT.

TABLE 3-4 Bonus Schedule

Overtime − $\frac{2}{3}$*Absent	Bonus Paid
>40 hours	$50
>30 but ≤40 hours	$40
>20 but ≤30 hours	$30
>10 but ≤20 hours	$20
≤10 hours	$10

gives a total of 48 hours, which entitles him/her to the full bonus of $50. For this employee the first test in step 2 succeeds. Thus, the "then" alternative is selected and PAYMENT is set to 50.0. Since the rest of this step is contained within the "else" alternative for the first test, control passes immediately to step 3. We have made a total of one test. If we had, instead, written a series of unnested "if" statements, we would still have to test each of the other categories, all of which we know would fail.

Suppose that a different employee worked 30 hours of overtime but was absent for 12 hours. For this employee the testing value (OVERTIME − (2 / 3.0) * ABSENT) is 22. The first test (>40) fails; consequently, the "else" alternative is chosen. This alternative begins with a second test (>30), which also fails; consequently, its "else" alternative is chosen. This alternative begins with a third test (>20), which succeeds. Since the ">30" test failed, we know now that this employee falls in the range >20 but ≤30. The "then" alternative of this particular "if" sets PAYMENT to 30.0. Control then passes to step 3, as before.

You can see that the nesting structure employed in this algorithm cuts down on the number of *unsuccessful* tests required to process the various cases. This is a very useful form of organization. A series of unnested "if" statements would require each case to be tested every time.

We could improve this algorithm further by requiring the calculation of the testing value (OVERTIME − (2 / 3.0) * ABSENT) to be performed only once. To do this we introduce a temporary variable (let us call it TEST_VALUE) and break step 2 into two parts.

2.1 [Compute test value]
    TEST_VALUE ← OVERTIME − (2 / 3.0) * ABSENT
2.2 [Compute bonus]
      If TEST_VALUE > 40
      then PAYMENT ← 50.0
      else If TEST_VALUE > 30
            then PAYMENT ← 40.0
            else If TEST_VALUE > 20
                  then PAYMENT ← 30.0
                  else If TEST_VALUE > 10
                        then PAYMENT ← 20.0
                        else PAYMENT ← 10.0

In this version the calculated test value is stored in the variable TEST_
VALUE. Rather than having it recalculated at every test point, it is simply reexamined— a much faster process. This improvement cuts down the amount
of computation required, and therefore the ultimate processing cost of the
algorithm. A good programmer is ever watchful for such improvements.

We close this section with a seemingly trivial extension to the algorithm
MAX_3 that, in fact, adds considerable complexity. The problem is once
again to read in three numbers, but this time to determine the minimum
value as well as the maximum value. Try this problem yourself. When you
are satisfied that you have a solution, examine the following algorithm,
MAXMIN_3.

Algorithm MAXMIN_3. This algorithm reads three numbers, A, B, and
C, and prints the largest value and the smallest value. Values are assumed to
be distinct.

1. [Input data values]
    Read(A, B, C)
2. [Determine largest and smallest values]
    If A < B
    then If A < C
        then MIN ← A
            If B > C
            then MAX ← B   (A < C < B)
            else MAX ← C   (A < B < C)
        else MIN ← C   (C < A < B)
            MAX ← B
    else If A > C   (A > B at this point)
        then MAX ← A
            If B > C
            then MIN ← C   (A > B > C)
            else MIN ← B   (A > C > B)
        else MAX ← C   (C > A > B)
            MIN ← B
3. [Output results]
    Write('LARGEST VALUE IS', MAX,', SMALLEST IS', MIN)
4. [Finished]
    Exit   □

3-2.3 SUMMARY

The "if" statement that we have presented in this section is one form of a
very useful and very powerful programming construct. Virtually all programming languages will have some form of conditional construct. The version
we discussed allows, in its most general form, a choice between two alternative courses of action. We have referred to these as the "then" alternative,
taken when the testing condition evaluates to true, and the "else" alternative,

taken when the testing condition is false. The "else" alternative may be omitted, in which case no special action is taken if the testing condition is false (see Fig. 3-3).

We also discussed the possibility of nesting, in which further tests are embedded within the alternatives. This allows us to test for a multiplicity of conditions as illustrated in the examples, but can quickly lead to confusion if abused. Great care must be taken to show clearly the totality of conditions in effect at any point. Appropriate comments and spacing can help considerably.

EXERCISES 3-2

1. (a) Modify the algorithm REPORT, given in Sec. 2-6, so that a special message is printed for failing students.
   (b) Modify the algorithm REPORT further, so that the student is given a division standing according to his/her final grade. Assume that the divisions are based on the following table:

Final grade	Division
80–100	1
70–79	2
60–69	3
50–59	4
Less than 50	5

2. Give an algorithm to read in the base and height of a triangle and print out the area of the triangle (area = $\frac{1}{2}$ * base * height). During the preparation of data for this algorithm, it is entirely possible for a mistake to be made that may inadvertently result in one of the values base or height being negative. Clearly, this is undesirable, since it will result in a negative area being printed. Design into your algorithm the capability to check for negative values on input. If one is encountered, you should print it out along with a message identifying it as the base or height value (this may make it easier for someone to correct the error). Test your algorithm thoroughly. Be particularly careful of the case where *both* input values are negative. This would produce a positive area that could result in the error going undetected.

3. Revise the algorithm MAX_3 to account for the possibility of input values being equal.

4. Design an algorithm to read the lengths of the three sides of a triangle (S1, S2, and S3) and determine what type of triangle it is, based on the following cases. Let A denote the largest of S1, S2, and S3, and B and C the other two. Then

If $A \geq B + C$	No triangle is formed
If $A^2 = B^2 + C^2$	A right-angled triangle is formed
If $A^2 > B^2 + C^2$	An obtuse triangle is formed
If $A^2 < B^2 + C^2$	An acute triangle is formed

5. The Department of the Environment maintains three lists of industries known to be heavy polluters of the atmosphere. Results from a number of measurements are combined to form what is called the "pollution index." This is monitored regularly. Normal values fall between .05 and .25. If at any time it should reach .30, the industries on list A are asked to suspend operations until such time as the values return to the normal range. If it should exceed .40, industries from list B are notified as well. If it should exceed .50, industries on all three lists are asked to suspend activities. Design an algorithm that reads the pollution index and indicates the appropriate notifications.

6. Design an algorithm to read an integer value and determine whether it is even or odd. (Generalization: Read m, n. Does n divide m?)

7. Modify the algorithm NOMINATIONS so that it does not require presorting the ballot results.

3-3 LOOPING

Computers are particularly well suited to applications in which an operation or series of operations is to be repeated many times. For this purpose, we introduce a new control structure. The *loop* is a fundamental programming construct and, as we shall see, can occur in a variety of forms.

Consider, for example, the calculation of the factorial function, an important mathematical function. For a given value, N, the factorial of N (usually written N!) is computed as the product $N(N - 1)(N - 2) \cdots (2)(1)$. This is clearly nothing more than a series of repeated multiplications, where the multiplier is reduced by 1 prior to each multiplication. The following algorithm, FACT_N, uses a loop to compute the factorial of an input value.

Algorithm FACT_N. This algorithm computes the factorial of a value read from the input stream. Variables N, PRODUCT, and MULTIPLIER are of type integer.
1. [Input]
   Read(N)
2. [Initialize variables]
   PRODUCT ← 1
   MULTIPLIER ← N
3. [Establish loop]
   Repeat thru step 5 while MULTIPLIER ≥ 1

4. [Compute partial product]
      PRODUCT ← PRODUCT * MULTIPLIER
5. [Reset multiplier]
      MULTIPLIER ← MULTIPLIER − 1
6. [Output result]
      Write('FACTORIAL OF', N, 'IS', PRODUCT)
7. [Finished]
      Exit □

The loop in this example is controlled by the "Repeat" statement in step 3, which tells us to repeat the steps down to and including step 5 as long as the value of MULTIPLIER remains greater than or equal to 1. This is a form of loop known as a *conditional loop*, since the process of looping is controlled by a stated condition.

3-3.1 CONDITIONAL LOOPS

In our algorithmic notation we express a conditional loop in the following manner:

Repeat thru step s_n while "condition"

We refer to the steps specified in the repeat statement (that is, the steps from the repeat up to and including step s_n) as the *range* of the loop. The steps in the range are to be repeated as long as the stated condition remains true. The condition is evaluated prior to each pass through the loop. If it is true, the loop is executed one more time; if it is false, the step following the last step in the range of the loop (that is, step s_n) is executed next. If the condition is false on the first entry into the loop, the loop will be bypassed altogether. Notice the use of the long bracket to emphasize the steps in the range of the loop. This is done for improved readability. In some applications it will not be necessary to include the "thru step s_n" specification in the "Repeat" statement. When this is omitted, the range of the loop will be clearly indicated. Examples will show when this is desirable.

As was the case with the if–then–else control structure of the previous section, we will consider the "repeat" control structure as an entity with a single entry point and a single exit point, as shown in Fig. 3-4. Entry and exit are both through the repeat statement itself—it is both the first statement executed and the last. That is, the evaluation of the condition expressed in the repeat statement is done before any steps in the range of the loop are executed, and we exit from the loop only when the condition evaluates to false. It is the responsibility of the programmer to ensure that this does happen eventually; otherwise, the loop may run "forever," a serious and costly error that is unfortunately all too common.

Let us trace the execution of algorithm FACT_N with some sample data, paying particular attention to the operation of the loop. Suppose that we supplied the data value 3. Table 3-5 shows the results of this trace. Note the

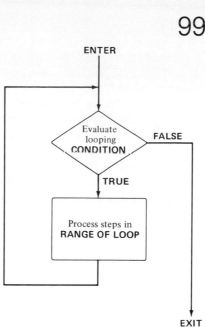

ENTER

FIGURE 3-4
General flowchart of the
loop construct.

introduction of a specific indication of the point at which the algorithm terminates.

Step 1 reads the input value (3) and assigns it to the variable N. Step 2 initializes a pair of variables that we will be using in the loop: PRODUCT gets the value 1, and MULTIPLIER gets the current value of N (here 3). We now come to the step controlling the loop. The first thing we do is evaluate the stated condition. Since the current value of MULTIPLIER (3) is in fact greater than 1, we enter into the range of the loop and proceed to step 4. In step 4 we multiply the current values of PRODUCT (1) and MULTIPLIER (3) and assign

TABLE 3-5 Trace of Algorithm FACT N

Step	N	PRODUCT	MULTIPLIER	Condition	Output
1	3	?	?		
2	3	1	3		
3	3	1	3	True	
4	3	3	3		
5	3	3	2		
3	3	3	2	True	
4	3	6	2		
5	3	6	1		
3	3	6	1	True	
4	3	6	1		
5	3	6	0		
3	3	6	0	False	
6	3	6	0		FACTORIAL OF 3 IS 6
7	Termination				

the result (3) to PRODUCT. In step 5 we reduce the value of MULTIPLIER by 1, so it is now 2. At this point we do not proceed ahead to step 6. Remember that we are in the range of a loop, under control of step 3 in this particular case. The only way we can get out of a loop is for the stated condition to evaluate to false. We now return to step 3 and reevaluate this condition. The current value of the variable MULTIPLIER is 2, which is still greater than 1, so we move once again to step 4. The current values of PRODUCT (3) and MULTIPLIER (2) are multiplied together, and the result (6) becomes the new value of PRODUCT. MULTIPLIER is reduced by 1 once again (to 1) and we return to reevaluate the looping condition once more. Again, the condition evaluates to true (MULTIPLIER is now equal to 1), so we proceed to step 4. Once again, the current values of PRODUCT (6) and MULTIPLIER (1) are multiplied together and the result (6) is assigned to PRODUCT. In this case no values are actually changed. The value of MULTIPLIER is then reduced from 1 to 0 in step 5 and we return to reevaluate the looping condition. This time the condition is false. The current value of MULTIPLIER (0) is neither greater than nor equal to 1. As a result we are ready to leave the loop. We now pass automatically to the step immediately following the last step of the loop. In this case it is step 6, which produces the output.

FACTORIAL OF 3 IS 6

This is happily the correct result. The algorithm then terminates in step 7.

Although this example is fairly simple, it illustrates several important points. It would be worthwhile to reread it. Of particular importance are the flow of control through the loop, and the manner in which the variables involved in the loop (PRODUCT and MULTIPLIER) receive a succession of values. This last point can often be a source of difficulty in more complex problems. It can be helpful to tabulate the sequence of values received, as shown in Table 3-6.

To check your understanding of the foregoing discussion, run through the algorithm FACT_N yourself. You might try 6 as the input data value. The succession of values taken on by the loop variables is shown in Table 3-7.

You might have observed by now that the value of the result in this particular algorithm is unaffected by the last pass through the loop. Since the last pass is redundant, we could change the looping condition from "while MULTIPLIER $\geq$ 1" to "while MULTIPLIER $>$ 1." This would not alter the final

TABLE 3-6   Loop Behavior of Algorithm FACT_N for 3!

Pass Through Loop	Value of PRODUCT	Value of MULTIPLIER
0*	1	3
1	3	2
2	6	1
3	6	0

* The 0th pass refers to conditions in effect before the loop is entered.

TABLE 3-7   Loop Behavior of Algorithm **FACT_N** for 6!

Pass Through Loop	Value of PRODUCT	Value of MULTIPLIER
0	1	6
1	6	5
2	30	4
3	120	3
4	360	2
5	720	1
6	720	0

result, since multiplication by 1 leaves the result unaffected. It is very important, when looping, to take special care with the first pass through a loop and the last pass. Errors often result from failure to account properly for initial conditions or end conditions. They often require special attention.

In order to illustrate this point, the initial values of **PRODUCT** (1) and **MULTIPLIER (N)** were carefully chosen so that the correct answer would result on certain special inputs. What happens, for example, if we input **0** as the data value? In this case the first test of the condition "**MULTIPLIER** $\geq$ 1" evaluates to false, since the value of **MULTIPLIER** is, by assignment, 0. Thus, the loop is not executed at all. Control passes immediately to step 6 and the following is printed:

FACTORIAL OF 0 IS 1

By definition, in mathematics 0! is 1.

3-3.2   LOOP-CONTROLLED INPUT
In all the examples presented thus far, the problem has been solved for a single set of data values. By means of an appropriate loop, it is possible to read additional sets of data values and repeat the algorithm. It is possible, for example, to process the grades for an entire class in one run, or to compute the payoffs for an entire day of racing. This cuts down considerably on the amount of manual intervention required.

To illustrate the technique, let us modify the algorithm **REPORT** from Sec. 2-6 to compute the grades for an entire class. For convenience, we reproduce the original algorithm here.

Algorithm REPORT. Given individual marks in three aspects of a year's work in a course (**LAB_WORK, MIDTERM_EXAM, FINAL_EXAM**) for a student named **NAME**, this algorithm computes the final grade with weighting of 20%, 30%, and 50%, respectively. The output is a printed line giving the student's name, his/her individual marks, and final grade. The variable **GRADE** is real.
1. [Get input data]
     Read(NAME, LAB_WORK, MIDTERM_EXAM, FINAL_EXAM)

2. [Compute final grade]
    GRADE ← 0.20 * LAB_WORK + 0.30 * MIDTERM_EXAM
    + 0.50 * FINAL_EXAM
3. [Display results]
    Write('STUDENT NAME:', NAME)
    Write('LABORATORY MARK:', LAB_WORK)
    Write('MIDTERM EXAMINATION:', MIDTERM_EXAM)
    Write('FINAL EXAMINATION:', FINAL_EXAM)
    Write('FINAL GRADE:', GRADE)
4. [Finished]
    Exit   □

The process of returning to read additional data (assuming input is in the form of punched cards) is no problem now that we know about loops. The difficulty is in determining when to stop. We will discuss three ways of doing this.

One approach is to introduce one additional data card at the front of the regular card deck. This card tells us how many sets of data values follow, giving us the information required to determine the number of times to loop. We will refer to this as *counter-controlled* input (see Fig. 3-5). Applying this approach to our algorithm REPORT gives the following improved version.

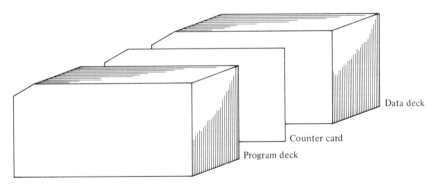

Data deck

Counter card

Program deck

FIGURE 3-5
Deck setup for
counter-controlled input.

Algorithm REPORT (revision 1). Given individual marks in three aspects of a year's work in a course (LAB_WORK, MIDTERM_EXAM, FINAL_EXAM) for a student named NAME, this algorithm computes the final grade with weighting of 20%, 30%, and 50%, respectively. The output is a printed line for each of a series of students giving the student's name, his/her individual marks, and final grade. The variable GRADE is real. The input is counter-controlled. The variable STUDENTS indicates the number of student records to be processed; the variable COUNT gives the number that have been processed. Both are of type integer.
1. [Input data counter, in this case the number of students]
    Read(STUDENTS)

2. [Initialize count of processed students]
   COUNT ← 0
3. [Engage the counter-controlled repetition]
   Repeat thru step 7 while COUNT < STUDENTS
4. [Get input data]
   Read(NAME, LAB_WORK, MIDTERM_EXAM, FINAL_EXAM)
5. [Compute final grade]
   GRADE ← 0.20 * LAB_WORK + 0.30 * MIDTERM_EXAM
   + 0.50 * FINAL_EXAM
6. [Display results]
   Write('STUDENT NAME:', NAME)
   Write('LABORATORY MARK:', LAB_WORK)
   Write('MIDTERM EXAMINATION:', MIDTERM_EXAM)
   Write('FINAL EXAMINATION:', FINAL_EXAM)
   Write('FINAL GRADE:', GRADE)
7. [Update count of students actually processed]
   COUNT ← COUNT + 1
8. [Finished]
   Exit   □

Study this algorithm carefully, paying particular attention to the material inserted to control the repetition. Two additional variables are used: STUDENTS and COUNT. STUDENTS gives the number of sets of data values to be read (in this case the number of students to be processed in total), and COUNT is used to compute the number of sets of data values actually read. The value of COUNT is increased by 1 each time through the loop. When COUNT is equal to STUDENTS, we know that we have read all the input values and the algorithm terminates. This is a good example of a counter-controlled input loop.

A second method of controlling an input loop involves the use of an extra set of data values at the end of the regular set of data values. Something special about this set of values makes it easily identifiable, so that when it is read you know that all the normal values must already have been read. Since an element of watching is involved, we refer to the extra card on which these special values are given (assuming punched card input again) as a *sentinel* card. This form of input control is referred to as *sentinel-controlled* input (see Fig. 3-6). Since the values on the sentinel card are not meant to be part of the calculation, but an indication that there are no more data values, care must be taken to ensure that they are not included in whatever calculations are taking place.

The following version of algorithm REPORT operates on sentinel-controlled input. Examine its structure carefully. There are several important ideas worthy of special note.

Algorithm REPORT (revision 2). Given individual marks in three aspects of a year's work in a course (LAB_WORK, MIDTERM_EXAM, FINAL_EXAM)

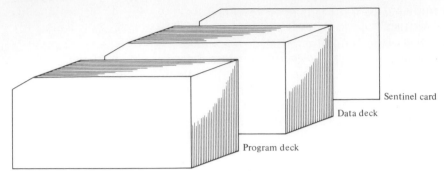

FIGURE 3-6
Deck setup for
sentinel-controlled input.

for a student named **NAME**, this algorithm computes the final grade with weighting of 20%, 30%, and 50%, respectively. The output is a printed line for each of a series of students giving the student's name, his/her individual marks, and final grade. The variable **GRADE** is real. The input loop is sentinel-controlled. A name of **'END'** (with dummy values for the other three variables) signals the end of the input.

1. [Read first card in data deck]
   Read(NAME, LAB_WORK, MIDTERM_EXAM, FINAL_EXAM)
2. [Engage the sentinel-controlled repetition]
   Repeat thru step 5 while NAME ≠ 'END'
3. [Compute final grade]
   GRADE ← 0.20 * LAB_WORK + 0.30 * MIDTERM_EXAM
   + 0.50 * FINAL_EXAM
4. [Display results]
   Write('STUDENT NAME:', NAME)
   Write('LABORATORY MARK:', LAB_WORK)
   Write('MIDTERM EXAMINATION:', MIDTERM_EXAM)
   Write('FINAL EXAMINATION:', FINAL_EXAM)
   Write('FINAL GRADE:', GRADE)
5. [Read next data card]
   Read(NAME, LAB_WORK, MIDTERM_EXAM, FINAL_EXAM)
6. [Finished]
   Exit ☐

The algorithm is designed so that the end of actual data is signaled by a data card on which a value of **'END'** is specified for the name of the student. This is not likely to be the name of any actual student; consequently, it serves as a good sentinel indicator. You will no doubt observe some changes in the location of the read statement over that of previous versions of the algorithm. This is necessary for two reasons: first, to ensure that the sentinel card itself is not processed as part of the computations in the loop, and second, to allow for the possibility (albeit remote) that the sentinel card may, in fact, be the first card read (that is, there are no actual data). For these reasons a read is performed immediately prior to entering the loop (step 1 in this case) to read the first card. If the values on the card constitute valid data, the looping condition will succeed and we will enter the loop to process the data on this particular card. If, however, the first card turns out to be the sentinel card, the looping condition will fail and we will bypass the loop entirely. In the event that the loop is entered, following the processing of the first data card we read the data card for the next pass through the loop. This second read statement is the last step in the range of the loop (step 5). Following execution of this step, control returns to step 2 and the looping condition is re-evaluated.

Notice that in this particular organization, the data read within pass i of the loop (here, in step 5) are actually intended for processing in pass i + 1. For this reason this is said to employ a "read-ahead" scheme. As described, a "read-ahead" scheme is essential whenever sentinel-controlled input is used.

To illustrate the operational differences between the two methods discussed for controlling input loops, let us take a small sample of input data and examine the workings of the two versions of the algorithm REPORT. Suppose that the following constitutes the data for a very small class:

'RALPH KRAMDEN', 58, 63, 72
'ED NORTON', 77, 51, 60

We process these data first using a counter-controlled input loop. This requires the introduction of a data card *before* the two actual data cards. On this card we enter the number of actual data cards to follow—here 2. The trace of this algorithm is given in Table 3-8.

Step 1 of the algorithm reads this first data card and assigns the value 2 to the variable STUDENTS. Step 2 initializes the data counter, COUNT, to 0. In step 3 we begin the loop. The looping condition (while COUNT < STUDENTS) is currently true, so we enter the range of the loop. In step 4 the first data card (data for Ralph Kramden) is read. Step 5 computes Ralph Kramden's final grade, and step 6 prints the results. In step 7, we prepare to repeat the loop by updating the data counter (from 0 to 1). Control then returns to step 3 for reevaluation of the looping condition. Since COUNT is still less than STUDENTS, our condition remains true, so we enter the loop a second time. In step 4 this time, we read the data card corresponding to Ed Norton. His final grade is computed in step 5 and printed in step 6. The data counter is updated in step 7, from 1 to 2, and control returns to step 3. This time, however, the value of COUNT is equal to the value of STUDENTS. Since the looping condition is now false, we proceed to step 8, at which point the algorithm terminates, having processed all the data.

To process these data using a sentinel-controlled input loop, we introduce a dummy card (no reflection on any students) *after* the two actual data cards. The important value of this card is the value of NAME, which must be 'END'. The other three values are not important but must be present nonetheless; otherwise, an input error will result when an attempt is made to read nonexistent values. We will assume that the following card has been inserted after the regular data deck:

'END', 0, 0, 0

The execution of this version of the algorithm is traced in Table 3-9.

Step 1 of this algorithm reads the first data card (corresponding to Ralph Kramden). The loop begins in step 2. Since this is an actual data card, the name is not 'END' (in fact, it is 'RALPH KRAMDEN'). As a result, we are allowed to proceed into the range of the loop. The final grade for Ralph Kramden is calculated in step 3 and printed in step 4. In step 5 we prepare for the next pass through the loop by reading the next data card (corresponding to Ed Norton). Control then returns to step 2 for reevaluation of the looping condition. Again, the value of the variable NAME (now, 'ED NORTON') is not equal to 'END', so we are allowed to proceed. The final grade for Ed Norton is computed and printed, and then, in step 5, the next card is read. This time it happens to be our sentinel card. Once again, control returns to step 2 for reevaluation of the looping condition. This time the value of NAME is 'END', so we proceed to step 6, where the algorithm terminates.

This discussion ought to convince you that even though they are structurally different, the two methods of controlling input loops produce the same effect. When do you choose one over the other? In many cases it is hard to see an advantage, and the choice is strictly one of personal preference. Sometimes you may not know how many data values you have, and you may not be prepared to count them. In this case you may opt for a sentinel-controlled input loop. The data counter employed in counter-controlled input loops, however, often has other useful functions in an algorithm. For example, it could be used in the calculation of the average grade in the entire class. However, such a data counter can also be employed very easily in a sentinel-controlled loop. You would be wise to become comfortable with both methods of controlling input loops and let the nature of the specific application dictate the choice.

Many programming languages provide yet another method of controlling input loops, which we will refer to as the "end-of-file" method. If the language you are using contains such a facility, it is possible to have a special end-of-data signal detected automatically, and a specified action taken accordingly. This operates something like the sentinel method without requiring the user to include any special data.

We will often express end-of-data tests as part of the read statement in our algorithmic notation. If there are data, the read statement will function normally. If, however, there are no more data, the indicated action will be taken. The following version of the algorithm REPORT uses this method.

Algorithm REPORT (revision 3). Given individual marks in three aspects of a year's work in a course (LAB_WORK, MIDTERM_EXAM, FINAL_EXAM) for a student named NAME, this algorithm computes the final grade with weighting of 20%, 30%, and 50%, respectively. The output is a printed line for each of a series of students giving the student's name, his/her individual marks, and final grade.

1. [Engage the repetition]
   Repeat thru step 4 while there are data elements to read
2. [Read next data card; exit on end of file]
   Read(NAME, LAB_WORK, MIDTERM_EXAM, FINAL_EXAM)
   If there is no more input data, then Exit
3. [Compute final grade]
   GRADE ← 0.20 * LAB_WORK + 0.30 * MIDTERM_EXAM
   + 0.50 * FINAL_EXAM
4. [Display results]
   Write('STUDENT NAME:', NAME)
   Write('LABORATORY MARK:', LAB_WORK)
   Write('MIDTERM EXAMINATION:', MIDTERM_EXAM)
   Write('FINAL EXAMINATION:', FINAL_EXAM)
   Write('FINAL GRADE:', GRADE)   □

For this application the third revision of the algorithm is very compact, owing largely to the use of the end-of-file signal to control the input loop. This particular algorithm has several interesting variations from previous algorithms we have seen. First, the loop control statement (step 1) has a less specific condition. Part of the loop control is given by the end-of-file test in step 2. The second item of interest is that this is the first algorithm we have seen in which the exit is not from the last step.

3-3.3   COUNTED LOOPS

For some applications the conditional loop control imposed by the "repeat-while" construct is unnecessarily complicated. In many cases you might want a loop executed a fixed number of times, where the number is already known. For applications such as these, a modified form of repetition construct is in order. This construct is of the following form.

   Repeat "range of loop" for "values of loop control variable."

Examples:

   1. Repeat thru step 11 for ID = 1, 2, . . ., 10
   2. Repeat for VAL = 1, 2, . . ., N
   3. Repeat thru step 10 for INDEX = −9, −8, . . ., −2
   4. Repeat for COUNT = 2, 4, . . ., 18

In each case we use what is called a *loop control variable* (ID, VAL, INDEX, and COUNT, respectively). Its purpose is to control the number of times the loop is executed. Each time through the loop, the loop control variable is given a new value automatically. The particular values it gets are specified as part of the repeat statement as shown, usually specified in the form of a

TABLE 3-8 Trace of Algorithm **REPORT** (revised for counter-controlled input)

Step	NAME	LAB_WORK	MIDTERM_EXAM	FINAL_EXAM	GRADE	STUDENTS	COUNT	Condition	Output
1	?	?	?	?	?	2	?		
2	?	?	?	?	?	2	0		
3	?	?	?	?	?	2	0	True	
4	RALPH KRAMDEN	58	63	72	?	2	0		
5	RALPH KRAMDEN	58	63	72	66.5	2	0		
6	RALPH KRAMDEN	58	63	72	66.5	2	0		As specified
7	RALPH KRAMDEN	58	63	72	66.5	2	1		
3	RALPH KRAMDEN	58	63	72	66.5	2	1	True	
4	ED NORTON	77	51	60	66.5	2	1		
5	ED NORTON	77	51	60	60.7	2	1		
6	ED NORTON	77	51	60	60.7	2	1		As specified
7	ED NORTON	77	51	60	60.7	2	2		
3	ED NORTON	77	51	60	60.7	2	2	False	
8	Termination								

TABLE 3-9 Trace of Algorithm REPORT (revised for sentinel-controlled input)

Step	NAME	LAB_WORK	MIDTERM_EXAM	FINAL_EXAM	GRADE	Condition	Output
1	RALPH KRAMDEN	53	63	72	?		
2	RALPH KRAMDEN	53	63	72	?	True	
3	RALPH KRAMDEN	58	53	72	66.5		
4	RALPH KRAMDEN	58	53	72	66.5		As specified
5	ED NORTON	77	51	60	66.5		
2	ED NORTON	77	51	60	66.5	True	
3	ED NORTON	77	51	60	60.7		
4	ED NORTON	77	51	60	60.7		As specified
5	END	0	0	0	60.7		
2	END	0	0	0	60.7	False	
6	Termination						

series of integer values. These values may be positive or negative, in increasing order or decreasing order. The loop terminates when the list of values has been exhausted.

To illustrate the use of counted loops, we present a simple algorithm to print out the first 20 powers of 2. Note carefully the structure of the loop and the use of the control variable **POWER**.

Algorithm POWERS. This algorithm prints the first 20 powers of 2. **POWER** is an integer variable.
1. [Compute and print]
   Repeat for POWER = 1, 2, . . ., 20
      Write(2 ↑ POWER)
2. [Finished]
   Exit ☐

The loop in this algorithm is quite short, and therefore its range is included in the single step. Notice that when the range of a loop is included in the same step as the loop control statement, the steps in the range will be indented as shown. Each time through the loop, the control variable **POWER** takes on a new value, beginning with 1 and increasing to 20 by 1 each time. This algorithm could be made more general by allowing the user to specify how many powers are wanted through the use of a data card.

Algorithm POWERS (more general). This algorithm prints the first N powers of 2. **POWER** and **N** are integer variables.
1. [Input number of values required]
   Read(N)
2. [Compute and print N values]
   Repeat for POWER = 1, 2, . . ., N
      Write(2 ↑ POWER)
3. [Finished]
   Exit ☐

Different programming languages handle the specification and testing of counted loops in different ways. In many languages, the specification consists of an initial value, an increment, and a final value. Each new value is computed by adding the increment (which in some languages may be negative). The loop terminates when the final value is reached or exceeded. In some languages (FORTRAN is an example) the testing is done *after* execution of the steps in the range of the loop, as shown in Fig. 3-7a. Such a loop is sometimes referred to as *bottom-tested*. In other languages (such as PL/I) the testing is done *before* execution of the steps in the range of the loop, as shown in Fig. 3-7b. With a bottom-tested loop the steps in the range will always be executed at least once, regardless of the initial conditions. *Top-tested* loops allow the possibility of skipping these steps entirely

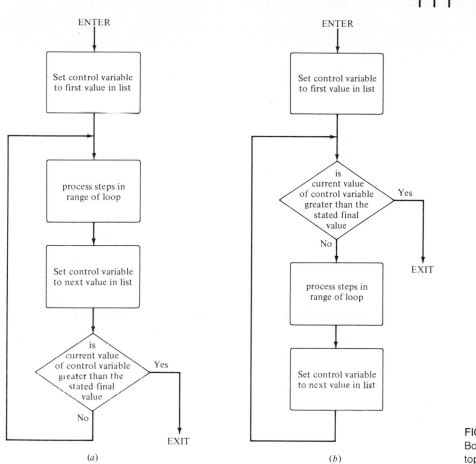

FIGURE 3-7
Bottom-tested and
top-tested loops.

and thus allow more generality. For this reason, we will assume that loops in our algorithmic language are top-tested.

As a final example of the use of a counted loop, let us return to the factorial calculation described earlier in this section. An examination of its operation (see Table 3-3, for example) reveals that although it uses a conditional "repeat-while" construct to control the looping, it actually executes like a counted loop, with **MULTIPLIER** as the loop control variable. In this case, however, the programmer has taken on the burden of controlling its values directly. This algorithm is perhaps more naturally expressed using a counted loop.

Algorithm FACT_N (revised). This algorithm computes the factorial of a value read from the input stream.

1. [Input]
   Read(N)

2. [Initialize product]
    PRODUCT ← 1
3. [Compute]
    Repeat for MULTIPLIER = N, N − 1, . . ., 1
        PRODUCT ← PRODUCT * MULTIPLIER
4. [Output result]
    Write('FACTORIAL OF', N, 'IS', PRODUCT)
5. [Finished]
    Exit   ☐

3-3.4 NESTED LOOPS

We close this section with some comments on nested loops. Just as it was possible to have an if–then–else construct within another if–then–else construct, it is also possible to have a loop within the range of another loop. The rules of nesting are similar in both cases. The inner construct must be completely embedded within the outer construct. There can be no overlapping Figures 3-8*a* and 3-8*b* show examples of proper and improper nesting, respectively.

As an example of the use of nested loops, let us modify our algorithm FACT_N for the calculation of the factorial function, to add to it an input loop. The factorial algorithm itself uses a loop to calculate the factorial; we will be enclosing this loop with a second loop to read a series of input values. The following algorithm FACTORIALS results.

Algorithm FACTORIALS. This algorithm reads in a series of input values (N) and computes and prints the factorial of each. The result is accumulated in the variable PRODUCT.

1. [Engage outer loop]
    Repeat thru step 5 while there are data items to read
2. [Read in next data value]
    Read(N)
    If there is no more input, then Exit
3. [Initialize PRODUCT for this particular calculation]
    PRODUCT ← 1
4. [Loop to calculate this factorial]
    Repeat for MULTIPLIER = N, N − 1, . . ., 1
        PRODUCT ← PRODUCT * MULTIPLIER
5. [Output result]
    Write('FACTORIAL OF', N, 'IS', PRODUCT)   ☐

Examine the structure of this algorithm carefully. You will notice that the loops are properly nested (the inner bracket is included here as well to show this). The range of the outer loop is steps 2 through 5; the range of the inner loop is included as part of step 4. Let us examine the loop behavior as the algorithm executes with the following two data values: 4, 5. The loop behavior is summarized in Table 3-10.

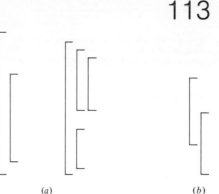

(a)   (b)

FIGURE 3-8
Examples of (a) proper and
(b) improper nesting.

Execution begins at step 1, which sets the outer loop in motion. The first data value (4) is read in step 2 and assigned to the variable N. Step 3 initializes the variable PRODUCT to 1. Step 4 then engages the second loop. Note, however, that we are still within the range of, and under control of, the first loop, and will be until step 5. Step 4 sets up a counted loop with MULTIPLIER as the control variable. The assignment is then executed repetitively, with MULTIPLIER receiving a succession of values 4, 3, 2, 1, as shown in Table 3-10. The value of PRODUCT grows accordingly. When the loop has been executed with the last specified value of MULTIPLIER (1), it terminates, and control passes to step 5, where the result

FACTORIAL OF 4 is 24

is printed. Step 5 is, however, the last step in the range of the outer loop. Control thus returns to step 1, the controlling statement. Since there are still data to read, we proceed to step 2, read the next data value (5), and assign it to the variable N. Step 3 resets the variable PRODUCT for the upcoming calculation, and step 4 engages the calculation loop once more. This time the control variable MULTIPLIER gets the succession of values 5, 4, 3, 2, 1, and PRODUCT is computed accordingly. On termination of this loop, control

TABLE 3-10   Loop Behavior in Algorithm FACTORIALS

Pass through Outer Loop	Pass through Inner Loop	Value of N	Value of MULTIPLIER	Value of PRODUCT
1	1	4	4	4
	2	4	3	12
	3	4	2	24
	4	4	1	24
2	1	5	5	5
	2	5	4	20
	3	5	3	60
	4	5	2	120
	5	5	1	120

passes to step 5 for printing of

FACTORIAL OF 5 IS 120

and then back to step 1 as dictated by the outer loop. This time when we try to read a data value, end of file is signaled, and as a result, the algorithm terminates.

The concept of nested loops is an important one and can, of course, be done with either type of loop. It would be well worth your time to reread the description above until you are completely comfortable with what is happening. Note in particular that the two loops operate at different rates, much like a system of gears. The inner loop moves faster than the outer loop, as is shown clearly in Table 3-10. Any values required in the inner loop must be reset (as in the case of **PRODUCT**) before the loop is reentered. If sufficient care is not taken, you may be processing the steps in a loop with the values from the previous pass through the loop. What would happen in the algorithm **FACTORIALS** if step 3 were removed and inserted before step 1?

3-3.5 SUMMARY

In this section we have presented the concept of a loop. We saw that there are basically two kinds of loops, distinguished by the manner in which the looping is controlled. A conditional loop is controlled by means of a condition expressed in a "while" clause of the repeat statement. Each time the flow of control returns to the repeat statement, the condition is reevaluated. Looping continues as long as the looping condition is true.

The second type of loop we referred to as the counted loop. In a counted loop, a specially designated variable (the control variable) takes on a specified sequence of values as the looping progresses. This type of loop makes possible the repetition of a group of statements a fixed number of times.

Once again, the choice of which type of repetition construct to use should be dictated by the nature of the application. Too many programmers force one type of loop on all programs. This results in programs which are unnecessarily difficult to understand. Unfortunately, some programming languages offer only one form. Where a choice is available, take advantage of it.

We also considered the use of nested loops, in which the range of a loop is embedded within the range of another loop. The inner loop is executed completely (that is, all passes) on each pass through the outer loop. With both the loop and the if statement, you should avoid nesting too deeply. Programs with more than two or three levels of nesting tend to be difficult to understand and, as a result, error-prone. One alternative to deeply nested structures is discussed in Sec. 3-4.

As a final comment on looping, we remind you that it is the responsibility of the programmer to ensure that loops will terminate. Loops that fail to do so can prove extremely costly, and embarrassing to the programmer. The frequency of loops failing to terminate is probably higher with conditional loops, because the mechanism for counted loops is largely automatic.

Nevertheless, it is well worth your time to double check all loops before you submit your programs for execution.

Many programming languages contain an additional control structure that we have not described—the *go to* statement. The "go to" statement forces an immediate transfer of control to some designated point in an algorithm. In so doing it has the power to override the actions of the other control structures described.

We have deliberately avoided discussing the "go to" statement in this section, although we will be dealing with it for specialized applications in later chapters. This seemingly innocuous little statement has been the object of an extensive controversy for the past several years. One camp advocates its abolition from programming; the other argues that it is a basic tool.

There is a reasonable amount of evidence to suggest that, in general, the quality of programs goes down as the number of go to's in them goes up. Although we prefer not to be fanatical, our feeling is that if you can get along without it, you should. Since the go to can undermine the actions of the other control structures, the resulting logic can be very difficult to follow and the overall structure, hard to grasp—a situation that often leads to errors that are quite hard to detect. The repetition and if–then–else constructs described can do a great deal for you if you let them. Their execution is fairly easy to understand. Unless it complicates the solution unnecessarily, we feel that it is best to avoid the use of the go to statement. Certainly, any use of it must be considered very carefully! More readings on this particular subject are found in the references listed at the end of Chap. 7.

EXERCISES 3-3

1. In each of the following algorithm segments, indicate whether or not the loop will terminate. If it will not, why not? Assume that all are integer variables.

    (a) 1. COUNTER ← 0
    TOTAL ← 0
    2. Repeat while COUNTER ≥ 0
    TOTAL ← TOTAL + 2

    (b) 1. COUNTER ← 0
    TOTAL ← 0
    2. Repeat while COUNTER ≤ 10
    TOTAL ← TOTAL + 2
    COUNTER ← COUNTER + 1

    (c) 1. TOTAL ← 0
    2. Repeat for INDEX = 1, 2, . . ., 10
    TOTAL ← TOTAL + 1
    INDEX ← INDEX − 1

2. In each of the following segments, give the value that will be printed for the variable **VAR**. Assume integer variables throughout.

(a) 1. VAR ← 0
   2. Repeat for INDEX = 1, 2, . . ., 10
       VAR ← VAR + 1
   3. Write(VAR)

(b) 1. VAR ← 0
   2. Repeat for INDEX = 4, 8, . . ., 36
       VAR ← VAR + 1
   3. Write(VAR)

(c) 1. VAR ← 0
   2. Repeat step 3 for INDEX1 = 1, 2, . . ., 15
   ⎡ 3. Repeat for INDEX2 = 1, 2, . . ., 8
   ⎣   VAR ← VAR + 1
   4. Write(VAR)

(d) 1. VAR ← 0
   2. Repeat for INDEX = 10, 9, . . ., 1
       VAR ← VAR + 1
       INDEX ← INDEX + 1
   3. Write(VAR)

3. Penny Programmer is worried about her performance in her computer science class. On her first program, she made one mistake; on her second, she made two; on the third, four; and so on. It appears that she makes twice the number of mistakes on each program as she made on the program before. The class runs for 13 weeks with two programming problems per week. Design an algorithm to compute the number of errors Penny can expect on her final program at her current rate of performance.

4. Design an algorithm to compute the sum of the following series to 100 terms:

$$1 - \tfrac{1}{2} + \tfrac{1}{4} - \tfrac{1}{6} + \tfrac{1}{8} - \tfrac{1}{10} + \tfrac{1}{12} - \cdots$$

5. The ancient Greek philosopher Zeno is perhaps best known for the paradox of Achilles and the tortoise. Achilles and the tortoise run a race. Achilles runs 10 times as fast as the tortoise, but the tortoise has a 100-yard headstart. It can be argued that Achilles can never pass the tortoise, since whenever he reaches the point where the tortoise *was*, the tortoise has moved ahead a small amount. Design an algorithm to use a loop to compute the time at which Achilles does, in fact, pass the tortoise.

6. Saskatchewan fishing regulations impose a limit on the total poundage of a day's catch. Suppose that you plan to take your portable computer terminal with you on your next fishing trip, and you require a program

to tell you when you have exceeded your limit. Design an algorithm that first reads the daily limit (in pounds), and then reads input values one by one (the weights of the fish recorded as they are caught) and prints a message at the point when the limit is exceeded. A weight of 0 indicates the end of input. After each fish is recorded, your algorithm should print the total poundage caught up to that time.

7. A number is defined to be a *prime* number if it has no divisors other than 1 and the number itself. Design an algorithm to read a number and determine whether or not the number is a prime number.

8. Manny Motorist has just returned from a recent motoring holiday. At each stop for gas he recorded his odometer reading and the amount of gas purchased (assume that he filled the tank each time). In addition, assume that gas was purchased immediately prior to leaving for the trip, and immediately upon return, taking odometer readings at each point. Design an algorithm to read first the total number of stops made (including the first and last), and then the data recorded for gas purchases, and compute
   (a) The gas mileage achieved between every pair of stops on the trip.
   (b) The gas mileage achieved through the entire trip.

9. Each team in the Canadian Football League has a roster of 33 players. Assume that each team in the league prepares a card for each of its players, with the format

   player's name, weight, age

   The data for the nine teams in the league are collected and sent to league headquarters for analysis. A single deck is prepared, with the data cards grouped by individual team; that is, the first 33 cards are for team 1, the next 33 are for team 2, and so on. Design an algorithm to read this deck and compute the following statistics:
   (a) Average weight and age for each of the nine teams.
   (b) Average weight and age for the entire league.

10. The Who-Do-You Trust Company plans to use a computer to prepare customer statements for their deposit accounts. For each customer a set of data cards is prepared, containing information on his/her deposits and withdrawals for that month. The data for each customer begin with a special card containing his/her name, address, and balance forwarded from the previous month. This is then followed by transaction cards, which contain the customer's name, a description of the transaction, and the amount of the transaction. Account withdrawals will have a negative amount of transaction. Typical input would be as follows:

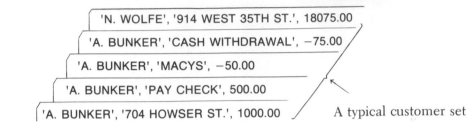

The last card is a "dummy"
card with customer name
of 'LASTCARD'

'N. WOLFE', '914 WEST 35TH ST.', 18075.00

'A. BUNKER', 'CASH WITHDRAWAL', −75.00

'A. BUNKER', 'MACYS', −50.00

'A. BUNKER', 'PAY CHECK', 500.00

'A. BUNKER', '704 HOWSER ST.', 1000.00   A typical customer set

Design an algorithm to produce a statement of account for each customer. These statements appear as follows:

WHO-DO-YOU TRUST COMPANY
416 FIFTH AVE.
NEW YORK

TO: A. BUNKER
   704 HOWSER ST.

ITEM	DEPOSITS	WITHDRAWALS	TOTAL
OPENING BALANCE	1,000.00		1000.00
PAY CHECK	500.00		1500.00
MACYS		50.00	1450.00
CASH WITHDRAWAL		75.00	1375.00
SERVICE CHARGE		.50	1374.50
INTEREST PAID	3.75		1378.25

The service charge must be calculated for each customer at a rate of 25¢ for each withdrawal. Interest is to be calculated by your program at 1% on any final balance over $1,000.00. A new customer's leading card is detected by a change in name.

3-4   USE OF COMPOUND CONDITIONS

In the previous two sections we presented a pair of programming constructs that involve the expression of conditons. These conditions were described in terms of simple relationships between values, using the relational operators in Table 3-1. For some problems simple relations are inadequate for describing the conditions required. The result can usually be achieved with nesting (see algorithm MAX_3 from Sec. 3-2, for example), but this may make the algorithms unnecessarily complicated and difficult to understand.

An alternative method is the use of compound conditions. These are derived from the standard simple relations using the logical connectives "and," "or," and "not." An "and" operation combines two simple conditions and yields a true result only if both components are true; an "or" operation on two simple conditions requires only one of them to be true. The "not" operation operates on a single simple condition and simply negates (or reverses) its value.

The meanings of these operators are best described using what is known as truth tables. Suppose that c_1 and c_2 represent conditions. Each may be either true or false. The truth tables in Table 3-11 define "and," "or," and "not." As we shall see, these definitions give natural interpretations for the compound conditions.

As an example, suppose that ONE is a variable whose value is 1. Consider the following simple conditions: "ONE < 2" and "ONE < 0". Clearly, the first relation is true and the second is false. Consider the following three compound conditions derived from these simple relationships.

1. (ONE < 2) and (ONE < 0)
2. (ONE < 2) or (ONE < 0)
3. Not (ONE < 2)

If we apply the truth table for the "and" operator to (1), it tells us that the value of this compound condition is false since the first operand is true but the

TABLE 3-11   Truth Tables for Logical
Connectives

(a) "and" (conjunction)

c_1	c_2	c_1 and c_2
True	True	True
True	False	False
False	True	False
False	False	False

(b) "or" (disjunction)

c_1	c_2	c_1 or c_2
True	True	True
True	False	True
False	True	True
False	False	False

(c) "not" (negation)

c_1	not c_1
True	False
False	True

second is false. This makes sense intuitively. Suppose that this were used as the condition in an if statement:

If (ONE < 2) and (ONE < 0)

Since ONE is less than 2 but is not less than 0, the test above must fail. The truth-table approach gives us the correct value, false.

Applying the appropriate truth tables to the other two examples, we get the values true for (2) and false for (3).

In the previous examples we bracketed the individual simple conditions. Although this clarifies the presentation, it is not necessary for unambiguous processing of the conditions. The precedences for the logical connectives are set so that the relational operators have higher precedence, and therefore tighter binding, than the connectives. As a result the required interpretation is achieved without the use of parentheses.

Compound conditions can be used either in "if" statements or in conditional repeats. The form and interpretation are the same in each case. As an example of an application using compound conditions, we will reformulate the algorithm MAX_3 from Sec. 3-2. You will recall that the problem was to determine the largest of three input values A, B, and C. Our solution then involved pairwise comparisons and nested ifs. The nesting can be eliminated through the use of compound conditions. The following algorithm results.

Algorithm MAX_3 (version 2). This algorithm reads three numbers A, B, and C and prints the value of the largest. Values are assumed to be distinct.
1. [Input data values]
    Read(A, B, C)
2. [Determine largest value]
    If A > B and A > C then MAX ← A
    If B > A and B > C then MAX ← B
    If C > A and C > B then MAX ← C
3. [Print largest value]
    Write('LARGEST VALUE IS', MAX)
4. [Finished]
    Exit □

You will notice that, in this case, by eliminating the nesting we produce an algorithm that is easier to understand. The conditions in effect are clearly evident, and no comments are required. As a parenthetical remark, even though this version is shorter and easier to understand, the previous version will probably execute faster because fewer tests are required. As a general rule, we will, however, lean toward algorithms and programs that are easier to read and understand. Programmer time is more valuable than computer time.

As an exercise you might try to rewrite the algorithm MAXMIN_3 to determine the largest and smallest of three values (see Sec. 3-2), using compound conditions.

The laws governing the interpretation of compound conditions are based on principles of formal mathematical logic, a subject into which we prefer not to venture. Just as extensive nesting can hinder understanding of an algorithm, so too can elaborate compound conditions. In some cases we may wish to unwind a nested structure using compound conditions; in others, the complexity of the conditional expression may dictate the use of nesting instead. Even though there are formal laws governing the interpretation of complex conditional expressions (see DeMorgan's laws, for example), we prefer to restrict ourselves to expressions that are easily read and understood. A useful test of understandability is the "telephone test." Can you read the expression to someone over a telephone and have them interpret it correctly? This is a valuable test to bear in mind for any complex expression.

EXERCISES 3-4

1. Suppose that I and J are integer variables with values 6 and 12, respectively. Which of the following conditions are true?
   (a) $2*I \leq J$
   (b) $2*I - 1 < J$
   (c) $I > 0$ and $I < 10$
   (d) $I > 25$ or $(I < 50$ and $J < 60)$
   (e) $I < 4$ or $J > 5$
   (f) not $I > 6$

2. Assume that A, B, C, and D are variables and S_1, S_2, S_3, and S_4 are statements.
   (a) Express the conditions necessary for the execution of S_1, S_2, S_3, and S_4 in the following statement as separate compound conditions.

   ```
   If A > B
   then If B ≤ C
        then If C ≠ D then S₁
                      else S₂
        else S₃
   else S₄
   ```

   (b) Express the following as a nested "if" structure using only simple conditions.

   If $(A < B$ and $C \neq D)$ and $(B > D$ or $B = D)$ then S_1

3. Students are recommended for graduate fellowships according to their overall undergraduate average. The nature of the recommendations is based on the following table:

Average	Recommendation
≥90%	Highest recommendation
≥80% but <90%	Strong recommendation
≥70% but <80%	Recommended
<70%	Not recommended

A card is prepared for each student applicant according to the following format:

student's name, overall average

Design an algorithm to read the deck of cards for the applicants and prepare a list giving the name of each student, his/her average, and recommendation. At the end of the list (denoted by a sentinel card with student name 'END OF LIST'), give the overall average of the applicants and a count of the number of recommendations of each type.

4. The Purple Nasties hockey team has had a good season and is rewarding its players with a salary increase for the coming season. Salaries are to be adjusted as follows:

Present Salary	Action
0–$9,000	20% increase
$9,001–$13,000	10% increase
$13,001–$18,000	5% increase
Over $18,000	No increase

The team roster contains 20 players. Design an algorithm to read the name and present salary for each player, and then print his name, present salary, and adjusted salary. At the end of this list, give the total payroll at present, and the total payroll given the proposed adjustments.

5. Students were given five examinations (A, B, C, D, E). Statistics are required to determine the number that
(*a*) Passed all exams.
(*b*) Passed A, B, and D, but not C or E.
(*c*) Passed A and B, C or D, but not E.
Design appropriate algorithms, using the end-of-file method to signal the end of input.

3-5 APPLICATIONS

We now present a number of problems in which the techniques of this chapter are applied. Considerable attention is paid to the design of the algorithms. The actual tracing of the algorithms is left as an exercise.

The first application deals with the problems faced by a college bookstore ordering textbooks for the coming term and the factors influencing its decisions. The second application is in the area of monthly mortgage payments. The third and final application concerns trying to reconcile checks cashed by employees of a company against those that were issued.

3-5-1   BOOKSTORE ORDERS

A campus bookstore would like to make an estimate of the number of books to order and its likely profit margin for the coming academic term. The number of students expected to purchase any given book is governed by a number of factors, such as the importance of the book to a course, the availability of used books, and the likelihood of groups of students sharing a single book. Since storage space is severely limited, and the bookstore must pay the shipping charges on books returned to the supplier, the desire not to overstock is very strong. On the other hand, as a campus service, it must ensure that all students who want to purchase a copy of the book have the opportunity to do so.

For each book to be ordered, a card is punched containing the following information:

1. The book identification number—a five-digit numeric code (for example, 38605)
2. Quantity currently in stock
3. Classification of the book—1 indicates a prescribed course text, 2 indicates a book recommended as supplementary reading
4. Estimated student enrollment in the course for which the book is being ordered
5. Whether the book is being used for the first time this term—1 indicates yes, 0 indicates no
6. Wholesale unit cost of the book

Bookstore studies have found that their market is dictated mainly by items 3, 4, and 5. For prescribed texts, sales are projected at 60% of estimated enrollment for books that have been used in a previous term, and 85% for new books. For books recommended as supplementary reading, the figures are 25% of estimated enrollment for books that have been used in a previous term, and 40% for books that have not. These projections are used to calculate the number of books required; the quantity to order is determined by subtracting the number currently in stock. A negative result indicates the amount of excess stock that is to be returned.

The profit margin is based on the wholesale cost of the book and the bookstore's markup. From this, all bookstore expenses must be paid. There is a 25% markup on all books costing $10 or less, and a 20% markup on all others.

The output required is a report of the following form:

IDENTIFICATION	ON HAND	TO ORDER	PROFIT MARGIN
38605	13	67	200.00
.	.	.	.
.	.	.	.
.	.	.	.
.	.	.	.

TOTAL PROFIT   25653.72

The algorithm to solve this problem requires a loop to read and process all the data cards, and "if" statements to process the individual cases. We can describe this in a general or "high-level" algorithm as follows:

1. Repeat thru step 4 for all data cards
2. Read information for next book
3. Determine number of copies required, the size of the order, and the profit margin on this book, and print
4. Add profit margin on this book to total profit margin
5. Print final total

We first introduce some variables. From the general algorithm and description of the input, the following are suggested (types in parentheses):

IDENT (integer)	Book identification number
STOCK (integer)	Quantity currently in stock
TYPE (integer)	Classification of the book as pre-scribed text or supplementary reading
ENROLLMENT (integer)	Estimated course enrollment
NEW (integer)	New text or used previously
COST (real)	Wholesale cost of the book
NUMBER_REQUIRED (integer)	Number of copies required
ORDER (integer)	Number of copies to be ordered
PROFIT (real)	Profit margin on the book
TOTAL_PROFIT (real)	Total profit margin

We can now proceed to the design of our final algorithm. We will first consider processing a single book (steps 2 through 4 of the general algorithm), and incorporate the details of the input loop afterward.

The input statement is straightforward, given the preceding list of variables. It can be written directly as

```
Read(IDENT, STOCK, TYPE, ENROLLMENT, NEW, COST)
```

Next we must determine the number of copies required of a given book. This is a function of three integer variables: TYPE, denoting the book's classification as a prescribed text or supplementary reading; ENROLLMENT, denoting the estimated enrollment for the coming session; and NEW, denoting whether or not this book is new this term. From these values the number of copies required (NUMBER_REQUIRED) is calculated as follows (notice that the result is rounded to the nearest integer through the use of the built-in function ROUND):

```
If TYPE = 1
then If NEW = 1
      then NUMBER_REQUIRED ← ROUND(0.85 * ENROLLMENT)
      else NUMBER_REQUIRED ← ROUND(0.60 * ENROLLMENT)
else  If NEW = 1
      then NUMBER_REQUIRED ← ROUND(0.40 * ENROLLMENT)
      else NUMBER_REQUIRED ← ROUND(0.25 * ENROLLMENT)
```

Two conditions are required before an assessment can be made. We need to know the book's classification and whether or not it is new this term. We could have used compound conditions (such as "If TYPE = 1 and NEW = 1"); however, a total of four tests would be required to handle the four possibilities. For this problem, the nesting approach is probably easier to understand.

Having determined the number of copies required (NUMBER_REQUIRED), it is a simple matter to determine the number to order. We use the integer variables STOCK, denoting the number of copies currently on hand, and ORDER, denoting the number of copies that will have to be ordered to fulfill the projected requirement, in the statement

```
ORDER ← NUMBER_REQUIRED − STOCK
```

Since any stock in excess of that required is to be returned to the supplier, we might issue a message to that effect if the value of ORDER turns out to be negative (note the use of the built-in function ABS).

```
If ORDER < 0
then Write('OVERSTOCKED:', ABS(ORDER),' COPIES TO RETURN')
```

The calculation of the profit margin is based on the wholesale cost of the book (the real variable COST). We use the real variable PROFIT to denote the projected profit on each copy of a particular book.

```
If COST ≤ 10.00
then PROFIT ← NUMBER_REQUIRED * 0.25 * COST
else  PROFIT ← NUMBER_REQUIRED * 0.20 * COST
```

Within the loop, we will accumulate the total profit margin in the real variable TOTAL_PROFIT.

For the report we display the values of four variables for each book: IDENT(book's identification number), STOCK, ORDER, and PROFIT.

To read the input values we will use a loop controlled by an end-of-file test. After all the data have been processed, we will display the value of TOTAL_PROFIT.

Putting this all together in accordance with the general structure developed earlier results in the following algorithm.

Algorithm BOOKSTORE. This algorithm reads data on inventory and sales projections and determines the profit margin (PROFIT, TOTAL_PROFIT) on book orders for an upcoming term. Sales projections for a given book are based on the estimated enrollment in the course (ENROLLMENT), the sales classification of the book (TYPE), and whether or not the book was used in the previous term (NEW). The stock on hand is given by STOCK and the wholesale unit cost by COST. Each book is identified by an identification number (IDENT). Variables IDENT, STOCK, TYPE, ENROLLMENT, NEW, NUMBER_REQUIRED, and ORDER are integer; COST, PROFIT, and TOTAL_PROFIT are real.

1. [Initialize]
      TOTAL_PROFIT ← 0.00
2. [Set up report headings]
      Write('IDENTIFICATION', 'ON HAND', 'TO ORDER', 'PROFIT MARGIN')
3. [Engage input loop]
      Repeat thru step 9 while data items remain to be read
4. [Read information for next book]
      Read(IDENT, STOCK, TYPE, ENROLLMENT, NEW, COST)
      If there are no more data items
      then Write('TOTAL PROFIT', TOTAL_PROFIT)
          Exit
5. [Determine number of copies required]
      If TYPE = 1
      then If NEW = 1   (prescribed text)
          then NUMBER_REQUIRED ← ROUND(0.85 * ENROLLMENT)
              (new this term)
          else NUMBER_REQUIRED ← ROUND(0.60 * ENROLLMENT)
              (used before)
      else If NEW = 1   (supplementary reading)
          then NUMBER_REQUIRED ← ROUND(0.40 * ENROLLMENT)
              (new this term)
          else NUMBER_REQUIRED ← ROUND(0.25 * ENROLLMENT)
              (used before)
6. [Determine size of order and, if necessary, issue overstocked notice]
      ORDER ← NUMBER_REQUIRED − STOCK
      If ORDER < 0
      then Write(IDENT, 'IS OVERSTOCKED:', ABS(ORDER),
          'COPIES TO RETURN')
7. [Determine profit on this book]
      If COST ≤ 10.00
      then PROFIT ← NUMBER_REQUIRED * 0.25 * COST
      else PROFIT ← NUMBER_REQUIRED * 0.20 * COST

8. [Update total profit statistic]
   TOTAL_PROFIT ← TOTAL_PROFIT + PROFIT
9. [Print line for this particular book]
   Write(IDENT, STOCK, ORDER, PROFIT) □

The input loop used in the algorithm is controlled by an end-of-data file signal. Is this method the best suited to the application? How can the algorithm be modified to incorporate a counter-controlled or sentinal-controlled input loop?

3-5.2 MORTGAGE PAYMENTS

The purchase of a home is probably the largest single investment that the average consumer will make. Many factors must be considered, not the least of which is the size of the monthly mortgage payment.

The monthly payment is based on three factors: the amount of principal owed, the interest rate for the mortgage, and the number of years for the mortgage (or the term of the mortgage). We want to develop an algorithm to compute the monthly mortgage payment from the following standard formula:

$$\frac{P \times i \times (i + 1)^n}{(i + 1)^n - 1}$$

where P is the amount of principal, i the monthly interest rate (in this formula interest is compounded monthly), and n the number of months in the term of the mortgage.

For our algorithm we will define the following variables:

PRINCIPAL (real)   Amount of principal
INT_RATE (real)   Yearly interest rate
TERM (integer)   Numbers of years for the mortgage
PAYMENT (real)   Monthly payment

We will assume that these variables represent the various factors as they are normally quoted to the consumer. This means that the interest rate and term must be converted from yearly figures to monthly figures for use in the standard formula. We introduce variables I (real) and N (integer) for this purpose. This will require the statements

I ← INT_RATE / 12.0
N ← TERM * 12

The calculation of the monthly payment is a straightforward translation of the standard formula,

PAYMENT ← (PRINCIPAL * I * (I + 1) ↑ N) / ((I + 1) ↑ N − 1)

In addition to the monthly payment, we will produce a table showing a breakdown, by month, of the amount of the monthly payment going to principal and the amount going to interest. At the end of each year we will print yearly totals of amount paid to principal and interest, and amount of principal outstanding. We will develop an algorithm to do this by generalizing on a particular example.

Suppose that we have a $25,000 mortgage at 10% with a term of 25 years. If we apply the computations above, we get a monthly interest rate (I) of .10/12 and a term in months (N) of 300. Thus, the value of **PAYMENT** is calculated as

$$ \frac{25{,}000 \times \dfrac{.10}{12} \times \left(\dfrac{.10}{12} + 1\right)^{300}}{\left(\dfrac{.10}{12} + 1\right)^{300} - 1} $$

which gives a monthly payment of $227.17 (assume rounding to two digits). The interest for the first month of the mortgage is easy to compute. It is simply .10/12 of $25,000, or $208.33. This means that of our total monthly payment of $227.17, $208.33 goes to interest and the remaining $18.84 to principal.

What happens in month 2? Our principal has been reduced by the payment in month 1, $18.84, leaving a total of $24,981.16. The calculation of the monthly interest/principal breakdown for this month proceeds as for the first month. Our interest payment is .10/12 of $24,981.16, or $208.18. Thus, in month 2 our monthly payment of $227.17 comprises a payment of $208.18 to interest and $18.99 to principal. We can continue in this fashion for the entire 25 years (or 300 months) of the mortgage.

Generalizing these calculations yields an algorithm with the following basic structure:

1. Read in quoted figures (amount of principal, yearly interest rate, term in years)
2. Compute basic monthly payment
3. Repeat thru step 5 for each year of mortgage
4. Repeat for each month in the year
   Compute payments to interest and principal for the month from quoted interest rate, amount of principal outstanding, and amount of monthly payment
5. Compute yearly totals and amount of outstanding principal for the next year

We now proceed to supply the details. First, we will introduce some additional variables:

MONTHLY_INT_PAYMENT (real)   Monthly payment of interest
MONTHLY_PRINC_PAYMENT (real)   Monthly payment of principal
YEARLY_INT_PAYMENT (real)   Yearly payment of interest
YEARLY_PRINC_PAYMENT (real)   Yearly payment of principal

The calculation of the monthly payment has already been described. The breakdown of a monthly payment (PAYMENT) is straightforward, as shown in the example above.

    MONTHLY_INT_PAYMENT ← I * PRINCIPAL
    MONTHLY_PRINC_PAYMENT ← PAYMENT − MONTHLY_INT_PAYMENT

For the year-end statistics, these components must be added to running totals YEARLY_INT_PAYMENT and YEARLY_PRINC_PAYMENT.

Since the term of the mortgage is fixed, we will repeat the calculations a fixed number of times. This application is perfectly tailored to a counted loop.

    Repeat thru step . . . for YEAR = 1, 2, . . ., TERM

What happens on each pass through this loop? We begin with an outstanding principal of PRINCIPAL, and make 12 payments to it, computing the payment breakdown for each month as we go. Before the next pass through this particular loop, the value of YEARLY_PRINC_PAYMENT is adjusted to reflect each of the 12 monthly payments, and the value of PRINCIPAL is also adjusted accordingly. For each month, then, the following calculation is made.

    MONTHLY_INT_PAYMENT ← I * PRINCIPAL
    MONTHLY_PRINC_PAYMENT ← PAYMENT − MONTHLY_INT_PAYMENT
    YEARLY_INT_PAYMENT ← YEARLY_INT_PAYMENT
                                        + MONTHLY_INT_PAYMENT
    YEARLY_PRINC_PAYMENT ← YEARLY_PRINC_PAYMENT
                                        + MONTHLY_PRINC_PAYMENT
    PRINCIPAL ← PRINCIPAL − MONTHLY_PRINC_PAYMENT

In addition, the following report line is printed each month:

    Write(YEAR, MONTH, MONTHLY_PRINC_PAYMENT,
    MONTHLY_INT_PAYMENT)

At the end of each year, a yearly summary is printed:

    Write('YEAR END SUMMARY:')
    Write('PRINCIPAL PAID:', YEARLY_PRINC_PAYMENT,
      'INTEREST PAID:', YEARLY_INT_PAYMENT,
      'OUTSTANDING PRINCIPAL:', PRINCIPAL)

Putting all of this together we get the following algorithm.

Algorithm MORTGAGE. This algorithm reads in the principal owing on a mortgage (PRINCIPAL), the yearly interest rate (INT_RATE), and the duration of the mortgage in years (TERM). It computes and prints the monthly payment and a breakdown by year of the monthly payments to interest and to the principal.

1. [Input]
    Read(PRINCIPAL, INT_RATE, TERM)

2. [Convert to monthly figures for use in standard formula and compute monthly payment]
    I ← INT_RATE / 12.0
    N ← TERM * 12
    PAYMENT ← (PRINCIPAL * I * (I + 1) ↑ N / ((I + 1) ↑ N − 1)
    Write('MONTHLY PAYMENT IS $', PAYMENT)

3. [Set up table headings for the monthly reports]
    Write('YEAR', 'MONTH', 'AMT PAID TO PRINCIPAL',
        'AMT PAID TO INTEREST')

4. [Engage loop to compute and print yearly payments]
    Repeat thru step 7 for YEAR = 1, 2, . . ., TERM

5. [Initialize yearly running totals]
    YEARLY_INT_PAYMENT ← 0
    YEARLY_PRINC_PAYMENT ← 0

6. [Compute and print each monthly breakdown]
    Repeat for MONTH = 1, 2, . . ., 12
        MONTHLY_INT_PAYMENT ← I * PRINCIPAL
        MONTHLY_PRINC_PAYMENT ← PAYMENT
                                    − MONTHLY_INT_PAYMENT
        YEARLY_INT_PAYMENT ← YEARLY_INT_PAYMENT
                                    + MONTHLY_INT_PAYMENT
        YEARLY_PRINC_PAYMENT ← YEARLY_PRINC_PAYMENT
                                    + MONTHLY_PRINC_PAYMENT
        Write(YEAR, MONTH, MONTHLY_PRINC_PAYMENT,
            MONTHLY_INT_PAYMENT)
        PRINCIPAL ← PRINCIPAL − MONTHLY_PRINC_PAYMENT

7. [Print yearly summary]
    Write('YEAR END SUMMARY:')
    Write('PRINCIPAL PAID:', YEARLY_PRINC_PAYMENT, 'INTEREST PAID:',
        YEARLY_INT_PAYMENT, 'OUTSTANDING PRINCIPAL:', PRINCIPAL)

8. [Finished]
    Exit ☐

3-5.3 CHECK RECONCILIATION

A computer card containing the check number and the amount of the check is prepared for every check issued to employees of the Red Nose Winery. At the end of each month, all the cashed checks are collected from the bank and a similar card is punched for each check, again containing the check number and the amount of the check. These two card decks are merged together manually, by check number, to form the input to a reconciliation algorithm that is to match the checks issued against those that were cashed.

The following is typical of the input that might occur and illustrates the cases that can arise.

23871	48.50
23871	48.50

← check 23871 issued and cashed for $48.50

23872   150.00   ← check 23872 issued for $150.00 but not yet cashed

23873	36.20
23873	236.20

← check 23873 issued and cashed for different amounts

The reconciliation algorithm is to list the checks issued but not yet cashed, and to print the check number and two amounts of any check which is issued and then cashed at a different amount. Also, it is to print the total amount of the checks cashed and the total amount of the outstanding checks (checks issued but not yet cashed).

Sample output:

CHECK NUMBER	AMOUNT	
23872	150.00	
23873	36.20	236.20

TOTAL CASHED = $123761.42
TOTAL OUTSTANDING = $8731.49

This problem is typical of many problems that arise in business data processing. The techniques involved in its solution are important. The organization of read statements is somewhat different from what we have seen previously, since two data cards are required to process each check. The first of the two cards gives us the number and amount of the check that was issued; the second card tells us what has happened to it. For example, most often the check will have been cashed normally, in which case the amount on the second card is the same as the amount on the first card (as was the case with check 23871). On rare occasions the check may have been issued and cashed for different amounts, in which case the amounts on the two cards will be unequal (check 23873, for example). Finally, the check may not have been cashed yet, in which case there is no second card (the case for check 23872). How do we detect that there is no second card? We do this by comparing the check numbers of the two cards actually read. If they differ, we know that there is no second card, and, in fact, we have probably read the first card of the next pair.

Referring to the sample input, we can illustrate how this reading must work. First we read two cards. Since they have the same check numbers (23871), we know that they constitute a valid pair and we proceed to process them accordingly. In this case, in fact, since the values match, the checks have been issued and cashed correctly. We then read the next two cards. In this case, however, the check numbers are different (23872 and 23873). This tells us immediately that, since there is no second card for it, check number 23872 has not yet been cashed. We have, in fact, read the first card of the next pair. We then read one more card and find that it has a check number of 23873, matching the other card we had just read. We can now process these two cards as a valid pair. In this instance the check has been issued and cashed for different amounts.

As a first attempt at setting up the general structure for an algorithm to solve this problem, we might get something like the following:

1. Repeat thru step 4 while data cards remain to be processed
2. Read in two records
3. If the check numbers are identical,
    then if the amounts are equal,
        then the check has been correctly cashed;
        else it has been issued and cashed at different amounts
4. If the check numbers differ,
    then the first card read is the record of a check yet outstanding

This algorithm works correctly if the cards continue to come in as pairs. Should we encounter a case, however, where a check has yet to be cashed, we run into difficulty. In this situation we require only one additional card; our algorithm delivers two.

To solve this particular problem, we need to introduce an additional variable that tells us which type of card is required. This will allow us to process abnormal cases as well as normal ones. For example, upon encountering an outstanding check, we will issue an appropriate message, then treat the second card as the first card of the next pair, and set our special variable to indicate that a card of type 2 is required. This will cause us to skip over the reading of a first card. The following algorithm operates in this fashion.

1. CARD_REQUIRED ← 1   (look for a "first card")
2. Repeat thru step 6 while data cards remain to be processed
3. If CARD_REQUIRED = 1,
    then read a card and treat it as a "first card"
4. Read in "second card"

5. If the check numbers are identical,
    then if the amounts are equal,
        then the check has been correctly cashed;
        else it has been issued and cashed at different amounts
        CARD_REQUIRED ← 1   (look for a "first card" again)
6. If the check numbers differ,
    then identify the first card as that of an outstanding check and
        consider the second card now as a "first card"
        CARD_REQUIRED ← 2   (look for a "second card")

Returning again to the sample input, since CARD_REQUIRED is 1 in step 3, the first card,

    23871   48.50

is read and treated as a "first card." Step 4 causes the next card,

    23871   48.50

to be read as a "second card." Since the check numbers are identical and the amounts are equal, the check is noted as correctly cashed by step 5. CARD_REQUIRED is then set to 1 to initiate a search for a new "first card." The test in step 6 fails, causing control to return to step 3. Since CARD_REQUIRED is 1, the next card,

    23872   150.00

is read and treated as a "first card." Step 4 causes the next card,

    23873   36.20

to be read as a "second card." Since the check numbers of these two cards are not equal, step 5 is bypassed in favor of step 6. Here, the card with check number 23872 is recognized as that of an outstanding check, and the card with check number 23873 is now considered as a "first card." CARD REQUIRED is set to 2, to indicate a search for a "second card." Control then returns to step 3. Now, since CARD_REQUIRED ≠ 1, step 3 is bypassed. In step 4 the next card,

    23873   236.20

is read as a "second card." Since the check numbers on the two cards match, but the amounts are unequal, step 5 notes this as a situation of checks being issued and cashed at different amounts. CARD_REQUIRED is then set to 1 to indicate a search for a new pair. This approach appears to function correctly.

An additional problem with the use of two read statements is the treatment of end of file. Since the input can run out on either read statement, we must give some consideration to the consequences of end of file at each point. If it occurs on an attempt to read a first card, no special action is required. If, however, it occurs on an attempt to read a second card, we infer that the first card of a pair has been read but there is no second card. From this we conclude that the last check has not yet been cashed, and print a message to this effect. We will incorporate the appropriate actions into our final solution.

We are now ready to specify the details of the solution in a complete algorithm. The purpose of additional variables not yet introduced is clear from their context.

Algorithm RECONCILE. This algorithm processes data on checks issued and checks cashed, noting discrepancies and computing total amount cashed and outstanding. The total amount of checks cashed is stored in the real variable CASHED; the amount outstanding, in the real variable OUTSTANDING. The data have been merged manually into one input stream. Each card contains a check number and amount. The first card of a pair was prepared at the time the check was issued. The second card, if present, was prepared after it was cashed.

1. [Initialize totals and card indicator]
      CASHED ← 0.0
      OUTSTANDING ← 0.0
      CARD_REQUIRED ← 1
2. [Engage major input loop]
      Repeat thru step 5 while data remain to be processed
3. [Read first card of a pair if not previously read]
      If CARD_REQUIRED = 1
      then Read(CHECK1_NO, CHECK1_AMT)
            If end of input, then Write('TOTAL CASHED = $', CASHED)
                              Write('TOTAL OUTSTANDING = $',
                                    OUTSTANDING)
                        Exit
4. [Read second card of a pair]
      Read(CHECK2_NO, CHECK2_AMT)
      If end of input, then Write(CHECK1_NO, CHECK1_AMT) (last check was
                  not chased)
                        Write('TOTAL CASHED = $', CASHED)
                        OUTSTANDING ← OUTSTANDING + CHECK1_AMT
                        Write('TOTAL OUTSTANDING = $', OUTSTANDING)
                        Exit
5. [Process the pair of cards]
      If CHECK1_NO = CHECK2_NO
      then If CHECK1_AMT = CHECK2_AMT
            then CASHED ← CASHED + CHECK1_AMT
                  (issued and cashed for same amount)
            else Write(CHECK1_NO, CHECK1_AMT, CHECK2_AMT)
                  (issued and cashed for different amounts)

```
                CARD_REQUIRED ← 1 (complete pair has been read)
        else  Write(CHECK1_NO, CHECK1_AMT)
                 (check not cashed yet)
              OUTSTANDING ← OUTSTANDING + CHECK1_AMT
              CHECK1_NO ← CHECK2_NO  (becomes first card of next pair)
              CHECK1_AMT ← CHECK2_AMT
              CARD_REQUIRED ← 2   (read second card of this next pair)    □
```

The first part of this algorithm is straightforward. We initialize the two variables used for totals and engage the loop. We then have the two read statements described earlier, with consideration for the consequences of end of file at each point. If it occurs while trying to read a first card, we simply print the final totals. Should it occur, however, while trying to read a second card, we deduce that the first card of a pair has been read but there is no second card. From this we conclude that the last check has not yet been cashed, print the required message to this effect, and print the final totals.

Most of step 5 has been described already. The comments indicate the various cases being processed. As you might expect, once again special action is required if there is no second card in the pair (that is, the check numbers on the two cards read are different). First, we print the required message and update the total of outstanding checks. We must then deal with the second card we did read. This is now to be treated as the first card of the next pair. Since "first card" data are to be stored in variables CHECK1_NO and CHECK1_AMT, we must effect this by copying the values currently in CHECK2_NO and CHECK2_AMT. This ought to be the end of the loop and we should return to read two more cards. As described, however, we already have the first card of our pair—we require only the second. The setting of the variable CARD_REQUIRED to 2 will cause us to bypass the reading of a first card in step 3. We will proceed instead to step 4, and we carry on processing in the normal fashion from this point. In particular, if a valid pair is encountered next, we must reset the variable CARD_REQUIRED to 1, to allow for regular processing of subsequent cards.

We will leave the complete tracing of this algorithm as an exercise.

CHAPTER EXERCISES

1. Design an algorithm to print the sum of the squares of the first 100 integers.

2. (a) In mathematics the following is always true:

$$|\sin(x)| \leq 1$$

Design an algorithm to verify this property for the built-in function **SIN** for values of x going from -10.0 to $+10.0$ in steps of .2.

(*b*) In mathematics, the following property holds:

$$\sqrt{x} \le \sqrt{y} \qquad \text{whenever } x \le y$$

Design an algorithm to check that this property holds for the built-in function **SQRT** for integral values of x and y in the range 0 to 100.

3. Commercial fishermen are required to report monthly information on their catch to the Department of Fisheries. These data are analyzed regularly to determine the growth or reduction of the various species of fish and to indicate any possible trouble. From the catch reports and previous data, a card is prepared containing the following information:

region fished (numerical code), species name (string), number caught this year (integer), number caught last year (integer)

e.g., 16, 'HALIBUT', 20485, 18760

This sample card indicates that in region 16, a total of 20,485 halibut were caught as compared to 18,760 in the same month last year. Design an algorithm to read the deck of cards prepared (terminated by a card with a negative region number) and flag any unusual growth or reduction in catches. An unusual growth or reduction is defined as one in which the percentage change exceeds 30%, where percentage change is defined as

$$\frac{(\text{this year}) - (\text{last year})}{(\text{last year})} \times 100\%$$

4. The present government of Gooba-Gooba Land instituted wage and price controls shortly after it was elected. The Prime Minister has received word from her economic advisors that future predictions, based on current trends, are that wages will continue to increase by 5%, the prices of essential goods and services by 10%, and personal taxes by 15% per year. Her political advisors tell her that if taxes plus the cost of essential goods and services consume more than 75% of their annual wage for a significant number of citizens, she will be in trouble at the next election (expected in three years' time), and that if this reaches 80%, she will be in real trouble. The Prime Minister has collected and placed on punched cards the present annual wage, taxes, and cost of essential goods and services for a sample taken of her constituents. Design an algorithm to determine the number and percentage of the total surveyed who will fall into the 75% and 80% categories described. The number in the sample is unknown. End of input is to be determined by the end-of-file method.

5. A toy distributor has made a bargain purchase of 10,000 small toys packaged in rectangular boxes of varying size. He intends to put the boxes into brightly colored plastic spheres and resell them as surprise packages. The spheres come in four diameters: 4, 6, 8, and 10 in. To place his order he needs to know how many of each diameter he will need. Since the diagonal of a rectangular box with dimensions A, B, and C, given by

$$D = \sqrt{A^2 + B^2 + C^2}$$

is its largest measurement, the distributor must calculate the diagonal lengths of the boxes and then determine the number that are 4 in or less, 4 to 6 in, and so on. The dimensions of each box are punched on a separate card. Design an algorithm to read these data and determine the number of spheres of each size needed to repackage the toys.

6. The marketing office of a publishing company is faced with the task of computing the break-even point for any book they propose to publish. The break-even point is the number of copies of the book that must be sold for the sales revenue to equal production costs. Production costs consist of a fixed cost for layout, typesetting, editing, and so on, plus a per copy cost for printing, binding, and other expenses. For each candidate for publication, a market analysis determines likely sales figures, as well as the production cost, based partly on the size of the book (number of pages) and the number of copies produced according to the formula

prod. cost = fixed prod. cost + no. produced × (pages × 0.0305)

An analysis of these projections is used to determine the price at which the book must be sold to break even. A card is prepared for each prospective book, containing the following information:

title of book (string), sales projections, fixed production cost, number of pages

For example, the card

'THE COMPUTER-PHILES', 5000, 7500, 365

indicates that a book entitled *The Computer-Philes*, with projected sales of 5,000 copies, has a fixed production cost of $7,500 and is 365 pages in length. For this particular book, the cost of producing 5,000 books would be

7,500 + 5,000 × (365 × .0305) = $63,162.50

To break even, this book must sell for

$$\frac{\$63,162.50}{5,000} = \$12.63$$

Design an algorithm to read the deck of cards prepared for the season's prospective books and produce a line for each, giving the title of the book, projected sales, and the computed break-even selling price. Assume that the deck of cards is terminated by a special card on which the book title is 'END OF DATA'.

7. The Saskatchewan Government Insurance Office has compiled data on all traffic accidents in the province over the past year. For each driver involved in an accident, a card has been prepared with the following three pieces of information:

> year driver was born (numeric), sex ('M' or 'F'), registration code (1 for Saskatchewan registration, 0 for everything else)

Design an algorithm to read the deck of data cards and print the following summary statistics on drivers involved in accidents:

(a) Percentage of drivers under 25.
(b) Percentage of drivers who are female.
(c) Percentage of drivers who are males between the ages of 18 and 25.
(d) Percentage of drivers with out-of-province registration.

Use the end-of-file method to signal the end of input.

8. The following table is taken from the 1976 Canadian income tax guide.

1976 Rates of Federal Income Tax

Taxable Income	Tax
$ 654 or less	6%
654	$ 39 + 18% on next $ 653
1,307	157 + 19% on next 1,307
2,614	405 + 20% on next 1,307
3,921	667 + 21% on next 2,614
6,535	1,216 + 23% on next 2,614
9,149	1,817 + 25% on next 2,614
11,763	2,470 + 27% on next 2,614
14,377	3,176 + 31% on next 3,921
18,298	4,392 + 35% on next 13,070
31,368	8,966 + 39% on next 19,605
50,973	16,612 + 43% on next 27,447
78,420	28,414 + 47% on remainder

Design an algorithm to read taxable income and determine the federal income tax to be paid according to this schedule of rates (given on page 138).

9. The city police department has accumulated information on speeding violations over a period of time. The department has divided the city into four quadrants and wishes to have statistics on speeding violations by quadrant. For each violation, a card is prepared containing the following information:

> vehicle registration number (numeric code), quadrant in which offense occurred (1–4), speed limit in miles per hour (integer), actual speed traveled in miles per hour (integer)

This set of cards is terminated by a special card with vehicle registration number of 0.

Design an algorithm to produce 2 reports. First, give a listing of speeding fines collected, where the fine is calculated as the sum of court costs ($20) plus $1.25 for every mile per hour by which the speed limit was exceeded. Prepare a table with the following headings:

SPEEDING VIOLATIONS

VEHICLE REGISTRATION	SPEED RECORDED(MPH)	SPEED LIMIT(MPH)	FINE

This report is to be followed by a second report, in which an analysis of violations by quadrant is given. For each of the four quadrants shown above, give the number of violations processed and the average fine.

10. (a) Design an algorithm to compute and tabulate the values for the function

$$f(x,y) = \frac{x^2 - y^2}{x^2 + y^2}$$
for $x = 2, 4, 6, 8$
$y = 6, 9, 12, 15, 18, 21$

(b) Design an algorithm to compute the number of points with integer-valued coordinates that are contained within the ellipse

$$\frac{x^2}{16} + \frac{y^2}{25} = 1$$

Notes:

1. Points on the ellipse are considered to be within it
2. Range of coordinate values is limited by the major and minor axes of the ellipse (that is, $-4 \le x \le 4$ and $-5 \le y \le 5$)

11. The We-Spray-Anything Company uses planes to spray crops for a variety of problems. The rates they charge the farmer depend on what he is spraying for and how many acres he wants sprayed according to the following schedule:

Type 1: Spraying for weeds, $1 per acre
Type 2: Spraying for grasshoppers, $2 per acre
Type 3: Spraying for army worms, $3 per acre
Type 4: Spraying for all of these, $5 per acre

If the area to be sprayed is greater than 1,000 acres, the farmer receives a 5% discount. In addition, any farmer whose bill is over $1,500 receives a 10% refund on the amount over $1,500. If both discounts apply, the acreage discount is taken first. Design an algorithm that will read in a series of cards containing the following information:

farmer's name, type of spraying requested (integer code 1–4), and number of acres to be sprayed (integer)

e.g., 'BROWN', 3, 950

For each card read, calculate the total cost to the farmer and print out the farmer's name followed by his bill. Use the end-of-file method to signal the end of input.

12. (*a*) Design an algorithm to compute the amount of savings you would have at the end of 10 years if you were to deposit $100 each month. Assume a constant annual interest rate of 6%, compounded every 6 months (that is, interest in the amount of 3% is awarded each 6 months).

(*b*) We want to invest a sum of money that will grow to be X dollars in Y years time. If the interest rate is R percent, then the amount we have to invest (the *present value* of X) is given by the formula

$$\frac{X}{(1 + .01 \times R)^Y}$$

Design an algorithm that will print out a table of the present value of $5,000 at 7.5% interest, for periods of 1 to 21 years, in steps of 2 years.

13. A computer dating service maintains a card file of its clients. Each card contains the following information:

> name, sex ('M' or 'F'), age, height (in inches), weight (in pounds), color of eyes (1 for blue, 2 for brown, 3 for any other), color of hair (1 for brown, 2 for blonde, 3 for any other)

Design an algorithm that will read through this card file and print the names of:
(a) All blonde-haired blue-eyed females between 5 feet and 5 feet 2 inches, weighing less than 115 pounds.
(b) All brown-eyed males over 6 feet tall, weighing between 180 and 220 pounds.
Use the end-of-file method to signal the end of input.

14. The sieve of Eratosthenes, named after a third-century Greek astronomer and geographer, is a technique for generating prime numbers. We begin by writing all the odd integers from 3 up to N, then crossing out every third element after 3, every fifth element after 5, and so on until multiples of all odd integers less than $\sqrt{N}$ have been eliminated. The integers remaining in the list are exactly the prime numbers from 3 to N. Design an algorithm to generate the prime numbers from 3 to 1,000 using the sieve technique.

15. Many professional sports teams use a computer to assist in the analysis of scouting reports on prospective players. Suppose that our professional hockey team has such a system. For each player scouted, a card containing the following information is prepared:

> player's name, age, height (in inches), weight (in pounds), goals scored last season, assists last season, penalty minutes last season, league factor (real number)

The players are evaluated according to the following formula:

> (goals + assists + (penalty minutes)/4
>   + (height + weight)/5 − age) * league factor

Design an algorithm to read the entire deck of scouting reports, listing for each player the information on his card and his evaluation figure. At the end of this list (denoted by a special card with player name 'END OF LIST'), give the name and evaluation figure for the player with the highest evaluation.

16. Assume that a particular store sells all of its merchandise for a price of $1 or less. Assume further that all customers pay for each purchase with a $1 bill. Design an algorithm that reads in the purchase price of an article and calculates the number of each type of coin to be given in change so that the smallest number of coins is returned. For example, if the purchase price is 63¢, the change will be 1 quarter, 1 dime, and 2 pennies.

17. More and more banks are moving toward computerized customer accounts systems. Develop an algorithm to process transactions from teller stations in the following hypothetical system. Consider the customer accounts to be maintained as a deck of cards (in increasing order of account number), each having the following format:

	customer name	account number	current balance
example:	'BARBARINO,VINCENT'	501865	298.13

As each of the day's customers is processed, the teller prepares a card describing the transaction that took place. The transaction cards have the following format:

	customer name	account number	action	amount
example:	'HORSHACK,ARNOLD'	308512	'DEPOSIT'	16.28

Other possible transactions are 'WITHDRAWAL' and 'NEW ACCOUNT'. At the end of the day the transaction cards are collected from the tellers, sorted manually in increasing order of account numbers, and used to update the information in the current customer accounts file.

The transaction cards are processed in the following manner. A 'DEPOSIT' transaction causes the amount of the transaction to be added to the indicated customer's balance; a 'WITHDRAWAL' transaction causes the amount to be subtracted from the customer's balance; a 'NEW ACCOUNT' transaction causes a new account to be created with the amount given as the opening balance. As each transaction card is processed, a line is printed giving the format of the new customer account card for this customer. These cards will be keypunched and inserted manually into the customer-accounts deck at a later time.

Note that the inclusion of both customer name and account number on the transaction card serves as a safeguard in this update process.

Although the information is processed by account number, if the name on the transaction card fails to match the name in the customer accounts file, that transaction is to be bypassed and an appropriate message printed.

Design an algorithm that will read the two decks of cards simultaneously from separate card readers (use **Read1** for the first reader and **Read2** for the second). Then process the transaction cards against the customer accounts file, as described, and produce the appropriate output (updated information and error messages). Use the end-of-file method to signal the end of the input on each of the two card readers.

4

Vectors and Arrays

Earlier we introduced the concept of a variable. Recall that a variable can have a single value such as an integer, a real number, or a string. In this chapter we extend the notion of a variable to a collection or set of variables which is called an array.

The first section of the chapter deals with the general concept of a data structure. Data structures are broadly classified as primitive and nonprimitive structures. One of the most commonly used nonprimitive data structures is the array. The algorithmic notation for one-dimensional arrays or vectors is introduced in Sec. 4-1.

Section 4-2 describes several operations that can be performed on one-dimensional arrays, such as read, write, add, and scalar multiply. The counted loop control statement introduced in Sec. 3-3.3 is used throughout this section. The technique of computing the location or index of an element within a vector is discussed by using a "rainfall report" application.

The third section describes the important operations of searching and sorting with vectors.

The fourth section extends the ideas of the previous two sections to the manipulation of higher-dimensional arrays.

Section 4-5 describes several applications of both one-dimensional and higher-dimensional arrays.

4-1 THE VECTOR AS A DATA STRUCTURE

In organizing a solution to a problem which is to be solved with the aid of a computer, we are confronted with several interrelated subproblems. One of these subproblems is to understand thoroughly the relationship between the data items that are relevant to the solution of the problem. To understand the relationship between the data items in the problem implies that we must understand the data. Data in a particular problem consist of a set of elementary items. An item usually consists of single elements such as integers, bits, characters, or a set of such items. A person solving a particular problem is concerned with structurally organizing the data. Choosing items of data is a necessary and key step in defining and then solving a problem. The possible ways in which the data items or atoms are logically related define different *data structures*.

Data structures can be broadly classified as primitive and nonprimitive. In Sec. 2-3 we introduced several primitive data structures: the integer, the real number, the character, the logical value, and the pointer. These structures are called *primitive* because the instruction set of a computer has instructions which will manipulate these structures (recall Sec. 2-3). We can perform the common arithmetic operations on numbers. It will also be seen in Chap. 5 that a string which contains a number of characters can be modified using a number of machine language instructions.

Let us concern ourselves next with a *nonprimitive*, yet very simple data structure—a complex number. A complex number is not considered to be primitive since very few computers, if any, have instructions which add, subtract, multiply, and divide complex numbers. Many higher-level languages, however, such as PL/I and FORTRAN, permit the handling of complex numbers.

Let $U = x + yi$ and $V = a + bi$ be two complex numbers, where i denotes $\sqrt{-1}$ and x, y, a, and b are real numbers. Then the complex arithmetic operations which can be performed on these numbers are defined as follows:

$$U \oplus V = (x + a) + (y + b)i$$
$$U \ominus V = (x - a) + (y - b)i$$
$$U \circledast V = (x * a - y * b) + (x * b + y * a)i$$
$$U \oslash V = \frac{(x * a + y * b) + (y * a - x * b)i}{(a * a + b * b)}$$

where the operators $\oplus$, $\ominus$, $\circledast$, and $\oslash$ denote complex addition, subtraction, multiplication, and division, respectively. From this it is clear that complex numbers can be used in complex arithmetic operations using the ordinary arithmetic operators. Whereas there is a single machine-language instruction for the multiplication of two real numbers, for example, note that the multiplication of two complex numbers consists of a sequence of many (six) real instructions (four multiplications, one subtraction, and one addition). The same applies to the remaining complex arithmetic operators. Also, a complex number is considered to consist of an ordered pair of real numbers. Each real number in this ordered pair is treated differently.

We now proceed to the introduction of another nonprimitive data structure, one that is more complicated, yet very common. Suppose that we are given a class list for an introductory computer science course. Each entry in this list consists of a student name and a grade. It is required to print the name of each student whose grade is greater than the average grade of the class.

To accomplish this task we must scan (or read) the class list twice. On the first scan we must determine the average grade obtained in the class. Once this average grade is known, a second scan of the class list is performed in which the grade of each student is compared with this average. The student's name is printed if that student's grade exceeds the average. How can we perform the required task? Assuming that each name and grade is placed on one card, one way is to place the resulting deck of cards for the class list in the card reader twice. It is read once to determine the average, and it is read again to generate the required output. Not only is this approach inefficient, but in many computing systems it is also not allowed. An alternative way of achieving the same goal is to store the class list in the memory of the computer.

Let us assume further that the class list contains the names and grades of five students. Let NAME1 and GRADE1 be variables which denote the name and grade, respectively, of the first student in the class list. In a similar manner, let variables NAME2 and GRADE2 be associated with the second student and, finally, let the variables NAME5 and GRADE5 correspond to the fifth student. The required report can then be generated by the following algorithm.

Algorithm CLASS_STAT. Given a class list of five students, this algorithm produces a report which contains the name of each student whose grade is greater than the average grade of the class. The variables NAME1, GRADE1, . . ., NAME5, GRADE5 are as described earlier. The real variable AVERAGE denotes the average grade of the class.

1. [Read the class list]
   Read(NAME1, GRADE1, NAME2, GRADE2, NAME3, GRADE3, NAME4, GRADE4, NAME5, GRADE5)
2. [Obtain the average grade]
   AVERAGE ← (GRADE1 + GRADE2 + GRADE3 + GRADE4 + GRADE5) / 5.0
3. [Check the first student]
   If GRADE1 > AVERAGE then Write(NAME1)
4. [Check the second student]
   If GRADE2 > AVERAGE then Write(NAME2)
5. [Check the third student]
   If GRADE3 > AVERAGE then Write(NAME3)
6. [Check the fourth student]
   If GRADE4 > AVERAGE then Write(NAME4)
7. [Check the fifth student]
   If GRADE5 > AVERAGE then Write(NAME5)
8. [Finished]
   Exit ☐

Note that this algorithm is only valid for five students. What would the algorithm be for a class of 100 students? Using the previous approach, we would require 100 variables, say NAME1, NAME2, . . ., NAME100 for student names and 100 variables for student grades, say, GRADE 1, GRADE2, . . ., GRADE100. The modified algorithm to handle this case would look like this:

Algorithm CLASS_STAT (revision 1). This algorithm is identical to the previous version except for the number of students being processed. The algorithm handles 100 students.

1. [Read the class list]
   Read(NAME1, GRADE1, NAME2, GRADE2, . . ., NAME100, GRADE100)
2. [Obtain the average grade]
   AVERAGE ← (GRADE1 + GRADE2 + . . . + GRADE100) / 100.0
3. [Check the first student]
   If GRADE1 > AVERAGE then Write(NAME1)
4. [Check the second student]
   If GRADE2 > AVERAGE then Write(NAME2)
   .
   .
   .
102. [Check the last student]
   If GRADE100 > AVERAGE then Write(NAME100)
103. [Finished]
   Exit   □

Note that we have used the symbol . . . to denote a continuation of the processing in the read statement in step 1 and the symbol : to denote the checking of the grades for the third through the ninety-ninth student (steps 5 thru 101).

This algorithm for handling 100 students requires 103 steps and is clearly absurd. If one were given a class list consisting of 200 students, a total of 400 distinct variable names would be required for the student names and grades. Furthermore, the algorithm to perform the required task would contain some 203 steps!

A more realistic approach to solving this problem is to associate one name with the entire set of student names and another name with the set of student grades. Let the variables NAME and GRADE correspond to the sets of student names and grades, respectively. If the class list contains five students, the variable NAME denotes an ordered set of five elements. Each element in this set represents a character string. Using this notation, how can we refer to or select a particular student's name? We can use a *subscript* to select a particular element in the ordered set. For example, the name of the third student is denoted by $NAME_3$ or NAME[3]. Although $NAME_3$ is usually the notational convention used in mathematics, we will use NAME[3] in our algorithmic notation because most character alphabets on computers do not contain subscript symbols. When this convention becomes unwieldy, we will revert to the familiar mathematical notation. The variable NAME[10] refers to the name of the tenth student in the class list and GRADE[10] selects that student's grade. When we append a subscript such as 3 or 10 to the name given to an ordered set, the resulting variable is called a *subscripted variable*. The subscript associated with the set name can also be a variable. For example, the term NAME[i] refers to the fifth name in the class list when i has a value of 5. Obviously, a variable which is used as a subscript to refer to a particular element in a collection or a set must yield an integer value.

A *vector* is defined as an ordered set which contains a *fixed* number of elements. An element in the vector can be, for example, an integer, a real number, or a string. Note, however, that all elements in the vector must be of the same type. A vector is also called a *one-dimensional array*.

Using this notation, the following algorithm is a reformulation of the earlier algorithm which handled a class of five students.

Algorithm CLASS_STAT (revision 2). Given a class list of five students whose names and grades are represented by the vectors NAME and GRADE, respectively, this algorithm produces the same report as the original version of the algorithm.

1. [Read the class list]
    Repeat for i = 1, 2, . . ., 5
        Read(NAME[i], GRADE[i])
2. [Obtain the average grade]
    SUM ← 0
    Repeat for i = 1, 2, . . ., 5
        SUM ← SUM + GRADE[i]
    AVERAGE ← SUM / 5.0
3. [Check each student in the class list]
    Repeat for i = 1, 2, . . ., 5
        If GRADE[i] > AVERAGE then Write(NAME[i])
4. [Finished]
    Exit □

The first step of the algorithm contains a loop which repeats the read statement five times. Each time that this statement is executed, two values are read from the input stream. For example, when i is 1, the name and grade on the first card are placed in variables NAME[1] and GRADE[1], respectively. When i is 2, the name and grade on the second card are copied into the variables NAME[2] and GRADE[2], respectively. The first statement of step 2 initializes the variable SUM to zero. The next loop statement proceeds to accumulate the sum of the grades in the class list. Upon the completion of this loop, the average grade is computed. Step 3 contains a loop statement which checks the grade of each student against the average grade of the class. When this is done, control in the algorithm passes to step 4 and an exit results.

Now, what changes are required in this algorithm to handle a class of 100 students? The only modifications which are required are to change the

last value of the loop control variable i in the repeat statements in steps 1
through 3 and the value of the divisor in step 2. In particular, the following
algorithm handles a class of 100 students.

Algorithm CLASS_STAT (revision 3). This algorithm is similar to its preceding version and handles a class of 100 students.
1. [Read the class list]
    Repeat for i = 1, 2, . . ., 100
      Read(NAME[i], GRADE[i])
2. [Obtain the average grade]
    SUM ← 0
    Repeat for i = 1, 2, . . ., 100
      SUM ← SUM + GRADE[i]
    AVERAGE ← SUM / 100.0
3. [Check each student in the class list]
    Repeat for i = 1, 2, . . ., 100
      If GRADE[i] > AVERAGE then Write(NAME[i])
4. [Finished]
    Exit   ☐

Finally, can we modify the preceding algorithm to handle an arbitrary
class size? This can be accomplished easily by having an extra card at the
front of the input deck. (See the counter-controlled input in Sec. 3-3.) This
card contains an integer value which denotes the number of students in the
class. The following revised algorithm, in which the variable N designates
the size of the class, handles this general case.

Algorithm CLASS_STAT (revision 4). Given the class size (N) and two
vectors NAME and GRADE which represent names and grades of students on
the class list, respectively, this algorithm produces the desired report. The
variable SUM is real.
1. [Read the size of the class]
    Read(N)
2. [Read the class list]
    Repeat for i = 1, 2, . . ., N
      Read(NAME[i], GRADE[i])
3. [Obtain the average grade]
    SUM ← 0.0
    Repeat for i = 1, 2, . . ., N
      SUM ← SUM + GRADE[i]
    AVERAGE ← SUM / N
4. [Check each student on the class list]
    Repeat for i = 1, 2, . . ., N
        If GRADE[i] > AVERAGE then Write(NAME[i])
5. [Finished]
    Exit   ☐

This algorithm is very similar to the previous two algorithms. A new step has been added to obtain the size of the class first. The variable N is also used in steps 2 through 4 to denote the number of times the loops are to be performed.

An important aspect of this algorithm concerns the possible values of N. Can it be arbitrarily large? If it cannot be, how large can its value be? The value for N must be finite and the precise limit on its value depends on how much storage the computer has at its disposal. Larger memories usually permit the use of a larger value for N.

Another important aspect concerning vectors is the way in which they are organized in the memory of the computer. The elements of a vector are stored in contiguous or consecutive sequential storage locations in computer memory. For example, a 100-element integer vector could be stored from storage position 1000 to storage position 1099 inclusive in this kind of memory representation. It is assumed that each integer would occupy one storage position.

We have at this point introduced the notion of a vector and its associated subscript. We can select any element in the vector by using this subscript notation. In the next section we shall discuss a number of operations which are performed on vectors.

4-2 OPERATING ON VECTORS

In the previous section the notion of a vector, or one-dimensional array, was introduced. A vector X of n elements was defined as the sequence X[1], X[2], ..., X[n]. Although the lower bound on the subscript in this instance is 1, this, in general, need not be the case. The general case is described at the beginning of this section. Several common operations that can be performed on vectors are also described. Such operations as assignment, input, and output are discussed. Also, by the use of a simple example, we introduce the important notion that a subscript referring to a particular element of a vector can have its value computed at run time.

In general, a vector X can be an ordered sequence of elements such as

X[l], X[l + 1], ..., X[h]

where l and h are integers (h ≥ l) that denote the lower and upper bounds of the subscript, respectively. For example, the vector consisting of the elements

X[0], X[1], X[2], X[3], X[4]

contains five elements and has a lower subscript bound of 0. In this vector

X[0] denotes its first element. Another example is the vector Y, which consists of the elements

Y[-2], Y[-1], . . ., Y[2]

It also contains five elements and its lower subscript bound is -2. So Y[-2] and Y[1] denote the first and fourth elements of the vector, respectively. Throughout the remainder of this book we will encounter several examples involving vectors where the lower subscript bounds are not 1. Many programming languages permit the use of vectors with an arbitrary lower bound on their subscript values. There are, however, certain languages where the lower bound of the subscript must always be 1.

Let us now consider the assignment of a value to a vector element. If we have a vector X consisting of the elements X[1], X[2], . . ., X[10], then the assignment X[5] ← 100 will assign a value of 100 to its fifth element. A subscripted variable can be used in the left part of an assignment statement in much the same way as an ordinary variable.

Processing arrays normally involves processing the individual elements of the array one at a time. You will find the counted loop particularly useful for this purpose. For example, often in a number of applications, it is required to initialize each element of a vector to the same value. For example, the vector A, consisting of the elements A[0], A[1], . . ., A[9], can be initialized to 0 by the following statement:

Repeat for i = 0, 1, . . ., 9
    A[i] ← 0

We will allow in our algorithmic notation an alternative way of assigning the same value to each element of the vector. In particular, the statement

A ← 0

yields the same result as the previous repeat statement. Many programming languages permit this generalization of the assignment operator to arrays.

Another aspect concerning vectors deals with their initialization by reading in values. For example, the statement

Repeat for i = 0, 1, . . ., 9
    Read(A[i])

will read 10 values from cards. The first value will be placed in A[0], the second in A[1], . . ., and the tenth value in A[9]. In our algorithmic notation, the same result occurs if we use the statement

Read(A)

It is assumed that the first value is copied into the first element of the vector, the second value is copied into its second element, and so on. Note that the number of items read is determined by the size of the vector, not by the number of input items.

An analogous situation arises for the output of vectors. The statement

```
Repeat for i = 0, 1, . . ., 9
    Write(A[i])
```

will print the 10 values of the vector A. The same result is accomplished by using the statement

```
Write(A)
```

As an example which involves vectors, let us formulate an algorithm to compute the standard deviation D of the elements of the vector x according to the formula

$$D = \sqrt{\frac{\sum_{i=1}^{N} (x_i - \bar{x})^2}{N - 1}}$$

where the symbols Σ, $\sqrt{\ }$, and $\bar{x}$ denote the summation operation, the square-root operation, and the average of x, respectively. The following algorithm accomplishes this task.

Algorithm STANDARD_DEVIATION. Given a vector X consisting of N elements, this algorithm computes the standard deviation, D, just described. The variable AVERAGE denotes the arithmetic average of the elements of the array. SUM and S are temporary variables. All variables contain real values.

1. [Read the size of the vector]
    Read(N)
2. [Read the vector X]
    Read(X)
3. [Compute the average]
    SUM ← 0
    Repeat for i = 1, 2, . . ., N
        SUM ← SUM + X[i]
    AVERAGE ← SUM / N
4. [Compute and write standard deviation]
    S ← 0
    Repeat for i = 1, 2, . . ., N
        S ← S + (X[i] − AVERAGE) ↑ 2
    D ← SQRT(S / N − 1)
    Write(D)
5. [Finished]
    Exit   □

An important concept in using vectors is the idea that a subscript can be computed. We shall illustrate this concept by considering the following problem.

Input:

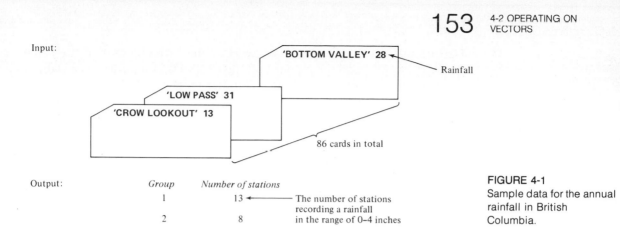

'BOTTOM VALLEY' 28 ← Rainfall

'LOW PASS' 31

'CROW LOOKOUT' 13

86 cards in total

Output:

Group	Number of stations	
1	13 ←	The number of stations recording a rainfall in the range of 0–4 inches
2	8	

FIGURE 4-1
Sample data for the annual rainfall in British Columbia.

The Department of Transport of British Columbia keeps weather records of annual rainfall (in inches) at 86 different weather stations (a matter of prime interest to anglers and water conservation authorities alike). The station name and total depth of rain recorded at that station for this past year have been punched on cards, as illustrated in Fig. 4-1. We wish to design an algorithm which will determine the number of stations recording rainfalls in each of the groups specified in Table 4-1.

We will assume that no station receives more than 49 in of rain in 1 year. Sample input data and the desired output are given in Fig. 4-1.

There are essentially two ways to attack this problem. The first method features the brute-force approach and the second method involves the computation of a subscript. In both approaches we will let the variables STATION_NAME and RAINFALL denote the station identification and rainfall, respectively. Also common to both approaches is the need for 10 counters; one counter is required to record the number of stations reporting rainfall within the same range. Let the vector RANGE denote this set of counters. The following algorithm is the result of the brute-force approach.

TABLE 4-1 Annual Rainfall Records

Group	Range, in
1	0–4
2	5–9
3	10–14
4	15–19
5	20–24
6	25–29
7	30–34
8	35–39
9	40–44
10	45–49

Algorithm RAIN. Given a series of cards, each of which contains the name (STATION_NAME) and the amount of rain (RAINFALL) reported by a station, this algorithm produces the desired summary report just described.

1. [Initialize range counters]
    RANGE ← 0
2. [Process the data]
    Repeat thru step 4 while there remains a weather record
3. [Read a station's statistics]
    Read(STATION_NAME, RAINFALL)
    If there are no more input data
    then Write('GROUP', 'NUMBER OF STATIONS')
        Repeat for i = 1, 2, . . ., 10
            Write(i, RANGE[i])
        Exit
4. [Update appropriate range counter]
    If 0 ≤ RAINFALL and RAINFALL ≤ 4 then RANGE[1] ← RANGE[1] + 1
    If 5 ≤ RAINFALL and RAINFALL ≤ 9 then RANGE[2] ← RANGE[2] + 1

                            .
                            .
                            .

    If 45 ≤ RAINFALL and RAINFALL ≤ 49 then RANGE[10] ← RANGE[10] + 1
    □

The algorithm is straightforward. The first step initializes the 10 range counters to 0. Steps 2 through 4 process all the data. A station's statistic is read in step 3. Note that the station name is read but never used. When the end of the input data is reached, the required report is produced. Step 4 updates the appropriate range counter. This is accomplished by exhaustively checking the rainfall to determine into which range it falls. This check requires 10 "if" statements! Clearly, this approach is inefficient. Step 4 can be changed to the following:

4. [Update the appropriate range counter]
    POSITION ← TRUNC(RAINFALL / 5) + 1
    RANGE[POSITION] ← RANGE[POSITION] + 1

Note that the variable POSITION is used to denote the particular counter which is to be updated. The value of POSITION is determined from the rainfall reported by the station. Note that the built-in function TRUNC has been used to obtain the desired truncated value. The reader should verify that this computation does indeed yield the correct subscript value of the counter to be updated. There are many applications in which this technique can be used.

EXERCISES 4-2

1. For a vector A of n real numbers, formulate an algorithm which will determine the largest and the second largest elements of this vector. Assume that these values are distinct.

2. Given a vector A of n real numbers, obtain the largest difference between two consecutive elements of this vector.

3. Repeat exercise 2 and obtain the smallest difference between two consecutive elements.

4. Formulate an algorithm to obtain the following statistics for a vector X of n elements:

$$\text{Mean deviation (MD)} = \frac{1}{n} \sum_{i=1}^{n} |x_i - \bar{x}| \qquad \text{where } \bar{x} = \frac{1}{n} \sum_{i=1}^{n} x_i$$

$$\text{Root mean square (RMS)} = \sqrt{\frac{1}{n} \sum_{i=1}^{n} x_i^2}$$

$$\text{Harmonic mean (HM)} = n \left/ \sum_{i=1}^{n} (1/x_i) \right.$$

$$\text{Range (R)} = \text{maximum}\{x_1, x_2, \ldots, x_n\} - \text{minimum}\{x_1, x_2, \ldots, x_n\}$$

$$\text{Geometric mean (GM)} = \sqrt{x_1 \times x_2 \times \ldots \times x_n}$$

5. Design an algorithm which reads an unsorted vector A of n integers and prints the vector in the same sequence after ignoring duplicate values found in the given vector. The number of remaining elements (m) is also required. For example, given the vector

A_1	A_2	A_3	A_4	A_5	A_6	A_7	A_8	A_9	A_{10}
15	31	23	15	75	23	41	15	31	85

of 10 integers, the compressed vector returned would be

A_1	A_2	A_3	A_4	A_5	A_6	A_7	A_8	A_9	A_{10}
15	31	23	75	41	85				

with $m = 6$.

6. Formulate an algorithm to convert decimal (base 10) integers to their octal (base 8) representations by successive divisions. Let **NUMBER** denote the integer to be converted and **BASE** the base to which the

integer is to be converted (8 in our case). For example, to compute the octal representation of 150, it is repeatedly divided by 8 and the resulting remainders are saved in order.

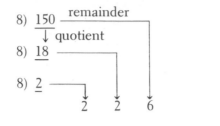

$$226_8 = 2 \times 8^2 + 2 \times 8^1 + 6 \times 8^0 = 150_{10}$$

7. Given a polynomial $p(x)$ of the form

$$p(x) = a_0 x^n + a_1 x^{n-1} + \cdots + a_{n-1} x + a_n$$

where $a_0, a_1, \ldots, a_n$ are real numbers which denote the coefficients of the polynomial, construct an algorithm which will input n followed by these coefficients and a sequence of values for x. For each value of x, the value of $p(x)$ is to be tabulated.

8. (*The "character-distance" problem.*) Examine an input stream of n characters (letters of the alphabet only), and form a result stream of numeric values, one for each input character. Each position of the result stream will be occupied by a number representing a count of the characters separating the character in the corresponding input position from the nearest similar character to its left in the input stream. No distance larger than 9 will be recorded. Any character not matching anything to its left within nine positions will have a zero in the result stream. For example, the given string **AABCDBEFFEABGBWB** would yield the result **0100030013960202**. Formulate an algorithm for this problem. For input purposes, however, assume that each character is in quotes. For example, the input of the preceding example string would have the following form:

16 ← number of characters
'A' 'A' 'B' 'C' 'D' 'B' 'E' 'F' 'F' 'E' 'A' 'B' 'G' 'B' 'W' 'B'

9. As the shaft concrete lining was poured at a nearby potash mine, samples of the concrete were taken and tested for maximum strength. The record book of shaft depth versus concrete strength has been keypunched as follows:

The first card contains the starting shaft depth and the total number of test results for consecutive one foot increments down the shaft

The following cards all contain 10 test results per card, but the last card may contain less than 10 results, depending on the total number of results taken

Following is an example input:

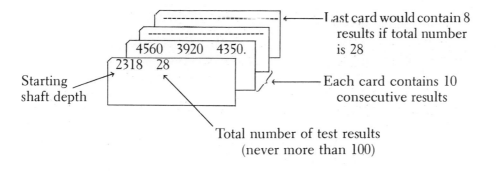

Last card would contain 8 results if total number is 28

4560 3920 4350.

2318 28

Starting shaft depth

Each card contains 10 consecutive results

Total number of test results (never more than 100)

A running average of eight results is used as an indication of the average concrete strength and would indicate any extremely weak section in the shaft. The running average is the average of the readings at that depth and the next 7 ft below. The running average for each of the last 7 ft is the average of the remaining readings. Obtain an algorithm to generate a running average table as follows:

DEPTH	TEST RESULT	RUNNING AVERAGE
2318	4560	4341
2319	3920	4256
.	.	.
.	.	.
.	.	.
2339	4820	4515
.	.	.
.	.	.
.	.	.
2345	4500	4500

The "length" of the running average, in this case 8 ft, varies with each application. After you have a solution, generalize your algorithm so that only one data card must be changed to change the length of the average.

10. The following formula gives the variance of the values in a vector x ($\bar{x}$ denotes the mean of the n values of the vector):

$$\sigma_x^2 = \frac{1}{n-1}\left[\sum_{i=1}^{n} x_i^2 - n(\bar{x})^2\right]$$

Design an algorithm to read the elements of a vector x and compute the mean and variance of the values using this formula for variance. Can you see any computational advantage in computing the variance in this way, as opposed to the more familiar formula

$$\sigma_x^2 = \frac{1}{n-1}\left[\sum_{i=1}^{n}(x_i - \bar{x})^2\right]$$

11. An important problem in statistics concerns the predictability of the value of one variable from the value of another variable. Two variables that can be used in this way with good chance of success are said to be *strongly correlated*. The strength of correlation is determined by the *correlation coefficient*. We wish to conduct an experiment to determine the strength of correlation between a student's final high school average and his/her performance on first-year university classes. Following final examinations, a card is prepared for each first-year student containing two real values: high school average (H) and first-year average (F). Assume that there are N students involved in this study. Design an algorithm to read these data into two vectors, H[i] and F[i], $i = 1, 2, \ldots, N$. Then compute the correlation coefficient r according to the following formula:

$$r = \frac{N\,\Sigma\,H[i]F[i] - \Sigma\,H[i]\,\Sigma\,F[i]}{\sqrt{(N\,\Sigma\,H[i]^2 - (\Sigma\,H[i])^2)(N\,\Sigma\,F[i]^2 - (\Sigma\,F[i])^2)}}$$

If the correlation coefficient exceeds .85, a message is to be printed saying that these variables appear to be strongly correlated.

12. In any experiment, there is a certain amount of error associated with any measurement. A technique known as *smoothing* can be used to reduce the effect of this error in the analysis of the results. Suppose that a series of real values has been recorded from N replications of a particular experiment. These values are stored in a vector V. Prior to the analysis of these experimental results, the following simple smoothing operation is to be applied to the values of V. For each value (except the first and the last, which are to remain unchanged), V_i is to be replaced by

$$\frac{V_{i-1} + V_i + V_{i+1}}{3}$$

Design an algorithm to read the initial measurements, and then print the observed values and the "smoothed" values. These smoothed values are to be stored into a separate vector SMOOTH.

13. The Hollywood movie factories are singularly unimaginative in their production of monster movies. Following the success of "Frankenstein" we soon get sequels, such as "The Son of Frankenstein" and "The Bride

of Frankenstein." When there are pairs of contending monsters, such as Frankenstein and Dracula, we get additional titles, such as "Frankenstein Meets Dracula."

(a) Develop an algorithm to generate possible monster movie titles from a list of popular monsters. Your algorithm will use as its basic "title template" the following four cases:

1. _____
2. THE SON OF _____
3. THE BRIDE OF _____
4. _____ MEETS _____

With this particular template, your algorithm will generate all possible titles from an input list of monster names.

(b) Test your algorithm by hand tracing with the following set of input data. Show clearly the action of your algorithm and the output that results from it.

'KING KONG'
'MOTHRA'
'GODZILLA'

Use the end-of-file method to detect the end of the data.

14. A real estate company has 25 salespersons on its staff. Each sale made by each salesperson is recorded on one card. This card contains

salesperson number, salesperson name, amount of sale

The number of input cards is unknown. The number of sales may vary from salesperson to salesperson. For example, one salesperson may have 12 sales and another 10 sales. The input cards are not in sequence. Formulate an algorithm which generates the total amount of sales for each salesperson. The report is to look as follows:

SALESPERSON NO.	SALESPERSON NAME	TOTAL AMOUNT
1	JOHN DOE	50240.75
2	SUE BROWN	71326.50
.	.	.
.	.	.
.	.	.
25	BILL SMITH	21375.00

The end-of-file method should be used to detect the end of the data. Special citations are to be given to the salespersons with the top *two* amounts. Design your algorithm so that these two persons are specially noted.

15. Construct an algorithm which generates a yearly sales report. The report is to give a breakdown of sales for each month of the year and a yearly total. Each sales transaction is recorded on a card as follows:

sales amount, month number

where each month of the year is numbered from 1 to 12. The number of input cards is unknown, and these cards are not in any sequence. Use the end-of-file method to detect the end of the data.

4-3 SORTING AND SEARCHING WITH VECTORS

Two very important operations in computer science are those of *sorting* and *searching*. Indeed, it is estimated that computers, on the average, perhaps spend up to half of their time performing these operations. Although these operations are not, by any means, used only with vectors, their use frequently occurs in conjunction with them.

Suppose that we are given an integer vector. Sorting such a vector is the operation of arranging its elements into sequential order according to an ordering criterion. For our example, numerical elements can be sorted into increasing or decreasing order. In numerical sorting, elements are arranged in ascending or descending order, according to the numerical value of the element. In sorting terminology, the element on which sorting is done is often called a *key*. Assuming that the vector K which contains the elements 73, 65, 52, 24, 83, 17, 35, 96, 41, 9 is to be sorted, the result of an *ascending* sort is the sequence

9, 17, 24, 35, 41, 52, 65, 73, 83, 96

while a *descending* sort would yield the sequence

96, 83, 73, 65, 52, 41, 35, 24, 17, 9

Searching is one of the most commonly performed operations in data processing. Basically, it involves the scanning of a set of items in order to locate a desired item. For example, in a list of names we may want to access or find an element with a particular name. There are many ways that can be used to search a vector. Some of these methods are more efficient than others. Invariably, the better searching methods require that the data to be searched be organized in special ways. Examples of such organizations are given in Chaps. 10 and 11.

In this section we discuss some elementary sorting and searching methods involving vectors. In particular, we describe the selection sort in Sec.

4-3.1. The next section describes searching and applies it to a problem connected with the operation of a post office. Finally, the important topic of merging and its application to sorting are discussed.

4-3.1 SELECTION SORT

One of the easiest ways to sort a vector is by *selection*. Suppose that we wish to sort a vector with numeric elements into ascending order. Beginning with the first element in the vector, a sequential search is performed to locate the element which has the smallest value. When this element is found, it is written out in the first position of another vector, which is to eventually have all the original elements in ascending order. In order to remember which element was selected, the smallest element in the unsorted vector is changed to some distinguished value. This value is assumed to be invalid as far as the given data are concerned. We choose 9999 as this value. A search for the second smallest value is then carried out. This is accomplished by examining the value of each element in the given vector. If the element value is equal to the special value 9999, it is ignored since it has already been written out. (Note what would happen here if the value of the smallest element had not been changed.) At the end of this searching process the second smallest value is copied into the second position of the partially sorted vector. The process of searching for the element with the next smallest value and placing it in its proper position in the partially sorted vector continues until all elements have been sorted in ascending order. The process of searching through the entire vector and selecting the element with the next smallest value is called a *pass*. If a vector K consisting of n elements is sorted in this manner, n passes are required. This is because each pass places one element in its proper location, and in the nth pass, the element just selected must be the largest.

A general algorithm for this sorting process is contained in these steps.

1. Repeat step 2 a total of n times
2. Examine each element in the unsorted vector and place the smallest of these into the next position of the partially sorted vector (note that elements having a special value of 9999 are not used in our search for the next smallest element); once this smallest element is found, its value is changed to 9999

We first introduce some variables. The following are suggested (types in parentheses).

N (integer)	Number of elements to be sorted
K (real)	Vector to be sorted
OUTPUT (real)	Vector which contains the sorted elements

MIN_INDEX (integer)　　　Position of smallest element in a particular pass

PASS (integer)　　　Current pass number

We can now proceed to the design of our final algorithm. Let us examine more closely the second step of the previous general algorithm. The smallest element in a particular pass can be obtained as follows:

```
MIN_INDEX ← 1
Repeat for I = 2, 3, . . ., N
    If K[I] < K[MIN_INDEX] and K[I] ≠ 9999
    then MIN_INDEX ← I
```

Note that at the end of the loop, MIN_INDEX contains the position of the smallest element of the current pass. Also, any element having a value of 9999 is ignored during the pass. Such an element has already been accounted for.

Now that we have the position of the smallest element, it can be placed in its proper output position in OUTPUT. This element is then set to a value of 9999. The following algorithm segment accomplishes the desired task:

```
OUTPUT[PASS] ← K[MIN_INDEX]
K[MIN_INDEX] ← 9999
```

Putting these segments together with step 1 of the general algorithm, we obtain the following detailed algorithm.

Algorithm SELECTION_SORT. Given a vector K which contains N elements, this algorithm sorts this vector in ascending order. The vector OUTPUT represents the sorted version of the given unsorted vector; that is, OUTPUT[1] ≤ OUTPUT[2] ≤ . . . ≤ OUTPUT[N]. The variable PASS denotes the pass index. The variable MIN_INDEX denotes the position of the smallest element encountered thus far in a particular pass. The variable I indexes elements K[2] to K[N] in a given pass. It is assumed that each element of K has a value which is less than 9999. All variables are of type integer.

1. [Loop on pass index]
    Repeat thru step 4 for PASS = 1, 2, . . ., N
2. [Initialize minimum index]
    MIN_INDEX ← 1
3. [Make a pass and obtain element with smallest value]
    Repeat for I = 2, 3, . . ., N
        If K[I] < K[MIN_INDEX] and K[I] ≠ 9999
        then MIN_INDEX ← I
4. [Copy current smallest element and change its value in K to 9999]
    OUTPUT[PASS] ← K[MIN_INDEX]
    K[MIN_INDEX] ← 9999
5. [Finished]
    Exit　　　　□

Note that at the beginning of the search for each pass, we assume that
K[MIN_INDEX] = K[1] is the current smallest element encountered to this
point. For this reason, I is always initialized to index the second element in
step 3. If element K[1] has already been selected and moved to its proper
position in the partially sorted vector OUTPUT, the initial test in step 3 still
works. This is because K[MIN_INDEX] has a value of 9999 and all values in the
given vector to be sorted are assumed to be less than this special value.

The behavior of the selection sort on some sample data is given in Table
4-2. Each encircled entry denotes the element with the smallest value as
selected in a particular pass. The trace of each pass is given before the selected element is replaced by the special value 9999.

There are a number of obvious inefficiencies in this approach. First, we
must use an additional vector in order to store the sorted vector. Second,
during each pass, we must always examine all elements in the unsorted
vector. Finally, the special value 9999 could create a problem. This problem
could arise if valid element values could be equal to this special value. We
turn now to a revised version of the previous algorithm which eliminates all
of the previous drawbacks.

Beginning with the first element in the vector, a search is performed to
locate the element which has the smallest value. When this element is
found, it is interchanged with the first element in the vector. This interchange places the element with the smallest value in the first position of the
vector. A search for the second smallest value is then carried out. This is
accomplished by examining the values of each element from the second
element onward. The element which has the second smallest value is interchanged with the element located in the second position of the vector. The
process of searching for the element with the next smallest value and placing
it in its proper position continues until all elements have been sorted in
ascending order. If a vector K consisting of n elements is sorted in this manner, $n - 1$ passes are required. This is because each pass places one element

TABLE 4-2   Behavior of a Selection Sort

	Pass Number (PASS)									
	1	2	3	4	5	6	7	8	9	10
1	73	73	73	73	73	73	73	(73)	9999	9999
2	65	65	65	65	65	65	(65)	9999	9999	9999
3	52	52	52	52	52	(52)	9999	9999	9999	9999
4	24	24	(24)	9999	9999	9999	9999	9999	9999	9999
5	83	83	83	83	83	83	83	83	(83)	9999
6	17	(17)	9999	9999	9999	9999	9999	9999	9999	9999
7	35	35	35	(35)	9999	9999	9999	9999	9999	9999
8	96	96	96	96	96	96	96	96	96	(96)
9	41	41	41	41	(41)	9999	9999	9999	9999	9999
10	(9)	9999	9999	9999	9999	9999	9999	9999	9999	9999

into its proper location, and after $n - 1$ passes, the last element in the vector must be the largest. Consequently, an nth pass is not required. Note that in the ith pass we concern ourselves only with the examination of elements K[i] to K[n] inclusive, since the others have already been sorted by that time.

A general algorithm for this sorting process is contained in these steps.

1. Repeat thru step 3 a total of $n - 1$ times
2. Examine the remaining unsorted elements and place the smallest of these into its proper position
3. Reduce the size of the unsorted vector by one element

The following algorithm formalizes this process.

Algorithm SELECTION_SORT (revision 1). Given a vector K of N elements, this algorithm rearranges the vector in ascending order; that is, its elements will be in the order $K[1] \le K[2] \le \ldots \le K[N]$. The sorting process is based on the techniques just described. The variable PASS denotes the pass index and the position of the first element in the vector which is to be examined during a particular pass. The variable MIN_INDEX denotes the position of the smallest element encountered thus far in a particular pass. The variable I indexes elements K[PASS] to K[N] in a given pass. All variables are of type integer.

1. [Loop on pass index]
   Repeat thru step 4 for PASS = 1, 2, . . ., N − 1
2. [Initialize minimum index]
   MIN_INDEX ← PASS
3. [Make a pass and obtain element with smallest value]
   Repeat for I = PASS + 1, PASS + 2, . . ., N
       If K[I] < K[MIN_INDEX] then MIN_INDEX ← I
4. [Exchange elements]
   If MIN_INDEX ≠ PASS then K[PASS] ⇔ K[MIN_INDEX]
5. [Finished]
   Exit   □

The first step in the algorithm controls the number of passes which are required. Steps 2 to 4 constitute a pass. The elements to be examined in a particular pass are K[PASS] to K[N]. Note that in step 3, MIN_INDEX will be changed to the value of I, if the condition K[I] < K[MIN_INDEX] is true. Unless the element K[MIN_INDEX] is already in its final position with respect to the ordering, step 4 will interchange the contents of K[PASS] and K[MIN_INDEX]. We have introduced the special symbol ⇔ to denote this exchange operation. This corresponds to the sequence of assignments

TEMP ← K[MIN_INDEX]
K[MIN_INDEX] ← K[PASS]
K[PASS] ← TEMP

An example of the selection sort is given in Table 4-3. Each circled entry denotes the element with the smallest value which is selected in a particular pass. The elements above the bar for a given pass are those elements that have been placed in order.

We now turn to an analysis of the performance of this algorithm. During the first pass, in which the element with the smallest value is found, $n - 1$ elements are compared. In general, for the ith pass of the sort, $n - i$ comparisons are required. The total number of comparisons is, therefore, the sum

$$\sum_{i=1}^{n-1} (n - i) = \frac{n(n - 1)}{2}$$

Therefore, the number of comparisons is proportional to n^2. This fact is often denoted $O(n^2)$ (that is, order n^2). The number of element interchanges depends upon how "unsorted" the vector is to begin with. Since, during each pass, no more than one interchange is required, the maximum number of interchanges for the sort is $n - 1$.

To get a better appreciation for the performance of this method, let us assume that we want to sort the 100,000 customers of a credit card company by customer number. If we assume that each comparison takes only 10^{-6}s, the average time to perform all the comparisons in the sort is

$$\frac{100,000 \times 99,999}{2} \times 10^{-6}s$$

which is slightly more than 83 minutes! Clearly, if the required sort was only one step in a complex algorithm, such a situation might become untolerable.

TABLE 4-3   Behavior of a Selection Sort

	Unsorted			Pass Number (PASS)						Sorted
j	$K[j]$	1	2	3	4	5	6	7	8	9
1	73←	9	9	9	9	9	9	9	9	9
2	65	65←	17	17	17	17	17	17	17	17
3	52	52	52←	24	24	24	24	24	24	24
4	24	24	(24)←	52←	35	35	35	35	35	35
5	83	83	83	83	83←	41	41	41	41	41
6	17	(17)←	65	65	65	65←	52	52	52	52
7	35	35	35	(35)←	52	(52)←	(65)	65	65	65
8	96	96	96	96	96	96	96	96←	73	73
9	41	41	41	41	(41)←	83	83	83	(83)	83
10	(9)←	73	73	73	73	73	73	(73)←	96	96

Up to this point, we have considered only the sorting of numerical elements. Sorting, however, can be performed on nonnumerical elements. This variation of sorting will be discussed in Chap. 5. Also, more efficient sorting techniques will be presented in Chaps. 10 and 11.

4-3.2 BASIC SEARCHING

In this section we examine two simple searching techniques. These techniques can be applied to many searching situations. An application containing an instance of searching is described.

The most straightforward method of finding a particular element in an unordered vector is the *linear search* technique. The technique simply involves the scanning of each element of the vector in a sequential manner until the desired element is found. An algorithm for such a search procedure is as follows:

Algorithm LINEAR_SEARCH. Given an unordered vector K consisting of $n(n \geq 1)$ elements, this algorithm searches the vector for a particular element having the value of x. Assume that all variables are of type real.

1. [Search the vector]
   Repeat for i = 1, 2, ..., n
     If K[i] = x
     then Write('SUCCESSFUL SEARCH')
         Exit
2. [Element not found]
     Write('UNSUCCESSFUL SEARCH')
3. [Finished]
     Exit   ☐

This algorithm sequentially scans each element of the vector for x. If this element is found, an appropriate message is printed, and i is the index to the desired element. In the case of an unsuccessful search, the message **UNSUCCESSFUL SEARCH** is produced and an exit results. It will take, on the average, $n/2$ comparisons to find a particular element in a vector containing n elements. The worst case will require n comparisons! This can be very time consuming when the number of elements is large. A more efficient way of performing a linear search involves the following modification to the previous algorithm. A sentinel element K[n + 1] is assumed. The purpose of this element is to receive the value of x.

Algorithm LINEAR_SEARCH (revision 1). Given an unordered vector K consisting of n + 1 (n $\geq$ 1) elements, this algorithm searches the vector for a particular element having the value x. Vector element K[n + 1] serves as a sentinel element and receives the value of x prior to the search. Assume that all variables are of type real.

1. [Initialize search]
   i ← 1
   K[n + 1] ← x
2. [Search the vector]
   Repeat while K[i] ≠ x
      i ← i + 1
3. [Successful search?]
   If i = n + 1
   then Write('UNSUCCESSFUL SEARCH')
   else Write('SUCCESSFUL SEARCH')
4. [Finished]
   Exit   ☐

The purpose of the sentinel element K[n + 1] is to hold the value of x. Using this approach, the search will always be successful. Note that because of this, the index i is never checked against $n + 1$, and therefore this algorithm is more efficient than the previous version. If the index i reaches $n + 1$, then the element x was not in the original vector, and consequently the search is unsuccessful.

If the elements have previously been sorted, much more efficient searches can be conducted. A relatively simple method of searching a sorted list is known as the *binary search* method. Assume that the elements in the vector are stored in ascending order. The selection sort of the previous section can be used to achieve this ordering. A search for an element with a particular value resembles the search for a name in a telephone directory. The approximate middle entry of the vector is located and its value is examined. If this value is too high, then the value of the middle entry element of the first half is examined and the procedure is repeated on the first half until the required element is found. If the value is too low, then the value of the middle entry element of the second half of the vector is tried and the procedure is repeated on the second half. This process continues until the desired element is found or the search interval becomes empty.

A general formulation of the algorithm is as follows:

1. Repeat thru step 3 while the search interval is not empty
2. Obtain the position of the midpoint element in the current search interval
3. If the values of the desired and midpoint elements are equal, then the search is successful
   If the value of the desired element is less than the value of the midpoint element,
   then reduce the search interval to the first half of the current search interval;

else reduce the search interval to the second half of the current search interval.

4. Unsuccessful search

The following algorithm is a formalization of this process.

Algorithm BINARY_SEARCH. Given a vector K whose elements are in ascending order, this algorithm searches the structure for a given element whose value is given by x. The variables LOW, MIDDLE, and HIGH denote the lower, middle, and upper limits of the search interval, respectively.

1. [Initialize]
   LOW ← 1
   HIGH ← n
2. [Perform search]
   Repeat thru step 4 while LOW ≤ HIGH
3. [Obtain index of midpoint of interval]
   MIDDLE ← (LOW + HIGH) / 2
4. [Compare]
   If x < K[MIDDLE]
   then HIGH ← MIDDLE − 1
   else If x > K[MIDDLE]
       then LOW ← MIDDLE + 1
       else Write('SUCCESSFUL SEARCH')
           Exit
5. [Unsuccessful search]
   Write('UNSUCCESSFUL SEARCH')
   Exit   □

As an example, assume that we have the ordered vector

61, 147, 197, 217, 309, 448, 503

The behavior of the previous algorithm for $x = 197$ and 503 is given in Table 4-4. This algorithm will work only if the size of the vector is $n = 2^m - 1$ for some integer value m. In the example used, $m = 3$. The algorithm, however,

TABLE 4-4 Binary Search Behavior

Search for 197				Search for 503			
Iteration	LOW	HIGH	MIDDLE	Iteration	LOW	HIGH	MIDDLE
1	1	7	4	1	1	7	4
2	1	3	2	2	5	7	6
3	3	3	3	3	7	7	7

can easily be generalized to any n. Such a generalization involves changing
step 3 to the following:

3. [Obtain index of midpoint of interval]
    MIDDLE ← TRUNC((LOW + HIGH) / 2)

The behavior of this modified algorithm for the sample vector

61, 147, 197, 217, 309, 448, 503, 577, 629, 701, 831

is given for $x = 197$ and 503 in Table 4-5.

An average of TRUNC($\log_2$ n) − 1 comparisons is required in order to
locate a specific element, where $\log_2 n$ denotes the logarithm to the base 2
of n. The worst case, however, will take at most TRUNC($\log_2$ n) + 1 comparisons.

From this analysis it is clear that the binary search method performs as
well as or better than the linear search method. However, owing to the complexities of the binary search strategy, additional overhead is required. Updating the lower, middle, and upper indices (i.e., LOW, MIDDLE and HIGH in
the previous algorithm) is an example of such overhead. Figure 4-2 shows a
comparison of the two search methods with the presence of overhead.

We now turn to an application of searching which involves the mailing
of packages in a postal system. Table 4-6 gives a recent table of costs of
parcel post within the province of Saskatchewan. To be accepted, a parcel
is subject to the following constraints:

1. Weight limit of 35 lb
2. The rates of Table 4-6 apply to parcels not exceeding 3 ft in length,
   width, or depth and with a combined length and girth not exceeding
   6 ft; parcels exceeding these dimensions are accepted providing no
   one dimension exceeds 6 ft and the combined length and girth does
   not exceed 10 ft, subject to a surcharge of 75¢

The girth of a package is the circumference of the package around its two

TABLE 4-5   Binary Search Behavior

Search for 197				Search for 503			
Iteration	LOW	HIGH	MIDDLE	Iteration	LOW	HIGH	MIDDLE
1	1	11	6	1	1	11	6
2	1	5	3	2	7	11	9
				3	7	8	7

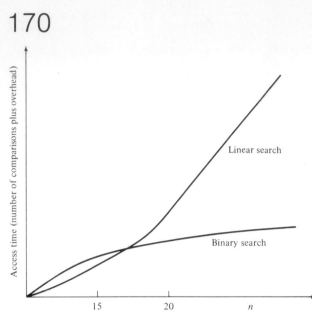

FIGURE 4-2
Comparison of linear
search and binary search.

smallest sides, as indicated in Fig. 4-3. Mathematically, the girth is given by the formula

$$2 * (D1 + D2 + D3 - \text{LARGEST})$$

where **LARGEST** is the largest of the three box dimensions **D1**, **D2**, and **D3**.

We wish to design an algorithm to process a transaction file containing one card for each box mailed during the week. Each card contains a transaction number, followed by the weight of the box, followed by its three dimensions (in no particular order). The end of file is signaled by a dummy card with transaction number **0**. The algorithm must print the transaction number, weight and postal charge for all accepted packages, and the transaction number, weight, and dimensions for all rejected packages. At the end of the report, we must print the number of packages processed and the number rejected.

TABLE 4-6 Parcel Post Package Rate Structure

	Packages up to and including:								
Weight	2 lb	3 lb	4 lb	5 lb	6 lb	7 lb	8 lb	9 lb	10 lb
Cost	.45	.65	.75	.85	.95	1.05	1.15	1.25	1.35
Weight	11 lb	12 lb	13 lb	14 lb	15 lb	20 lb	25 lb	30 lb	35 lb
Cost	1.50	1.65	1.80	1.95	2.10	2.30	2.55	2.80	3.05

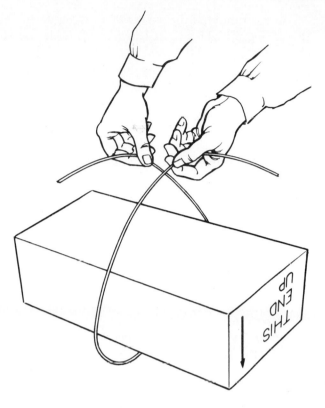

FIGURE 4-3
The girth of a package.

This problem can be solved by storing Table 4-6 as two vectors: WEIGHT and COST. The postal cost of each parcel can then be determined by searching the WEIGHT vector to find the required entry for a given parcel. The corresponding element in the COST vector can then be selected as the basic postal cost. A general algorithm for this problem is as follows:

1. Read a parcel description
2. Repeat thru step 4 while the last card has not been encountered
3. If the parcel is too large,
   then reject the parcel and update rejection statistics;
   else if box is large,
        then assess a penalty of 75¢;
        else assess no penalty
        Search the postal table for the basic postal cost
        Obtain total postal cost by adding the penalty cost to the basic
           postal cost
        Accept parcel and update acceptance statistics
4. Read another parcel description
5. Print acceptance and rejection statistics

We assume that the weight of a parcel, W, has already been rounded up to the appropriate weight category of Table 4-6. For example, an actual parcel weight of 2.4 lb would be recorded as 3 lb for computational purposes. It is further assumed that the dimensions of the parcel are given to the nearest inch.

We now recap the variables introduced thus far and introduce additional variables which we require. The updated list is as follows (type in parentheses):

WEIGHT (integer)	Vector which represents table of parcel weights
COST (real)	Vector corresponding to WEIGHT which contains table of mailing costs
NUMBER (integer)	Identification number of package
W (integer)	Weight of parcel
D1 (integer)	First dimension of package (nearest inch)
D2 (integer)	Second dimension of package (nearest inch)
D3 (integer)	Third dimension of package (nearest inch)
LARGEST (integer)	Largest dimension of package (inches)
GIRTH (integer)	Girth of package
PENALTY (real)	Penalty associated with parcel
POSTAL_COST (real)	Cost of mailing parcel
NR (integer)	Number of rejected packages
NA (integer)	Number of accepted packages
LOW (integer)	Minimum position of table segment in binary search
MIDDLE (integer)	Position of middle entry in table segment being searched
HIGH (integer)	Maximum position of table segment in binary search

We can now proceed to the design of the final algorithm. Let us examine more closely the third step of the previous general algorithm. The following algorithm segment computes the girth of the package.

```
LARGEST ← D1
If LARGEST < D2 then LARGEST ← D2
If LARGEST < D3 then LARGEST ← D3
GIRTH ← 2 * (D1 + D2 + D3 − LARGEST)
```

From this girth computation, we are in a position to either accept or reject the given parcel. If the parcel is acceptable, then a binary search of the postal table (along with a possible penalty for an oversized parcel) yields the postal cost of mailing this parcel; otherwise, the parcel is rejected. The statement for accepting or rejecting the parcel is as follows:

```
If W > 35 or LARGEST + GIRTH > 120 or LARGEST > 72
then Write('REJECTED PACKAGE', NUMBER, W, D1, D2, D3)
     NR ← NR + 1
else  If LARGEST + GIRTH > 72 or LARGEST > 36
      then PENALTY ← 0.75
      else PENALTY ← 0.0
      Search postal table for basic postal cost and print acceptance
         message
```

The statements which realize the binary search for determining initially
the basic postal cost and then the total postal cost of a given parcel are:

```
LOW ← 1
HIGH ← 18
Repeat while LOW ≤ HIGH
   MIDDLE ← TRUNC((LOW + HIGH) / 2)
   If W < WEIGHT[MIDDLE]
   then HIGH ← MIDDLE − 1
   else  If W > WEIGHT[MIDDLE]
         then LOW ← MIDDLE + 1
         else  POSTAL_COST ← PENALTY + COST[MIDDLE]
               Write('ACCEPTED PACKAGE', W, POSTAL_COST)
               NA ← NA + 1
               LOW ← HIGH + 1
```

The details of the remaining steps of the general algorithm are simple in
comparison to the previous step. In particular, the input of a parcel description (steps 1 and 4) becomes

```
Read(NUMBER, W, D1, D2, D3)
```

The loop-control statement (step 2), which handles a series of parcels, is

```
Repeat thru step ... while NUMBER ≠ 0
```

The last step of the algorithm, which summarizes the acceptance and rejection statistics, becomes

```
Write('NUMBER ACCEPTED', NA, 'NUMBER REJECTED', NR)
```

We are finally in a position to formulate a detailed algorithm. This
algorithm, which also reads the postal table, follows.

Algorithm POSTAL. Given the postal information of Table 4-6, which is
represented by the vectors **WEIGHT** and **COST**, and the parcel specifications,
consisting of identification number (**NUMBER**), weight (**W**), and parcel
dimensions D1, D2, D3, this algorithm generates the required report just
described. The variables **LARGEST** and **GIRTH** denote the largest dimension

and the girth of the parcel, respectively. PENALTY represents the penalty cost for a large parcel, and POSTAL_COST denotes the total cost of mailing the parcel. NA and NR are used to count the number of accepted and rejected parcels, respectively.

1. [Initialize]
    NA ← 0
    NR ← 0
2. [Read the postal table]
    Repeat for i = 1, 2, . . ., 18
        Read(WEIGHT[i], COST[i])
3. [Read the first parcel description]
    Read(NUMBER, W, D1, D2, D3)
4. [Process the parcel]
    Repeat thru step 8 while NUMBER ≠ 0
5. [Find the largest dimension of the parcel]
    LARGEST ← D1
    If LARGEST < D2 then LARGEST ← D2
    If LARGEST < D3 then LARGEST ← D3
6. [Compute the girth of the parcel]
    GIRTH ← 2 * (D1 + D2 + D3 − LARGEST)
7. [Process the parcel]
    If W > 35 or LARGEST + GIRTH > 120 or LARGEST > 72
    then Write('REJECTED PACKAGE', NUMBER, W, D1, D2, D3)
        NR ← NR + 1
    else If LARGEST + GIRTH > 72 or LARGEST > 36
        then PENALTY ← 0.75
        else PENALTY ← 0.0
        (search postal table for basic postal cost)
        LOW ← 1
        HIGH ← 18
        Repeat while LOW ≤ HIGH
          MIDDLE ← TRUNC((LOW + HIGH) / 2)
          If W < WEIGHT[MIDDLE]
          then HIGH ← MIDDLE − 1
          else If W > WEIGHT[MIDDLE]
              then LOW ← MIDDLE + 1
              else POSTAL_COST ← PENALTY + COST[MIDDLE]
                    Write('ACCEPTED PACKAGE', W, POSTAL_COST)
                    NA ← NA + 1
                    LOW ← HIGH + 1
8. [Read another parcel description]
    Read(NUMBER, W, D1, D2, D3)
9. [Print summary]
    Write('NUMBER ACCEPTED', NA, 'NUMBER REJECTED', NR)
    Exit

The first step of this algorithm initializes the summary counters to 0. Step 2 reads in the postal table, which consists of the two 18-element vectors WEIGHT and COST. The third step reads in the description of the first parcel. Step 4 controls the loop for processing all the parcel data. The largest dimension and girth of a parcel are calculated in steps 5 and 6, respectively. Step 7 checks the parcel as to its acceptability. If it is acceptable, the postal cost of the parcel is determined. The binary search technique is used in performing this task. Step 8 reads the parcel description of another parcel. Finally, the appropriate summaries are printed by step 9.

4-3.3   MERGING AND MERGE SORTING

In the early days of data processing, merging was performed on cards with the aid of a machine called a *collator*. The collator had as input two separate decks of cards, each of which was sorted, and it proceeded to combine these two decks and to output a single sorted deck of cards. Such a process is called *merging*. In this subsection we examine more closely the operation of merging and its application to sorting, based on successive merges.

First, let us examine the merging of two sorted vectors which are to be combined to produce a single sorted vector. We can accomplish this process easily by successively selecting the element with the smallest value or key in either of the vectors and placing this element in a new vector, thereby creating a sorted vector of elements. For example, from the vectors

Vector 1:	13	21	39
Vector 2:	7	28	

we obtain the behavior given in Table 4-7, where vector 3 is a vector which

TABLE 4-7   Sample Behavior of Merging

Vector					
1	13	21	39		
2	⑦	28			
3	7				
1	⑬	21	39		
2	28				
3	7	13			
1	㉑	39			
2	28				
3	7	13	21		
1	39				
2	㉘				
3	7	13	21	28	
1	㉟				
2					
3	7	13	21	28	39

contains a trace of the desired results. This process is formalized in the following algorithm.

Algorithm SIMPLE_MERGE. Given two sorted vectors A and B containing n and m elements, respectively, this algorithm merges these vectors and produces the sorted vector C, which contains n + m elements. The variables i, j, k, and r are used as indices to the vectors.

1. [Initialize]
   $i \leftarrow j \leftarrow k \leftarrow 1$

2. [Compare corresponding elements and output the smallest]
   Repeat while $i \leq n$ and $j \leq m$
       If $A[i] \leq B[j]$
       then $C[k] \leftarrow A[i]$
           $i \leftarrow i + 1$
           $k \leftarrow k + 1$
       else $C[k] \leftarrow B[j]$
           $j \leftarrow j + 1$
           $k \leftarrow k + 1$

3. [Copy the remaining unprocessed elements into output area]
   If $i > n$
   then Repeat for $r = j, j + 1, \ldots, m$
           $C[k] \leftarrow B[r]$
           $k \leftarrow k + 1$
   else Repeat for $r = i, i + 1, \ldots, n$
           $C[k] \leftarrow A[r]$
           $k \leftarrow k + 1$

4. [Finished]
   Exit □

Note in step 1 the introduction of a multiple assignment statement $i \leftarrow j \leftarrow k \leftarrow 1$, in which the variables i, j, and k are all initialized to a value of 1.

Let us apply this merging concept to a familiar application. An advertising firm receives two separate alphabetically ordered mailing lists from separate sources. These lists contain information such as names and addresses of potential customers. The problem, however, is that the same customer name can be on both mailing lists. For economic reasons (and to avoid annoyance to the customer) the firm wants a single mailing list in which duplicate name entries do not occur. For our purposes, we assume that each entry on the mailing lists consists of only a customer name.

The basic notion of algorithm SIMPLE_MERGE applies here. We shall use the relational operators < and > with strings. The behavior of these operators when used with strings is very similar to that encountered when used with numbers. These notions will be formalized in Sec. 5-2. We can scan both mailing lists for a matching pair of names. If such a match is found, then only one of these names is written out. This modification of the previous algorithm follows.

Algorithm MAILING_LIST. Given two alphabetically ordered mailing lists which are represented by vectors A_LIST and B_LIST containing n and m customer names, respectively, this algorithm constructs a new mailing list. This mailing list is to contain no duplicate names. The vector NEW_LIST represents the resulting list. The variables i, j, k, and r are used as indices to the vectors.

1. [Initialize]
   $i \leftarrow j \leftarrow k \leftarrow 1$

2. [Compare corresponding names and output only one in case of a match]
   Repeat while $i \le n$ and $j \le m$
       If A_LIST[i] < B_LIST[j]
       then NEW_LIST[k] ← A_LIST[i]
           $i \leftarrow i + 1$
           $k \leftarrow k + 1$
       else If A_LIST[i] > B_LIST[j]
           then NEW_LIST[k] ← B_LIST[j]
               $j \leftarrow j + 1$
               $k \leftarrow k + 1$
           else NEW_LIST[k] ← A_LIST[i]
               $i \leftarrow i + 1$
               $j \leftarrow j + 1$
               $k \leftarrow k + 1$

3. [Copy the remaining names into new mailing list]
   If $i > n$
   then Repeat for $r = j, j + 1, \ldots, m$
           NEW_LIST[k] ← B_LIST[r]
           $k \leftarrow k + 1$
   else Repeat for $r = i, i + 1, \ldots, n$
           NEW_LIST[k] ← A_LIST[r]
           $k \leftarrow k + 1$

4. [Finished]
   Exit   □

Algorithm SIMPLE_MERGE can be generalized to merge k sorted vectors into a single sorted vector. Such a merging operation is called *multiple merging* or *k-way-merging*.

Multiple merging can equivalently be accomplished by performing a simple merge repeatedly. For example, if we are to merge eight vectors of one element each, we can first merge them in pairs using the algorithm SIMPLE_MERGE. The result of this first step yields four vectors, which are again merged in pairs to give two vectors. Finally, these two vectors are merged to give the required sorted vector. A pictorial trace of this process is given in Fig. 4-4. In this example, three separate passes are required to yield a sorted vector. In general, k separate passes are required to merge 2^k separate vectors of an element each into one sorted vector. This strategy can be applied to sorting. Given an unsorted vector of n elements in which n is an

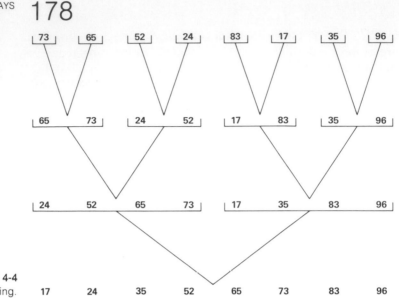

FIGURE 4-4
Two-way merge sorting.

integral power of 2 (say 2^m), one considers this original vector to be a set of n vectors, each of which contains a single element. Clearly, a vector which contains a single element is sorted. The sort requires m passes.

We now develop an algorithm based on this successive merging strategy. We assume, however, that the number of elements which are to be sorted in this manner is an integral power of 2 (that is, $n = 2^m$). This assumption simplifies the required algorithm. This restriction on the number of elements to be sorted can be lifted. The more general algorithm is more complicated (but not drastically so) than the one which we develop. We leave this modification as an exercise.

Returning to the development of the simplified version of the desired algorithm, we note that a merge sort requires an auxiliary storage area equal in size to that of the vector which is to be sorted. As mentioned previously, the sort requires m passes on the data, where $m = \log_2 n$. Each pass requires a number of subpasses. This number is a function of the pass number. A subpass is essentially the same as a simple merge for which an algorithm already exists.

In our current example three passes are required, since we want to sort eight elements. On the first pass we have four subpasses or simple merges to perform. Each of these subpasses involves the merging of two single elements. In the second pass we must do two subpasses. Each of these subpasses involves the merging of two sets of two elements each. The final pass requires only one subpass; each of the two sets contains four elements. An informal algorithm to perform a merge sort is as follows.

1. Repeat thru step 5 for m passes where $m = \log_2 n$
2. Set up control to keep track of next pass

3. Repeat thru step 5 until one pass is complete
4. Set up control to keep track of next subpass
5. Perform a subpass (that is, do a simple merge)
6. Determine which area (the original or auxiliary) the sorted vector
is in

Steps 2 and 4 in this algorithm perform a very important function. Step 2 must keep track of the number of elements which are to be merged in a particular subpass. Recall that this number is a function of the pass number (**PASS**). Specifically, if **SIZE** represents the desired quantity, then its value is given by

SIZE $\leftarrow$ 2 $\uparrow$ (PASS $-$ 1)

The second step must also determine the number of subpasses which are required in a particular pass. This quantity (**SUBPASS**) depends on the total number of elements (n) to be sorted and the pass number (**PASS**):

SUBPASS $\leftarrow$ n / (2 $*$ PASS)

Step 4 must keep track of the elements which are to be merged in a particular subpass or simple merge. For example, using the sample unsorted vector in Fig. 4-4, the second pass of the merge sort with **PASS** = 2, **SIZE** = 2, and **SUBPASS** = 2 can be represented as follows:

(a) state at beginning of first subpass during second pass

now merged

(b) state at beginning of second subpass during second pass

where variables i and j keep track of the elements which are to be simply merged and variables p and q keep track of the merging process during a particular pass. We have corresponding elements to compare in the two subvectors being merged in a particular subpass (or simple merge) if i + 1 $-$ p $\leq$ SIZE and j + 1 $-$ q $\leq$ SIZE both hold. If, however, one of these conditions fails, then the remaining elements in the other subvector are copied into the output area, as was the case in algorithm **SIMPLE_MERGE**.

Clearly, this sorting technique requires a temporary or output area whose size is the same as that of the original unsorted vector. Consequently, for our example vector of eight elements, an output area of the same size is required. During the first pass of the merge sort, the elements of the original vector are rearranged and written out in the output area. During the second pass, however, the elements in the output area are merged in groups, with the results being copied in the original vector area. This "swapping" between the original vector and output areas continues until the original vector is in order (that is, until all passes are complete).

We are now in a position to give a detailed algorithm for a merge sort.

Algorithm MERGE_SORT. Given an unsorted vector K which contains n elements, this algorithm performs a sort according to the method just described. The vector C represents the output area. The integer variables PASS, SIZE, and SUBPASS are as previously described. The integer variables p, q, i, and j were just defined. The variables r, s, and t are index variables.

1. [Perform sort]
   Repeat thru step 7 for PASS = 1, 2, . . ., $\log_2 n$

2. [Initialization for this pass]
   SIZE $\leftarrow 2 \uparrow (\text{PASS} - 1)$
   p $\leftarrow 1$
   q $\leftarrow$ p + SIZE
   SUBPASS $\leftarrow n / (2 * \text{PASS})$

3. [Perform one pass]
   Repeat thru step 7 for s = 1, 2, . . ., SUBPASS

4. [Initialize simple merge]
   i $\leftarrow$ p
   j $\leftarrow$ q
   t $\leftarrow$ p

5. [Compare corresponding elements and output smallest]
   Repeat while $i + 1 - p \leq \text{SIZE}$ and $j + 1 - q \leq \text{SIZE}$
   If PASS is odd
   then If $K_i \leq K_j$
        then $C_t \leftarrow K_i$
             i $\leftarrow$ i + 1
             t $\leftarrow$ t + 1
        else $C_t \leftarrow K_j$
             j $\leftarrow$ j + 1
             t $\leftarrow$ t + 1
   else If $C_i \leq C_j$
        then $K_t \leftarrow C_i$
             i $\leftarrow$ i + 1
             t $\leftarrow$ t + 1
        else $K_t \leftarrow C_j$
             j $\leftarrow$ j + 1
             t $\leftarrow$ t + 1

6. [Copy the remaining unprocessed elements from a subvector into output area]

    If $i + 1 - p > $ SIZE

    then If PASS is odd

        then Repeat for $r = j, j + 1, \ldots, q + $ SIZE $- 1$

            $C_t \leftarrow K_r$

            $t \leftarrow t + 1$

        else Repeat for $r = j, j + 1, \ldots, q + $ SIZE $- 1$

            $K_t \leftarrow C_r$

            $t \leftarrow t + 1$

    else If PASS is odd

        then Repeat for $r = i, i + 1, \ldots, p + $ SIZE $- 1$

            $C_t \leftarrow K_r$

            $t \leftarrow t + 1$

        else Repeat for $r = i, i + 1, \ldots, p + $ SIZE $- 1$

            $K_t \leftarrow C_r$

            $t \leftarrow t + 1$

7. [Update subvector indices]

    $p \leftarrow q + $ SIZE

    $q \leftarrow p + $ SIZE

8. [Recopy, if required]

    If $\log_2 n$ is odd

    then Repeat for $i = 1, 2, \ldots, n$

        $K_i \leftarrow C_i$

9. [Finished]

    Exit   □

    The first step controls the number of passes which are required in the sort. Step 2 computes the size of each subvector (SIZE) which is used in a particular subpass or simple merge. This step also computes the number of simple merges or subpasses (SUBPASS) required during a particular pass. Finally, the initialization of the variables p and q, which keep track of the subvectors being merged during each subpass, takes place in this step. The third step controls the number of subpasses required during a particular pass. Step 4 initializes the variables i, j, and t. These variables index through the pair of subvectors which are to be simply merged. Steps 5 through 7 perform a simple merge and these steps are a rewrite of the algorithm SIMPLE_ MERGE. Note that if the variable PASS is odd, then the output area is the vector C; otherwise, the output area is the original vector K. Step 8 determines which area contains the sorted vector. If an odd number of passes was required in the sort, then the output area contains the results. In this case the elements in the output area C are copied into K.

    A sample behavior for this algorithm is given in Table 4-8. The vertical strokes delineate the subtables being merged in a particular pass. This sort-

TABLE 4-8   Behavior of Two-Way Merge Sorting

Pass	K_1	K_2	K_3	K_4	K_5	K_6	K_7	K_8	C_1	C_2	C_3	C_4	C_5	C_6	C_7	C_8
	73|	65|	52|	24|	83|	17|	35|	96|	—	—	—	—	—	—	—	—
1	73	65	52	24	83	17	35	96	65	73|	24	52|	17	83|	35	96|
2	24	52	65	73|	17	35	83	96|	65	73	24	52	17	83	35	96
3	24	52	65	73	17	35	83	96	17	24	35	52	65	73	83	96
	17	24	35	52	65	73	83	96	17	24	35	52	65	73	83	96

ing method is quite efficient. Since $\log_2 n$ passes are required in the sort, the total number of comparisons required is proportional to $n \times \log_2 n$. This quantity represents the worst case, as well as the average case. An obvious disadvantage in using this method is that an output area equal in size to the original unsorted vector is required.

In closing, we reemphasize that algorithm MERGE_SORT only works if $n = 2^m$. This algorithm can be changed to handle the general case. This extension, however, is left as an exercise.

EXERCISES   4-3

1. Consider a vector X which contains 10 elements. Assume that the distribution of requests for each element in this vector is as shown in Table 4-9.

    (a) Compute the average number of comparisons for a sequential search of this vector.
    (b) Suggest a more efficient arrangement of the data entries. Based on this arrangement, recompute the average number of comparisons for a sequential search.

TABLE 4-9

i	X_i	Distribution of Requests (%)
1	X_1	10
2	X_2	5
3	X_3	20
4	X_4	5
5	X_5	25
6	X_6	10
7	X_7	5
8	X_8	5
9	X_9	5
10	X_{10}	10

2. Given a vector X of n real elements where n is odd, design an algorithm to calculate the median of this vector. The median is the value such that half the numbers are greater than that value and half are less. For example, given the vector X

$$
\begin{array}{ccccccccc}
X_1 & X_2 & X_3 & X_4 & X_5 & X_6 & X_7 & X_8 & X_9 \\
17 & -3 & 21 & 2 & 9 & -4 & 6 & 8 & 11
\end{array}
$$

containing nine elements, the execution of your algorithm should give a value of 8.

3. Algorithm **MERGE_SORT** given in the text only works if $n = 2^m$. Extend this algorithm so that it can handle any value of n. A trace of this more general two-way merge process for a vector of 10 elements is given in Fig. 4-5.

4. Another familiar sorting method is the *bubble sort*. It differs from the selection sort in that, instead of finding the element with the smallest value and then performing an interchange, two elements are interchanged immediately upon discovering that they are out of order. Using this approach, at most $n - 1$ passes are required. During the first pass, K_1 and K_2 are compared, and if they are out of order they are interchanged; this process is repeated for elements K_2 and K_3, K_3 and K_4, and so on. This method will cause elements with small values to "bubble up." After the first pass, the element with the largest value will be in the nth position. On each successive pass, the elements with the next largest value will be placed in position $n - 1$, $n - 2$, . . ., 2, respectively.

After each pass through the vector, a check can be made to determine whether or not any interchanges were made during that pass. If no interchanges have occurred during the last pass, then the vector must be sorted and, consequently, no further passes are required. A sample behavior of this sorting process is given in Table 4-10. Formulate an algorithm for this sorting process.

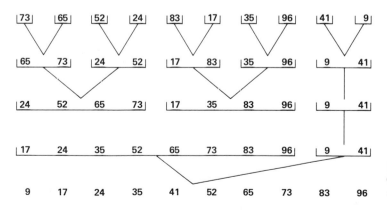

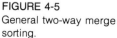

FIGURE 4-5
General two-way merge sorting.

TABLE 4-10   Behavior of a Bubble Sort

j	K_j	1	2	3	4	5	6	7	8	9
	Unsorted			Pass Number						Sorted
1	73	65	52	24	24	17	17	17	17	9
2	65	52	24	52	17	24	24	24	9	17
3	52	24	65	17	35	35	35	9	24	24
4	24	73	17	35	52	41	9	35	35	35
5	83	17	35	65	41	9	41	41	41	41
6	17	35	73	41	9	52	52	52	52	52
7	35	83	41	9	65	65	65	65	65	65
8	96	41	9	73	73	73	73	73	73	73
9	41	9	83	83	83	83	83	83	83	83
10	9	96	96	96	96	96	96	96	96	96

5. Alter the algorithm obtained in exercise 4 to take advantage of the fact that all elements below and including the last one to be exchanged must be in the correct order; consequently, those elements do not have to be examined again.

6. Modify the algorithm obtained in exercise 5 such that alternate passes go in opposite directions. That is, during the first pass, the element with the largest value will be at the end of the vector and during the second pass the element with the smallest value will be in the first position of the vector, and so on. A sample behavior of such a modified approach is given in Table 4-11.

7. A large firm has plants in five different cities. The firm employs a total of

TABLE 4-11   Behavior of a Modified Bubble Sort

j	K_j	1	2	3	4	5	6	7	8	9
	Unsorted			Pass Number						Sorted
1	73	65	9	9	9	9	9	9	9	9
2	65	52	65	52	17	17	17	17	17	17
3	52	24	52	24	52	24	24	24	24	24
4	24	73	24	65	24	52	35	35	35	35
5	83	17	73	73	65	65	52	52	52	41
6	17	35	17	17	73	35	65	41	41	52
7	35	83	35	35	35	41	41	65	65	65
8	96	41	83	41	41	73	73	73	73	73
9	41	9	41	83	83	83	83	83	83	83
10	9	96	96	96	96	96	96	96	96	96

n employees. Each employee record contains (in part) the following fields:

employee name, city, employee number

These records are not kept in any order. Assume that the city field is coded with an integer value of 1 to 5. The information on the employees can be represented by three vectors: NAME, CITY, and NUMBER. Construct an algorithm which sorts all the employee records such that they are printed by increasing employee number within each city. That is, the format is as follows:

First city:

name	number
name	number

.
.
.

| name | number |

Second city:

name	number
name	number

.
.
.

| name | number |

.
.
.

Fifth city:

name	number
name	number

.
.
.

| name | number |

8. Management information systems are becoming more and more common. They allow an administrator to type a request into a computer and obtain the answer to the request. In this problem we will consider one such request: given the name of an employee, find the department in which the employee works. These requests come in the form of the keyword 'DEPARTMENT' followed by the name of the employee. The last of these requests has a keyword of 'FINISHED'.

In order to respond to such requests, the following information is available. First, there is a file of employee information. This file contains the employee's name and the name of his supervisor. This information is in alphabetical order by employee name, and the last record in the file has a sentinel employee name of 'ZZZZZ'. The company has a large number of employees; the current number (which changes from time to time) is 134. Since the number of employees is large and the file is in alphabetical order by employee name, when seeking the record for a specific employee, a binary search should be used. It is known that every supervisor is also the manager of a department. Thus, in order to determine the department in which an employee works, we must determine the name of the department that the employee's supervisor manages. This information can be determined from a second file, the department file. This file contains the name of each department and the name of the manager of the department. This file is ordered by department name. Note there is always less than 50 departments.

Formulate an algorithm to respond to this type of request. For your data, assume that the department file comes first, preceded by a number specifying the number of records (departments) in the department file. The employee file comes next. Last come the requests for information.

4-4   ARRAYS

4-4.1   TWO-DIMENSIONAL ARRAYS

Each week the management of a local appliance store records the sales of the individual items in its stock. At the end of each month, these weekly summaries are sent to the head office, where they are analyzed. In a typical month, the sales might be as shown in Table 4-12.

This report is in the form of a table, with the rows denoting the weekly sales summaries and the columns denoting the sales figures for the individual appliances. Any individual figure in the report can be acquired simply by referring to the row and column in which it appears. For example, the number of stoves sold in the second week is given in the second row and the fourth column, and is, in fact, 3.

To represent structures of this form, we introduce a new data structure that we will call an *array*. Like a vector, an array is a subscripted variable;

TABLE 4-12   Monthly Sales Summary

Week	Washers	Dryers	Refrigerators	Stoves	Freezers
1	6	4	8	9	3
2	7	7	10	3	5
3	5	3	7	8	2
4	8	10	15	12	5

however, more than one subscript is allowed. The sales summary table shown in Table 4-12 is conveniently represented as an array with two subscripts: the first specifying the row and the second, the column. The subscripts are separated by commas. If the summary report shown is implemented as a two-dimensional array, called REPORT, the figure for the number of stoves sold in the second week is stored in REPORT[2, 4].

Each subscript of an array references a *dimension* of the array. Two-dimensional structures, such as tables, require two subscripts. It is possible to visualize three-dimensional structures, which could be represented by arrays with three subscripts. Although it is possible to have arrays with more than three subscripts, it is hard to illustrate them with realistic examples. Arrays of two dimensions, because they are so common, are often referred to by a special term, *matrices*.

You might have speculated that a vector is really a special case of an array, having only one dimension. This is, in fact, the case, and suggests that certain restrictions are the same for both. For example, as was the case with vectors, we will insist that all elements of an array be of the same type.

Returning to Table 4-12, let us suppose that the head office wishes to accumulate the following statistics from the sales summary reports. First, the total number of appliances sold each week, and second, the total number of each type of appliance sold in the month. This requires totaling the values in each row of the table to get the weekly totals as in the following:

Repeat for each week
    total the appliances sold in this week

and totaling the values in each column to get the monthly totals by appliance type as in the following:

Repeat for each type of appliance
    total the sales for the 4 weeks

To do this we can use nested counted loops as shown in the following algorithms, WEEKLY_TOTALS and UNIT_TOTALS.

Algorithm WEEKLY_TOTALS. This algorithm computes the total of sales of appliances in each week. The sales are stored in a two-dimensional array REPORT. All values are assumed to be integer.
1. [Repeat the computation for each week]
    Repeat thru step 4 for WEEK = 1, 2, 3, 4
2. [Initialize row total]
    TOTAL ← 0
3. [Total the sales for the five appliances]
    Repeat for APPLIANCE = 1, 2, . . ., 5
        TOTAL ← TOTAL + REPORT[WEEK, APPLIANCE]

4. [Print total for this week]
    Write('SALES FOR WEEK', WEEK, 'WERE', TOTAL, 'UNITS')
5. [Finished]
    Exit □

Algorithm UNIT_TOTALS. This algorithm computes the total sales of each type of appliance in a month. The sales are stored in a two-dimensional array REPORT. All values are assumed to be integer.
1. [Repeat the computation for each type of appliance]
    Repeat thru step 4 for APPLIANCE = 1, 2, . . ., 5
2. [Initialize column total]
    TOTAL ← 0
3. [Total the sales of this appliance for the 4 weeks]
    Repeat for WEEK = 1, 2, 3, 4
        TOTAL ← TOTAL + REPORT[WEEK, APPLIANCE]
4. [Print total for this appliance]
    Write('SALES FOR APPLIANCE', APPLIANCE, 'WERE', TOTAL, 'UNITS')
5. [Finished]
    Exit □

TABLE 4-13

Value of Row Subscript, WEEK	Value of Column Subscript, APPLIANCE	Value of Array Element Considered, REPORT[WEEK, APPLIANCE]
1	1	6
	2	4
	3	8
	4	9
	5	3
2	1	7
	2	7
	3	10
	4	3
	5	5
3	1	5
	2	3
	3	7
	4	8
	5	2
4	1	8
	2	10
	3	15
	4	12
	5	5

Although the loops in the algorithms WEEKLY_TOTALS and UNIT_TOTALS are short, it is easy to be confused by their execution. Let us examine them carefully, paying particular attention to the order in which the individual array elements are processed.

In the algorithm WEEKLY_TOTALS, the application requires that we process the array one row at a time (that is, along the rows). For the first row, then, we consider each of its columns in turn. This process continues for the second row, the third row, and finally, the fourth row. As you can see, the subscript referring to the column (the second subscript) must pass through all its values for each value of the subscript referring to the row (the first subscript). Consequently, the inner loop controls the column subscript, while the outer loop controls the row subscript. Table 4-13 shows the order in which the array elements are considered.

Turning to the second algorithm, UNIT_TOTALS, here we must process the array one column at a time (that is, down the columns). For each of the columns we look in turn at each of its rows. Thus, the inner loop in this case must control the values of the row subscript, while the outer loop controls the values of the column subscript. Table 4-14 shows the order in which the array elements are considered in this instance.

TABLE 4-14

Value of Column Subscript, APPLIANCE	Value of Row Subscript, WEEK	Value of Array Element Considered, REPORT[WEEK, APPLIANCE]
1	1	6
	2	7
	3	5
	4	8
2	1	4
	2	7
	3	3
	4	10
3	1	8
	2	10
	3	7
	4	15
4	1	9
	2	3
	3	8
	4	12
5	1	3
	2	5
	3	2
	4	5

Notice that we do not change the array itself, simply the *order* in which the elements are considered. The value of REPORT[3, 4] is always 8, regardless of the processing order. This order is determined by the manner in which values are assigned to the row and column subscripts.

This process can be easily extended to arrays of still higher dimension. The visual representation, however, is difficult for this type of array. Also, we can no longer use the terms "row" and "column" which refer to a two-dimensional table.

The input and output of arrays can be handled in an element-by-element fashion using loops like those in the two previous algorithms. For example, to write out the entire array REPORT by rows, we could give the following statements:

```
Repeat for ROW = 1, 2, 3, 4
  Repeat for COLUMN = 1, 2, . . ., 5
    Write(REPORT[ROW, COLUMN])
```

As mentioned earlier, it is possible in many programming languages to specify the reading or printing of an entire array in one operation. In the case of two-dimensional arrays, the reading or writing is done in many languages by row. For example, the statement

```
Write(REPORT)
```

would write out the elements of the array REPORT in row order. If we required a print out in column order, we would be forced to resort to element-by-element printing.

4-4.2 ARRAYS OF MORE THAN TWO DIMENSIONS

The Antique Car Club of Saskatoon is conducting a survey of the antique cars in the province of Saskatchewan. For each car, a data card is punched with the following information: the manufacturer (an integer code from 0 to 30), the year of the car (from 1900 to 1950), and the car's condition (integers 1 through 4 for poor, fair, good, excellent, respectively). To enable various analyses to be conducted, these data are stored in a three-dimensional array, as shown in Fig. 4-6.

The data structure shown in Fig. 4-6 could be represented as an array CARS with three subscripts: the first subscript gives the manufacturer, the second gives the year, and the third gives the condition. The value stored is the number of cars of that type found. For example, the entry shown in Fig. 4-6 is denoted by CARS[26, 1904, 2], and shows that six cars were found with these characteristics.

To illustrate the use of this three-dimensional array, we give the following algorithm, which reads in a year and prints out the number of cars found for that year and their average condition.

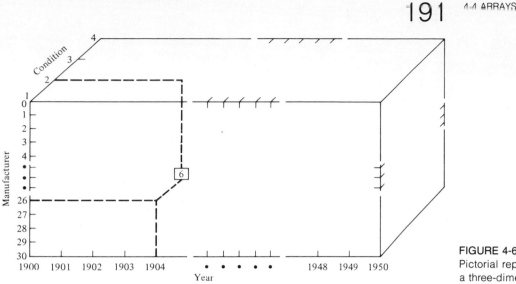

FIGURE 4-6
Pictorial representation of
a three-dimensional array.

Algorithm CAR_REPORT. Assuming the existence of the three-dimensional array **CARS** just described, the algorithm reads in a year (**YEAR**), counts the number of cars found (**COUNT**) of that year, and computes their average condition (**AVG_COND**). Variables **YEAR** and **COUNT** are of type integer; **AVG_COND** is real.

1. [Input]
      Read(YEAR)
2. [Initialize counters]
      COUNT ← 0
      AVG_COND ← 0.0
3. [Process all makes of car for that year]
      Repeat for MAKE = 0, 1, 2, . . ., 30
         Repeat for COND = 1, 2, 3, 4
            NUM ← CARS[MAKE, YEAR, COND]
            If NUM ≠ 0
            then AVG_COND ← AVG_COND + COND * NUM
               (weighted average)
               COUNT ← COUNT + NUM
4. [Output computed statistics]
      Write('YEAR:', YEAR,
            'CARS RECORDED:', COUNT,
            'AVERAGE CONDITION:', AVG_COND / COUNT)   □

This example illustrates the use of higher-dimensional arrays. As you can see, the techniques introduced earlier in the chapter generalize easily. We will leave the tracing of this example as an exercise.

EXERCISES 4-4

1. Given a matrix A of the form

$$
\begin{pmatrix}
a_{11} & a_{12} & \cdots & a_{1m} \\
a_{21} & a_{22} & \cdots & a_{2m} \\
\cdot & \cdot & & \cdot \\
\cdot & \cdot & & \cdot \\
\cdot & \cdot & & \cdot \\
a_{n1} & a_{n2} & \cdots & a_{nm}
\end{pmatrix}
$$

the *transpose* A′ of A is given by

$$
\begin{pmatrix}
a_{11} & a_{21} & \cdots & a_{n1} \\
a_{12} & a_{22} & \cdots & a_{n2} \\
\cdot & \cdot & & \cdot \\
\cdot & \cdot & & \cdot \\
\cdot & \cdot & & \cdot \\
a_{1m} & a_{2m} & \cdots & a_{nm}
\end{pmatrix}
$$

That is, the transpose of a matrix is obtained by interchanging its rows and columns. Formulate an algorithm to input a matrix and obtain its transpose.

2. A matrix A of the form

$$
\begin{pmatrix}
a_{11} & a_{12} & \cdots & a_{1m} \\
a_{21} & a_{22} & \cdots & a_{2m} \\
\cdot & \cdot & & \cdot \\
\cdot & \cdot & & \cdot \\
\cdot & \cdot & & \cdot \\
a_{n1} & a_{n2} & \cdots & a_{nm}
\end{pmatrix}
$$

is *symmetric* if $n = m$ and

$$a_{ij} = a_{ji} \quad \text{for } 1 \le i \le n \quad \text{and} \quad 1 \le j \le m$$

For example, the matrix

$$
\begin{pmatrix}
1 & 4 & 7 \\
4 & 2 & 9 \\
7 & 9 & 3
\end{pmatrix}
$$

is symmetric. Construct an algorithm which inputs a matrix and determines whether or not it is symmetric.

3. Given the two matrices A and B, where

$$A = \begin{pmatrix} a_{11} & a_{12} & \cdots & a_{1m} \\ a_{21} & a_{22} & \cdots & a_{2m} \\ \cdot & \cdot & & \cdot \\ \cdot & \cdot & & \cdot \\ \cdot & \cdot & & \cdot \\ a_{n1} & a_{n2} & & a_{nm} \end{pmatrix} \quad \text{and} \quad B = \begin{pmatrix} b_{11} & b_{12} & \cdots & b_{1r} \\ b_{21} & b_{22} & \cdots & b_{2r} \\ \cdot & \cdot & & \cdot \\ \cdot & \cdot & & \cdot \\ \cdot & \cdot & & \cdot \\ b_{m1} & b_{m2} & \cdots & b_{mr} \end{pmatrix}$$

the *product* C of A and B is given by

$$C = \begin{pmatrix} c_{11} & c_{12} & \cdots & c_{1r} \\ c_{21} & c_{22} & \cdots & c_{2r} \\ \cdot & \cdot & & \cdot \\ \cdot & \cdot & & \cdot \\ \cdot & \cdot & & \cdot \\ c_{n1} & c_{n2} & \cdots & c_{nr} \end{pmatrix}$$

where

$$c_{ij} = \sum_{k=1}^{m} a_{ik} \times b_{kj}$$

For example, given

$$A = \begin{pmatrix} 1 & 2 & 3 \\ 4 & 5 & 6 \end{pmatrix} \quad \text{and} \quad B = \begin{pmatrix} 1 & 4 \\ 2 & 5 \\ 3 & 6 \end{pmatrix}$$

the product is

$$\begin{pmatrix} 14 & 32 \\ 32 & 77 \end{pmatrix}$$

Note that one can only multiply matrices A and B if the number of columns in A is equal to the number of rows in B. Obtain an algorithm which inputs matrices A and B and computes their product.

4. Develop an algorithm to compute Pascal's triangle of binomial coefficients. Such a triangle has the form

where, in general, a row is obtained by noting that each element in this row is the sum of the two elements immediately above it. Generate the first 10 rows of this triangle.

5. Given a matrix **GRADE** whose element **GRADE**[I, J] contains the Jth score of the Ith student and a weight vector **WEIGHT** whose element **WEIGHT**[J] denotes the weight of the Jth problem, formulate an algorithm which ranks the students. The ranks are to be recorded into a vector **RANK** such that **RANK**[I] denotes the position of the Ith student in the class. The algorithm is to read input data for **N** students and **M** problems. For example, the following arrays contain the grade, problem, and rank information for a class of five students and three problems. In this example the average of the first student is

$$\frac{3 \times 65 + 2 \times 80 + 1 \times 85}{6} = 73.3$$

and this student is ranked fourth in a class of 5 students.

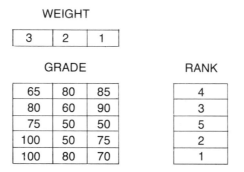

6. A final examination contains 100 multiple-choice questions. Each question has five choices and only one choice can be the correct one. The examination results and student information can be represented as follows:

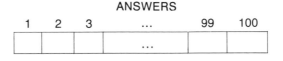

SCORE

	1	2	...	99	100		NAME
1			...				
2			...				
.	.	.		.	.		.
.	.	.		.	.		.
.	.	.		.	.		.
n			...				

where **ANSWERS** contains the correct answers to the examination questions, **SCORE** is a matrix whose row contains the answers to the 100 multiple-choice questions, and the vector **NAME** contains the names of the students in the class. The answer to each question is coded as **1, 2, 3, 4,** or **5.** If more than one choice is marked, this possibility is recorded as a **6.** Construct an algorithm which outputs the names of the students who passed. A minimum grade of 60 is required to pass.

7. Design an algorithm which calculates the salary for a group of 100 salespersons. The salary of each salesperson is paid on a commission basis. Each salesperson sells 50 items. The input data consist of a set of cards each of which contains a salesperson number, item number, unit price, and units of this item which are sold. The cards are not in any particular sequence. Use an end-of-file test to detect the end of the data. Assume a commission rate of 5% on all items.

8. An analysis is underway of traffic accidents in midtown Manhattan. For convenience the streets and avenues are represented by a grid as follows:

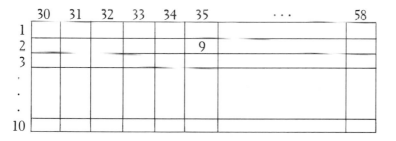

where the row headings denote the avenues from First Avenue to Tenth Avenue, and the column headings denote the streets from 30th Street to 58th Street. The entries in the array denote the number of accidents occurring in that vicinity in the most recent observational period. For example, in the case shown nine accidents occurred near the intersection of Second Avenue and 35th Street. An unknown number of accident data are to be read in. Each accident is given by a pair of numbers describing its location. For example, the pair 7,42 denotes an accident occurring in the vicinity of Seventh Avenue and 42nd Street. Formulate an algorithm that reads in this information and prepares an array of the form shown. Use an end-of-file test to determine the end of the data. Be sure to incorporate a range check to ensure that all data values lie within the range of streets and avenues under study. Reject any invalid data with an appropriate message. Once the data have been read and stored, produce a list of the ten most dangerous intersections.

9. Using the three-dimensional array CARS described in this section, design algorithms to compute the following statistics:
    (a) The number of cars made before 1910 with condition rated good or excellent.
    (b) The most popular make of car, as judged by the number recorded of all the manufacturers.
    (c) Identify the manufacturer whose cars appear to be in the best average condition.

4-5 APPLICATIONS OF VECTORS AND ARRAYS

In this section we present four applications involving arrays. We begin with a simple problem of subscript calculation to determine the amount of family allowance benefits to be paid based on income and size of family. The second problem deals with more elaborate processing of actual array elements as survey statistics are computed for a group known as Overweights Anonymous. The third problem deals with the preparation of league standings for a hypothetical hockey league. Finally, an application of arrays to the problem of a computer-dating service is described.

4-5.1 FAMILY ALLOWANCE PAYMENTS

In the kingdom of Fraziland, family allowance benefits are paid to the citizenry each month, according to the yearly earnings of the family and the number of children in the family. Monthly payments are made according to Table 4-15. Assume that a card is prepared for each family in Fraziland, containing the yearly income of the family and the number of children in the family. We require an algorithm to determine the amount of the monthly family allowance payment.

This problem requires that the schedule of payments be stored in an array, with the rows corresponding to income level, and the columns to number of children. For any family, the subscript values must be determined from the values on the data card. The subscript values corresponding to number of children can be taken almost directly from the input value. Notice that this means that the first column subscript will be 0. This presents no particular problems in the algorithmic language, although it might

TABLE 4-15   Family Allowance Payments

Yearly Income	Number of Children						
	0	1	2	3	4	5	6 and up
Less than $3,000	0	17	19	20	22	24	25
$3,000–$3,999	0	16	18	19	21	23	24
$4,000–$4,999	0	15	17	18	20	22	23
$5,000–$5,999	0	14	16	17	19	21	22
$6,000–$6,999	0	13	15	16	18	20	21
$7,000–$7,999	0	12	14	15	17	19	20
$8,000–$8,999	0	11	13	14	16	18	19
$9,000–$9,999	0	10	12	13	15	17	18
$10,000 and over	0	9	11	12	14	16	17

not be possible in some programming languages (for example, FORTRAN). The calculation of the row subscript, however, requires an appropriate transformation of the input value. The algorithm BENEFITS does the job.

Algorithm BENEFITS. This algorithm computes the monthly family allowance payments according to the schedule shown in Table 4-15. Assume that this information is stored in the integer array SCHED. The integer variables R and C are used to determine the appropriate row and column subscripts, respectively, from the values read into the integer variables INCOME and CHILDREN.

1. [Repeat for all families]
    Repeat thru step 5 while there still remains a family
2. [Input data for next family]
    Read(INCOME, CHILDREN)
    If there is no more input then Exit
3. [Determine appropriate row subscript]
    If INCOME < 3000
    then R ← 1
    else If INCOME ≥ 10000
        then R ← 9
        else R ← TRUNC((INCOME − 1000) / 1000)
4. [Determine appropriate column subscript]
    If CHILDREN ≥ 6
    then C ← 6
    else C ← CHILDREN
5. [Select correct payment]
    Write('PAYMENT IS', SCHED[R, C]).   □

Let us consider the execution of this algorithm on some sample data. Suppose that the following cards are submitted:

7600	3
3950	4
2500	0
12700	8

Card number 1 is read, with the result that INCOME and CHILDREN receive values 7600 and 3, respectively. In step 3, the appropriate row subscript is calculated as TRUNC ((7600 − 1000) / 1000), which is 6. The column subscript is calculated in step 4 as 3. Thus, the correct payment is SCHED[6, 3], or $15.

For the second data card, the correct payment is SCHED[2, 4], or $21.

The income value on the third data card falls in the special case of less than $3,000. Thus, in step 3, R is set to 1. C is set to 0 in step 2. The correct payment, then, is SCHED[1, 0], or 0.

Both the income and number of children on the fourth data card fall into special case regions. Since the income value exceeds $10,000, R is set to 9 in step 3. Since the number of children is more than 6, C is set to 6 in step 4. Thus, the correct payment is SCHED[9, 6], or $17.

Since there are no more data cards, execution then terminates in step 2.

4-5.2 OVERWEIGHTS ANONYMOUS

The local branch of Overweights Anonymous is conducting a study of the effectiveness of its weight-reducing programs. Fifty members were selected at random as subjects for the study. Over the past 12 months, a record has been made of the weight of each subject at the start of the month. For each of these subjects, a card is prepared containing this set of 12 readings (assume that they are all rounded to the nearest pound). The branch now requires an algorithm to analyze this information to determine the following:

1. The average weight change for all subjects over the entire 12-month period
2. The number of subjects whose total weight change exceeded the average
3. The average monthly weight change per subject
4. The number of instances during the year in which a subject lost more than the average monthly weight change during a single month

The solution of this problem requires an integer array to store the data recorded for the 50 subjects over the 12-month period. To this end, we define an integer array POUNDS with 50 rows and 12 columns, as shown in Fig. 4-7.

An algorithm to compute the required statistics has the following general form:

1. Repeat for each subject in the sample
   Read 12-month history
   Add weight change for the year to running total

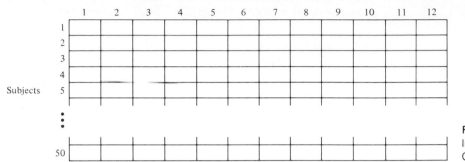

FIGURE 4-7
Information layout in
Overweights Anonymous.

2. Compute average weight change per subject over the 12-month
   period
3. Repeat for each subject in sample
   If total weight change exceeds average,
   then increment count
   Compute average monthly weight change
   Add to running total
4. Compute average monthly weight change per subject
5. Repeat for months 2 thru 12
   Calculate number of instances where a subject experienced a
   weight change in the past month that exceeds the figure from
   step 4

This algorithm requires that we move through the elements of the array
POUNDS a total of 3 times (steps 1, 3, and 5). Let us consider each of the
steps of this algorithm in more detail.

In step 1, we perform two important functions. We read the entire set of
data values, but as we do so, we simultaneously compute the total weight
change in the year. For each subject this is done by subtracting the value for
month 12 from the value for month 1. In step 2, we divide this total by 50 to
determine the per subject average. In step 3, we also perform two functions
as we read through the array. (In both cases, this is done to cut down on the
number of times we process each array element; this can have a noticeable
effect on the execution time.) First, we count the number of times that a
subject's weight change for the year exceeds the per subject average. At the
same time, we compute for each subject the average monthly weight change
(this requires an inner loop) and add this to a running total. In step 4 this is
divided by the number of subjects to get a per subject, per month value.
Finally, in step 5 we count the number of subjects whose weight change in
any month exceeds the per subject, per month figure computed in step 4.
Notice that the form of this step suggests columnwise processing of the array
rather than the rowwise processing of the previous steps.

This process is formalized in the algorithm **OVERWEIGHTS**.

Algorithm OVERWEIGHTS. This algorithm calculates the statistics described for a sample of 50 subjects. The integer array POUNDS, with 50 rows and 12 columns, is used to hold the input data. PY_AVG is a real variable used to compute the average weight change per subject per year; PM_AVG is a real variable used to compute the average weight change per subject per month. Y_COUNT and M_COUNT are integer variables for the number of subjects exceeding the yearly average change and the monthly average change, respectively. TEMP is a temporary integer variable. Integer variables SUBJECT and MONTH are used as loop indices.

1. [Initialize counts and totals]
   PY_AVG ← Y_COUNT ← PM_AVG ← M_COUNT ← 0
2. [Input data and compute total yearly weight change]
   Repeat for SUBJECT = 1, 2, . . ., 50
     Repeat for MONTH = 1, 2, . . ., 12
       Read(POUNDS[SUBJECT, MONTH])
     PY_AVG ← PY_AVG + (POUNDS[SUBJECT, 1]
       − POUNDS[SUBJECT, 12])
3. [Compute per subject yearly average]
   PY_AVG ← PY_AVG / 50
   Write('PER SUBJECT, PER YEAR WEIGHT CHANGE IS', PY_AVG)
4. [Count number exceeding average and determine total average monthly weight change]
   Repeat for SUBJECT = 1, 2, . . ., 50
     If (POUNDS[SUBJECT, 1] − POUNDS[SUBJECT, 12]) > PY_AVG
     then Y_COUNT ← Y_COUNT + 1
        TEMP ← 0
     Repeat for MONTH = 2, 3, . . ., 12
       TEMP ← TEMP + (POUNDS[SUBJECT, MONTH − 1]
         − POUNDS[SUBJECT, MONTH])
     PM_AVG ← PM_AVG + TEMP / 11.0
   Write('NUMBER OF SUBJECTS EXCEEDING THIS FIGURE IS', Y_COUNT)
5. [Compute per subject monthly average]
   PM_AVG ← PM_AVG / 50
   Write('PER SUBJECT, PER MONTH WEIGHT CHANGE IS', PM_AVG)
6. [Count number exceeding this average]
   Repeat for MONTH = 2, 3, . . ., 12
     Repeat for SUBJECT = 1, 2, . . ., 50
       If (POUNDS[SUBJECT, MONTH − 1] − POUNDS[SUBJECT, MONTH])
         > PM_AVG
       then M_COUNT ← M_COUNT + 1
   Write('NUMBER OF TIMES THIS FIGURE EXCEEDED IS', M_COUNT)
7. [Finished]
   Exit   □

   We leave the tracing of this algorithm on appropriate data values as an exercise.

4-5.3 GLOBAL HOCKEY LEAGUE

Inspired by the surge in popularity of international hockey tournaments, a group of entrepreneurs has formed the Global Hockey League. A worldwide campaign for franchises has netted the following 12 teams (listed alphabetically):

The Burma Shaves

The Chile Beans

The Hammond Eggs

The Kentucky Derbies

The Labrador Retrievers

The Louisiana Purchase

The Mobile Homes

The Peking Ducks

The Scotland Yards

The Toronto Island Ferries

The Trafalgar Squares

The Vichy Ssoise

To enable more efficient processing of game results, which are coming from all corners of the globe, the league organizers intend to use a computer. After completion of each league game, the result is sent to league headquarters as in the following example:

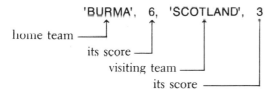

A team is awarded 2 points for each win, 1 point for each tie, and no points for a loss.

League standings are published regularly in the following form:

TEAM	GAMES PLAYED	WINS	LOSSES	TIES	POINTS
—	—	—	—	—	—
—	—	—	—	—	—

These statistics are computed from the most recent published standings, updated according to the recently arrived unprocessed game results. The teams are listed in decreasing order of points. We wish to design an algorithm to prepare and print the league standings.

To solve this problem we require two different arrays, since we have two types of values. The first will, in fact, be a vector of elements of type string. This will contain the names of the teams as they appear in the published standings (and in the individual game reports). The second array is a two-dimensional array with integer elements. Each row of this array contains the statistics for one of the teams in the league. The columns contain the number of games played, the number of games won, the number of games lost, the number of games tied, and the total points. (Another approach would be to use five separate vectors, one for each category, and process them in parallel. We have chosen to use a two-dimensional array instead.) Figure 4-8 shows the standings after the first round of play. The order is significant. The team from Labrador is currently in first place with a record of 6 wins, 2 losses, and 3 ties for a total of 15 points. Cold climates obviously breed a hockey tradition. The Labrador team is followed closely by the team from Toronto Island with a record of 6 wins, 3 losses, and 2 ties for 14 points. At the present time the teams from Scotland and Peking trail the field with identical records of 3 wins, 7 losses, and 1 tie for 7 points.

After each run of the program to compute and print the league standings, data cards are punched giving the team names in their current order, and their records. This serves as input to the next run, along with the results of the subsequent games.

The algorithm must operate as follows. First, the current standings are read in and used to construct the two essential arrays (team names and statistics). Attention then turns to the processing of the individual game results. The game cards are read, one at a time, and the statistics of the teams involved are updated accordingly. After all the game results have been proc-

Team		Games played	Wins	Losses	Ties	Points
LABRADOR	← →	11	6	2	3	15
TORONTO ISLAND	← →	11	6	3	2	14
CHILE	← →	11	5	3	3	13
TRAFALGAR	← →	11	6	4	1	13
HAMMOND	← →	11	6	5	0	12
VICHY	← →	11	5	4	2	12
KENTUCKY	← →	11	5	5	1	11
LOUISIANA	← →	11	5	5	1	11
MOBILE	← →	11	4	6	1	9
BURMA	← →	11	4	7	0	8
SCOTLAND	← →	11	3	7	1	7
PEKING	← →	11	3	7	1	7

FIGURE 4-8
Global Hockey League: standings after the first round of play.

Array of type string

Array of type integer

essed, the total points for each team are recalculated. Finally, a sort into descending order of total points must be performed and the standings printed. More formally, our algorithm has the following general structure:

1. Read and store current standings
2. Repeat for each new game result
   Process the game result
   Update team statistics
3. Recalculate total points for all teams
4. Sort arrays by points
5. Print out new standings

Let us consider each of these steps of the algorithm in more detail. The first step, the input and storing of current standings, is actually a straight-forward application of read statements. We will use a string vector **TEAMS**, with 12 elements, to hold the team names, and a two-dimensional integer array **STATS**, with 12 rows and 5 columns, to hold the statistics. Using loops, this can be expressed as

Repeat for ROW = 1, 2, . . ., 12
  Read(TEAMS[ROW]) (read the team name)
  Repeat for COLUMN = 1, 2, . . ., 5
    Read(STATS[ROW, COLUMN])   (read the data for this team)

Alternatively, if the data were punched in the appropriate order, our input statement could be simply

Read(TEAMS, STATS)

We will opt for the more complete representation. The processing of new game results requires some further elaboration. For each data card we must determine the teams involved in the game and the appropriate adjustments to make to their statistics. From the score given, we determine the winner of the game and the loser (or possibly it was a tie). We then find the appropriate rows in the statistics table by searching the list of names. For example, if the game result were

'BURMA', 6, 'SCOTLAND', 3

an examination of the scores tells us that Burma has won this game and Scotland has lost. We then search the list of names (Fig. 4-8) to find a match on 'BURMA', and discover that it is in row 10. We then add one more win to the statistics for this team by adding 1 to the array element STATS[10, 2]. We also increase STATS[10, 1] by 1 to indicate another game played. For 'SCOTLAND' we perform a similar process, updating STATS[11, 1] and STATS[11, 3] by one. We repeat this process until all game results have been read.

1. Repeat thru step 5 while game results remain to be processed
2. [Read game result]
   Read(TEAM1, SCORE1, TEAM2, SCORE2)
3. [Find positions of teams]
   Repeat for ROW = 1, 2, . . ., 12   (linear search)
     If TEAMS[ROW] = TEAM1
     then ROW1 ← ROW
     else If TEAMS[ROW] = TEAM2
        then ROW2 ← ROW
4. [Update games played]
   STATS[ROW1, 1] ← STATS[ROW1, 1] + 1
   STATS[ROW2, 1] ← STATS[ROW2, 1] + 1
5. [Determine game outcome]
   If SCORE1 > SCORE2
   then STATS[ROW1, 2] ← STATS[ROW1, 2] + 1   (team 1 wins)
        STATS[ROW2, 3] ← STATS[ROW2, 3] + 1   (team 2 loses)
   else If SCORE2 > SCORE1
      then STATS[ROW2, 2] ← STATS[ROW2, 2] + 1
           STATS[ROW1, 3] ← STATS[ROW1, 3] + 1
      else STATS[ROW1, 4] ← STATS[ROW1, 4] + 1   (tie game)
           STATS[ROW2, 4] ← STATS[ROW2, 4] + 1

The next step is to recompute the point totals for all the teams. For this, we simply run through each row of the array STATS, awarding 2 points for each win and 1 point for each tie. The following loop will do this:

Repeat for ROW = 1, 2, . . ., 12
  STATS[ROW, 5] ← 2 $*$ STATS[ROW, 2] + STATS[ROW, 4]
                      _____   _____
                         wins   ties

After the current point totals have been computed, the teams must be sorted into descending order of points. We will use the selection sort described earlier. We have an important additional requirement, however. Because the statistics relate to particular teams, we must keep all the rows intact and preserve the correspondence between elements of the array TEAMS and rows of the array STATS. This means that, even though our sorting is done on the total points alone (the fifth column of the STATS array), when it comes time to exchange elements (note the use of the exchange operator " ⇔ ") we will be moving not only all elements in that row of the STATS array, but also the corresponding element of the NAMES array.

1. [Loop on pass index]
   Repeat thru step 4 for PASS = 1, 2, . . ., 11
2. [Initialize minimum index]
   TOP ← PASS   (current top of unsorted vector)

3. [Find smallest key]
    Repeat for ROW = PASS + 1, PASS + 2, . . ., 12
      If STATS[ROW, 5] > STATS[TOP, 5] then TOP ← ROW
4. [Exchange rows]
    If PASS ≠ TOP then STATS[PASS, 1] ⇔ STATS[TOP, 1]
                   STATS[PASS, 2] ⇔ STATS[TOP, 2]
                   STATS[PASS, 3] ⇔ STATS[TOP, 3]
                   STATS[PASS, 4] ⇔ STATS[TOP, 4]
                   STATS[PASS, 5] ⇔ STATS[TOP, 5]
                   TEAMS[PASS] ⇔ TEAMS[TOP]

The final step is to print the new standings. This is no more than setting up the appropriate write statements. The analysis of this problem is now complete and we are ready to assemble the entire algorithm from the individual pieces that we have designed.

Algorithm STANDINGS. This algorithm computes and prints team standings for the Global Hockey League. Input consists of the most recently published standings and the results of games played since then. The output is the revised standings sorted in decreasing order of total points.

1. [Read in most recent standings]
    Repeat for ROW = 1, 2, . . ., 12
      Read(TEAMS[ROW])   (read the team name)
      Repeat for COLUMN = 1, 2, . . ., 5
        Read(STATS[ROW, COLUMN])   (read data for this team)
2. [Engage major loop for reading and processing new results]
    Repeat thru step 6 while there still remain game results
3. [Input game result]
    Read(TEAM1, SCORE1, TEAM2, SCORE2)
4. [Find positions of teams involved in game]
    Repeat for ROW = 1, 2, . . ., 12   (linear search)
      If TEAMS[ROW] = TEAM1
      then ROW1 ← ROW   (position of team 1)
      else If TEAMS[ROW] = TEAM2
        then ROW2 ← ROW   (position of team 2)
5. [Update games played for appropriate teams]
    STATS[ROW1, 1] ← STATS[ROW1, 1] + 1
    STATS[ROW2, 1] ← STATS[ROW2, 1] + 1
6. [Determine game outcome and adjust statistics accordingly]
    If SCORE1 > SCORE2
    then STATS[ROW1, 2] ← STATS[ROW1, 2] + 1   (team 1 wins)
        STATS[ROW2, 3] ← STATS[ROW2, 3] + 1   (team 2 loses)
    else If SCORE2 > SCORE1
      then STATS[ROW2, 2] ← STATS[ROW2, 2] + 1   (team 2 wins)
        STATS[ROW1, 3] ← STATS[ROW1, 3] + 1   (team 1 loses)

else STATS[ROW1, 4] ← STATS[ROW1, 4] + 1   (tie)
    STATS[ROW2, 4] ← STATS[ROW2, 4] + 1

7. [After processing all game results, compute current point totals]
    Repeat for ROW = 1, 2, . . ., 12
        STATS[ROW, 5] ← 2 * STATS[ROW, 2] + STATS[ROW, 4]

8. [Begin sorting by total points]
    Repeat thru step 11 for PASS = 1, 2, . . ., 11

9. [Find current top of unsorted elements]
    TOP ← PASS

10. [Find largest key among unsorted elements]
    Repeat for ROW = PASS + 1, PASS + 2, . . ., 12
        If STATS[ROW, 5] > STATS[TOP, 5]
        then TOP ← ROW

11. [Exchange rows]
    If PASS ≠ TOP
    then STATS[PASS, 1] ⇔ STATS[TOP, 1]
        STATS[PASS, 2] ⇔ STATS[TOP, 2]
        STATS[PASS, 3] ⇔ STATS[TOP, 3]
        STATS[PASS, 4] ⇔ STATS[TOP, 4]
        STATS[PASS, 5] ⇔ STATS[TOP, 5]
        TEAMS[PASS] ⇔ TEAMS[TOP]

12. [Output headings]
    Write('TEAM', 'GAMES PLAYED', 'WINS', 'LOSSES', 'TIES', 'POINTS')

13. [Output new standings]
    Repeat for TEAM = 1, 2, . . ., 12
        Write(TEAMS[TEAM], STATS[TEAM, 1], STATS[TEAM, 2],
            STATS[TEAM, 3], STATS[TEAM, 4], STATS[TEAM, 5])

14. [Finished]
    Exit   □

A final comment on this algorithm concerns the control of the major loop in steps 2 through 6. The intention as expressed in the loop control statement (step 2) is to perform this loop for each game result read. Once all the game results have been read and processed by repeating steps 3 through 6, control is to pass to step 7. The actual manner in which such a requirement is effected, using an end-of-file test, depends largely on the particular programming language being used. Once again, we do not wish to concern ourselves with such details in this book. As a result we will assume that the loop as it is expressed in this algorithm operates as required; specifically, that once the end of file is detected, control transfers automatically to the step immediately following the loop (step 7) and execution resumes from that point.

You will notice that this is different from the manner in which end-of-file tests were handled in Chap. 3. When algorithms become more elaborate, such that the end-of-file wrap-up operations become more involved, it is convenient to be able to flow through in this way and split up what must be done into separate steps that can be individually written and, thus, individually explained. Not only does this make the algorithm cleaner to write, it also makes it much easier to read and understand. We will employ similar techniques elsewhere in the book.

4-5.4   COMPUTER DATING SERVICE

Universal Dating Inc., a nationwide mate-matching organization, has gathered information on some romantically inclined young men and women.

Currently, each candidate fills out a questionnaire which contains several factors, such as sports, dancing, movies, classical music, parties, and world affairs. Each candidate is required to assess the degree of his/her feelings of like and dislike toward these factors. The rating of each factor is accomplished by associating a number with every factor, according to the following code of values:

1—intense dislike	5—mild like
2—moderate dislike	6—moderate like
3—mild dislike	7—intense like
4—neutral	

The input data for each candidate are recorded on one card according to the following format:

Name

Sex (1 = male, 2 = female)

Ratings of the factors by increasing factor number (that is, the first rating is for factor 1, the second rating is for factor 2, and so on)

The first card of the data is a header card which contains three data items:

Date

Number of candidates

Number of factors surveyed

The following sample input data describe the ratings obtained on three factors from five candidates.

Date	Number of Candidates	Candidate Name	Sex	Number of Factors	Ratings for Factors 1, 2, and 3, Respectively		
'MARCH 1, 1979'	5			3			
		'MARY MATCH'	2		1	6	5
		'TIM TALL'	1		2	4	5
		'FRED FUN'	1		2	5	7
		'GARY GALLANT'	1		1	6	7
		'LINDA LOVE'	2		2	4	6

Several applicant cards follow the header and candidate data cards. Each applicant card represents a date-matching inquiry. Such an inquiry involves the searching of the candidate data for those individuals who are "most suited" to the applicant. There are many ways of specifying what is meant by "most suited." One way of doing this is to compare the applicant's rating of each factor with that of each candidate. The square of the difference of the ratings for each factor can then be summed. A most suited individual could be a candidate with the least sum. This criterion is often referred to as *least squares*.

As an example, using the previous candidate data and an applicant with the following particulars

'BARBARA BEAUTY'   2   2   5   6

the least-squares sum for each candidate when compared with this applicant would be:

For **TIM TALL**: $(2 - 2)^2 + (5 - 4)^2 + (6 - 5)^2 = 2$

For **FRED FUN**: $(2 - 2)^2 + (5 - 5)^2 + (6 - 7)^2 = 1$

For **GARY GALLANT**: $(2 - 1)^2 + (5 - 6)^2 + (6 - 7)^2 = 3$

Since the smallest sum of squares is obtained for **FRED FUN**, he is deemed the individual most suited for **BARBARA BEAUTY**. Note that, in general, it is possible to have more than one most suited individual. The general algorithm for this problem follows.

1. Read in the date, number of candidates, and number of factors
2. Repeat for each candidate
     Read in the candidate's profile
3. Repeat thru step 8 for each applicant

4. Read in applicant's profile
5. Repeat for each candidate
    Compute the candidate's sum-of-squares statistic
6. Determine the minimum sum-of-squares statistics
7. Repeat for each candidate
    Select those candidates who have the minimum statistics
8. Output the desired report for current applicant

The factor ratings given by the candidates can be stored in a two-dimensional array, RATING, each row of which represents the ratings of a particular candidate. Each element in a particular row contains the rating for a certain factor. Furthermore, we store the candidate names and sexes in vectors NAME and SEX, respectively. Using these representations, our sample candidate data given earlier become

NAME	SEX	RATINGS		
'MARY MATCH'	2	1	6	5
'TIM TALL'	1	2	4	5
'FRED FUN'	1	2	5	7
'GARY GALLANT'	1	1	6	7
'LINDA LOVE'	2	2	4	6

The factor data on an applicant's card is stored as a vector called FACTOR. In the previous applicant, the representation is

    FACTOP
2   5   6

The previous comments are incorporated in the following detailed algorithm.

Algorithm DATING. Given the date, DATE, the number of candidates, N; the number of factors, M; the candidate ratings, name, and sex, which are represented by an array RATINGS and vectors NAME and SEX, respectively, as just described, this algorithm generates from the candidates those that are most suited to an applicant's profile. Each applicant's profile contains a name (APPLICANT_ID), sex (A_SEX), and set of ratings (vector FACTOR). The vector STAT represents the sum-of-squares statistics for the candidates as described earlier. The positions (that is, row numbers) of those candidates who have the smallest sum-of-squares statistics are stored in the integer vector BEST. The integer variable MIN contains the least sum of squares. I and J are index variables. The integer variable ACCEPT contains the number of candidates which have the least sum of squares.

1. [Input date, number of candidates, and number of factors]
    Read(DATE, N, M)

2. [Input the candidate profiles]
   Repeat step 3 for I = 1, 2, . . ., N

3. [Input the name, sex, and ratings of one candidate]
   Read(NAME[I], SEX[I])
   Repeat for J = 1, 2, . . ., M
       Read(RATINGS[I, J])

4. [Process all applicants]
   Repeat thru step 9 while there still remains an applicant

5. [Read in applicant's profile]
   Read(APPLICANT_ID, A_SEX)
   If there is no more data then Exit
   Repeat for J = 1, 2, . . ., M
       Read(FACTOR[J])

6. [Compute the candidate's sum-of-squares statistics]
   Repeat for I = 1, 2, . . ., N   (initialize statistics)
       STAT[I] ← 0
   Repeat for I = 1, 2, . . ., N
       If A_SEX ≠ SEX[I]
       then Repeat for J = 1, 2, . . ., M
               STAT[I] ← STAT[I] + (RATINGS[I, J] − FACTOR[J])↑2

7. [Determine the minimum sum-of-squares statistic]
   I ← I
   Repeat while A_SEX = SEX[I]
       I ← I + 1
   MIN ← STAT[1]
   Repeat for I = 2, 3, . . ., N
       If A_SEX ≠ SEX[I]
       then If STAT[I] < MIN then MIN ← STAT[I]

8. [Select those candidates with smallest statistic]
   ACCEPT ← 0
   Repeat for I = 1, 2, . . ., N
       If A_SEX ≠ SEX[I]
       then If STAT[I] = MIN
               then ACCEPT ← ACCEPT + 1
                   BEST[ACCEPT] ← I

9. [Output the desired report for this applicant]
   Write('DATE IS', DATE)
   Write('APPLICANT'S NAME IS:' APPLICANT_ID)
   Write('THE BEST MATCHED CANDIDATES ARE:')
   Repeat for I = 1, 2, . . ., ACCEPT
       Write(NAME[BEST[I]])   □

The first step of the algorithm inputs the first card, which contains the date, number of candidates, and the number of factors which were rated. Steps 2 and 3 input all candidate names, sex, and associated factor ratings. The fourth step engages the loop, which controls the processing of all applicants. Step 5 reads in an applicant's name, sex, and factor ratings. Step 6 computes the sum-of-squares statistic associated with a candidate of the opposite sex. These statistics are stored in the vector STAT. We select the smallest statistic (MIN) in step 7. Step 7 scans the statistic of the opposite sex (that is, STAT[I]) and compares this value against the value of MIN. If the candidate's statistic is the smallest value, then the position or index (that is, row number) of this particular candidate is copied into the integer vector BEST. Step 9 displays the desired results.

CHAPTER EXERCISES

1. Design an algorithm which inputs a class list of student names and grades and outputs a list of student names and grades in descending order of grades.

2. (*Radix sort.*) The radix sort is a method of sorting which predates any digital computer. This was performed and is still performed on a mechanical card sorter. Such a sorter usually processes a standard card of 80 columns, each of which may contain a character of some alphabet. When sorting cards on this type of sorter, only one column at a time is examined. A metal pointer on the sorter is used to select any one of the 80 columns. For numerical data the sorter places all cards containing a given digit into an appropriate pocket. There are 10 pockets, corresponding to the 10 decimal digits. The operator of the sorter combines in order the decks of cards from the 10 pockets. The resulting deck has the cards of pocket 0 at the bottom and those of pocket 9 on top. In general, numbers consisting of more than one digit are sorted. In such a case an ascending-order sort can be accomplished by performing several individual digit sorts in order. That is, each column is sorted in turn, starting with the lowest-order (rightmost) column first and proceeding through the other columns from right to left. As an example, consider the following sequence of numbers (one number on each card):

   42, 23, 74, 11, 65, 57, 94, 36, 99, 87, 70, 81, 61

   After the first pass on the unit-digit position of each number, we have

		61								
		81			94			87		
	70	11	42	23	74	65	36	57		99
Pocket:	0	1	2	3	4	5	6	7	8	9

   Now by combining the contents of the pockets such that the contents of the 0 pocket are on the bottom and the contents of the 9 pocket are on the top, we obtain

   70, 11, 81, 61, 42, 23, 74, 94, 65, 36, 57, 87, 99

On the second pass, we sort on the higher-order digit, yielding

					65	74	87	99
	11	23	36	42	57	61	70	81	94

Pocket: 0 1 2 3 4 5 6 7 8 9

By combining the 10 pockets in the same order as in the first pass, we complete the sort. This type of sort is called a *radix sort*.

This mechanical method of sorting can be implemented on a computer. If the maximum number of digits in a key is m, then m successive passes, from the unit digit to the most significant digit, are required in order to sort the numbers. Formulate an algorithm which performs the radix sort. Assume that a key K contains m digits of the form $b_m b_{m-1} \cdots b_2 b_1$ and that a selection mechanism is available for selecting each digit.

3. It has been discovered by the leaders of two international espionage organizations (called Control and Kaos) that a number of employees are on the payrolls of both groups! A secret meeting is to be held for loyal employees of Control and Kaos (that is, excluding those on both payrolls) to determine a suitable course of action to be taken against the "double agents." Design an algorithm which will accomplish the following task. Read as input two alphabetically ordered lists of names, one name per data card, the first list containing the names of agents on the Control payroll and the second containing names of agents on the Kaos payroll. (Each of the two lists is followed by a card with the name 'ZZZZ'.) Then scan the two lists together and print in alphabetical order the names of those agents who should be invited to the proposed meeting (that is, all those whose name appears on one list but not on both).

The following is an example input:

Example input:

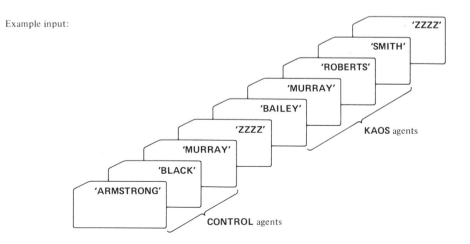

Corresponding example output:

ARMSTRONG

BAILEY

BLACK

ROBERTS

SMITH

4. At any large school or university the task of drafting an exam timetable is both difficult and time consuming. An aid to the development of an exam timetable is an algorithm which would "check out" all students against a tentative exam timetable and determine if any exam conflicts exist (an exam conflict means the student writes more than one exam at any one time). Input to the algorithm consists of the tentative exam timetable and student records indicating the classes taken by each student. This input is punched on cards as follows:

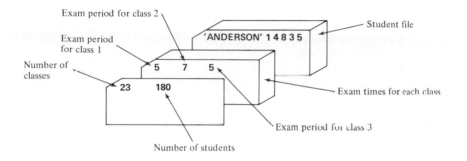

The first card contains two numbers: the number of classes and the number of students. The next set of cards indicates the exam period for each class, with the first card indicating the exam periods for classes 1 to 10; the second card indicating the exam periods for classes 11 to 20, and so on. The final set of cards consists of the student's name and the number of the classes this student takes. Each student always takes five classes. In the example illustrated, Anderson takes classes numbered 1, 4, 8, 3, and 5. Both classes 1 and 3 have been scheduled for exam period 5; therefore, Anderson has an exam conflict. Your algorithm must print the names of all students who have exam timetable conflicts.

5. Let **NAME** and **SEX** be vectors that contain the name and sex of each member of a youth club. Male and female are denoted by **'M'** and **'F'**, respectively. Formulate an algorithm which creates two new vectors, **MALES** and **FEMALES**, such that **MALES** contains the names of all males in alphabetic order and **FEMALES** contains the names of all females in alphabetic order.

6. Assume that the Saskatchewan Real Estate Board has conducted a survey of each of its licensees. Each licensee filled in a questionnaire of the following form:

Item	Answer Code
License type	1 = broker
	2 = salesperson
Residence town	840 towns coded from 1 to 840
Age	Age in years
Sex	1 = male
	2 = female
Education	1 = less than high school diploma
	2 = high school diploma
	3 = technical institute or community college
	4 = college degree

Develop an algorithm which analyzes these questionnaires. In particular, calculate the following for the group of respondents:
(a) Total number of respondents.
(b) Percentage brokers and percentage salespersons.
Calculate the following separately for brokers and salespersons:
(a) Number of respondents from each town.
(b) Average age.
(c) Percentage male and percentage female.
(d) Number of respondents in each educational classification.
The input data for each questionnaire consist of five questionnaire answer codes, representing license type, residence town, age, sex, and education. Use an end-of-file test to determine the end of the data.

7. Universal Exports keeps a computer file of its inventory. Each record contains the following items:

    product number,   product name,   supplier name,   unit price

These records are kept in a product number order. A number of queries are made concerning these inventory records. Each query takes the form

    product name,   product number

For each query the corresponding inventory record must be found, and the product number, product name, and supplier name must be reported. The query data are kept in no particular order. The data are organized as follows:

inventory data

'**' ← end of inventory data

query data

Formulate an algorithm which processes these data and generates the required report. Use an end-of-file test to detect the end of the query data.

8. Students at the University of Saskatchewan are classified by class and college as follows:

Class	Code	College	Code
Freshman	1	Agriculture	1
Sophomore	2	Arts and science	2
Junior	3	Commerce	3
Senior	4	Dentistry	4
Graduate	5	Education	5
		Engineering	6
		Graduate studies and research	7
		Home economics	8
		Law	9
		Medicine	10
		Nursing	11
		Pharmacy	12
		Physical education	13
		Veterinary medicine	14

Each student is assigned a class code and a college code. Construct an algorithm which inputs the student data and determines how many students of each class have enrolled in each college. An end-of-file test is to be used to detect the end of the data.

9. Students from 50 New Brunswick high schools take college entrance examinations. A tabulation is desired showing the scores of people from each high school that took the examination. The input has a student identification number, his/her high school code (from 1 to 50), and his/her examination score. The output is to be a tabulation showing the scores of the students from each high school. The scores from each high school should be sorted in descending order. The input is arranged by high school code. No more than 100 students per school took the examination. A sample of the output format follows:

HIGH SCHOOL	STUDENT NUMBER	SCORE	
1	—	—	descending order
	—	—	
	.	.	
	.	.	
	.	.	
	—	—	
2	—	—	
	—	—	
	.	.	
	.	.	
	.	.	
	—	—	
	.	.	
	.	.	
	.	.	
	—	—	
50	—	—	
	.	.	
	.	.	
	.	.	
	—	—	

10. A profile of student attitudes towards a certain course is being determined by a method that requires each student in the course to assess the degree of his/her feelings of like or dislike toward certain factors. Each student is requested to rate each factor by associating a number with every factor, according to the following scale of values:

1—intense dislike   5—mild like

2—moderate dislike   6—moderate like

3—mild dislike   7—intense like

4—neutral

It is required to formulate an algorithm which will perform a simple analysis of these data and produce a report giving:

(a) The average rating of each factor.

(b) Information on the students whose ratings are "closest" to the average ratings. (More will be said about what is meant by "closest" shortly.)

The input data are punched on cards and consist of one header card followed by a number of survey cards, one for each student surveyed. The header card contains three data items:

name of the course, number of students surveyed, number of factors surveyed

Each survey card contains a student's identification number and this student's ratings of the factors by increasing factor number. That is, the first rating is for factor 1, the second rating is for factor 2, and so on. The sample input data describe the ratings obtained on three factors from five students in course **CMPT 180A**.

Course Name	Number of Students	Student Number	Number of Factors	Ratings for Factors 1, 2, and 3, Respectively		
'CMPT 180A'	5		3			
		10175		1	6	5
		12791		2	4	5
		9981		2	5	7
		38005		1	6	7
		27091		2	4	6

As mentioned earlier, the algorithm must produce the average rating for each factor in the course and output the student or students whose ratings are "closest" to the average ratings. The "closeness" is measured by the statistic S_j, which is computed for the jth student from his/her ratings for the various factors and the average for each of these factors. The lower the S_j value for a student, the "closer" the student's ratings are to the averages. S_i is defined as follows:

$$S_i = \sqrt{\sum_{i=1}^{n} (r_{ij} - \bar{r}_j)^2}$$

where S_i = "closeness" statistic for person i

n = number of factors surveyed

r_{ij} = ith student's rating of the jth factor
$\bar{r}_j$ = average rating of the jth factor

The output report is to consist of:

(a) Name of the course.
(b) The "smallest" (that is, the smallest S_j) statistic for the course.
(c) The average rating for each factor.
(d) The identification numbers of those students "closest" to the averages.
(e) For each of these "closest" students, the difference between each of their ratings and the corresponding averages.

A sample report for the previous input data follows:

ATTITUDE REPORT FOR CMPT 180A

SMALLEST S: 1.077

FACTOR	AVERAGE VALUE
1	1.60
2	5.00
3	6.00

CLOSEST STUDENTS

STUDENT NUMBER: 9981

FACTOR	DIFF. FROM AVG.
1	0.40
2	0.00
3	1.00

STUDENT NUMBER: 27091

FACTOR	DIFF. FROM AVG.
1	0.40
2	−1.00
3	0.00

11. The College of Arts and Science wishes to determine the age distribution of the faculty members in its various departments. In particular, they want to know for each department how many faculty members are in each of the following categories:

<20	40–49
20–29	50–59
30–39	>59

The following data have been prepared for the program. The first card gives the number of departments in the college. This is followed by the names of the departments in alphabetical order. These names are in quotes. After all the department names comes the information on the individual faculty members. This information consists of the faculty member's name (in quotes), the name of the department in which the faculty member is located, and the faculty member's age. This information is in alphabetical order by the name of the faculty member. The following is a set of sample data:

```
33
'ANATOMY'
'ANTHROPOLOGY'
        .
        .
        .

'SOCIOLOGY'
'ABBOTT'          'HISTORY'          37
'ACKERMAN'        'PSYCHOLOGY'       53
     .                 .              .
     .                 .              .
     .                 .              .

'ZOOK'            'ART'              42
'END'             'DATA'             0
```

Give an algorithm that will use these data to output the age distribution for each department.

The output should have the following format:

DEPARTMENT	AGE CATEGORIES					
	<20	20–29	30–39	40–49	50–59	>59
ANATOMY	0	2	5	4	2	2
ANTHROPOLOGY	0	1	2	4	2	1
.						
.						
.						
SOCIOLOGY	0	4	5	6	4	1

5

Strings and Things

Computers have traditionally been associated with the solution of numerical problems such as summations, the calculation of the roots of an equation, and so on. In general, the primary objective of numerical computations is to obtain a numerical answer. In recent years, however, a good deal of interesting work has been done using computers for essentially nonnumeric problems such as sorting, translation of languages, text editing, pattern recognition, and symbolic manipulation of mathematical equations. Numerical answers occur infrequently in such problems. The major attributes of the computer used for these applications are decision-making and storage capability rather than its ability to do arithmetic.

Although the concept of a string was brought up in Chap. 2, this chapter formally introduces the basic notions of string processing. The first section deals with character information. A character is considered to be a primitive structure on most machines. The internal representation of characters in the computer's memory has important consequences in computing. Section 5-2 formally introduces the concepts of a string in terms of characters. The assignment of a string value to a variable and the rules for comparing strings are examined. In Sec. 5-3 our algorithmic notation is expanded to include a number of primitive string manipulation functions. The need for these functions is motivated by considering certain string-oriented applications. Finally, several simple string applications are discussed.

5-1 CHARACTER INFORMATION

In Sec. 4-1 we introduced the notion of a primitive data structure. Recall that a data structure was said to be primitive if the computer had machine-level instructions which permitted the manipulation of such a structure. The integer and real number are two familiar examples of primitive data structures. Another, perhaps not so familiar, is the character. Since a string is an ordered sequence of characters, the character is the fundamental entity in string manipulation.

This section is therefore concerned with the character, or more generally with a finite set of characters—an *alphabet*. An example of an alphabet is the set of English letters. Another common example is the set of decimal digits. We will give examples of certain popular alphabets which have been used in the computing industry. Furthermore, the representation (or coding) of characters within an alphabet will be described. Such representations are important to the understanding of the comparison of two strings.

The first computers introduced in Chap. 1 were actually sophisticated calculators in the sense that they were only capable of handling numeric data. Even the first computer programs were themselves written in a strictly numeric form (that is, machine code). It was soon realized that machine-code programming was cumbersome and the programs were hard to read and correct. To overcome this problem, the use of character data eventually led to the development of modern problem-oriented languages such as FORTRAN and PL/I.

Many character sets (or alphabets) have been designed for computer use over the years. Two of the most popular and largest character sets are those known as EBCDIC (Extended Binary Coded Decimal Interchange Code) and ASCII (American Standard Code for Information Interchange). EBCDIC is a character-coding system used primarily on the IBM/360–370 series of computers. ASCII was developed as a standard coding scheme for the computer industry and is used on many non-IBM machines. The character sets provided by these two coding schemes are as follows:

ASCII character set:

1. English alphabet in both lowercase and capital letters {a, b, c, . . ., z, A, B, C, . . ., Z}
2. Decimal number characters {0, 1, 2, 3, 4, 5, 6, 7, 8, 9}
3. Operation and special characters {+, −, *, /, >, =, <, |, space (SP), !, ", #, $, %, &, ', (,), ,, ., :, ;, ?, @, [,\,], ↑, →, ", {,}, ~}
4. Control characters such as DEL (Delete or rub out), STX (Start of text), ETX (End of text), ACK (Acknowledge), HT (Horizontal tab), VT (Vertical tab), LF (Line feed), CR (Carriage return), NAK (Negative acknowledge), SYN (Synchronous idle for synchronous transmission), ETB (End of transmission block), FS (File separator), GS (Group separator), and RS (Record separator)

EBCDIC character set:

(1) through (3) above, plus control characters which, although having different mnemonics and names than those given in (4), perform the same control functions.

Let us examine these character sets in terms of the function they commonly perform. English characters, decimal number characters, and special characters can be combined to form English text. Computer applications involving natural language text are both wide-ranging and numerous. Of course, numbers are most often used in computations. Characters which are operational in nature (for example, +, −, *, /, =) are commonly used in programs to represent operations in the programming language such as addition, subtraction, multiplication, and division. A discussion of control characters is beyond the scope of this book.

Character sets have been created for special-purpose computer applications. Many computer graphic systems use operational-type characters for the manipulation of points and lines on a cathode-ray-tube display device. For example, special characters have been used to rotate, translate, enlarge, or contract pictures on the screen. Another example of a special character set is the character set used in the APL (A Programming Language) programming system. APL, a mathematically oriented programming language designed originally by K. Iverson, is very effective for the manipulation of mathematical items such as vectors and arrays. The APL character set includes the capital letters from the English alphabet, the decimal numerals, the special characters which are included in the EBCDIC and ASCII character sets, some Greek letters (α, Δ, ϵ, ι, ρ, ω), and a number of mathematically oriented symbols ($\subset$, $\supset$, $\cap$, $\cup$, $[,]$, $\perp$, $\top$, $\downarrow$, $\rightarrow$, $\div$, $\times$, $\ddot{,}$, $\leqslant$, $\geqslant$, $\neq$, $\circ$, $\square$, $\circ$), which include special operators on vectors and arrays. Many other "nonstandard" alphabets exist for special-purpose applications, but it is beyond the scope of this text to document these cases.

We now turn to the storage representation of character data. A character can be represented in memory as a sequence of bits (that is, a sequence of 0's and 1's), where a distinctive bit sequence is assigned to each character in the character set by the coding convention chosen. It is a generally adopted policy by computer manufacturers that character sets are encoded in fixed-length bit sequences. That is, every character in a character set is represented by the same number of bits.

It can be shown that a bit sequence of length n could be used to represent x unique characters, where $\log_2 x = n$. Therefore, 2^n characters can be represented using bit sequences of length n. Consequently, with $n = 7$ and 8, we can encode up to 128 and 256 characters, respectively. These correspond to the ASCII and EBCDIC character sets, respectively.

Earlier in this section, we introduced a number of character sets which have been designed for use on computers. Most of these character sets include the decimal numerals, the letters from a natural language alphabet, punctuation characters, and arithmetic operators. Unfortunately, the bit codes for representing identical or nearly identical character sets have not been standardized in spite of a considerable effort toward standardization in the computer industry. Some of the more commonly used codes for the FORTRAN character set are listed in Table 5-1 (we have chosen this particular character set since the FORTRAN language, and hence its character set, has been standardized and is available on a wide variety of computers). The Hollerith code has become a standard code for representing information on punched cards. This code can be handled on a wide variety of machines.

Let us now examine the EBCDIC bit representation of a number of characters more closely. For example, the representation of the character 'A' is the bit sequence.

$$1 \quad 1 \quad 0 \quad 0 \quad 0 \quad 0 \quad 0 \quad 1$$
$$2^7 \quad 2^6 \quad 2^5 \quad 2^4 \quad 2^3 \quad 2^2 \quad 2^1 \quad 2^0$$

where the weight of each binary digit is given below that digit. This bit representation can be interpreted as a binary number. The decimal equivalent of the bit representation for 'A' is therefore

$$1 \times 2^7 + 1 \times 2^6 + 0 \times 2^5 + 0 \times 2^4 + 0 \times 2^3 + 0 \times 2^2 + 0 \times 2^1 + 1 \times 2^0$$
$$= 128 + 64 + 1 = 193$$

In a similar manner the decimal number associated with the character 'B' in the EBCDIC code is 194. Continuing in this fashion the decimal equivalent of 'Z' is 233. For character comparison purposes we have 'A' < 'B' < 'C' < ... < 'Y' < 'Z'. It is also easily seen that the decimal number equivalent of 0 (240) is less than that of 1 (241), and so on. Therefore, for comparison purposes the number associated with any letter is less than the number associated with any digit. Finally, the blank character has a weight (64) which is less than any other character in the character set. For comparison purposes, the ordering of the FORTRAN character set given in Table 5-1 is blank.(+$*) −/,=A through Z, 0 through 9. This ordering of a character set is called the *collating sequence* of the character set and is important in applications involving sorting. This ordering on characters will be extended to strings in the next section.

In the earlier discussion in this section, it was pointed out that the introduction of character data was a necessary step toward the development of high-level languages. However, it is interesting to note that many of the first languages, such as FORTRAN and ALGOL 60, accommodate character data only to the limited extent of allowing the programmer to annotate his/her output with literal text. In languages which were developed later, such as SNOBOL, PL/I, and ALGOL 68, character strings, together with instructions to manipulate string data, were provided.

While it is worthwhile to investigate characters as primitive data structures, they are in many ways too primitive to be useful in expressing much of the nonnumeric information which can be processed by a computer. It was not purely coincidental that the designers of languages such as PL/I chose the character string as the basic data type rather than the character. The reasoning behind this design decision will be further illustrated in the next section.

5-2 STRING CONCEPTS AND TERMINOLOGY

In this section we want to extend the notion of strings introduced in Chap. 2. Initially, we draw an analogy between the natural number system and a string system. In so doing the operation of concatenation on strings will be introduced. The assignment statement is then expanded to incorporate the assignment of string values. Finally, the comparison of strings with respect to the relational operators introduced in Chap. 2 is then completed.

TABLE 5-1   Some Character Codes in Common Use for the
FORTRAN Character Set

Character	370 EBCDIC (Bit Representation)	ASCII	Hollerith Punch Positions
A	11000001	1000001	12-1
B	11000010	1000010	12-2
C	11000011	1000011	12-3
D	11000100	1000100	12-4
E	11000101	1000101	12-5
F	11000110	1000110	12-6
G	11000111	1000111	12-7
H	11001000	1001000	12-8
I	11001001	1001001	12-9
J	11010001	1001010	11-1
K	11010010	1001011	11-2
L	11010011	1001100	11-3
M	11010100	1001101	11-4
N	11010101	1001110	11-5
O	11010110	1001111	11-6
P	11010111	1010000	11-7
Q	11011000	1010001	11-8
R	11011001	1010010	11-9
S	11100010	1010011	0-2
T	11100011	1010100	0-3
U	11100100	1010101	0-4
V	11100101	1010110	0-5
W	11100110	1010111	0-6
X	11100111	1011000	0-7

Throughout this section and the next, we are concerned with what kind of operations can be performed on strings. We certainly expect those operations to be drastically different from the familiar arithmetic operations on numbers. Many string operations, however, are similar to some of their arithmetic counterparts. Let us examine some interesting properties for arithmetic operations over the natural numbers. To refamiliarize ourselves with some of the properties associated with arithmetic operations, let us consider the operation of addition on the natural numbers. This operation can be represented in general, by a functional system in two variables:

$$f(x,y) = x + y$$

where x and y are natural numbers. This system is well known to us and it exhibits certain interesting properties. First, the sum of any two numbers is a natural number. This property is called *closure*. Closure is a necessary prop-

TABLE 5-1   Some Character Codes in Common Use for the
FORTRAN Character Set (Continued)

Character	370 EBCDIC (Bit Representation)	ASCII	Hollerith Punch Positions
Y	11101000	1011001	0-8
Z	11101001	1011010	0-9
0	11110000	0110000	0
1	11110001	0110001	1
2	11110010	0110010	2
3	11110011	0110011	3
4	11110100	0110100	4
5	11110101	0110101	5
6	11110110	0110110	6
7	11110111	0110111	7
8	11111000	0111000	8
9	11111001	1011001	9
+	01001110	0101011	12-6-8
−	01100000	0101101	11
*	01011100	0101010	11-8-4
/	01100001	0101111	0-1
Blank	01000000	0100000	Space
(	01001101	0101000	12-5-8
)	01011101	0101001	11-5-8
$	01011011	0100100	11-8-3
=	01111110	0111101	8-8
,	01101011	0101100	0-8-3
.	01001011	0101110	12-8-3

erty for a system (that is, a set and an operation on that set) to be classified as an algebra or algebraic system. Second, expressions such as $(x + y) + z = x + (y + z) = x + y + z$ are equivalent when x, y, and z are natural numbers; the operation of addition is said to be *associative*. Third, there exists a number i such that for every natural number x, $x + i = x$. This number is zero and is called the unit element or *identity* of the additive system. There are many other important properties, such as distributivity and cummutativity, which exist when arithmetic operations such as addition and multiplication are applied to the set of natural numbers.

We have already introduced the notion of a string in Chap. 2. A string can be defined more formally in terms of an alphabet, a notion which was the topic of discussion in the previous section. A string is merely an ordered sequence of characters, each of which is a member of an alphabet. Examples of strings over the alphabet [X, Y, Z] are 'X', 'XY', 'XXYYZ', and 'ZYX'. A string can contain no characters at all. Such a string is denoted by the string

" and is called the *empty string* or *null string*. We shall use the symbol □ to denote the space character when it is not otherwise clear that it is part of a given string. Note that the string '□' is a string which contains the blank character and is not to be confused with the empty string.

Let us now turn to the manipulation of character strings. Here the operations may be less familiar than the normal numeric operations, but as we will see, they are not difficult. A very basic operation on character strings is to take two character strings and join them together to make one string. This operation is known as *concatenation*. We will denote the concatenation operator by the symbol "○". Thus, if we have two strings, 'SASKA' and 'TOON', the result of the operation. 'SASKA'○'TOON' is the new character string 'SASKATOON'. This operation is something like combining two piles of leaves on the lawn to make one bigger pile.

The empty string acts as the identity with respect to concatenation, that is, for any string x over an alphabet, x○"="○x = x. Associativity is another property of strings with respect to concatenation; that is, for any strings x, y and z, (x○y)○z = x○(y○z) = x○y○z. Finally, it is obvious that for any strings x and y, x○y will yield a string. So the system of strings under concatenation is closed. Therefore, the system of strings under concatenation behaves in a manner similar to the set of natural numbers under addition.

The concatenation operator, like the arithmetic operators, can be applied several times in one expression. Thus, the result of

'MIC' ○ 'KEY' ○ '□MOUSE'

is 'MICKEY MOUSE'. Similarly, the expression 'EDGAR□' ○ 'ALAN□' ○ 'POE' gives the string 'EDGAR ALAN POE'.

We now turn to the assignment of a string value to a variable. If CITY is a string variable, then the statement

CITY ←'SASKA' ○ 'TOON'

will assign the string value 'SASKATOON' to CITY. Any string variable can assume a string value whose length is a finite number. The *length* of a string is the number of characters in that string. Some programming languages require that some maximum length be specified for each string variable declared. In our algorithmic notation we will not abide by this constraint. In general, a string expression can contain string variables as well as string constants such as 'SASKA' or 'TOON'. For example, the sequence of assignment statements

A ←'COMPUTE'
B ←'SCIENCE'
C ← A ○ 'R□' ○ B

shows each type of element where A, B, and C are string variables, and results in a value for C of 'COMPUTER SCIENCE'.

We now turn to the comparison of strings. In the earlier chapters the notion of testing strings for equality and inequality was introduced in an informal manner. We now wish to elaborate further on these two operators and extend the comparison of strings to the other relational operators. A test for string equality, for string variables $x = x_l \ldots x_n$ and $y = y_l \ldots y_m$, is of the form $x = y$. The condition $x = y$ is considered to be true if the following holds:

1. The number of characters in x and y are identical (that is, $n = m$)
2. $x_i = y_i$ for all $1 \le i \le n$

For example, 'JOHN' = 'JOHN' is true while 'BILL' = 'BILLY' and 'BILLY' = 'BULLY' are false. The inequality relation is the negation of equality. For example, 'BILL' $\neq$ 'BILLY' and 'JOHN' $\neq$ 'JOHN□' are both true.

It is easy to expand this comparison feature to include the other relational operators, such as $<$, $\le$, $>$, and $\ge$. The meaning of these comparisons is based on the collating sequence of a character set which was introduced in the previous section. In our notation we shall use the character set which is applicable to most card readers. The collating sequence for this character set is assumed to be: □¢.<(+|&!$*);-/.%_>?:#@'="A through Z, 0 through 9. This sequence is based on the internal representation of the characters, which was discussed in the previous section. The ordering on strings is similar to the one found in a dictionary or a telephone directory. For example, 'BILL' $<$ 'BILLY', 'ANN' $<$ 'JOAN', and 'JONES' $\ge$ 'BUNT' are all true, while 'TREMBLAY' $<$ 'BUNT', 'COMPUTER' $>$ 'SCIENCE', and 'BOB' $\le$ 'ALAN' are all false. From these examples, it is obvious that the condition is tested by making a sequence of character comparisons in a left-to-right manner. Note that the presence of any character (even a space) is always considered to be greater than the omission of a character. For example,

'SCIENCE□' $>$ 'SCIENCE'

is true.

The comparison of character strings is very important in the sorting of character data. Such a sorting operation is required in many data-processing and string-manipulation applications. The selection sort algorithm of Sec. 4-3.1 can be used directly with a vector of strings such as a list of student names.

So far, the only string operation that we have introduced is that of concatenation. Clearly, if we are going to solve string-manipulation problems, we need a greater variety of string operators than this. The next section is concerned with the introduction of additional string operations.

5-3 BASIC STRING OPERATIONS

In the previous section we discussed the operations of concatenating two strings and testing two strings for equality. In addition to these operations, a close examination of the basic string-handling facilities required of any string-manipulation system would probably include the following list of primitive operations:

1. Create a string of text
2. Compute the length of a string
3. Search and replace (if desired) a given substring within a string

In this section we discuss the importance of each of these three operations in a string-handling system and incorporate into our algorithmic notation operations which effect these operations.

The creation of a string implies not only the ability to construct a representation for a string, but also the ability to retain the value of a string in a variable (or memory cell location). The ability to create a string must be present in any string-handling system.

For many applications it is useful for us to be able to find the length of a given string. This operation, which we will denote by the built-in function LENGTH, gives us this information. For example, the result of LENGTH('FOR WHOM THE BELL TOLLS') is 23. At this point it is interesting to note one important fact about this function. Although the argument to this function is a string, the result is an integer. In all previous cases the results of an operation had the same type as the operands. Since the result is numeric it can be used as part of an arithmetic expression. For example, the expression 3 + 2 + LENGTH('LES FEUX FOLLETS') has the result 21.

Another important string operation is one which allows us to extract a specified portion of a given string. In a way this is the reverse of the concatenation operation that combines strings to make a larger string. This new operation allows us to take smaller strings from a larger string.

The operation is known as the *substring* operation. There are several possible formats for this operation. We will adopt the following three argument built-in function format:

SUB(a_1, a_2, a_3)

where

a_1 is the string from which we want the piece to be taken

a_2 is the number in the original string of the position at which the desired piece begins

a_3 is the length of the desired piece

Note that a_1 is a string argument and a_2 and a_3 are integers. Any or all of these may be expressions themselves.

As an example, consider the following expression:

SUB('EDMONTON, ALBERTA', 6, 3)

Note the three arguments:

a_1 is the string 'EDMONTON, ALBERTA'

a_2 is the number 6, indicating that we wish to extract a substring beginning at the sixth position (the character 'T')

a_3 is the number 3, indicating that our substring will be three characters in length

Thus, the result of the indicated operation is the string 'TON'. Try one yourself. What is SUB('VANCOUVER, B.C.', 3, 5)?

As a special case of this operation it is possible to omit the third argument. If this is done, it is assumed that the desired substring begins at the position indicated by the second argument and runs through to the end of the original string. This may save us some writing in a number of cases. For example, the result of the expression SUB('ENDING', 4) is the string 'ING'. The same result would come from the expression SUB('ENDING', 4, 3).

Just as we can develop interesting and useful arithmetic expressions by combining the numeric operations (for example, $6.0 / 3 + 5 * 4$), so, too, can we combine string operations into one expression. Consider the following expression, for example:

SUB('EDMONTON', 3, 4) ∘ SUB('PORT CREDIT', 7, 2) ∘ SUB('CALGARY',
    2, 2)

This expression has three separate substring components or subexpressions, the results of which are concatenated to form the final result of the expression. Let us take the subexpressions one at a time

    length 4
SUB('EDMONTON', 3, 4)
    ↑
    position 3

The value of this subexpression is the string 'MONT'.

     length 2
SUB('PORT CREDIT', 7, 2)
     ↑
     position 7

The value of this subexpression is the string 'RE'.

length 2

SUB('CÂLGARY', 2, 2)

↑
position 2

The value of this subexpression is the string 'AL'.

To get the final result of the original expression, we then concatenate these three intermediate values, 'MONT' ∘ 'RE' ∘ 'AL'. We can see that our final result is the string 'MONTREAL'.

For more interesting applications, this sort of combination will be very common. Do not become flustered by the apparent complexity of the expression. Instead, concentrate on identifying and processing the various subexpressions that form the terms of the expression as you do with numeric expressions. If you take things a step at a time, there should be no major difficulty.

To complete the definition of SUB, some additional cases must be handled.

1. If $a_3 \leq 0$ (regardless of a_2), then the empty string is returned
2. If $a_2 \leq 0$ (regardless of a_3), then the empty string is returned
3. If $a_2 > k$, where $k = \text{LENGTH}(a_1)$, then the empty string is returned
4. If $a_2 + a_3 > k + 1$, where $k = \text{LENGTH}(a_1)$, then a_3 is assumed to be $k - a_2 + 1$

The examples SUB('ABCDEF', 0, 4) and SUB('ABCDEF', 7) both return ''.

As another example, consider the problem of transforming the string

'EDGAR□ALAN□POE'

to the string

'POE,□E□A'

This task can be accomplished by first scanning the name for the leftmost blank. Once we know where this blank is, the next character gives us the middle initial. By noting the position of the second blank, we can then obtain the last name. Clearly, we know the position of the first initial. Therefore, we have all the pieces required to generate the desired output.

Algorithm EDIT_NAME (version 1). Given a string variable NAME which represents the name of an individual in the format just discussed, it is required to generate an equivalent name, called DESIRED_NAME, in the form of last name followed by the first and middle initials. Let the string variables FL, MI, and LAST denote the first initial, middle initial, and last name of the individual, respectively. The variable I is used as temporary integer variable.

1. [Obtain the first initial]
    FI ← SUB(NAME, 1, 1)

2. [Scan the name for the first blank]
   I ← 1
   Repeat while SUB(NAME, I, 1) ≠ '□'
      I ← I + 1
3. [Obtain the second initial]
   MI ← SUB(NAME, I+1, 1)
4. [Get rid of first name and blank]
   NAME ← SUB(NAME, I+1)
5. [Scan the name for the second blank]
   I ← 1
   Repeat while SUB(NAME, I, 1) ≠ '□'
      I ← I + 1
6. [Obtain the last name]
   LAST ← SUB(NAME, I+1)
7. [Output the desired name]
   DESIRED_NAME ← LAST ∘ ', □' ∘ FI ∘ '□' ∘ MI
   Write(DESIRED_NAME)
   Exit   □

Note that in step 4 we got rid of the first name and the leftmost blank which follows it. By rescanning the reduced name, we were then able to locate the position of the second blank. Another approach to solving this problem could have been to replace step 4 by a new step which stored the position of the first blank in the original name. Step 5, in the search for the second blank, would thus start to scan the original name by looking at the character following the first blank. Clearly, both approaches will yield the same result.

The function SUB can also be used on the left-hand side of an assignment statement. For example, if S = 'ABCDEFG', then the statement

   SUB(S, 4, 1) ← 'G'

would change the value of S to 'ABCGEFG'. That is, the fourth character of S, or 'D' is changed to the letter 'G'. The statement

   SUB(S, 1, 3) ← "

would change the original value of S to 'DEFG'. That is, the first three characters of S are replaced by the empty string. The use of a function such as SUB on the left-hand side of an assignment statement is an example of what is often called a *pseudo-variable*. Such a functional reference involves two steps:

1. The location of the affected substring within the string must be determined
2. This affected substring is changed to the value of the expression on the right-hand side of the assignment statement

In order to make the definition of SUB complete when it is used as a pseudo-variable, a number of unusual cases must be specified. If SUB(a_1, a_2, a_3) appears on the left-hand side of an assignment statement and $a_2 \leq 0$ or $a_3 \leq 0$, then the assignment is not executed. If $a_2 > k$, where k is the length of a_1 or $a_2 + a_3 > k + 1$, then characters are assigned to positions beyond the right-hand end of the string a_1. Intermediate character positions which are unassigned are set to blank characters. For example, the statement

SUB(S, 8, 3) ←'HIJ'

where S has the value 'ABCDEFG', will change the value of S to 'ABCDEFGHIJ'. The statement

SUB(S, 10, 3) ←'HIJ'

for the same initial value of S, on the other hand, will change the value of S to 'ABCDEFG□□HIJ'.

In the previous algorithm, we were required to find the positions of each blank in the given name string. Although we were able to determine these positions by using the SUB function, such an approach can be somewhat tedious for the programmer. This type of phenomenon occurs in so many applications that we introduce a new operation to handle it. We denote this operation in our algorithmic notation by the built-in function INDEX. For example, the invocation of INDEX in

INDEX('EDGAR□ALAN□POE', '□')

will return the position of the leftmost blank in the string 'EDGAR□ALAN□POE' (that is, a value of 6). The function, in its general form, is

INDEX(S, P)

where S denotes the string which is to be examined for the leftmost occurrence of the substring given by P. The string S is quite often called the *subject string* while P is called the *pattern string*. This process of searching for a pattern string in a subject string is commonly called *pattern matching*. The INDEX function is our first example of a pattern-matching operation. Many pattern matching operations exist, and other such operations will be illustrated in Chap. 9. If the pattern in an INDEX function call is not found in the subject string, then the pattern matching process has failed. Such a failure is reported by returning a value of 0. For example, the value of INDEX('ABCD', '□') is 0. Of course, the pattern can be a string as in the case of 'INDEX('PQXYZZXY', 'XYZ'), which returns a value of 3. In this case, the position of the leftmost character of 'XYZ' in the subject string is returned. We can now reformulate algorithm EDIT_NAME.

Algorithm EDIT_NAME(version 2). Given a string variable NAME, which represents the name of an individual in the format previously discussed, it is required to generate an equivalent name, called DESIRED_NAME, in the form of the last name followed by the first and middle initials. Let the string variables FI, MI, and LAST denote the first initial, middle initial, and last name of the individual, respectively. The integer variable I is used to denote the position of a blank within the given string.

1. [Obtain the first initial]
   FI ← SUB(NAME, 1, 1)
2. [Obtain the position of the first blank]
   I ← INDEX(NAME, '□')
3. [Obtain the second initial]
   MI ← SUB(NAME, I + 1, 1)
4. [Delete first name and blank]
   NAME ← SUB(NAME, I + 1)
5. [Obtain the position of the second blank]
   I ← INDEX(NAME, '□')
6. [Obtain the last name]
   LAST ← SUB(NAME, I + 1)
7. [Output the desired name]
   DESIRED_NAME ← LAST ○ ', □' ○ FL ○ '□' ○ MI
   Write(DESIRED_NAME)
   Exit □

This second version of the algorithm is very similar to the first version with steps 2 and 5 changed to incorporate the INDEX function. Note the absence of loops in this version.

Many string-oriented applications also involve the use of arrays. Two simple examples of this will now be discussed. The first example involves the scanning of some English text to obtain the relative frequency of occurrence of each letter in the English alphabet.

The basic approach is to scan the piece of text for each letter while counting the number of times that each individual letter occurs. Also, as we scan the text, we can also count the total number of letters which occur. The relative frequency of a particular letter can then be obtained by dividing the number of times that letter occurs by the total number of letters in the text. The number of occurrences for each letter can be accumulated in a 26-element vector. We will associate the letter 'A' with the first element of the vector, 'B' with the second element, and so on. A general algorithm for this task follows.

1. Initialize the frequency counters for the letters and the counter which represents the total number of letters in the text
2. Scan the text for the next character until the entire text is processed; if this character is a letter, then update both the frequency counter for this letter and the counter which keeps track of the total number of letters in the text

3. Obtain the relative frequency of occurrence for each letter and write the results

One problem which arises concerns the convenient update of the particular counter which corresponds to the letter just scanned. This goal can be achieved by using the INDEX function. Let the string variable ALPHABET contain the letters of the English alphabet, in order. Also, let the vector FREQUENCY represent the frequency of occurrence of each letter so that FREQUENCY[1] is associated with the letter 'A', FREQUENCY[2] with the letter 'B', and so on. If CHARACTER contains the letter just scanned, then the statement

J ← INDEX(ALPHABET, CHARACTER)

returns the position of the current character in the English alphabet. For example, the letter 'C' will return a value of 3 for J. This information can be used in the statement

FREQUENCY[J] ← FREQUENCY[J] + 1

to update the appropriate counter. We can now proceed to a formal algorithm.

Algorithm LETTER_FREQUENCY. Given a string variable TEXT, it is required to compute the relative frequency of occurrence of each letter in the text. The string variables ALPHABET and CHARACTER and the vector FREQUENCY have already been described. The real variable TOTAL_LETTERS represents the total number of letters in the given text and the integer variable TEXT_LENGTH denotes the length of the input string. The integer variables I and J are temporary variables.

1. [Initialize]
    ALPHABET ← 'ABCDEFGHIJKLMNOPQRSTUVWXYZ'
    TOTAL_LETTERS ← 0
    TEXT_LENGTH ← LENGTH(TEXT)
    Repeat for I = 1, 2, ..., 26
        FREQUENCY[I] ← 0
2. [Process the input text]
    Repeat thru step 4 for I = 1, 2, ..., TEXT_LENGTH
3. [Scan the next character]
    CHARACTER ← SUB(TEXT, I, 1)
4. [If the character is a letter, then update the appropriate counter]
    J ← INDEX(ALPHABET, CHARACTER)
    If J ≠ 0
    then FREQUENCY[J] ← FREQUENCY[J] + 1
        TOTAL_LETTERS ← TOTAL_LETTERS + 1
5. [Calculate the relative frequencies of occurrence]
    Repeat for I = 1, 2, ..., 26
        FREQUENCY[I] ← FREQUENCY[I] / TOTAL_LETTERS

6. [Output the results]
    Write(FREQUENCY)
    Exit   □

As a final example, let us consider the generation of a monthly expense report for a typical household. The input data for this problem consist of an unordered sequence of cards. Each card contains two items: the first item is a string which represents the type of expenditure and the second item is a real number which specifies the amount of the expenditure. For example, the card containing

    'GROCERY'   31.57

represents an expenditure of \$31.57 for grocery items. Other expenditure types might be 'RENT', 'INSURANCE', 'CAR EXPENSES', and so on. As previously mentioned, the input card deck contains a number of such cards in no particular order. There may be several cards dealing with grocery expenditures. In the desired monthly report, it is required to summarize each type of expenditure. Since we do not know how many different types of expenditure are present in any given month, we must build a table which contains all the expenditure types for that month. Initially, this table is empty. As the input cards are processed, the expenditure type on each card is checked against the table entries obtained prior to reading the current card. If the current expenditure type is not already in the table, then the new type is added to the existing table. The table can be represented by a vector. When a new expenditure type is being processed, this new type can be added to the end of the existing type list. The linear search technique of Sec. 4-3.2 can be used to determine whether or not the new expenditure type is in the table. We will assume in this problem that the end of the input card deck can be determined by an end-of-file condition.

A general algorithmic formulation of this problem follows:

1. Repeat thru step 4 while there is still an expenditure
2. Read an expenditure card; if there are no more, print the desired report and exit
3. If the description type just read is not in the description table, insert the new type in the existing table
4. Update the total expenditure quantity with the current expenditure just read

We can use a pair of vectors to represent the expenditure type and the total amount of money spent on that type so far. Let the vectors TYPE and TOTAL_EXPENSE represent the expenditure type and the amount spent on that item, respectively. TYPE is a vector of strings and TOTAL_EXPENSE is a vector of reals. We shall assume that the size of these vectors is greater than the maximum number of distinct expenditure types that is likely to be encountered in any given month. The variable NEXT will denote the next elements in the vectors which are free. Steps 3 and 4 of the general algorithm, where ITEM and AMOUNT represent the expenditure type and expense on the current card, respectively, would look like this:

```
TYPE[NEXT] ← ITEM
I ← 1
Repeat while TYPE[I] ≠ ITEM
    I ← I + 1
If I = NEXT then NEXT ← NEXT + 1
TOTAL_EXPENSE[I] ← TOTAL_EXPENSE[I] + AMOUNT
```

This algorithm segment is a mere reformulation of the algorithm for LINEAR_SEARCH given in Sec. 4-3.2. We can now proceed to a detailed algorithm.

Algorithm MONTHLY_EXPENSE. Given an input card deck as previously described, it is required to generate a monthly expense report. The previously described vectors TYPE and TOTAL_EXPENSE, each containing 100 elements, are used. The variables NEXT (type integer), ITEM (type string), and AMOUNT (type real) are as previously described. I is a temporary integer variable.

1. [Initialize]
    NEXT ← 1
    Repeat for I = 1, 2, . . ., 100
        TOTAL_EXPENSE[I] ← 0
2. [Generate required report]
    Repeat thru step 5 while there is an expenditure
3. [Read an expenditure card]
    Read(ITEM, AMOUNT)
    If there are no more cards
    then Write('MONTHLY EXPENSE REPORT')
        Write('EXPENSE DESCRIPTION', 'TOTAL')
        Repeat for I = 1, 2, . . ., NEXT − 1
            Write(TYPE[I], TOTAL_EXPENSE[I])
        Exit
4. [Search the expenditure table for the current expenditure type; if unsuccessful, insert the new type in the table]
    TYPE[NEXT] ← ITEM
    I ← 1
    Repeat while TYPE[I] ≠ ITEM
        I ← I + 1
    If I = NEXT then NEXT ← NEXT + 1
5. [Update current item]
    TOTAL_EXPENSE[I] ← TOTAL_EXPENSE[I] + AMOUNT   □

In this section we have introduced a number of primitive string operations. More complex operations will be discussed in Chap. 9. We now proceed to the use of these operations in simple string applications.

EXERCISES 5-3

1. Give the results of the following expressions:
    (a) LENGTH('ALPHA' ○ 'BET' ○ '□SOUP')
    (b) SUB('HARPO' ○ 'CHICO' ○ 'GROUCHO' ○ 'ZEPPO', 11, 7) ○
        SUB('HARPO' ○ 'CHICO' ○ 'GROUCHO' ○ 'ZEPPO', 6, 5) ○
        SUB('HARPO' ○ 'CHICO' ○ 'GROUCHO' ○ 'ZEPPO', 18) ○
        SUB('HARPO' ○ 'CHICO' ○ 'GROUCHO' ○ 'ZEPPO', 1, 5)
    (c) SUB('SUFFIX', LENGTH('SUFFIX') − 3)

2. Formulate an algorithm for INDEX which uses other primitive functions (for example, SUB). That is, construct your own version of INDEX.

3. (a) You are given two cards with names written on them. Each name is separated by a comma, for example, JOHN, SUE, ..., JIM. Assume that a name appears only once on any one card. Construct an algorithm that will read these two cards and print out the union of the names on both cards. The union is the set of all the names that appear on one list or the other list, or both lists (if on both, only print once). Typical input and output are as follows:

    JOHN, MARY, JIM, JERRY, SUE, BOB, BARB
    BILL, BARB, JILL, BOB, SUE, JOHN
    UNION IS JOHN, MARY, JIM, JERRY, SUE, BOB, BARB, BILL, JILL

    (b) Repeat part (a) and print the intersection of the names on both cards. The intersection is the set of all names that appear on both lists. For the input data just given, the output is

    INTERSECTION IS JOHN, BOB, SUE, BARB

    Note that the output suggested in this problem is one possible form of the result.

4. Design an algorithm that inputs the name of a person, which is punched on one card in the form

    'EMILE□JEAN□PAUL□TREMBLAY'

    and outputs

    TREMBLAY, E.J.P.

    Your algorithm should handle an arbitrary number of names before the surname.

5. Construct an algorithm which inputs a string and replaces all occurrences of 'MRS.□' or 'MISS□' by 'MS□' and all occurrences of 'CHAIRMAN' by 'CHAIRPERSON'.

6. (*The "character-distance" problem.*) Examine an input stream of n characters (letters of the alphabet only), and form a result stream of numeric values, one for each input character. Each position of the result stream will be occupied by a number representing a count of the characters separating the character in the corresponding input position from the nearest similar character to its left in the input stream. No distance larger than 9 will be recorded. Any character not matching anything to its left within nine positions will have a zero in the result stream. Sample:

Input: 'AABCDBEFFEABGHIJKXYLMNOPQRSTUBWB'

Result: 01000300139600000000000000000002

Formulate an algorithm for this problem that does not use arrays.

7. Devise an algorithm that deletes all occurrences of trailing blanks in a given string. For example, the string 'R.B.□BUNT□□□□' should be transformed to the string 'R.B.□BUNT'.

8. Construct an algorithm which inputs a string, S, and a replication factor, N, and replicates the given string N times. For example, the results for the input

'HO!', 3

would be

'HO!HO!HO!'

9. Construct an algorithm which deletes all occurrences of each character contained in one given string from another given string. The two strings are as follows:
   (*a*) STR the string from which deletions are to be made.
   (*b*) LIST the string providing the characters whose occurrences in STR should be deleted.
   For example, if STR = 'THEX□EZNZZXDX' and LIST = 'XZ', the required answer is STR = 'THE□END'.

5-4   BASIC STRING APPLICATIONS

In this section we give a number of nonnumerical applications. These applications are a small representative set which involve string manipulation. More advanced applications of strings are given in Chap. 9. Section 5-4.1 discusses the processing and analysis of textual material. In Sec. 5-4.2 we examine one aspect of text editing and typesetting. Finally, we discuss a simplified form-letter generation application.

5-4.1   ANALYSIS OF TEXTUAL MATERIAL

A familiar application involving strings concerns the analysis of natural language text. Let us assume that the input text to be analyzed consists of a sequence of English words, each of which is separated by a single blank. We also assume that a blank follows the last word of text. In general, of course, words are also delimited from each other by various punctuation symbols such as !, ;, ., and : but, for the sake of simplicity, we will assume that such is not the case. In this particular application it is required to generate a frequency table which contains each distinct word in the text along with the number of times that word has occurred. This table is to be ordered in alphabetical order. For example, the sample text

    'YOU CAN FOOL SOME OF THE PEOPLE ALL OF THE TIME AND ALL
    OF THE PEOPLE SOME OF THE TIME BUT NOT ALL OF THE PEOPLE
    ALL OF THE TIME□'

would yield the following table:

WORD	FREQUENCY
ALL	4
AND	1
BUT	1
CAN	1
FOOL	1
NOT	1
OF	6
PEOPLE	3
SOME	2
THE	6
TIME	3
YOU	1

    We further assume that the number of cards which contain the narrative text is unknown and that words are not hyphenated or broken by card boundaries.

    Since we do not know in advance how many different distinct words are present in the given narrative text, we must build a table which contains all the distinct words in the given text. Initially, this table is empty. As the input cards are processed, each word is checked against the table entries obtained prior to the scanning of the current word. If the current word is not already in the table, this word is added to the existing table. The table can be represented by a vector. The linear search technique of Sec. 4-3.2 can be used to determine whether or not the current word being examined is in the table. If the search fails, then we must insert this new word at the end of the existing table. Once the table is complete, it can be sorted alphabetically with one of the sorting methods discussed in Chap. 4.

A general algorithmic formulation of this problem follows:

1. Read the narrative text
2. Repeat thru step 5 while there still remain words of text
3. Scan the next word of text
4. If the word just scanned is not in the word table, then insert this new word at the end of the existing word table
5. Update the frequency count for the word just scanned
6. Sort the word table and associated frequency tables
7. Print the desired results

We can use a pair of vectors to represent the distinct words and their respective frequencies encountered in the narrative text so far. Let the vectors WORD and FREQUENCY represent the distinct word table and the associated frequency table, respectively. WORD is a vector of strings and FREQUENCY contains integers. We assume that the size of these vectors is greater than the maximum number of distinct words that is likely to be encountered in any passage of narrative text. Step 4 of the general algorithm can be handled in a manner very similar to that used in the monthly expense report of Sec. 5-3. Also, step 6 involves the use of a sort algorithm. We choose the selection sort of Sec. 4-3.1 to perform this task.

We are now in a position to formulate a detailed algorithm.

Algorithm WORD_FREQUENCY. Given a passage of narrative text as described earlier, which is stored in the string variable TEXT, it is required to generate a report which contains each distinct word in the given text along with that word's frequency of occurrence. This report is to be in alphabetical order. The previously described vectors WORD and FREQUENCY, each containing 100 elements, are used. The variable NEXT denotes the next element in each vector, which is free or unoccupied. The string variable CARD represents a data card of narrative text. The integer variable P contains the position of the leftmost blank in the remaining narrative text. I denotes an index variable while the selection sort uses variables PASS and MIN_INDEX. The integer variable WORD_COUNT denotes the number of distinct words in the input. Finally, NEW_WORD is a string variable which contains the current word being scanned.

1. [Initialize]
    NEXT ← 1
    Repeat for I = 1, 2, . . ., 100
        FREQUENCY[I] ← 0

2. [Read in passage of narrative text]
    TEXT ← "
    Read(CARD)
    Repeat while there is no end-of-file
       TEXT ← TEXT ○ CARD
       Read(CARD)

3. [Process text]
    P ← INDEX(TEXT, '□')
    Repeat thru step 7 while P ≠ 0

4. [Scan the next word of text]
    NEW_WORD ← SUB(TEXT, 1, P − 1)
    TEXT ← SUB(TEXT, P + 1)

5. [Search and update word table for the word just scanned]
    I ← 1
    WORD[NEXT] ← NEW_WORD
    Repeat while WORD[I] ≠ NEW_WORD
      I ← I + 1
    If I = NEXT then NEXT ← NEXT + 1

6. [Update frequency count of the word just scanned]
    FREQUENCY[I] ← FREQUENCY[I] + 1

7. [Obtain position of next blank in the remaining narrative text]
    P ← INDEX(TEXT, '□')

8. [Using a selection sort, sort the word table and associated frequency count]
    WORD_COUNT ← NEXT − 1
    Repeat for PASS = 1, 2, . . ., WORD_COUNT − 1
      MIN_INDEX ← PASS
      Repeat for I = PASS + 1, PASS + 2, . . ., WORD_COUNT
        If WORD[I] < WORD[MIN_INDEX] then MIN_INDEX ← I
      If MIN_INDEX ≠ PASS
      then WORD[PASS] ⇔ WORD[MIN_INDEX]
         FREQUENCY[PASS] ⇔ FREQUENCY[MIN_INDEX]

9. [Print desired report]
    Write('WORD', 'FREQUENCY')
    Repeat for I = 1, 2, . . ., WORD_COUNT
      Write(WORD[I], FREQUENCY[I])

10. [Finished]
    Exit   □

The first step of the algorithm sets each element of the word frequency vector to zero. Moreover, the variable NEXT is set to 1. This value indicates that the word table is empty and, consequently, the first word of the narrative text is to be placed in the first position of the vector WORD.

    Step 2 reads all the input cards which make up the narrative text. Initially, the variable TEXT is set to the empty string. After the reading of each input card, the contents of this card are appended to the end of the current value of TEXT.

The third step engages a repetition loop which is to control the processing of the narrative text. This processing is to be performed on a word-by-word basis. The variable P denotes the position of the leftmost blank in the narrative text which remains to be processed. In this step P contains the location of the leftmost blank in the original text.

Step 4 copies the current word being scanned into the variable NEW_WORD. This word and the following blank are deleted from the narrative text.

Step 5 searches the word table, WORD, for the occurrence of NEW_WORD. If this word is not in the table, it is inserted at the end of the existing table. In this case NEXT is incremented by 1 to denote this insertion. This linear search technique is the same as that used in Sec. 4-3.2.

The next step of the algorithm adds 1 to the frequency count for the word just scanned.

Step 7 obtains the position of the leftmost blank in the remainder of the narrative text. Control is then returned to the repeat statement of Step 3. If the while test succeeds, the next word is processed. Otherwise, control passes on to step 8.

Step 8 performs an alphabetical selection sort on the vector WORD and its associated frequency vector FREQUENCY.

Finally, step 9 prints the desired results.

In this algorithm we have kept an unordered table of distinct English words and have performed the required sort after all words are examined. An alternative approach is to keep the word table ordered. A binary search technique (see Sec. 4-3.2) can then be used to search the existing table for the presence of the current word being scanned. An insertion in such an ordered table, however, involves the movement of data. This alternative approach is left as an exercise.

Another restriction in this algorithm involves the use of a single blank to separate the words of text. Typically, any delimiting character such as ;, !, :, ?, and , or even multiple blanks, can be used to perform this function. This variation of the preceding algorithm is also left as an exercise.

Another interesting problem in the analysis of English text is to obtain basic statistics for a certain passage of textual material. For example, statistics for the following may be required:

1. The number of sentences in the passage of text
2. The average number of words in a sentence
3. The average number of symbols per word

In order to bound the problem in terms of its complexity, it is convenient to make a number of assumptions concerning the textual material which is to be analyzed. Specifically, we assume the following:

1. Each word in the text is delimited by one blank space
2. Commas, colons, semicolons, question marks, exclamation marks, hyphens, and periods are considered to be valid punctuation symbols; furthermore, these symbols are not to be counted as textual characters in obtaining the required statistics for items 2 and 3 just given
3. The end of each sentence is denoted by a period; no periods are to occur within a sentence (for example, a real number such as 17.5 would not be allowed)
4. The last sentence in the text is to be indicated by a blank symbol followed by the symbol @ followed by a blank symbol

A general algorithmic formulation of this problem follows:

1. Read in the textual material to be analyzed
2. Repeat thru step 7 while there still remains text to be analyzed
3. Scan the next word of text based on the occurrence of a blank symbol
4. If the word just scanned is the symbol @, then compute the desired statistics and print the required summary report
5. If the word just scanned contains a period, then update the sentence counter
6. Update the word counter
7. Update the symbol counter and ignore any punctuation symbol in the symbol count of the word just scanned

We can now formulate a detailed algorithm.

Algorithm TEXT_ANALYSIS. Given a passage of textual material as previously described, it is required to generate, for this passage, a report which contains:

Number of sentences (SENTENCE_CTR, real)

Average number of words per sentence (AVG_WORDS, real)

Average number of symbols per word (AVG_SYMBOLS, real)

The passage of text is stored in the string variable TEXT. The integer variable P denotes the position of the leftmost blank in the remaining text to be processed. The real variables SYMBOL_CTR and WORD_CTR represent the number of symbols and words in the textual material, respectively.

1. [Initialize]
   WORD_CTR ← SYMBOL_CTR ← SENTENCE_CTR ← 0
2. [Read in the textual material]
   TEXT ← "

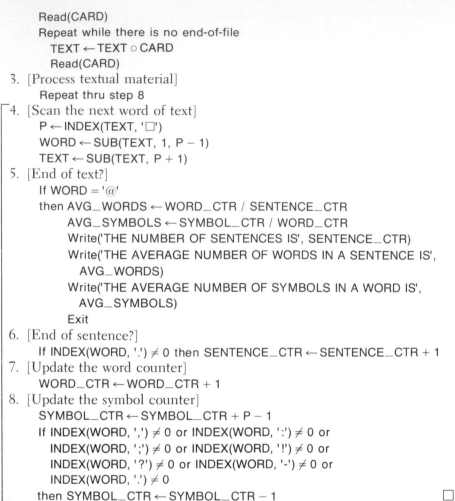

```
        Read(CARD)
        Repeat while there is no end-of-file
            TEXT ← TEXT ○ CARD
            Read(CARD)
  3. [Process textual material]
        Repeat thru step 8
  4. [Scan the next word of text]
        P ← INDEX(TEXT, '□')
        WORD ← SUB(TEXT, 1, P − 1)
        TEXT ← SUB(TEXT, P + 1)
  5. [End of text?]
        If WORD = '@'
        then AVG_WORDS ← WORD_CTR / SENTENCE_CTR
             AVG_SYMBOLS ← SYMBOL_CTR / WORD_CTR
             Write('THE NUMBER OF SENTENCES IS', SENTENCE_CTR)
             Write('THE AVERAGE NUMBER OF WORDS IN A SENTENCE IS',
                AVG_WORDS)
             Write('THE AVERAGE NUMBER OF SYMBOLS IN A WORD IS',
                AVG_SYMBOLS)
             Exit
  6. [End of sentence?]
        If INDEX(WORD, '.') ≠ 0 then SENTENCE_CTR ← SENTENCE_CTR + 1
  7. [Update the word counter]
        WORD_CTR ← WORD_CTR + 1
  8. [Update the symbol counter]
        SYMBOL_CTR ← SYMBOL_CTR + P − 1
        If INDEX(WORD, ',') ≠ 0 or INDEX(WORD, ':') ≠ 0 or
           INDEX(WORD, ';') ≠ 0 or INDEX(WORD, '!') ≠ 0 or
           INDEX(WORD, '?') ≠ 0 or INDEX(WORD, '-') ≠ 0 or
           INDEX(WORD, '.') ≠ 0
        then SYMBOL_CTR ← SYMBOL_CTR − 1                          □
```

The first step of the algorithm initializes the symbol, word, and sentence counters to zero.

Step 2 is identical to the second step of algorithm WORD_FREQUENCY; consequently, this step requires no further comment.

The third step engages a loop which is to process the entire passage of text. This potentially infinite loop is terminated in step 5.

Step 4 scans the remaining input text until the position of its leftmost blank is detected. This position is copied into the integer variable P. The current word of text is then extracted from the string variable TEXT. This word is then deleted from the current narrative text.

The next step determines whether or not the current word is the end-of-text delimiter (@). If so, the required averages are computed and the desired results are produced.

Step 6 examines the current word to determine whether or not it contains a period. If a period is present, then the sentence counter is updated.

Step 7 updates the word counter.

Finally, the eighth step of the algorithm updates the symbol counter. In doing so it checks the current word for the presence of punctuation symbols. The presence of such a symbol requires that the symbol counter be decremented by 1, since punctuation symbols are not to be counted as textual characters.

5-4.2   JUSTIFICATION OF TEXT

The typesetting of books, magazines, and newspapers has become very automated in recent years. Computers are used in the editing and formatting of text through the use of a text editor. Such an editor has an editing language associated with it. This language contains commands which enable the user to specify the typesetting of textual material. Such a language will be described in Chap. 9.

Right justification is a major problem in the typesetting process. By right justification of text we mean that the printed form of the text is such that the right margin is aligned for all lines in the output. In text that is typeset, such as most books, right justification is achieved by first attempting to split words across lines, then by leaving a certain amount of space between words, and finally, if necessary, by expanding the space between letters.

Word splitting can be handled in most instances by storing the syllables of most common words along with certain simple rules which govern syllable separation. To split an "uncommon" word may require operator intervention. Once such an uncommon word is split by manual intervention, the system can store this word and its split location. This may enable the system to split the word automatically the next time it overlaps the end of a line which is to be printed.

For textual material which is printed on a line printer or teletypewriter (the most popular types of hard-copy device used in computers), the spacing of words must be handled in a more primitive manner. The spacing between letters is impossible to adjust at present, since both printers and teletypes are fixed-print devices. That is, the amount of space between adjacent characters in such devices cannot be altered. Consequently, the spacing between words must be handled by allowing more than one blank character between words. In this subsection we discuss a solution to this problem.

We assume that words are not to be split between lines and that each line is to be both left- and right-justified (except for the last line of text). Any extra blank characters which are required in the justification of text are to be distributed as uniformly as possible between the words of a line. Furthermore, we assume that there are no paragraphs and that pagination and indentation are not required. Finally, each word in the textual material is separated from every other word by a blank and each punctuation symbol is followed by a blank.

As an example, let us consider the justification of the following sample text:

'THE□BUSINESS□WORLD□IS□RAPIDLY□CHANGING□AND□OUR□CORPORATION□ HAS□BEEN□KEEPING□PACE□WITH□THE□NEW□REQUIREMENTS□FORCED□UPON□ OFFICE□MACHINERY.□WE□ARE□GIVING□YOU,□MR.□DOE,□AS□A□KEY□FIGURE□ IN□THE□SASKATOON□BUSINESS□COMMUNITY,□AN□OPPORTUNITY□TO□BECOME□ FAMILIAR□WITH□THE□LATEST□ADVANCEMENTS□IN□OUR□EQUIPMENT.□'

Assuming 50 character positions per line, a right-justified equivalent of the previous text is

THE □ BUSINESS □ WORLD □ IS □□ RAPIDLY □□ CHANGING □□ AND □□ OUR □ CORPORATION□HAS□ BEEN□KEEPING □□ PACE □□ WITH □□ THE □□ NEW □ REQUIREMENTS □ FORCED □ UPON □ OFFICE □ MACHINERY. □ WE □□ ARE □ GIVING □ YOU, □ MR. □□ DOE, □□ AS □□ A □□ KEY □□ FIGURE □□ IN □□ THE □ SASKATOON □ BUSINESS □ COMMUNITY, □□ AN □□ OPPORTUNITY □□ TO □ BECOME □ FAMILIAR □ WITH □ THE □□ LATEST □□ ADVANCEMENTS □□ IN □ OUR □ EQUIPMENT

Note that the last line of the previous paragraph is not justified. The choice of output format for this line seems to be the most reasonable one to make.

A general algorithmic formulation of the text-justification problem follows:

1. Input paragraph of text and the number of characters per line of output
2. Repeat thru step 4 while the current line is not the last
3. If the current line of text is immediately right justifiable,
   then copy this line into output area;
   else determine the number of blanks to be inserted in current line
         Establish a loop to insert blanks into current line
             Distribute blanks in current line in a right-to-left manner
         Copy edited line in output area
4. Write current line
   Delete this line from paragraph
   If the leftmost character in the remaining text is a blank,
   then delete this character
5. Write last line and exit

We can now formulate a list of variable names (type in parentheses) for this problem. Such a list contains the following:

TEXT (string)	Input paragraph of English text which is to be right-justified
RMARGIN (integer)	Number of characters per printed line

LINE (string)   Edited line of text

BLANKS (integer)   Number of blanks to be inserted between
words in a line of text

BFIELD (string)   String of blank characters which separate
words of text

We next develop the details of our general algorithm. The step to read
in a paragraph of text and the number of characters per output line is

Read(TEXT, RMARGIN)

The statement

Repeat thru step . . . while LENGTH(TEXT) > RMARGIN

controls the output of full lines of right-justified text. The last output line of
text is not controlled by this statement.

The third step of the general algorithm represents the most significant
portion of the desired algorithm. Before developing the details of this step,
however, let us examine the right justification of a line more closely.

A line of text is immediately right-justifiable if the rightmost character
in the line is a nonblank character and the next character is a blank. For
example, assuming a line width of 25 and a text of

'THIS□BOOK□WAS□AUTHORED□BY□J.□E.□DOE.□HE□DISCUSSED□
THE□ART□OF□DEER□HUNTING.□'

the first line of the justified text can be printed as is. In this case the twenty-
fifth character is the 'Y' of 'BY' and the twenty-sixth character is a blank.
Consequently, no editing is required. That is, no blanks need be inserted in
the current line of text. Therefore, we can write

THIS⊔BOOK□WAS□AUTHORED□BY

In the next line of output, however, the case is somewhat more complex.
The remaining text to be printed is

'□J.□E.□DOE.□HE□DISCUSSED□THE□ART□OF□DEER□HUNTING.□'

Since this text is to be left-justified, we eliminate the leftmost blank. This
operation yields the string

'J.□E.□DOE.□HE□DISCUSSED□THE□ART□OF□DEER□HUNTING.□'

The twenty-fifth character in this string is the letter 'T'. The word 'THE'
must be printed on the next line. Consequently, we must insert two blanks
between the words in the second line. If these blanks are inserted in a right-
to-left manner between the words, then the second line of output becomes

'J.□E.□DOE.□□HE□□DISCUSSED'

The remaining text is now

'□THE□ART□OF□DEER□HUNTING.□'

Again, we delete the leftmost blank in this string. The last output line of text is

'THE□ART□OF□DEER□HUNTING.□'

Since the remaining text contains 25 characters, we need not perform any right justification on this line.

From these notions we can formulate the details of editing a line of output. The statement

If SUB(TEXT, MARGIN, 1) ≠ '□' and SUB(TEXT, RMARGIN + 1, 1) = '□'
then LINE ← SUB(TEXT, 1, RMARGIN)   (no justification required)
else (blanks must be inserted in the output line)

represents a skeleton of the required statement. To fill in the specifics of the else part of the previous statement, we must first determine the number of blanks that are to be inserted in the output line. The following algorithm segment produces this result:

i ← RMARGIN − 1
If SUB(TEXT, RMARGIN, 1) ≠ '□'
then Repeat while SUB(TEXT, i, 1) ≠ '□'
        i ← i − 1
i ← i − 1
BLANKS ← RMARGIN − i

The next part of the development is to distribute the required number of spaces (BLANKS) in the output line. As previously mentioned, the distribution of blanks is done by scanning the line from right to left and inserting extra blanks one at a time between the words. We require a loop to control the insertion of blanks. Assuming that the variable i carries over from the previous segment, the following statements distribute the desired blanks:

BFIELD ← '□'
Repeat for j = 1, 2, . . ., BLANKS
  Repeat while INDEX(SUB(TEXT, i, LENGTH(BFIELD)), BFIELD) = 0
     i ← i − 1
     If i = 0
     then i ← RMARGIN − BLANKS + j − 1
          BFIELD ← BFIELD ∘ '□'
  SUB(TEXT, i, LENGTH(BFIELD)) ← BFIELD ∘ '□'
  i ← i − 1
LINE ← SUB(TEXT, 1, RMARGIN)

This algorithm segment requires further comment. In general, we may have to scan the entire output line more than once in order to distribute all the blanks. Initially, we want to replace (in a right-to-left scan) every blank (initial value of BFIELD) by a sequence of two blanks. If more than one entire scan is required, however, we must be careful to detect the left end of the line. Such a check is made by testing the condition i = 0. If this condition holds, we must instigate a new scan starting from the right end of the line. Note that the position of the rightmost character in the line must be calculated at this point. Before starting the second scan we must also pad BFIELD with an extra blank. So, on the second scan of the line, an occurrence of a double blank (the revised value of BFIELD) is to be replaced by three blanks. This multiple scanning process continues until all blanks are distributed. A trace of the blank-distribution process is given following the detailed algorithm.

The fourth step of the algorithm outputs the right-justified line of text and deletes a leading blank from the next line, if necessary. The following statements accomplish this task:

```
Write(LINE)
TEXT ← SUB(TEXT, RMARGIN + 1)
If SUB(TEXT, 1, 1) = '□' then TEXT ← SUB(TEXT, 2)
```

Step 5 of the general algorithm is

```
Write(LINE)
Exit
```

We now give a complete algorithm for this problem.

Algorithm JUSTIFICATION. Given a character string TEXT, which contains English textual material beginning with a nonblank character, and the number of characters per printed line (RMARGIN), this algorithm right-justifies the given text. BLANKS is an integer variable which specifies the number of blanks to be inserted between the words in a line, and BFIELD is a character string of blank characters equal in size to the number of blanks which separate the words. Initially, only one blank separates these words. The integer variable i is used to index the characters of a line (in the right-to-left scan) when blanks are being inserted. The string variable LINE contains an edited line of text.

1. [Read in English text and characters per line]
   Read(TEXT, RMARGIN)
2. [Justify text]
   Repeat thru step 4 while LENGTH(TEXT) > RMARGIN
3. [Justify current line of text]
   If SUB(TEXT, RMARGIN, 1) ≠ '□' and SUB(TEXT, RMARGIN + 1, 1) = '□'
   then LINE ← SUB(TEXT, 1, RMARGIN)   (no justification required)
   else (blanks must be inserted in the output line)
        (check to see if position RMARGIN contains a nonblank character)
        i ← RMARGIN − 1

If SUB(TEXT, RMARGIN, 1) $\neq$ '□'
then Repeat while SUB(TEXT, i, 1) $\neq$ '□'
     i ← i − 1
     i ← i − 1
     (establish a loop for inserting blanks)
     BLANKS ← RMARGIN − i
BFIELD ← '□'
Repeat for j = 1, 2, . . ., BLANKS
     (successively add blanks to the blank fields separating the
     words)
     Repeat while INDEX(SUB(TEXT, i, LENGTH(BFIELD)), BFIELD)
     = 0
     i ← i − 1
     If i = 0
     then i ← RMARGIN − BLANKS + j − 1
         BFIELD ← BFIELD ∘ '□'
     SUB(TEXT, i, LENGTH(BFIELD)) ← BFIELD ∘ '□'
     i ← i − 1
     LINE ← SUB(TEXT, 1, RMARGIN)
4. [Output justified line]
    Write(LINE)
    TEXT ← SUB(TEXT, RMARGIN + 1)
    If SUB(TEXT, 1, 1) = '□' then TEXT ← SUB(TEXT, 2)
5. [Output last line]
    Write(TEXT)
    Exit   □

Recall that the third step of the algorithm performs the required editing. A test is made to determine whether or not blanks must be inserted in order to achieve right justification. If the character in position RMARGIN of TEXT is nonblank and the next character is a blank, this line can be written out directly. If the test fails, however, we must first determine the number of blanks which must be inserted in order to right-justify the line. If the character in position RMARGIN is a blank, then only one blank need be inserted. If the character in this position is nonblank, however, we must scan the text from the end of the line from right to left for a blank. The variable i is used as an index in this search. Once this blank is found, the variable BLANKS is assigned the number of blanks which must be distributed between the words in the line being prepared for output. The remainder of this step distributes the blanks between the words.

As an example, assume that TEXT has the value

'THIS□BOOK□WAS□AUTHORED□BY□J.□E.□DOE.□HE□DISCUSSED□
THE□ART□OF□DEER□HUNTING.□STILL□HUNTING□IS□DESCRIBED□AS□A . . .'

and RMARGIN is 25. Since this string contains more than 25 characters, we proceed to step 3. Because SUB(TEXT, 25, 1) = 'Y' and SUB(TEXT, 26, 1) =

'□', no right justification is required for the first line. Consequently, we proceed to step 4, where the first line written is

THIS□BOOK□WAS□AUTHORED□BY

The remaining text after deleting the leftmost blank becomes

'J.□E.□DOE.□HE□DISCUSSED□THE□ART□OF□DEER□HUNTING.□STILL□
HUNTING□IS□DESCRIBED□AS□A . . .'

A return to step 3 indicates that SUB(TEXT, RMARGIN, 1) = 'T'. Therefore, the word 'THE' cannot be printed this time. A search for the rightmost blank in the second line is successful in position 24. As a result, BLANKS receives a value of 2 and BFIELD is set to '□'. We then engage a loop to distribute these two blanks by scanning the line from position 23 from right to left. The while condition

INDEX(SUB(TEXT i, LENGTH(BFIELD)), BFIELD) = 0

for i = 23 reduces to the condition

INDEX(SUB(TEXT, 23, 1), '□') = 0

This condition fails since we are trying to match 'D' with '□'. We then continue to decrement i until the match succeeds. This situation occurs when i = 14. At that point we stop the search and exit from the loop. The blank in position 14 is replaced by two blanks. Now we have

'J.□E.□DOE.□HE□□DISCUSSED□THE□ART□OF□DEER□HUNTING.□STILL□
HUNTING□IS□DESCRIBED□AS□A . . .'

We now want to continue scanning the line because one more blank must be inserted. Before doing this, however, we reset i to a position which skips over the blank field just changed. This reset operation is performed by simply decrementing i by 1, which in our case yields a value of 13. We then resume the scan from position 13 leftward. The search for a blank is successful in position 11. We replace this blank by two blanks. The revised string now becomes

'J.□E.□DOE.□□HE□□DISCUSSED□THE□ART□OF□DEER□HUNTING.□STILL□
HUNTING□IS□DESCRIBED□AS□A . . .'

We have now inserted the required number of blanks for the second line. So the leftmost 25 characters of TEXT are copied into LINE and in step 4 the second line is written out as

J.□E.□DOE.□□HE□□DISCUSSED

In step 4 this line and the next blank are deleted from TEXT. As a result the remaining string is

'THE□ART□OF□DEER□HUNTING.□STILL□HUNTING□IS□DESCRIBED□AS□A . . .'

A return to step 3 indicates that SUB(TEXT, RMARGIN, 1) = '□'. Consequently, only one blank need be inserted into the third line in order to achieve justification. Thus, the third line written is

THE□ART□OF□DEER□□HUNTING.

After the deletion of this line and the next blank from TEXT we get

'STILL□HUNTING□□IS□DESCRIBED□AS□A . . .'

In step 3 SUB(TEXT, RMARGIN, 1) = 'E'. The word 'DESCRIBED' cannot be written out on this (the fourth) line. Nine blanks must be inserted in this line; that is, BLANKS = 9 and BFIELD = '□'. A scan for the rightmost blank from position 16 is successful in position 14. The substitution of two blanks here yields

'STILL□HUNTING□□IS□DESCRIBED□AS□A . . .'

At this point the scan for another blank continues and a successful match occurs in position 6. The replacement of this blank by a pair of blanks gives

'STILL□□HUNTING□□IS□DESCRIBED□AS□A . . .'

If we resume the scan for the next rightmost blank, the search fails. This condition (i = 0) is detected in step 3. When this situation arises, we want to repeat the scan from the right side of the line. The reset position of the scan is given by the statement

i ← RMARGIN − BLANKS + j − 1

In our case j = 3 since we are presently trying to add the third blank. Therefore, i is set to 25 − 9 + 3 − 1, or 18. Also note that the value of BFIELD is '□□'. Essentially, we want to replace a field of two blanks by a string of three blanks. If we perform the entire scan, we obtain the string

'STILL□□□HUNTING□□□IS□DESCRIBED□AS□A . . .'

Again, the condition i = 0 becomes true. Since we have only distributed four blanks, we must reset the scan so that the additional blanks can be inserted. The reset position is

i ← 25 − 9 + 5 − 1, or 20

The value of BFIELD changes to '□□□'. We now want to replace a field of three blanks by a string of four blanks. By repeating the entire scan, we get the string

'STILL□□□□HUNTING□□□□IS□DESCRIBED□AS□A . . .'

After two more scans of the line, we finally arrive at the string

'STILL□□□□□HUNTING□□□□□□IS□DESCRIBED□AS□A . . .'

and the fifth output line is

STILL☐☐☐☐☐HUNTING☐☐☐☐☐☐IS

The remaining text is then

'DESCRIBED☐AS☐A . . .'

Assuming that this string contains less than 26 characters, we can go directly to step 5, where the last line is written out.

In this subsection we have examined only the basics of text editing. A number of important editing considerations, such as pagination, indentation, and underlining, have been ignored. We will return to this topic in Chap. 9.

5-4.3   FORM-LETTER GENERATION

A business machine manufacturing corporation, which sells equipment throughout Canada, controls sales from its head office but has salespeople who work out of offices in each province. The company sends letters annually to prospective buyers, informing them that the provincial sales representative will call on them in the near future. The letters are mass-produced, but an effort is being made to personalize them by utilizing a computer.

In this subsection we wish to develop an algorithm for generating personalized form letters. An example of the input data for a form letter is given in Fig. 5-1. Each line in the letter represents a card. Note that blank cards (for blank lines) have been used to space out portions of the letter. Also present in this letter are a number of patterns or keywords such as

```
#40 187 MAIN STREET
#40 WINNIPEG 1, MANITOBA
#40 *DATE*

*X*
*ADDRESS*
*CITY*, *PROVINCE*

DEAR *Z*,
     THE BUSINESS WORLD IS RAPIDLY CHANGING AND OUR CORPORATION HAS BEEN
KEEPING PACE WITH THE NEW REQUIREMENTS FORCED UPON OFFICE MACHINERY.
WE ARE GIVING, YOU, *Z*, AS A KEY FIGURE IN THE *CITY* BUSINESS
COMMUNITY, AN OPPORTUNITY TO BECOME FAMILIAR WITH THE LATEST
ADVANCEMENTS IN OUR EQUIPMENT.  A REPRESENTATIVE OF OUR CORPORATION IN
*PROVINCE* WILL BE SEEING YOU WITHIN *N* WEEKS.  HE WILL TAKE SEVERAL
MACHINES TO *CITY* WHICH ARE INDICATIVE OF A WHOLE NEW LINE OF OFFICE
MACHINES WE HAVE RECENTLY DEVELOPED.
     OUR SALES REPRESENTATIVE IS LOOKING FORWARD TO HIS VISIT IN *CITY*.
HE KNOWS THAT THE MACHINES HE SELLS COULD BECOME AN INTEGRAL PART OF YOUR
OFFICE ONLY A FEW DAYS AFTER IMPLEMENTATION.

#40 SINCERELY,

#40 ROGER SMITH, MANAGER
#40 OFFICE DEVICES CORPORATION
```

FIGURE 5-1
Form letter.

```
*DATE*

*ADDRESS*

*CITY*

*PROVINCE*

*N*

*X*

*Z*
```

These patterns are to be replaced by actual and personalized information when a particular letter is generated. For example, the patterns *DATE* and *X* might be changed to JANUARY 1, 1978 and MR. R. B. BROWN, respectively. Also, certain lines in the form letter contain the indentation code, #. The form of this code is '#number□' and it applies to only one card. The code is used to indent that particular line by the specified number of spaces.

In the interest of simplicity we will not require the right justification of the personalized letters. Clearly, the notions introduced in the previous subsection could be used to achieve this goal, if desired. A personalized letter for the example form letter is given in Fig. 5-2. We have assumed that each line of the form letter after the substitution of personalized information (such as name, city, etc.) corresponds to one line of output in the personalized letter. Again,

```
                                187 MAIN STREET
                                WINNIPEG 1, MANITOBA
                                JANUARY 1, 1978

        MR. R.B. BROWN
        1712 ELK DRIVE
        JASPER, ALBERTA

        DEAR MR. BROWN,

            THE BUSINESS WORLD IS RAPIDLY CHANGING AND OUR CORPORATION HAS BEEN
        KEEPING PACE WITH THE NEW REQUIREMENTS FORCED UPON OFFICE MACHINERY.
        WE ARE GIVING, YOU, MR. BROWN, AS A KEY FIGURE IN THE JASPER BUSINESS
        COMMUNITY, AN OPPORTUNITY TO BECOME FAMILIAR WITH THE LATEST
        ADVANCEMENTS IN OUR EQUIPMENT.  A REPRESENTATIVE OF OUR CORPORATION IN
        ALBERTA WILL BE SEEING YOU WITHIN FOUR WEEKS.  HE WILL TAKE SEVERAL
        MACHINES TO JASPER WHICH ARE INDICATIVE OF A WHOLE NEW LINE OF OFFICE
        MACHINES WE HAVE RECENTLY DEVELOPED.
            OUR SALES REPRESENTATIVE IS LOOKING FORWARD TO HIS VISIT IN JASPER.
        HE KNOWS THAT THE MACHINES HE SELLS COULD BECOME AN INTEGRAL PART OF YOUR
        OFFICE ONLY A FEW DAYS AFTER IMPLEMENTATION.

                                SINCERELY,

                                ROGER SMITH, MANAGER
                                OFFICE DEVICES CORPORATION
```

FIGURE 5-2
Personalized letter.

this assumption simplifies our problem. Finally, we assume that several personalized letters must be produced by the algorithm.

The data are to consist of the following:

form letter

'**' an end-of-letter delimeter

first customer
{
'Date'
'MR. (or MRS., etc.)□Initials□Surname'
'Street Address'
'City,□Province'
'number of weeks before salesperson will visit'
}

second customer {

last customer {

For example, the customer data for the personalized letter of Fig. 5-2 would be

'JANUARY 1, 1978'
'MR. R. B. BROWN'
'1712 ELK DRIVE'
'JASPER, ALBERTA'
'FOUR'

A general algorithm for the form-letter problem is as follows:

1. Input the form letter

2. Repeat thru step 4 for every customer
3. Read the personalized data for a customer; if there are no more data, then exit
4. Repeat for every line of the form letter
    Get the next line of the form letter
    Repeat for each keyword in this line
        Replace keyword with personalized data
    If this line contains an indentation code, then insert the indicated number of indentation spaces
    Write out the next line of the personalized letter

Since we must produce a personalized letter for each customer, the form letter has to be saved. To preserve the one-to-one correspondence between each line of the form letter and the personalized letter, it is advisable to store the form letter line by line. A vector is a convenient structure for representing the form letter. Each string element of the vector represents one line of the letter. In step 1, the number of lines in the form letter can easily be determined as each line of the letter is read. On the detection of a line which contains the string '**' in its two leftmost character positions, a transfer to step 2 is made.

A list of variable names (type in parentheses) for this problem would include the following:

LETTER (string)	Vector, each of whose elements represents an output line in the letter
NUM_LINES (integer)	Number of lines in the form letter
DATE (string)	Date which is to appear on a personalized letter
NAME (string)	Name which is to appear on a personalized letter
LAST_NAME (string)	Last name of recipient of personalized letter
ADDRESS (string)	Street address which is to appear on a personalized letter
CITY_PROV (string)	City and province which is to appear on a personalized letter
WEEKS (string)	Number of weeks before a salesperson will visit an individual
PROVINCE (string)	Province which is to appear on a letter
CITY (string)	City which is to appear on a letter
RESULT (string)	Personalized data which are substituted for a keyword in the form letter

NUMBER (string)   Digit string which specifies the indentation spacing

VALUE (integer)   Integer which specifies the indentation spacing to be inserted in an indented line

We next develop the details of our general algorithm. The specifics of the first step for entering the form letter are as follows:

```
Read(LINE)
NUM_LINES ← 0
Repeat while SUB(LINE, 1, 2) ≠ '**'
    NUM_LINES ← NUM_LINES + 1
    LETTER[NUM_LINES] ← LINE
    READ(LINE)
```

Note that the end of the form letter is signaled by a line which contains the string '**' in its leftmost two positions.

In order to perform the indicated indentation of an output line, we need a sequence of blanks. The next statement realizes this need.

```
BLANKS ← '□□□□...□□□'      (a string of 80 blanks)
```

The second step of the algorithm which controls the generation of personalized letters has the form

```
Repeat thru step . . . while there are still customers
```

Step 3 reads and controls the input of personalized data for customers. The following statement performs this task.

```
Read(DATE, NAME, ADDRESS, CITY_PROV, WEEKS)
If there are no more customers then Exit.
```

Once the personalized data for a particular customer are read, we must examine these data so that the customer's last name, city, and state or province can be obtained. These items are inserted into certain lines of the personalized letter. The details of this task include

```
I ← INDEX(NAME, '□')
J ← INDEX(SUB(NAME, I + 1), '□')
LAST_NAME ← SUB(NAME, 1, I − 1) ○ '□' ○ SUB(NAME, I + J + 1)
I ← INDEX(CITY_PROV, ',')
PROVINCE ← SUB(CITY_PROV, I + 2)
CITY ← SUB(CITY_PROV, 1, I − 1)
```

The fourth and final step of the general algorithm controls and generates a personalized letter. The control function becomes the statement

```
Repeat thru step . . . for CC = 1, 2, . . ., NUM_LINES
```

We generate the letter line by line. The statement

```
LINE ← LETTER[CC]
```

obtains the next line of the form letter.

We must now perform all indicated substitutions of personalized information in the line which is to be output. Essentially, this involves performing a left-to-right scan on the line. This scan searches for the pattern '*KEY*', where KEY is any of the keywords 'DATE', 'X', 'ADDRESS', 'PROVINCE', 'CITY', 'Z', and 'N'. Each *KEY* pattern is replaced by the corresponding personalized data. More than one occurrence of a keyword can occur in a particular line. The following algorithm segment realizes this function.

```
I ← 0
J ← 0
K ← INDEX(LINE, '*')
Repeat while K ≠ 0
  I ← K + J + I
  J ← INDEX(SUB(LINE, I + 1), '*')
  KEY ← SUB(LINE, I, J + 1)
  If KEY = '*DATE*'
  then RESULT ← DATE
  else  If KEY = '*ADDRESS*'
        then RESULT ← ADDRESS
        else  If KEY = '*CITY*'
              then RESULT ← CITY
              else  If KEY = '*PROVINCE*'
                    then RESULT ← PROVINCE
                    else If KEY = '*N*'
                         then RESULT ← WEEKS
                         else If KEY = '*X*'
                              then RESULT ← NAME
                              else If KEY = '*Z*'
                                   then RESULT ← LAST_NAME
  LINE ← SUB(LINE, 1, I − 1) ∘ RESULT ∘ SUB(LINE, I + J + 1)
  J ← LENGTH(RESULT)
  K ← INDEX(SUB(LINE, I + J + 1), '*')
```

Finally, we must check for any indicated indentation. If the indentation code, #, is present in a particular line, a scan is made for the leftmost blank in that line. This search isolates the string which indicates the number of indentation spaces required. This blank specification string (NUMBER) is then converted to an integer (VALUE). The required number of indentation blanks is extracted from the string variable BLANKS. The following statements implement this approach.

```
If SUB(LINE, 1, 1) = '#'
```

```
then I ← INDEX(LINE, '□')
     VALUE ← 0
     NUMBER ← SUB(LINE, 2, I − 2)
     If LENGTH(NUMBER) = 2
     then CHAR ← SUB(NUMBER, 1, 1)
          VALUE ← INDEX('0123456789', CHAR) − 1) ∗ 10
          NUMBER ← SUB(NUMBER, 2)
     VALUE ← VALUE + INDEX('0123456789', NUMBER) − 1
     LINE ← SUB(BLANKS, 1, VALUE) ∘ SUB(LINE, I + 1)
```

All that remains to be done is to output the edited line of the personalized letter. The previous details are contained in the following algorithm.

Algorithm FORM_LETTER. Given a form letter and personalized customer data according to the formats just described, this algorithm generates a personalized letter for each set of customer data. The string vector, LETTER, contains the form letter. Each element in this vector represents a line of the form letter. The number of lines in the form letter is given by the integer variable NUM_LINES. As previously discussed, the string variables DATE, NAME, ADDRESS, CITY_PROV, WEEKS, LAST_NAME, PROVINCE, and CITY represent the personalized data in the form letter. I, J, K, and CC are integer variables. The string variable RESULT contains the personalized data which are substituted for a keyword in the form letter. The integer variable VALUE represents the number of indentation spaces to be inserted in an indented line. NUMBER is a digit string which specifies the indentation spacing.

1. [Input form letter]
   ```
   Read(LINE)
   NUM_LINES ← 0
   Repeat while SUB(LINE, 1, 2) ≠ '∗∗'
       NUM_LINES ← NUM_LINES + 1
       LETTER[NUM_LINES] ← LINE
       Read(LINE)
   ```
2. [Engage the loop to write personalized letters]
   ```
   BLANKS ← '□□□□...□□□'      (a string of 80 blanks)
   Repeat thru step 9 while there are still customers
   ```
3. [Read personalized data for one customer]
   ```
   Read(DATE, NAME, ADDRESS, CITY_PROV, WEEKS)
   If there are no more customers then Exit
   ```
4. [Split up the personalized data for one customer]
   ```
   I ← INDEX(NAME, '□')
   J ← INDEX(SUB(NAME, I + 1), '□')
   LAST_NAME ← SUB(NAME, 1, I − 1) ∘ '□' ∘ SUB(NAME, I + J + 1)
   I ← INDEX(CITY_PROV, ',')
   PROVINCE ← SUB(CITY_PROV, I + 2)
   CITY ← SUB(CITY_PROV, 1, I − 1)
   ```

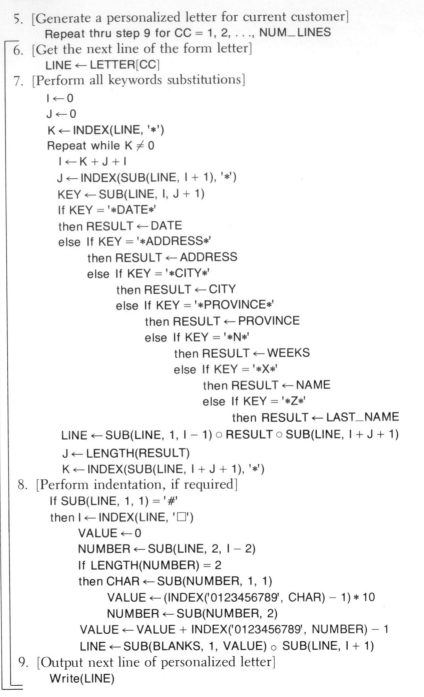

5. [Generate a personalized letter for current customer]
      Repeat thru step 9 for CC = 1, 2, . . ., NUM_LINES
6. [Get the next line of the form letter]
      LINE ← LETTER[CC]
7. [Perform all keywords substitutions]
      I ← 0
      J ← 0
      K ← INDEX(LINE, '*')
      Repeat while K ≠ 0
        I ← K + J + I
        J ← INDEX(SUB(LINE, I + 1), '*')
        KEY ← SUB(LINE, I, J + 1)
        If KEY = '*DATE*'
        then RESULT ← DATE
        else If KEY = '*ADDRESS*'
              then RESULT ← ADDRESS
              else If KEY = '*CITY*'
                    then RESULT ← CITY
                    else If KEY = '*PROVINCE*'
                          then RESULT ← PROVINCE
                          else If KEY = '*N*'
                                then RESULT ← WEEKS
                                else If KEY = '*X*'
                                      then RESULT ← NAME
                                      else If KEY = '*Z*'
                                            then RESULT ← LAST_NAME
        LINE ← SUB(LINE, 1, I − 1) ∘ RESULT ∘ SUB(LINE, I + J + 1)
        J ← LENGTH(RESULT)
        K ← INDEX(SUB(LINE, I + J + 1), '*')
8. [Perform indentation, if required]
      If SUB(LINE, 1, 1) = '#'
      then I ← INDEX(LINE, '□')
            VALUE ← 0
            NUMBER ← SUB(LINE, 2, I − 2)
            If LENGTH(NUMBER) = 2
            then CHAR ← SUB(NUMBER, 1, 1)
                  VALUE ← (INDEX('0123456789', CHAR) − 1) * 10
                  NUMBER ← SUB(NUMBER, 2)
            VALUE ← VALUE + INDEX('0123456789', NUMBER) − 1
            LINE ← SUB(BLANKS, 1, VALUE) ∘ SUB(LINE, I + 1)
9. [Output next line of personalized letter]
      Write(LINE)   □

It might be desirable to right-justify each personalized letter. Depending on the particular personalized information given for a certain customer, a one-to-one correspondence between each line of the form letter and the personalized letter may not always be suitable or acceptable. For example, for certain substitution possibilities, particular lines of the personalized letter may become too long. The ideas on right justification introduced in the previous subsection can be used to rectify this problem. Such a modification, however, is left as an exercise.

CHAPTER EXERCISES

1. Many programming languages such as FORTRAN permit the programmer to use blanks anywhere within certain parts of each statement of a program. The compiler of such a program would probably remove all the unnecessary blanks. Usually in these languages, there is a label field associated with each statement that is processed in a different manner. In the case of FORTRAN, for example, the first six characters of each statement represent the label or number of that statement. Furthermore, the remaining part of the statement must lie within a fixed field, such as characters 7 through 72 in FORTRAN. Formulate an algorithm that will:
    (a) Read in a text of 80 characters.
    (b) Delete the last eight of them.
    (c) Preserve unchanged the first six characters of the input.
    (d) Remove all blanks from the next 66 characters.
    (e) Print out the modified text.

2. Construct an algorithm for converting Roman numerals to Arabic numerals. The input consists of a sequence of Roman numerals. For each of these Roman numerals the corresponding Arabic numeral is to be generated. Table 5-2 gives the correspondence between the two number systems.

3. Formulate an algorithm for converting Arabic numerals to their corresponding Roman form (that is, do the inverse of Exercise 2).

TABLE 5-2

Roman Symbol	Arabic Equivalent
I	1
V	5
X	10
L	50
C	100
D	500
M	1,000

4. The usual way of writing a check requires five fields to be filled in; the date, the person being paid, the amount as a number, the amount in words, and the signature of the issuer. The amount is written twice for consistency and protection. Computer-generated checks sometimes do not generate the amount in words, since it is believed by some that a machine-printed amount is more difficult to change than its handwritten counterpart. Many companies have learned, however, that such is not the case; consequently, these companies do print on each check the amount in words. Formulate an algorithm which, given an integer amount, will print that amount in words. The integer amount lies in the range 100 through 99,999 (in pennies). Table 5-3 contains examples of numbers and their corresponding outputs.

5. A coded message is received on punched cards in groups of five letters separated by a blank. The last group of letters is followed by five 9's. The initial step in the decoding process is to replacc each letter by another, according to a table which changes each day. This table precedes the coded message information and occupies one card. For example, the string

    ABCDEFGHIJKLMNOPQRSTUVWXYZ
    'DEFGHIJKLMNOPQRSTUVWXYZABC'

represents a coding table in which D replaces A, E replaces B, F replaced C, . . ., B replaces Y, and C replaces Z. Using this code, the encoded message

    WKHZR□UOGLY□FRPLQ□JWRDK□HKGCC□99999

will be decoded as

    THEWORLDISCOMINGTOANENDZZ

Formulate an algorithm which inputs the given data and decodes the message.

6. In the algorithm WORD_FREQUENCY an unordered table of distinct words was kept and the required sort was performed after all words were examined. An alternative approach is to keep the word table ordered. Using this approach, reformulate WORD_FREQUENCY.

TABLE 5-3

Input (Pennies)	Amount in Figures	Amount in Words
17573	$175.73	ONE HUNDRED SEVENTY-FIVE AND 73/100
2900	29.00	TWENTY-NINE AND 00/100
48050	480.50	FOUR HUNDRED EIGHTY AND 50/100
1362	13.62	THIRTEEN AND 62/100

7. Generalize the algorithm **WORD_FREQUENCY** so that any delimiting character such as ;, !, and : can be used to separate words of text. Also, allow multiple blanks between words.

8. Generalize the algorithm **FORM_LETTER** so that the personalized letter which is produced is right-justified. State any assumptions which you make.

6

Subalgorithms: Functions and Procedures

Time and again throughout this book, we have maintained that a complex problem is best tackled by first breaking it up into a number of subproblems. Problem decomposition is an important aspect of problem solving. The subalgorithm provides us with an important tool for the development of algorithms and programs. In this chapter we will look at two different forms of subalgorithm and see how they can assist us in the solution of complex problems.

6-1 MOTIVATION

It has been estimated that the cost of producing a standard automobile if every part had to be constructed from basic units would exceed $100,000. Automobile assembly lines take full advantage of stocks of prefabricated components to reduce the cost as well as the time required to produce a car.

Similarly, in programming, we can significantly ease the production process through the use of previously written program components. Over the years, vast libraries of application programs have been compiled for applications in engineering, mathematical programming, and data processing, to name just a few areas. For example, various precoded sort routines (such as the selection sort given in Chap. 4) are available for data-processing applications, as are routines for such problems in mathematics as matrix inversion and the solution of differential equations. By having routines such as these available in a general form, the programmer of a particular problem is able instead to focus his/her attention on those aspects unique to his/her own problem.

In this chapter we discuss the concept of a subalgorithm. Subalgorithms can be loosely thought of as building blocks that can be assembled together in the process of constructing an algorithm or program. They may be precoded, as in the case of the library routines, or coded as part of the development of a suitable algorithm for the solution of a particular problem. In either case the use of subalgorithms allows the programmer to focus his/her attention selectively, as in the divide-and-conquer philosophy, and consequently helps reduce the overall complexity of the task.

In this chapter we will deal with two different types of subalgorithm, the function and the procedure, and their use in some typical applications.

6-2 FUNCTIONS

Users of most programming language systems have at their disposal a supply of system-defined or "built-in" functions. Trigonometric functions such as SIN (sine), COS (cosine), and TAN (tangent) are commonly found in scientific languages, as are other standard mathematical functions, such as ABS (absolute value) and SQRT (square root). A list of common built-in functions was given in Chap. 2. In Chap. 5 built-in functions SUB, INDEX, and LENGTH were introduced for the purpose of string handling. Built-in functions can be used directly in expressions, with the expected interpretation. For example, the statement

HYPOTENUSE ← SQRT(LEFTLEG ↑ 2 + RIGHTLEG ↑ 2)

computes the length of the hypotenuse of a right-angled triangle as the square root of the sums of the squares of the two legs. For a triangle with legs of 3 and 4 units, respectively, the value assigned to the variable HYPOTENUSE is SQRT(25), or 5. The statement

VAL ← SIN(ANGLE1) + COS(ANGLE2)

assigns to the variable VAL the sum of the sine of ANGLE1 and the cosine of ANGLE2. If the values of ANGLE1 and ANGLE2 are $\pi/2$ (radians) and 0 (radians), respectively, the value assigned to VAL is 1 + 1 or 2.

All these are examples of functions, in this case functions defined by the system for the use of the programmer. They are used in expressions as if they were simply regular variables. Like variables, they have a single value. It is the responsibility of the programmer to supply the particular *argument* (or, in some cases, arguments) on which the function is to operate for the specific computation at hand. In the examples the arguments were LEFTLEG ↑ 2 + RIGHTLEG ↑ 2, ANGLE1, and ANGLE2. Arguments can be any expression, the value of which is the value on which the named function operates.

The use of subalgorithms (in this case, functions) affects the flow of control in an algorithm. When a function is invoked, control passes to the instructions defining the function. Following execution of the function with the supplied arguments, control returns to the point of invocation in the calling (or main) algorithm, with whatever values have been computed in the function. This flow of control is illustrated schematically in Fig. 6-1.

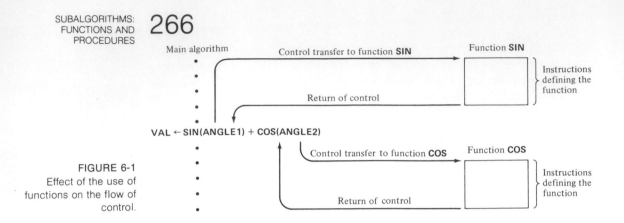

FIGURE 6-1
Effect of the use of
functions on the flow of
control.

For obvious reasons, no system can hope to provide *all* possible functions. In many problems, you would like to use an operation that is, unfortunately, not supplied by the system as a built-in function. It is important, too, that the programmer have the ability to define his/her own functions. We will now see how this is done.

In mathematics we write a function in a general form as follows:

$$f(x) = x^2 - 3x + 2$$

This particular function f has been defined in terms of a *parameter* x. Should we wish to evaluate this function for a particular value or argument of x, say, $x = 3$, we substitute that value for each occurrence of the parameter x in the general definition as in the following examples:

$$f(3) \ = 3^2 - 3 \times 3 + 2 = 2$$
$$f(7) \ = 7^2 - 3 \times 7 + 2 = 30$$
$$f(-2) = (-2)^2 - 3 \times (-2) + 2 = 12$$

Many functions will have more than one parameter. An important part of the definition of any function is a listing of all parameters in terms of which the function is defined. For example,

$$g(x,y) \ = x^2 - y^2$$
$$h(x,y,z) = x^2 + 3xy + z^2$$

are functions with two and three parameters, respectively. A corresponding number of arguments is required for each evaluation. Thus,

$$g(3,2) \ = 3^2 - 2^2 = 5$$
$$h(1,2,9) = 1^2 + 3 \times 1 \times 2 + 9^2 = 88$$

A correspondence is set up between the parameters of the definition and the arguments supplied, so that in the first case, 3 is substituted for each occurrence of *x* and 2 is substituted for each occurrence of *y*. Notice that the order is crucial; $g(3,2)$ is *not* the same as $g(2,3)$.

It is useful to view a function (and, as we will see later, a procedure) as an independent component of a program or algorithm, and for this reason we will define its instructions separately from those of the main part of the program or algorithm. The purpose of a function is to perform some computation when we require it, under control of the main algorithm (hence, the term "*sub*algorithm"). This computation may be required, in fact, at several places in the algorithm. An analogy might be a song sheet, in which the chorus, although written only once, is meant to be sung after every verse. It is usually given separately from the verses, and quite often, in fact, even has a different tune.

We will use the following format to define subalgorithms in the algorithmic notation. The example is a function to compute the average of three supplied values.

Function AVERAGE(VALUE1, VALUE2, VALUE3). The purpose of this function is to compute the average of the three values supplied. We assume all variables to be real.
1. [Compute average]
   AV ← (VALUE1 + VALUE2 + VALUE3) / 3.0
2. [Return result]
   Return(AV) □

Although this function is very simple, it serves to illustrate some key points. In terms of definition format the word "function" replaces the word "algorithm" in the standard algorithmic format. Following the name of the function is a list of *parameters* (in parentheses)—here **VALUE1**, **VALUE2**, and **VALUE3**. The function itself is defined in terms of operations on these parameters.

Any time the function is used in an algorithm, or *called*, a correspondence is automatically established between the function parameters and the arguments of the particular function call. The nature of this correspondence is discussed more fully in Sec. 6-4; suffice it for now to say simply that the values of the arguments somehow become the values of the parameters. The value that the function returns is given as an expression in parentheses immediately following Return; Return denotes the point at which control returns to the main algorithm, where the value returned is then used as part of the expression in which the function call appears.

To illustrate, we give the following algorithm, which uses the function **AVERAGE**.

Algorithm TEST_FUN. This algorithm illustrates the use of the function AVERAGE with various arguments.

1. [Initialize test values]
    A ← 2.0
    B ← 6.1
    C ← 7.5
2. [Use function in an expression]
    D ← AVERAGE(A, B, C)
3. [Display results]
    Write(D)
4. [Use function in a more elaborate expression]
    E ← AVERAGE(C, A, B) + AVERAGE(B, 3.2, A + 7)
5. [Display results]
    Write(E)
6. [Finished]
    Exit   □

You will notice that the function is called three times in this algorithm; once in step 2 and twice in step 4, with different arguments each time. The first call is with arguments A, B, and C. These have the values 2.0, 6.1, and 7.5, respectively. As the first function call is processed, the correspondence between the arguments and parameters is established, so that the parameters of the function, VALUE1, VALUE2, and VALUE3 are given the values 2.0, 6.1, and 7.5, respectively. The steps of the function itself are then executed with these values. The variable AV in the function definition is assigned the value 15.6/3.0, or 5.2, which is then returned as the value of the function in the next step. Back in the calling algorithm now, the returned value (5.2) is assigned to D and printed in step 3.

In step 4, the function is called again, twice in fact. The first of these calls is with a different ordering of the arguments. This results in VALUE1, VALUE2, and VALUE3, getting the values 7.5, 2.0, and 6.1, respectively. In this particular case, this results again in the value 5.2 being returned. The second call to the function in step 4 shows the versatility with which arguments can be expressed. The first argument (B) is a single variable; the second argument is a constant, 3.2; the third is an expression, (A + 7). The correspondence between arguments and parameters in this instance results in VALUE1, VALUE2, and VALUE3 being given the values 6.1, 3.2, and 9.0, respectively. Within the function, AV is assigned the value 18.3 / 3.0, or 6.1, which is returned as the value of the function for the second call.

To complete the processing of step 4 in the algorithm TEST_FUN, the two function values 5.2 and 6.1 are then added together and assigned to the variable E. Figure 6-2 summarizes the evaluation of this expression in step 4 of TEST_FUN. The value of E, 11.3, is then printed in step 5.

Although simple, this example illustrates several key points concerning the use of functions. Functions are defined as entities separate from the main

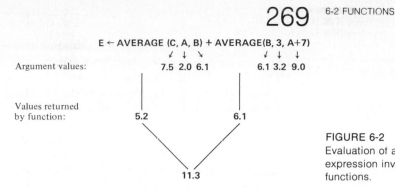

FIGURE 6-2
Evaluation of an
expression involving
functions.

algorithm, with the function parameters given in parentheses immediately following the function name. A function returns a single value given a particular set of arguments. The value to be returned is specified in parentheses in a Return statement in the function definition.

Most variables used in the definition of any subalgorithm (including parameters) are said to be *local* variables. This means that their names have meaning only inside the subalgorithm itself. For example, the variables VALUE1, VALUE2, VALUE3, and AV used in the definition of the function AVERAGE are local variables and thus they are unknown outside the function. Only the name of the function is known outside, as it must be used in any statement invoking the function. In fact, local variables can be given completely different meanings outside the subalgorithm without affecting the execution of the subalgorithm. From time to time, we will be assuming that certain variables of an algorithm are also known to its subalgorithms. Such variables are known as *global* variables and will always be clearly indicated as such in our algorithms. The distinction between global variables and local variables is important in many programming languages.

A function invocation or call is indicated by the presence of the function name (with appropriate arguments) in an expression. As the call is processed, a correspondence is established between the arguments of the call and the parameters specified in the function definition. In this way the values to be used in this particular execution of the function are established. Following execution of the statements in the function using these values, the *function value* is returned to the point of call.

The functions illustrated in this section all resulted in numerical values being returned to the point of call. We can refer to such functions as "numeric-valued." Since a function serves in an expression just as a regular variable, it is equally important that it have an associated type. More precisely, we may have "integer-valued" functions and "real-valued" functions, although we have not been that specific in these examples. We might also have applications requiring "string-valued" functions (such as the function SUB from Chap. 5) or even "logic-valued" or "pointer-valued" functions. Notice that the type of a function refers to the value returned, not to the type of its parameters. The function LENGTH from Chap. 5, for example, has a string parameter but is of type integer; INDEX is also of type integer, despite having two string parameters, whereas **SUB**, with one string and two integer parameters, is of type string.

The use of subalgorithms allows the programmer certain useful capabilities. He/she is able, for one thing, to extend the set of nonprimitive operations (see Sec. 4-1). Also, in the design of algorithms, a programmer is able to specify *what* to do with a value (by simply indicating that a subalgorithm will be called) before concerning himself/herself with *how* that value is to be obtained. That decision can be deferred until the appropriate subalgorithm is actually designed. This ability to postpone commitments, as we will see later, can be an important factor in managing the complexity of the programming task.

EXERCISES 6-2

1. Design a function that takes a parameter $x(x \neq 0)$ and returns the following value:

$$\frac{1}{x^5\left(\frac{e^{1.432}}{x} - 1\right)}$$

2. Design a function with two parameters, x and n, that returns the following:

$$\begin{cases} x + \dfrac{x^n}{n} - \dfrac{x^{n+2}}{n+2}, & \text{if } x \geq 0 \\[2ex] -\dfrac{x^{n-1}}{n-1} + \dfrac{x^{n+1}}{n+1}, & \text{if } x < 0 \end{cases}$$

3. (*a*) In Chap. 2 built-in functions **TRUNC** and **ROUND** were introduced. Although convenient for programming, **ROUND** is actually redundant. Design a function to perform the action of **ROUND** using **TRUNC**.

   (*b*) The function **FLOOR(x)** is defined as the largest integer value not exceeding x. This is not quite the same as the **TRUNC** function. For example, FLOOR(4.72) is 4, but FLOOR(-16.8) is -17. Use the built-in functions listed in Chap. 2 to design a function to compute **FLOOR(x)**.

   (*c*) The function **MOD(x, y)** is defined as the remainder that results from the division of x by y (x and y are both integers, as is the remainder). For example, MOD(8, 3) is 2. Formulate a function to compute **MOD(x, y)**.

4. (*a*) Design a function **SIGMA** to compute the summation of the n elements of a vector **X** (**X** and n are to be parameters of the function).

$$\sum_{i=1}^{n} x_i = x_1 + x_2 + \cdots + x_n$$

(b) Design a function PROD to compute the product of the n elements of a vector X (again, X and n are parameters of the function).

$$\prod_{i=1}^{n} x_i = x_1 \cdot x_2 \cdot \ldots \cdot x_n$$

5. Design functions MEAN and STD to compute the mean and standard deviation, respectively, of the n elements of a vector X, according to the following formulas:

$$\text{MEAN(X)} = \frac{1}{n} \sum_{i=1}^{n} x_i$$

$$\text{STD(X)} = \sqrt{\frac{1}{n-1} \sum_{i=1}^{n} [x_i - \text{MEAN(X)}]^2}$$

6. Design a function that takes as a parameter a card image (as a string) and returns to the point of call the number of distinct words on the card. Assume that words are separated by exactly one blank.

7. Design a string-valued function that takes as a parameter a card image (as a string) on which there appears some text. The purpose of the function is to remove all blanks from the text and return the string that results.

8. An expression involving a variable X such as

A(1)∗X¹ + A(2)∗X² + A(3)∗X³ + CONSTANT

can be evaluated for many values of X. For example, if

A(1) = 3, A(2) = 1, A(3) = 0.5, and CONSTANT – 5.2,

then the value of the expression for X = 2 is

(3)∗(2) + 1∗(2)² + 0.5∗(2)³ + 5.2 = 19.2

Desired is a function that would calculate

A(1)∗X¹ + A(2)∗X² + A(3)∗X³ + ... + A(N)∗Xᴺ + CONSTANT,

given the value of N, a particular set of coefficients (a vector A), a CONSTANT, and the value of X. Design a function EVAL that will accept the value of N, the CONSTANT, the coefficients (vector A), and a value for X. It will then evaluate the given expression for the value of X and return the resulting value to the point of call.

9. The special constant π plays an important role in mathematics. It is not surprising that there exist many methods of achieving numeric approxi-

mations to π. Many of these involve operations on an infinite series. Three such methods are the following:

1. $\pi = 4 \sum\limits_{i=0}^{\infty} \dfrac{(-1)^i}{2i+1} = 4(1 - \tfrac{1}{3} + \tfrac{1}{5} - \tfrac{1}{7} + \cdots)$

2. $\pi = \sqrt{\sum\limits_{i=1}^{\infty} \dfrac{6}{(i)^2}} = \sqrt{6 + \dfrac{6}{2^2} + \dfrac{6}{3^2} + \dfrac{6}{4^2} + \dfrac{6}{5^2} + \cdots}$

3. $\pi = 4 \cdot \dfrac{2}{3} \cdot \dfrac{4}{3} \cdot \dfrac{4}{5} \cdot \dfrac{6}{5} \cdot \dfrac{6}{7} \cdot \dfrac{8}{7} \cdots$

For practical computations, infinite-series calculations must be terminated after a finite number of terms, at the expense of accuracy in the result.

Design functions to calculate π according to each of the methods above. Each function is to accept, as a parameter, the value N indicating the number of terms to take part in the computation.

6-3   PROCEDURES

Although the function is a very useful programming tool, its abilities are somewhat restricted. In some situations, we may wish to specify an operation that is not conveniently stated as part of an expression. Such operations as the sorting of an array or the solution of a system of simultaneous linear equations have a much broader scope, and the returning of a single value for use in an expression would be somewhat artificial.

For these reasons, we introduce a second form of subalgorithm, which we will call a *procedure*. Although similar in most respects to the function, there are two important differences.

1. A procedure is invoked by means of a special statement, the *Call* statement. The execution of the Call statement results in execution of the calling routine being suspended, and control passing to the called routine or procedure. Following execution of the steps of the procedure, control returns to the calling routine at the statement *immediately following* the Call statement. Execution of the calling routine then continues from this point.
2. There is no single returned value as in the case of a function. Any values that are to be returned by a procedure are returned through the parameter list. Any number of values can be returned.

The following example illustrates the definition and use of a simple procedure. The procedure performs a division given two integer values, and returns as results the quotient and the remainder, again integers.

Procedure DIVIDE(DIVIDEND, DIVISOR, QUOTIENT, REMAINDER). This procedure divides the DIVIDEND by the DIVISOR giving the QUOTIENT and REMAINDER. Assume all numbers to be integer.
1. [Perform integer division]
   QUOTIENT ← DIVIDEND / DIVISOR
2. [Determine remainder]
   REMAINDER ← DIVIDEND − QUOTIENT ∗ DIVISOR
3. [Return to point of call]
   Return   □

Algorithm TEST_PROC. This algorithm illustrates the use of the procedure DIVIDE. All variables are integer.
1. [Initialize test values]
   A ← 5
   B ← 3
2. [Invoke procedure for division]
   Call DIVIDE(A, B, X, Y)
3. [Display results]
   Write(X, Y)
4. [Invoke procedure a second time]
   Call DIVIDE(A∗B − 1, B + 1, V, W)
5. [Display results]
   Write(V, W)
6. [Finished]
   Exit   □

An examination of the code for the procedure DIVIDE reveals that of its four parameters, two (DIVIDEND and DIVISOR) cause information to be transferred from the point of call to the procedure, and two (QUOTIENT and REMAINDER) cause information to be transferred in the opposite direction—from the procedure to the point of call. We will soon see that some parameters can, in fact, be vehicles for information transfer in both directions. Our mechanism for parameter/argument correspondence must be sufficiently flexible to handle all possibilities.

Let us examine what happens on the first call to procedure DIVIDE (step 2 of algorithm TEST_ PROC). The first two arguments here are A and B, which correspond positionally to DIVIDEND and DIVISOR, respectively, in the procedure definition. In a sense, these are *input* arguments, since the procedure requires them to arrive with values so that step 1 can be successfully executed. For this reason, they must have values at the time of the call, which they do (5 and 3, respectively). The last two arguments of the call statement,

X and Y, will receive values from the procedure (via QUOTIENT and REMAINDER). We can refer to these as *output* arguments.

As the first call is processed, the correspondence between arguments and parameters is effected as shown in Table 6-1. The steps of the procedure are executed accordingly and on execution of the Return statement in step 3, control returns to the point of call. The subsequent Write statement then causes the following line to be printed:

1   2

The procedure is called again in step 4 of the algorithm TEST_PROC, with a different pair of input arguments, A ∗ B − 1 and B + 1, and different output arguments, V and W. Although the input arguments are somewhat more elaborate in this case, the correspondence with parameters proceeds in the same fashion. This time the correspondence established between arguments and parameters is as shown in Table 6-2. This time the following line is printed after control returns from the procedure:

3   2

As was suggested, parameters in procedures are not always exclusively input or output. In many cases the same parameter will act in both capacities. As an input parameter it transfers information to the procedure but, in fact, it receives a new value from the procedure that it transfers back to the point of call, thus serving also as an output parameter. Consider the following procedure to interchange the values of two variables.

Procedure EXCHANGE(X, Y). The purpose of this procedure is to interchange the values given to variables X and Y. Assume all variables to be integer.
1. [Save the X value]
    TEMP ← X
2. [X takes on the value of Y]
    X ← Y
3. [Y takes on the previous value of X]
    Y ← TEMP
4. [Return to point of call]
    Return   ☐

TABLE 6-1

Argument		Parameter
A	corresponds to	DIVIDEND
B	corresponds to	DIVISOR
X	corresponds to	QUOTIENT
Y	corresponds to	REMAINDER

TABLE 6-2

Argument		Parameter
A * B − 1	corresponds to	DIVIDEND
B + 1	corresponds to	DIVISOR
V	corresponds to	QUOTIENT
W	corresponds to	REMAINDER

Clearly, X and Y serve as both input parameters and output parameters in this example. They bring the original values to the procedure, but also take the new values back.

The following algorithm uses the procedure EXCHANGE to sort three numbers in ascending order. Examine it carefully.

Algorithm SORT_3. This algorithm sorts three input values in ascending (increasing) order. All variables are integer.
1. [Input the values]
   Read(FIRST, SECOND, THIRD)
2. [Sort by pairwise comparisons]
   If FIRST > SECOND
   then Call EXCHANGE(FIRST, SECOND)
   If SECOND > THIRD
   then Call EXCHANGE(SECOND, THIRD)   (largest value is now in THIRD)

   If FIRST > SECOND
   then Call EXCHANGE(FIRST, SECOND)   (smallest value is now in FIRST)
3. [List sorted values]
   Write(FIRST, SECOND, THIRD)
4. [Finished]
   Exit   □

Let us trace the execution of this algorithm on some sample data, paying particular attention to the action of the procedure EXCHANGE. Assume that we have the following three sets of input data:

1. 3, 7, 5
2. 10, 8, 2
3. 6, 8, 12   (note that this is already sorted)

Table 6-3 shows the succession of values taken by the variables FIRST, SECOND, and THIRD as these three data sets are processed. The actual exchanges are indicated by arrows. Case 2 shows the maximum possible number of exchanges: case 3, the minimum number.

TABLE 6-3

Case	FIRST	SECOND	THIRD	Output
1	3	7	5	
	3	5	7	
				3 5 7
2	10	8	2	
	8	10	2	
	8	2	10	
	2	8	10	
				2 8 10
3	6	8	12	
				6 8 12

Let us examine in detail the action of the procedure on case 2, involving the greatest number of exchanges. We leave the other cases as an exercise. The initial values of FIRST, SECOND, and THIRD are 10, 8, and 2, respectively. Since FIRST > SECOND, the first call to the EXCHANGE procedure is with argument values of 10 and 8. These become associated with parameters X and Y, respectively. The procedure then interchanges these values so that immediately prior to returning, X has the value 8 and Y, the value 10. Since X and Y correspond to arguments FIRST and SECOND, respectively, upon return to the point of call, FIRST has the new value 8 and SECOND, the new value 10. As a result of these actions SECOND > THIRD now (10 > 2); thus a second call is required to interchange these values. On return from the second call, SECOND has the value 2 and THIRD, the value 10. The largest of the values is now in the variable THIRD. Finally, since FIRST > SECOND now (8 > 2), a third call is required to complete the sort. On return, FIRST has the value 2 and SECOND, the value 8. The action of these calls, showing the argument-parameter correspondence and the transfer of values between calling routine and procedure, is summarized in Fig. 6-3.

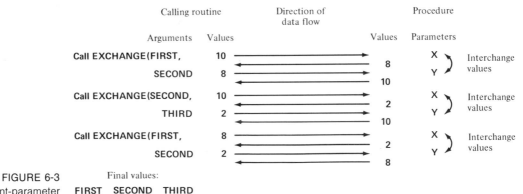

FIGURE 6-3
Argument-parameter
correspondence.

Final values:

FIRST	SECOND	THIRD
2	8	10

The sort method employed in this example differs from the method shown in Sec. 4-3.1 (selection sort). The method used here is sometimes referred to as the *exchange sort*, but also may be known as the *bubble sort* (see Exercises 4-3). The algorithm given is easily generalized to an algorithm that will sort a vector of any length. As a final illustration of the use of procedures, we will develop a general procedure to sort a vector of length n using the exchange sort method.

The basic approach is to examine all the elements of the vector, comparing successive pairs. If the vector is to be sorted, say, in ascending order, clearly all pairs must be in the correct relative order. If any pair is found to be out of order, that is, the first element of the pair is greater than the second, the elements are exchanged. This operation is performed repeatedly until no more exchanges are required, at which point the vector is in the correct order.

As was the case in the selection sort method, a maximum of $n - 1$ passes is required to sort a vector of n elements. In this case we can quit early should no exchanges be made on any pass, meaning that sorting is complete. Further, each pass over the elements results in at least one more element (the "bottom" one, since we are sorting in ascending order) being placed in its correct position, as will be shown. Thus, the number of comparisons can be reduced on each pass. A general algorithm for this sort would look like the following:

1. Repeat thru step 3 a total of $n - 1$ times
2. Compare each pair of consecutive elements in the unsorted vector and exchange their values if the first of the pair exceeds the second
3. If no exchanges are performed on this pass,
   then return to the point of call;
   else reduce the size of *unsorted* vector by 1 element
4. Sorting is complete, so return

The following procedure results from formalizing this process.

Procedure EXCHANGE_SORT(K, N). Given a vector K of N elements, this procedure sorts the elements into ascending (increasing) order using the method just described. The procedure EXCHANGE is used to interchange the elements when required. The variables PASS and LAST denote the pass counter and position of the last unsorted element, respectively. The variable i is used to index the vector elements. The variable EXCHS is used to count the number of exchanges made on any pass. All variables are integer.

1. [Initialize]
   LAST ← N   (entire list assumed unsorted at this point)
2. [Loop on pass index]
   Repeat thru step 5 for PASS = 1, 2, ..., N − 1
3. [Initialize exchanges counter for this pass]
   EXCHS ← 0

4. [Perform pairwise comparisons on unsorted elements]
    Repeat for i = 1, 2, . . ., LAST − 1
    If K[i] > K[i + 1]
    then Call EXCHANGE(K[i], K[i + 1])
        EXCHS ← EXCHS + 1
5. [Were any exchanges made on this pass?]
    If EXCHS = 0
    then Return   (mission accomplished; quit early)
    else LAST ← LAST − 1   (reduce size of unsorted list)
6. [Finished]
    Return   (maximum number of passes required)   □

Notice the use of a vector (K) as a parameter in this procedure. When entire vectors (or arrays) are used as arguments and parameters in any subalgorithm, they are denoted simply by their name. The arguments and parameters must, of course, agree in type and format. For example, we cannot pass a two-dimensional argument into a one-dimensional parameter. When individual elements are being used, however (see the call in step 4), they are treated as single variables.

To illustrate the use of this procedure, we give the following algorithm, which reads a list of recorded temperatures and prints them in ascending order.

Algorithm RECORD. This algorithm reads a list of 10 recorded temperatures, sorts them using the procedure EXCHANGE_SORT, and then prints the sorted list. The temperatures are stored in a vector TEMPS. All variables are integer.
1. [Read the temperatures]
    Repeat for i = 1, 2, . . ., 10
        Read(TEMPS[i])
2. [Perform the required sort]
    Call EXCHANGE_SORT(TEMPS, 10)
3. [Print the sorted list]
    Repeat for i = 1, 2, . . ., 10
        Write(TEMPS[i])
4. [Finished]
    Exit   □

You will notice that in this example we have a main algorithm (RECORD) that calls a procedure (EXCHANGE_SORT) that itself calls another procedure (EXCHANGE). This automatically establishes a chain of linkages as shown in Fig. 6-4. This should cause no special difficulties for the programmer.

Let us illustrate the action of the exchange sort by tracing the algorithm RECORD with a set of 10 temperatures. Step 1 reads the 10 values into the vector TEMPS, shown as the first column in Table 6-4. This vector is then

TABLE 6-4   Behavior of First Pass over Vector by EXCHANGE_SORT

Original Unsorted Vector							State at End of Pass One
73	65	65	65	65	65	65	65
65	73	52	52	52	52	52	52
52		73	24	24	24	24	24
24			73	73	73	73	73
83	no exchange → 83			17	17	17	17
17				83	35	35	35
35					83	83	83
96	no exchange				→ 96	41	41
41						96	9
9							96

passed to the procedure EXCHANGE_SORT, along with the number of elements (10), in step 2. Following the sort, the sorted vector is printed in step 3. We will concern ourselves with the action of the procedure EXCHANGE_SORT as it operates on the vector of temperatures. Table 6-4 shows the detailed behavior of the first pass over the vector, with arrows indicating the values being compared. A new column begins each time an exchange is made by the procedure EXCHANGE. Notice that at the end of pass 1, the largest element is in its rightful place at the bottom of the list (since we are sorting in increasing order). At the end of each subsequent pass, the next largest item will also assume its rightful place, until all are where they belong, at which time the list is sorted. Table 6-5 summarizes the behavior of the entire algorithm, showing the state of the list after each complete pass. Elements below the dash in each column are in their correct position. As you can see, a total of nine passes was required to sort this particular set of values.

The subalgorithm concept is a fundamental programming concept and is available in one form or another in most programming languages, although the terminology used may change. We have considered two forms of subalgorithm, which we have called the function and the procedure, that differ essentially in the manner they are called and in the manner by which values are returned to the point of invocation. Other aspects, such as control flow, argument-parameter correspondence, and the concept of local and global variables (see Sec. 6-2) are common to both.

Some programming languages, such as FORTRAN, PL/I, and ALGOL, provide the capability of defining subalgorithms of each type. Other lan-

FIGURE 6-4
Hierarchy of procedure calls.

TABLE 6-5 Complete Behavior of EXCHANGE_SORT

Original Unsorted Vector	Pass Number								Sorted
	1	2	3	4	5	6	7	8	9
73	65	52	24	24	17	17	17	17	9
65	52	24	52	17	24	24	24	9	17
52	24	65	17	35	35	35	9	24	24
24	73	17	35	52	41	9	35	35	35
83	17	35	65	41	9	41	41	41	41
17	35	73	41	9	52	52	52	52	52
35	83	41	9	65	65	65	65	65	65
96	41	9	73	73	73	73	73	73	73
41	9	83	83	83	83	83	83	83	83
9	96	96	96	96	96	96	96	96	96

guages are more restrictive. APL, LISP, and SNOBOL, for example, offer only the function capability. Some languages, such as COBOL and BASIC, offer only the equivalent of the procedure capability. As with other concepts we have seen, where a choice is available the particular requirement of any given problem should guide your choice.

In the next section we examine in more detail the correspondence between arguments and parameters. In Sec. 6-5 we consider the application of subalgorithms to the solution of a number of interesting problems.

EXERCISES 6-3

1. An algorithm was developed in Sec. 3-5.2 to compute and print monthly mortgage payments. Design a procedure with three input parameters

    PRINCIPAL (real)   Amount of the principal

    TERM (integer)   Number of years for the mortgage

    INT_RATE (real)   Yearly interest rate

    The procedure is to do two things:

    1. Compute and return to the point of call the monthly mortgage payment, calculated as in Sec. 3-5.2.
    2. Produce a complete table of monthly payments, giving the breakdown into payments of interest and principal.

2. One of the earliest applications of computers was the calculation of shell trajectories. If a shell is fired with an initial velocity V (ft/s) at an angle of

inclination B (radians), its position in the vertical x,y plane at time t (seconds) is calculated from the following:

$$x = (V \cos \theta)t$$
$$y = (V \sin \theta)t - \tfrac{1}{2}gt^2$$

where $0 < \theta < \pi/2$ and $g = 32$ ft/s^2.

Design a procedure with parameters θ and V that will list the x,y coordinates at intervals of .01 s for a particular firing, terminating the list when the shell hits the ground.

3. A square matrix is said to be symmetric if entry (i,j) = entry (j,i) for all i and j within the bounds of the matrix: that is, if A is a 4×4 matrix, then if $A(3,4) = 17$, $A(4,3)$ must $= 17$ for A to be symmetric. The following is an example of a symmetric matrix:

$$\begin{pmatrix} 0 & 3 & 5 & 1 \\ 3 & 7 & 6 & 9 \\ 5 & 6 & 2 & 4 \\ 1 & 9 & 4 & 2 \end{pmatrix}$$

Design a procedure that will decide whether or not a matrix is symmetric. Parameters of this procedure should include N, the matrix size, and A, the $N \times N$ matrix itself.

4. Design a procedure to center a title. The procedure takes as an input parameter a card image string, somewhere in which appears a title. The procedure is to produce a print line image string (of length 120) in which the input text (excluding leading and trailing blanks) is to be centered as nearly as possible within the string; that is, within one blank, the number of blanks before the title is the same as the number after. This print line image is then to be returned to the point of call.

5. Design a procedure to accept as a parameter a vector which may contain duplicate entries. The procedure is to replace each repeated value by -1 and return to the point of call the altered vector and the number of altered entries.

6. Design a procedure TRIM to accept as a parameter an arbitrary character string and return a string containing only the alphabetic characters in the original string. All blanks, punctuation marks, numbers, and special characters are to be removed.

7. Design a procedure REVERSE to accept as a parameter an arbitrary character string and return a string of the same length in which the order of the characters is reversed. That is, the first character of the output string is the last character of the input string, and so on.

8. A popular method of computing depreciation is known as the "declining balance" method. The first-year depreciation is applied to the initial cost of the item; in subsequent years, the amount of depreciation is based on the declining book value of the item. Assume that the cost of a particular item is given by the real variable ITEM_COST, and that its anticipated lifetime is given by the integer variable LIFETIME. The method proceeds as follows. In the first year

$$\text{DEPRECIATION}_1 = \frac{\text{ITEM\_COST}}{\text{LIFETIME}}$$

$$\text{BOOK\_VALUE}_1 = \text{ITEM\_COST} - \text{DEPRECIATION}_1$$

In the second year

$$\text{DEPRECIATION}_2 = \frac{\text{BOOK\_VALUE}_1}{\text{LIFETIME}}$$

$$\text{BOOK\_VALUE}_2 = \text{BOOK\_VALUE}_1 - \text{DEPRECIATION}_2$$

In general, for the ith year

$$\text{DEPRECIATION}_i = \frac{\text{BOOK\_VALUE}_{i-1}}{\text{LIFETIME}}$$

$$\text{BOOK\_VALUE}_i = \text{BOOK\_VALUE}_{i-1} - \text{DEPRECIATION}_i$$

where it is understood that

$$\text{BOOK\_VALUE}_0 = \text{ITEM\_COST}$$

For example, suppose that a small computer system, purchased for $250,000, is to be depreciated over 5 years using the declining balance method. The amount of depreciation for each of the 5 years is as follows:

Year	Depreciation	Book Value of Item
		$250,000
1	$50,000	200,000
2	40,000	160,000
3	32,000	128,000
4	25,600	102,400
5	20,480	81,920

Design a procedure to accept as parameters the initial cost of an item and its anticipated lifetime, and print the depreciation table over the lifetime of the item. The procedure should return to the point of call, the book value of the item at the end of its estimated lifetime.

6-4   ARGUMENT-PARAMETER CORRESPONDENCE

On each invocation of a procedure or function, a correspondence is established between the arguments of this particular invocation and the parameters of the procedure or function. In this section we discuss two ways this correspondence is effected, and comment on the advantages and disadvantages of each.

The most straightforward method of associating arguments with parameters is *pass by value* (sometimes known as "call by value"). As the invocation of the procedure or function is processed, each of the arguments is evaluated. The individual values are then, in effect, assigned to the respective parameters. It makes no difference whether an argument is a variable, a constant, or an expression; all that matters is that it have a value. This value becomes the value of the corresponding parameter.

Figure 6-5a illustrates the pass-by-value mechanism on a hypothetical procedure with three parameters. At the time the call statement is processed, each of the arguments of the call is evaluated. The value of the first argument, the variable A, is 3. The value of the second argument, the constant 17, is 17. The value of the third argument, the expression $B \uparrow 2 - 1$, is 35. These then become the values of the parameters X, Y, and Z, respectively, as execution of the procedure begins.

Although conceptually simple, pass by value has a major limitation. Although the parameters receive the values of the arguments, they are in no other way connected to the actual arguments themselves. Changes to the parameters within the procedure do not result in corresponding changes to the arguments. In a sense, the subalgorithm retains no knowledge of the original arguments. Thus, it is impossible to transfer values back to the point of call. In effect, all parameters are input only.

We have already seen cases where we require certain parameters to serve as output parameters. Consider, for example, the procedure EXCHANGE from the previous section. With pass by value it would not be possible for the results to be returned in the same two variables. The procedure EXCHANGE_ SORT would have to be totally redesigned since the same vector could not be used to return results. For this reason there are other mechanisms available for associating arguments with parameters. Among these is one that

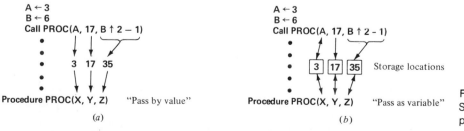

FIGURE 6-5
Schemes for parameter passing.

we will refer to as *pass as variable* (also sometimes known as "call by address" or "call by reference").

Rather than passing a value for each argument, pass as variable, in effect, passes the actual variable itself, potentially to be modified within the function or procedure. This can be done by the compiler in one of two ways with essentially the same effect. One way is to copy in the initial value of the argument at the time the call is processed and then copy out the final value prior to the return. A second method (which is, in fact, the method used by FORTRAN) passes the address of the argument. This address refers to a storage area into which the value of the argument is placed as the call is processed. This storage area can be used to pass information bidirectionally. For example, if the argument is a simple variable, the address passed is simply that of the storage area for which the variable stands.

With pass as variable, any reference to the corresponding parameter within the procedure is also a reference to the argument itself, which is, as we shall soon see, a mixed blessing. Pass as variable is illustrated in Fig. 6-5b. Although the problem of "input-only" parameters is solved with pass as variable, a new danger is introduced. Consider the following very simple function, for which we assume that processing of the parameter is done as pass as variable.

Function FUN(I).
1. I ← 5
2. Return(I)   □

We give the following algorithm that invokes the function FUN. Once again, assume everything to be integer.

Algorithm WHATS_UP.
1. K ← 3
2. J ← FUN(K) * 3
3. L ← FUN(3) * 3
4. Write(J, L)
5. Exit   □

What values are printed for J and L? Although the two calls look to be the same, the values printed are surprisingly different. During the processing of the function call of step 2, since pass as variable is used, the assignment of the value 5 to the parameter I simultaneously changes the value of its corresponding argument. Thus, the value of variable K is changed by the function from 3 to 5. The function returns the value 5, which is multiplied by 3 to give the value 15, which is assigned to L.

During the processing of the function call of step 3, however, something very strange happens. The argument in this case is the constant 3.

This constant is stored somewhere in the memory of the computer. Since pass as variable is used, a reference to the parameter is now effectively the same as a reference to the memory area in which the constant is stored. Thus, the assignment in step 1 of the function itself has the effect of changing the value of the constant 3 to 5. The function returns the value 5 which is again multiplied by the constant 3. But, since the constant 3 now has the value 5, the result of the multiplication is 25. What a mess!

We refer to changes made by a function or procedure to items outside of the function or procedure itself as *side effects*. In some cases, such as the procedures EXCHANGE and EXCHANGE_SORT given in the previous section, the side effects worked to our advantage. In the case of our function FUN, the side effects were disasterous.

The use of pass by value prevents this type of side effect from occurring since the function or procedure has no way of getting at the argument itself—just at its value. In some programming languages the programmer can specify for each parameter how it is to be processed—for example, as pass as variable or pass by value. The programmer must ensure that appropriate specifications are made to eliminate possible undesirable side effects. If the programming language being used offers only pass as variable (as, for example, in FORTRAN), the programmer must avoid undesirable side effects by exercising caution in the specification of his/her arguments. For example, the second function call in the algorithm WHATS_UP could be written using a nonsignificant temporary variable, as

```
M ← 3
L ← FUN(M) * 3
```

Any change made to M as a side effect will have no effect on the rest of the algorithm.

Pass by value and pass as variable are two examples of schemes by which parameters are made to correspond to actual arguments. Pass by value limits the subalgorithm to "read only" access to the supplied argument, whereas pass as variable permits both read and write access. This means that the value of the argument can, in fact, be changed by the subalgorithm, which is not possible with pass by value. Although pass as variable allows the capability of transferring values back to the point of call through the parameter list, it does raise the question of side effects. Where side effects are felt to be potentially dangerous, extreme care must be taken. If possible, the programmer may elect to specify pass by value in such situations.

EXERCISES 6-4

1. A procedure is desired to accept as a parameter an arbitrary vector of numeric elements. The procedure is to sort the elements "in place" and

return the sorted vector via the same parameter. What type of parameter passing is required? Why?

2. A function is required with two parameters; the first is the name of a vector, and the second is a value to be searched for. The function is to return a 1 if the value is found in the vector, and a 0 if it is not. Neither the vector nor the value are to be altered. What type of parameter passing is preferable? Why?

3. Consider the following procedure SWAP, with parameters X and Y. Assume T to be a local variable.

Procedure SWAP(X, Y).
1. T ← X
2. X ← Y
3. Y ← T
4. Return ☐

The procedure is used in the following statements, where A is a two element vector.

1. I ← 1
   A[1] ← 2
   A[2] ← 0
2. Call SWAP (I, A[I])
3. Write(I, A[I], A[2])

Give the results of the write statement under the following circumstances, explaining in detail what happens in each case:
(a) The argument-parameter correspondence is pass by value.
(b) The argument-parameter correspondence is pass as variable.

6-5 APPLICATIONS

In this section we present three applications to illustrate the use of procedures and functions in solving practical problems. The first application concerns the design of routines to insert and remove elements from a table used to define the symbols used in the translation of a computer program. The second application relates to the interesting problem of musical transposition. The important field of graph theory is introduced in the final application, which deals with the problem of finding paths in a graph.

6-5.1 PROCESSING SYMBOL TABLES

During the translation of many programming languages, a special table is constructed to contain relevant information about the variables encountered in processing the statements of the source programs. This table is known as a *symbol table* and contains information such as the name of a variable, its type, and its location in storage.

An entry is made into the symbol table whenever a new variable is first introduced. When the variable is subsequently used, the symbol table is interrogated for the information required by the compiler to complete the translation into executable machine language. In this section we investigate the design and use of routines to manage a symbol table. We shall defer until later chapters string processing considerations such as the identification and extraction of the variable names (or symbols) themselves from the statements of the program.

We turn first to the problem of placing a new entry into the symbol table. This occurs whenever a new variable is first introduced into a program. We will assume that a specification of the variable's type accompanies its introduction. The importance of the type as part of the symbol table information will be seen shortly.

When a new variable is introduced, the compiler has the following responsibilities. First, it must acquire a main storage area sufficient to hold a value of the specified type. The address of this main storage area becomes the address of the variable that must be specified in all subsequent instructions involving the variable. Then it must search the symbol table to ensure that no attempt has been made previously to define this same variable. Multiple definitions of a variable are not permitted in any programming language and attempts to redefine a variable must be treated as a programmer error. Should this attempt to define the variable prove to be legal, a new entry is made into the symbol table, consisting of the variable's name, the address of the memory area assigned to it, and an indication of its type.

Before we can design the appropriate symbol table routines, we must concern ourselves with representational issues. For the symbol table itself we will use arrays. In fact, we require more than one array, since the entries have elements of different types: string for the name of the variable and its type, and numeric for the address. We will employ the same solution used in Sec. 4-5.3 when similar requirements necessitated the use of "parallel" arrays. In this case we will use three vectors to represent our symbol table: two vectors of type string for the name and type, respectively, and a third vector of type integer for the address. This layout is shown in Fig. 6-6.

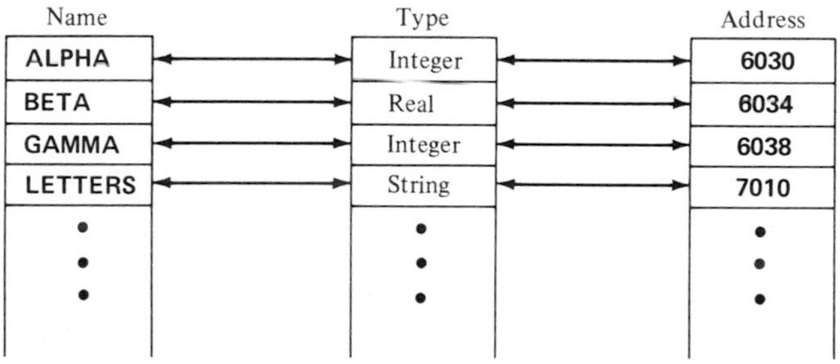

Name	Type	Address
ALPHA	Integer	6030
BETA	Real	6034
GAMMA	Integer	6038
LETTERS	String	7010
•	•	•
•	•	•
•	•	•

FIGURE 6-6
Symbol-table organization.

Following the specifications just given, a general algorithm for the insertion of an element into the symbol table would proceed as follows:

1. Search the symbol table for an element with the same name; if found, then issue an error message and return immediately
2. If no match is found, then using the next available table position, insert the name, type, and address of the symbol
3. Return

We shall not concern ourselves in this section with the acquisition of main storage space. Nor shall we be concerned with sorting the symbol table. Instead, we will assume the table to be unordered. For this reason we will be forced to rely on a linear search as described in Sec. 4-3.2. In practical compilers the number of variable names actually processed may force the designer to a more efficient search method, such as the binary search, in which case the table would have to be sorted prior to any retrieval attempts. The following search function is based on the algorithm LINEAR_SEARCH given in Sec. 4-3.2.

Function SEARCH(LIST, N, ELEMENT). This function will search (via a linear search) an unordered vector LIST of N elements for the first occurence of the particular element ELEMENT. The value returned to the point of call is an integer. If the element is found in the table, its position will be returned. If it is not found, 0 is returned.

1. [Search the vector]
   Repeat for i = 1, 2, . . ., N
       If LIST[i] = ELEMENT
       then Return(i)
2. [Match not found]
   Return(0)   □

We now give the entire insertion procedure. Its parameters are the name of the variable to be defined (VAR_NAME), its type (VAR_TYPE), and its address (VAR_ADDRESS). As mentioned, we defer until later chapters considerations of how these values are determined in practice. Since we require only that these pieces of information be inserted into the symbol table, there is no need for their values to be changed; therefore, we will assume that these parameters are passed by value. For reasons soon apparent we introduce an additional parameter SIZE, which gives the current number of entries in the symbol table. We will assume that the compiler sets SIZE to 0 prior to the first insertion. Since the value of SIZE will be altered, we must assume that it is passed as variable.

Procedure INSERT(VAR_NAME, VAR_TYPE, VAR_ADDRESS, SIZE). This procedure enters the variable defined by these parameters into a symbol table, organized as shown in Fig. 6-6, by the method described. The three vectors comprising the symbol table (NAME, TYPE, and ADDRESS) are global to this procedure. The table has a maximum of N positions and currently contains SIZE entries.

1. [Is this variable already defined?]
   If SEARCH(NAME, SIZE, VAR_NAME) ≠ 0
   then Write('*** ERROR – VARIABLE', VAR_NAME,
       'HAS BEEN PREVIOUSLY DEFINED')
   Return
2. [Compute the next available position]
   SIZE ← SIZE + 1
   IF SIZE > N   (table is full)
   then   Write('*** ERROR – TOO MANY VARIABLES DEFINED. ')
       Return
3. [Make insertion into the position found]
   NAME[SIZE] ← VAR_NAME
   TYPE[SIZE] ← VAR_TYPE
   ADDRESS[SIZE] ← VAR_ADDRESS
4. [Mission accomplished]
   Return   □

Notice the use of the function SEARCH to check for a previous definition. Rather than searching all positions of the table, we need only search those in which entries have been made. This number is given by the parameter SIZE. Notice also the error messages printed by this procedure. No doubt by now you appreciate the value of good error messages from a compiler.

We turn now to the problem of retrieving the information needed for translation each time a variable name is used in a source program. Imagine that you are a compiler attempting to translate the following statement:

ALPHA ← BETA * GAMMA

Since this is an assignment statement, the translation must cause the following machine-level operations to be performed. First, the values of the variables BETA and GAMMA must be retrieved from the appropriate memory areas (*load* operations). Since an arithmetic operation (multiplication) is indicated, a check should be made to see if the variables involved are of a numeric type (integer or real). These values are then to be multiplied together and the result (in the appropriate type) is to be placed in the memory area corresponding to the variable ALPHA (a *store* operation). Again, a check must be made on the type of ALPHA, and, if required, the result must be converted before being placed in the indicated memory area.

This sequence of actions requires three interrogations of the symbol table. The information needed consists of the addresses of the memory areas corresponding to the variables ALPHA, BETA, and GAMMA, and the types of these variables. Note that we make no mention of the values of the variables. We are concerned here only with the specification in executable machine language instructions of the steps required to perform the operations described above. The actual execution of these instructions on supplied data values will be done after the translation is complete.

For each variable name to be processed, the interrogation would be of the following form:

1. Has the variable been defined? If not, issue an appropriate error message and return immediately.
2. If the variable is found, then return to the calling program the information needed: specifically, the type and address.

We will defer discussion of what is done by the calling routine with the information returned by this procedure.

We now give the complete retrieval procedure. Its input parameter is the name of the variable being used (VAR_NAME). Its output parameters are the type and address of this variable (VAR_TYPE and VAR_ADDRESS) as recorded in the symbol table. For VAR_NAME, we will assume pass by value. For VAR_TYPE and VAR_ADDRESS, since their values are given by the procedure itself, we require pass as variable. Again, SIZE denotes the current number of entries and is also passed as variable.

Procedure RETRIEVE (VAR_NAME, VAR_TYPE, VAR_ADDRESS, SIZE). This procedure retrieves the required information (VAR_TYPE and VAR_ADDRESS) from a symbol table of N positions organized as shown in Fig. 6-6. The vectors comprising the symbol table (NAME, TYPE, and ADDRESS) are global to this procedure. The variable POS is local to this procedure.

1. [Has the variable been defined?]
    POS ← SEARCH(NAME, SIZE, VAR_NAME)
    If POS = 0
    then Write('*** ERROR – VARIABLE', VAR_NAME,
            'HAS NOT BEEN DEFINED.')
        Return
2. [If the variable is found, supply the required information]
    VAR_TYPE ← TYPE[POS]
    VAR_ADDRESS ← ADDRESS[POS]
3. [Mission accomplished]
    Return   □

The operations described here are indicative of an important field of computer science known as *information retrieval*. In general, there is a *data base* consisting of information of interest in some application area. The operators of the data base may acquire particular information by means of a retrieval operation such as that described, may add new information to the data base by means of an insertion operation similar to that given earlier, or may modify existing information by means of an update operation. Information retrieval systems are used, for example, in warehouses for parts inventories, in banks for depositors' accounts, in libraries for the whereabouts of books, or in department stores for credit card records. Properly designed, they can eliminate the need for a lot of manual clerical work, such as filing and unfiling, that is tedious, monotonous, and error-prone.

6-5.2   THE TRANSPOSITION OF MUSICAL SCORES

The ability to transpose music from one key to another is useful to anyone interested in playing music. Since the theory of transposition has a straightforward mathematical basis, the problem of transposing music presents an interesting application. Because of the alphabetic representation of musical keys and chords, it also provides us with an opportunity to exercise some of the string operations of Chap. 5.

Basically the problem is this: we have a musical score written in one key; we want the score shifted to a different key. To simplify matters we will restrict ourselves to the chords normally provided for guitar or piano accompaniment.

Musical transposition is really a problem in linear transformations, within a very limited universe. For example, the four most common chords in the key of C are C, Am (A minor), F, and G. If we transpose to the key of G, these become G, Em, C, and D, respectively. The amount of shift between the two primary keys, C to G, defines the amount of shift applied in all cases. Variations of the basic chords, such as minors and sevenths, pose no problem. For example, G minor 7th (Gm7) in the key of C transposes to D minor 7th (Dm7) in the key of G, just as G transposes to D.

For reasons soon to become clear, we will represent the possible keys by the character string

'A□□B□C□□D□□F□F□□G□□A□□B□C□□D□□E□F□□G□'

in which each of the basic notes of the scale appears twice. This will afford us an easy solution to the problem of "wraparound"; that is, A follows G. We have also included the character □ to denote the tones separating the respective notes. Two □'s indicate a full tone; where there is only a semitone difference (such as between B and C) only a single □ appears. This occurs a total of four times. We have also placed a single □ at the end of the string.

To perform a transposition, we must be given the original key and the key to which the score is to be transposed. From this we determine the amount of the "shift." Then, given any chord in the original key, we simply apply this shift to get the appropriate chord in the new key.

Let us suppose that we have been given the following input:

The original key (as a string)   e.g., 'G'

The desired key (as a string)　　　　　　　e.g., 'D'

The sequence of chords comprising
　　the score (as strings)　　　　　　　e.g., 'G', 'Em', 'G', 'C', 'D7'

Note the representation we are using for minor chords (for example, Em) and sevenths (for example, D7). We can use similar standard representations for other variations (for example, Adim, Gm7, Caug) since they will not affect our conversions. Sharps and flats, since they involve a semitone shift, will require special consideration. Note that in every case, the chord begins with one of our standard seven letters A through G.

An algorithm to perform the required transposition would have the following general form:

1. Read the original key and the key desired
2. Compute the amount of "shift" necessary to transform from one to the other
3. Repeat for the complete score
　　Read each chord
　　Determine transposed chord

The presence of sharps (denoted by #, as in 'F#') and flats (which we will denote by !, as in 'E!') requires semitone shifts in the transposition process. The sharps or flats could be part of either the original key or the required key or any of the chords to be converted, or any of the resulting chords. The actions are similar. For example, to convert a chord that is a sharp or flat (such as the chord E!), requires the addition or subtraction of 1, respectively, to the amount of shift for the particular chord to account for the semitone difference. This is reflected in the function TRANSPOSE, which will be given shortly. For example, suppose that we are converting from the key of F to the key of C. From our character string of notes we see that this represents a shift of 11. To transpose B! in the key of F, we first must subtract 1 from the shift distance. We then shift 10 notes to obtain the transposed chord F in the key of C. Transposing the chord F in the opposite direction (that is, from the key of C to the key of F—a shift of eight notes), we arrive at the semitone immediately before B. This tells us that the desired chord is B!. Similar shifts are required if either the original key or the key desired begins on a sharp or flat (such as the key F# or the key B!).

In the design of our algorithm we will assume, for the time being, the existence of a function TRANSPOSE with two parameters: the chord to be transposed, and the amount of shift necessary. The use of the double scale described earlier simplifies matters by making all shifts "forward," or positive, to avoid the problem of wraparound. To make matters even simpler, we will assume that the double scale is global to both the mainline algorithm and the function TRANSPOSE. The algorithm CHANGE_KEYS follows from the previous considerations.

Algorithm CHANGE_KEYS. The purpose of this algorithm is to transpose the chords of a musical score. The original key (ORIGINAL_KEY) and the key required (KEY_DESIRED) are read as input (in string format), followed by the chords of the score to be transposed. The string variable NOTES contains each of the notes of the scale twice, in the format described. The variables SHIFT and POSITION are of type integer. The function TRANSPOSE is called to determine the transpose of each chord.

1. [Initialize]
    NOTES ← 'A□□B□C□□D□□E□F□□G□□A□□B□C□□D□□E□F□□G□'
2. [Read keys]
    Read(ORIGINAL_KEY, KEY_DESIRED)
3. [Compute shift]
    POSITION ← INDEX(NOTES, SUB(ORIGINAL_KEY, 1, 1))
    SHIFT ← INDEX(SUB(NOTES, POSITION+1), SUB(KEY_DESIRED, 1, 1))
4. [Adjustment required?]
    If SUB(ORIGINAL_KEY, 2, 1) = '#'
    then SHIFT ← SHIFT − 1
    If SUB(ORIGINAL_KEY, 2, 1) = '!'
    then SHIFT ← SHIFT + 1
    If SUB(KEY_DESIRED, 2, 1) = '#'
    then SHIFT ← SHIFT + 1
    If SUB(KEY_DESIRED, 2, 1) = '!'
    then SHIFT ← SHIFT − 1
5. [Transpose chords of complete score]
    Write('ORIGINAL CHORD', 'TRANSPOSED CHORD')
    Repeat while there is input data
      Read(CHORD)
      If end of input then Exit
      Write(CHORD, TRANSPOSE(CHORD, SHIFT))   □

We turn now to the design of the function TRANSPOSE. The parameters of TRANSPOSE are the complete chord to be transposed (for example, D, Em, F#m7) and the distance of the shift between the original key and the required key. The following steps describe the general action of this function (with reference to the examples just given):

1. Determine position of basic chord (for example, D, E, or F) in the string of notes
2. If necessary, adjust for the presence of a sharp or flat
3. By shifting the specified amount, find the transpose of the basic chord
4. Determine if the transposed chord is a sharp or flat
5. Construct the final chord by concatenating that portion of the input chord not yet considered (for example, 'm' or 'm7' in the case of Em or F#m7)

The following function formalizes these steps.

Function TRANSPOSE (IN_CHORD, DIST). This function has two parameters: IN_CHORD , a string containing the chord to be transposed, and DIST, an integer giving the distance of the shift. Both are passed by value. The string variable NOTES is assumed to be global. Integer variables P1 and P2 are local to the function and are used to index into the string of notes. OUT_CHORD is a string variable representing the value to be returned.

1. [Find position of basic chord, shifting if necessary for sharp or flat]
 P1 ← INDEX(NOTES, SUB(IN_CHORD, 1, 1))
 IN_CHORD ← SUB(IN_CHORD, 2)   (strip off character just considered)
 If SUB(IN_CHORD, 1, 1) = '#'
 then P1 ← P1 + 1
   IN_CHORD ← SUB(IN_CHORD, 2)
 If SUB(IN_CHORD, 1, 1) = '!'
 then P1 ← P1 − 1
   IN_CHORD ← SUB(IN_CHORD, 2)

2. [Determine transposed chord]
 P2 ← P1 + DIST
 OUT_CHORD ← SUB(NOTES, P2, 1)
 If OUT_CHORD = '□'   (will be a flat or sharp)
 then If SUB(NOTES, P2 + 1, 1) ≠ '□'
   then OUT_CHORD ← SUB(NOTES, P2 + 1, 1) ○ '!'
   else If SUB(NOTES, P2 − 1, 1) ≠ '□'
     then OUT_CHORD ← SUB(NOTES, P2 − 1, 1) ○ '□'
     else Write('*** ERROR ***')

3. [Return transposed chord]
 OUT_CHORD ← OUT_CHORD ○ IN_CHORD
 Return(OUT_CHORD)   □

Because some of the steps in both the algorithm CHANGE_KEYS and the function TRANSPOSE are rather complex, we have given a trace in Table 6-6. To reduce clutter in this table, we have given a value under a particular variable only when a change is made to the variable. Both the algorithm and the function are shown: the algorithm on the left, and the function on the right. The transfers of control from algorithm to function and function to algorithm are clearly marked.

Traces are given for the following three input streams:

Stream 1:   'C ' (original key)   'F ' (desired key)
       'C '
       'Am '
       'F '   } chords in score
       'G7 '
       'E! '
       'D#m7'

Stream 2:   'E!'(original key)   'B '(desired key)
              'F ' }chord in score

Stream 3:   'E!'(original key)   'B!'(desired key)
              'F ' }chord in score

The first stream gives a typical chord sequence in the original key, followed by two somewhat more exotic chords. The second and third streams are included to illustrate the behavior of the algorithm on keys with flats. Keys with sharps will be processed in a similar fashion.

We will not explain the trace in complete detail, but we will cover some highlights of the processing of stream number 1. First, the original and desired keys are read in step 2 into variables ORIGINAL_KEY and KEY_DESIRED, respectively. The position of the original key (C) is determined through an INDEX operation on the string variable NOTES. The string sought is the first character of the string ORIGINAL_KEY, in this case simple 'C'. This is found in position 6. (Note that it also occurs in position 25; however, INDEX as defined in Chap. 5 always finds the *first* occurrence.) The value of SHIFT is then determined by looking for the first occurrence in the *remaining* portion of NOTES (given by SUB(NOTES, POSITION+1)) of the first character of KEY_DESIRED. This ensures that we always have a forward shift. 'F' is found at position 8 in the remaining string. Since neither the original key nor the desired key is a sharp or flat, no action is performed in step 4. Step 5 begins the processing of the chords of the original score.

The first chord read (into variable CHORD) is C. The function TRANSPOSE is then called with arguments CHORD ('C') and SHIFT (8). In the function, these become the value of IN_CHORD and DIST, respectively. Step 1 of the function looks for the first character of IN_CHORD ('C') in the string NOTES (which is global), and assigns its position (6) to the integer variable P1. The character C is then stripped off the string IN_CHORD, leaving it as the empty string, which we denote as ''. Since there is no sharp or flat on this chord, the rest of step 1 is bypassed. Step 2 begins by computing the position (P2) of the transposed chord, by shifting the required distance from P1. The value resulting is 14. Using the SUB function, the single character of NOTES at this position ('F') is extracted and assigned to the variable OUT_CHORD. Since it is not a space character ('□'), the rest of step 2 is ignored. Step 3 forms the final chord by concatenating to OUT_CHORD, the characters in IN_CHORD after the first. This is to account for variations such as minors, sevenths, and so on. In this case there is only the empty string, so OUT_CHORD remains as 'F', which is then returned to the point of call. At this point in the calling algorithm the values 'C' and 'F' are printed (not shown).

The processing of the next chord (Am) illustrates the action taken on a minor chord. Initially, the chord is processed as if it were simply an A. With actions like those just described, this is transposed to a D. The 'm' is then

TABLE 6-6 Trace of Algorithm CHANGE_KEYS (including Function TRANSPOSE)

Algorithm CHANGE_KEYS

Step	ORIGINAL_KEY	KEY_DESIRED	POSITION	SHIFT	CHORD	TRANSPOSE(CHORD,SHIFT)
2	C	F	?	?	?	
3			6	8	?	
5					C	Call
						F
5					Am	Call
						Dm
5					F	Call
						B!
5					G7	Call
						C7
5					E!	Call
						A!
5					D#m7	Call

Function TRANSPOSE

Step	IN_CHORD	DIST	P1	P2	OUT_CHORD
1	C	8	6	?	?
2	"			14	F
3					F Return
1	Am	8	1	?	?
2	m			9	D
3					Dm Return
1	F	8	14	?	?
2	"			22	□
					B!
3					B! Return
1	G7	8	17	?	?
2	7			25	C
3					C7 Return
1	E!	8	12	?	?
	!				
	"				
2			11	?	?
				19	□
					A!
3					A! Return
1	D#m7	8	9	?	?
	#m7				
2	#m7		10	18	□

	Algorithm CHANGE_KEYS							Function TRANSPOSE				
Step	ORIGINAL_KEY	KEY_DESIRED	POSITION	SHIFT	CHORD	TRANSPOSE(CHORD,SHIFT)	Step	IN_CHORD	DIST	P1	P2	OUT_CHORD
5	Termination					G#m7	3					G# / G#m7 Return
2	E!	B	?	?	?		1	F	12	14	?	?
3			12	11	?		2	;		26	□	C#
4				12	?		3					C# / C# Return
5					F	Call						
5	Termination					C#						
2	E!	B!	?	?	?		1	F	11	14	?	?
3			12	11	?		2	;		25	C	
4				12	?		3					C / C Return
5					F	Call						
5	Termination					C						

concatenated in step 3 of the function TRANSPOSE, to yield the correct result, Dm.

During the processing of the next chord (F), the character selected after the shift is found to be a space (step 2 of the function TRANSPOSE). A check is then made to determine if the character after it (that is, at position P2+1) or the character before it (that is, at position P2−1) is a nonspace. In this case the character at position P2+1 is 'B'. As a result, OUT_CHORD is assigned this character, concatenated with a flat symbol ('!'). In step 3 the correct chord (B!) is returned to the point of call.

The processing of the E! chord requires some special action because of the flat. First, the position of 'E' is found in the string NOTES to be 12. Then, because of the flat, this is backed up one place (P1 ← P1 − 1). The remaining steps proceed as previously described. The character at the shifted position is once again a space. This is adjusted to yield the chord A!

Detailed tracing of the remaining cases is left as an exercise.

6-5.3 FINDING PATHS IN A GRAPH

Graph theory is an important branch of mathematics or theoretical computer science that finds application in many diverse areas of computing. In this section we offer a short introduction to the field through a specific application which will, at the same time, illustrate the use of subalgorithms. Graphs themselves will be discussed in more depth in Chap. 11.

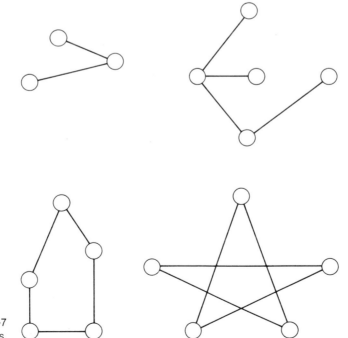

FIGURE 6-7
Examples of graphs.

The diagrams in Fig. 6-7 represent graphs. Very simply, a graph consists of a nonempty set of *nodes* or *vertices* represented as circles in our diagrams, along with a set of *edges* or *arcs*, shown as lines connecting certain pairs of nodes. Either set, or both, may have labels associated with their elements. Geometric details, such as the position of the nodes and the shape or lengths of the edges are unimportant at present.

A graph is actually a way of representing information and thus is yet another form of data structure, somewhat more elaborate perhaps than those we have seen to date. Suppose, for example, we represent a set of cities by nodes of a graph as in Fig. 6-8. A set of edges could represent major highways connecting these cities (Fig. 6-8a). For the same set of nodes, a different set of edges could represent air routes (Fig. 6-8b).

Before we can deal with graph problems, we must devise some scheme for representing the information contained in a graph in a form suitable for computer processing. For this purpose we will define a special type of matrix (or array). For a given graph G its *adjacency matrix*, $A(G)$, is a square matrix with one row and one column for each of the nodes of G. Entry a_{ij} of this matrix is defined to be a 1 if there is an edge connecting nodes i and j, and a 0 if there is not. Thus, for the graph depicted in Fig. 6-8a, the adjacency matrix is by definition

	HAMILTON	TORONTO	OTTAWA	KINGSTON	MONTREAL
HAMILTON	0	1	0	0	0
TORONTO	1	0	0	1	0
OTTAWA	0	0	0	1	1
KINGSTON	0	1	1	0	1
MONTREAL	0	0	1	1	0

An array such as this is known as a *binary* (or Boolean) array since it contains only 0 or 1 as elements.

Since we have not specified direction for the edges, this matrix is symmetric. That is, an edge connecting Hamilton to Toronto is the same as an edge connecting Toronto to Hamilton. If we were representing streets in a city, the presence of one-way streets would require an additional consideration of direction. We defer this extension to Chap. 11.

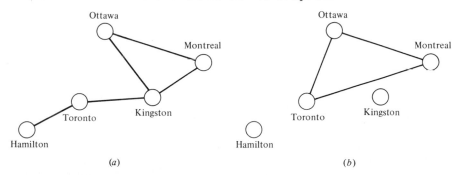

(a)   (b)

FIGURE 6-8
Graphical representations
of routes.

Now that we have a representation of a graph appropriate for computer processing, graph problems are simply problems of array manipulation.

An important problem in graph theory is the identification of paths in the graph. A *path* is defined as a sequence of edges that begins at one node and ends at another, with no breaks in between. For example, in Fig. 6-8a there is a path (of length 3 since three edges are involved) from Hamilton to Ottawa; in Fig. 6-8b, however, there is no path from Hamilton to Ottawa. This says that it is possible to drive from Hamilton to Ottawa, but not to fly. This sort of information is clearly important to travellers. For this section we will design an algorithm to determine if there exists a path (of any length) between two specified nodes of a given graph.

The elements of the adjacency matrix were defined to be 0 and 1 deliberately, to allow *logical* operations to take place. This is entirely appropriate to this problem; for example, if there exists a path from Hamilton to Toronto and a path from Toronto to Kingston, it follows that there exists a path from Hamilton to Kingston. We define logical operations $\wedge$ (AND) and $\vee$ (OR) in Table 6-7 by enumerating all possible calculations.

The extension of these operations to binary arrays is reasonably straightforward. Let the result of the matrix operation $A \wedge A$ be denoted by A^2, and its i,j entry be written as $a_{ij}^{(2)}$. Then

$$a_{ij}^{(2)} = a_{i1} \wedge a_{1j} \vee a_{i2} \wedge a_{2j} \vee \cdots \vee a_{in} \wedge a_{nj}$$

or

$$a_{ij}^{(2)} = \bigvee_{k=1}^{n} a_{ik} \wedge a_{kj}$$

By definition, if $a_{ij}^{(2)} = 1$, then there exists a path of length 2 between nodes i and j. Note that since no direction is specified, $a_{ji}^{(2)} = 1$ as well. Similarly $A^3 (= A^2 \wedge A)$ denotes the paths of length 3, A^4 the paths of length 4, and so on. For a graph of n nodes, the longest path that can be present is of length at most $n - 1$ (assuming that we do not allow nodes to be included more than once in a path).

With this background, our search for the existence of any path between two specified nodes i and j of a given graph whose adjacency matrix is denoted by A becomes the following:

1. Repeat thru step 3 for $l = 1, 2, \ldots, n$
2. Compute A^l
3. If $a_{ij}^{(l)} = 1$
   then Write('A PATH OF LENGTH ', l, 'EXISTS '),
       Exit
4. Write('NO PATH EXISTS')
5. Exit

Observe that if more than one path exists between the specified nodes, this algorithm will find the shortest.

In the formalization of this general algorithm we will design a procedure to take in two binary arrays, A and B, and return as its result $A \wedge B$. Clearly, such a procedure can be used to compute A^2 by passing as arguments A and A; in fact, for any n ($n \geqslant 1$), it can be used to compute A^n by passing as arguments A^{n-1} and A (assume that $A^0 = A^1 = A$). The following procedure POWER results from our earlier definition.

Procedure POWER(A, B, C, N). This procedure has three boolean array parameters. The fourth parameter, N, is an integer giving the size of each of these arrays ($N \times N$). A, B, and N are passed from the calling program; C is calculated as $C = A \wedge B$ according to the method just described. Assume pass by value for A, B, and N and pass as variable for C. We will also assume the existence of logical operators $\wedge$ and $\vee$ defined as in Table 6-7.

1. [Initialize result array]
   $C \leftarrow 0$
2. [Compute row elements]
   Repeat thru step 4 for i = 1, 2, ..., N
3. [Compute column elements]
   Repeat step 4 for j = 1, 2, ..., N
4. [Compute i,j element]
   Repeat for k = 1, 2, ..., N
   $c_{ij} \leftarrow c_{ij} \vee (a_{ik} \wedge b_{kj})$
5. [Finished]
   Return   ☐

To verify that this procedure is correct, let A be the adjacency matrix for the graph in Fig. 6-8a. The element corresponding to Toronto-Montreal in A should be 0, since there is no path of length 1 from Toronto to Montreal. There is, however, a path of length 2 (Toronto-Kingston-Montreal). Thus, the element corresponding to Toronto-Montreal in A^2 should be 1. Check it out. What about the same element in A^3?

We are now ready to formalize our complete path-finding algorithm using the procedure POWER. We will assume that the data are punched on cards in the following format. First a value giving the number of nodes in the

TABLE 6-7   Tables for $\wedge$ and $\vee$

$\wedge$	$\vee$
$1 \wedge 0 = 0$	$1 \vee 0 = 1$
$0 \wedge 1 = 0$	$0 \vee 1 = 1$
$1 \wedge 1 = 1$	$1 \vee 1 = 1$
$0 \wedge 0 = 0$	$0 \vee 0 = 0$

graph, followed by the elements of the adjacency matrix in row order, followed by two numbers identifying the nodes between which a path is being sought. The algorithm is to say whether or not such a path exists.

Algorithm PATHS. This algorithm determines whether or not a path exists between two specified nodes of a given graph according to the method just described. The graph itself has N nodes (N is an integer variable) and is described by the N × N adjacency matrix A. AL and T denote temporary arrays which, like A, are binary N × N arrays. I and J are integer variables denoting the two nodes between which a path is sought. L is an integer variable used to indicate the power of the adjacency matrix being computed.

1. [Input data values]
    Read(N, A, I, J)
2. [Initialize]
    AL ← A   (current power)
3. [Determine all possible paths in the graph]
    Repeat thru step 5 for L = 1, 2, . . ., N
4. [Check for a path]
    If AL[I, J] = 1
    then Write('A PATH OF LENGTH', L, 'EXISTS')
        Exit
5. [Compute next highest power of adjacency matrix]
    Call POWER(AL, A, T, N)
    AL ← T
6. [All possibilities checked—no success]
    Write('NO PATH EXISTS')
7. [Finished]
    Exit   □

Step 1 of this algorithm reads the input data according to the format described previously. In step 2 the temporary array AL is introduced. This is used to represent the "current" or most recently computed power of the adjacency matrix A. In step 2 it is initialized to A.

The search for paths is initiated in step 3. As described, there can be no path longer than $n - 1$ in a graph with n nodes, assuming that no nodes are visited more than once. Thus, we will examine powers of the adjacency matrix from 1 (the adjacency matrix itself) to N, with the value of L giving the path length being sought in each case.

In step 4 we test the appropriate entry of the most recently computed power of the adjacency matrix. If it is 1, then we have found a path. A message to this effect is printed and the algorithm terminates. Otherwise, we proceed to step 5, in which the procedure POWER is called to compute the next highest power of the adjacency matrix. Upon return from the procedure, the result is assigned to AL and the loop repeats.

If all N possibilities are tested and no path is found, control passes to step 6. A message indicating no paths is printed and the algorithm terminates.

CHAPTER EXERCISES

1. The *scalar product* (also called the *inner product* or the *dot product*) of two vectors A and B of length n is defined as

$$A \cdot B = \sum_{i=1}^{n} a_i b_i = a_1 b_1 + a_2 b_2 + \cdots + a_n b_n$$

   (a) Design a function with three parameters, A, B, and N, that computes the scalar product according to this formula.
   (b) If the scalar product of two vectors is zero, the vectors are said to be orthogonal. Design an algorithm that calls the function from part (a). If the value returned is zero, the ORTHOGONAL VECTORS message is printed.

2. Design a function FACTORIAL(N) that computes the factorial of the argument N (sometimes written $N!$). For an integer N, $N!$ is, by definition,

$$N! = N \times (N - 1) \times (N - 2) \times \cdots \times 1$$

   Incorporate in your function the special case

$$0! = 1$$

3. The number of combinations of n different objects selected r at a time (without regard to order) is given by

$$\binom{n}{r} = \frac{n!}{r!(n - r)!}$$

   where ! denotes the *factorial* function. The number $\binom{n}{r}$ is referred to as a *binomial coefficient* and has many applications in mathematics and engineering.

   Design a function with parameters n and r that computes $\binom{n}{r}$. You may assume the existence of the function FACTORIAL from Exercise 2.

4. A large number of important mathematical functions have infinite series approximations. In each case the accuracy of the approximation increases as more terms of the series are considered. Three series of this type are the following:

$$e^x = \sum_{i=0}^{\infty} \frac{x^i}{i!} = 1 + x + \frac{x^2}{2!} + \frac{x^3}{3!} + \cdots$$

$$\cos(x) = 1 + \sum_{i=1}^{\infty} (-1)^i \frac{x^{2i}}{(2i)!} = 1 - \frac{x^2}{2!} + \frac{x^4}{4!} - \frac{x^6}{6!} + \cdots$$

$$\sin(x) = \sum_{i=0}^{\infty} (-1)^i \frac{x^{2i+1}}{(2i+1)!} = x - \frac{x^3}{3!} + \frac{x^5}{5!} - \cdots$$

Design functions to compute approximations for each of these cases. Each function has a single argument x. Each approximation is to be obtained by adding new terms to the series until the absolute difference between two successive values is less than 10^{-3}; that is,

$$|\text{approx}_i - \text{approx}_{i+1}| < .001$$

Again, you may assume the existence of the function FACTORIAL developed in exercise 2.

5. A palindrome is a string that reads the same both forward and backward. For instance, 'REFER' and 'OTTO' are examples of word palindromes. In more complicated text, blanks and punctuation are ignored. Thus, 'I ROAMED UNDER IT, A TIRED NUDE MAORI' is an example of a phrase palindrome. Design a procedure to accept an arbitrary string as a parameter and determine whether or not this string is a palindrome. (*Hint:* You may wish to employ the procedures TRIM and REVERSE developed in the exercises for Sec. 6-3.)

6. Design a procedure to accept as a parameter an arbitrary string containing a series of words separated by one or more blanks, and return to the point of call the average number of letters in each word.

7. Design a procedure to accept as a parameter an adjacency matrix for a graph (see Sec. 6-5.3), and return to the point of call the number of edges in the graph.

8. Design a procedure to accept as a parameter an adjacency matrix for a graph, and a length k, and return to the point of call the number of paths of *exactly* length k in the graph.

9. The regional director for a large firm has prepared a vector of cities and towns in which the firm has an office.

'EDMONTON'	'SASKATOON'
'CALGARY'	'CHICAGO'
'MINNEAPOLIS'	'DENVER'
'REGINA'	'SPOKANE'

Periodically, he is required to visit these sites. He has at his disposal a graph showing the air routes connecting the sites. Design a function to accept as parameters the adjacency matrix of this graph, the vector of cities and towns, the name of the city or town he is currently in, and the name of the city or town he wishes to visit. The function is to return a value 1 if a direct flight exists between these two, and a 0 otherwise.

10. The Police Department requires a program to assist them in determining the identities of criminals from filed descriptions supplied by their victims. The police have cards describing known criminals. These cards have the following format:

	name	height (in inches)	weight (in pounds)	address
example:	'BUGSY MALONE'	53	119	'68 TOWN ST.'

Design an algorithm that first reads in the deck of cards giving the descriptions of known criminals and prepares a table of information on "known criminals." This set of cards is terminated by a special card of the form

'***' 0 0 '***'

A second set of cards follows, containing descriptions of criminals participating in unsolved crimes. These cards have the following format:

	description of crime	estimated height of criminal	estimated weight
example:	'21 JULY : MUGGING'	68	155

This second set of cards is terminated by a special card of the form

'***' 0 0

For each of the unsolved crimes, call a procedure (which you must also write) to determine possible suspects for the crime. This determination is based on the estimated height and weight of the criminals as given by the victims of the crimes. If the height is within 2 in *and* the weight is within 10 lb, the person is to be listed as a possible suspect for the crime involved. The parameters of this procedure are to include the table of "known criminals" and the card image on which the current crime is described.

7

Programming Style

There is no fixed set of rules according to which clear, understandable, and provable programs can be constructed. There are guidelines, of course, and good ones at that; but the individual programmer's style (or lack of it), his clarity of thought (or lack of it), his creativity (or lack of it), will all contribute significantly to the outcome.

Peter J. Denning, ACM Computing Surveys,
vol. 6, December 1974.

A computer programmer is in the business of designing solutions to problems and implementing them as computer programs. It is important that a beginning programmer recognize the importance of style in the practice of his/her craft, and develop habits of style that will carry him/her in good stead in professional life. Just as a good writing style does not come simply from a thorough knowledge of rules of English grammar, neither does a good programming style come from a thorough knowledge of the syntax of a programming language. This chapter is an attempt to bridge this gap.

7-1 THE IMPORTANCE OF STYLE

Creative people, such as artists, composers, writers, or architects, work very hard during their early training period to master the skills of their craft. At the same time, there develops a style that is unique and identifiable to each individual. This style is not incidental to their ultimate success at their craft; for if they are to be successful, it must be preferred in the marketplace to that of their competitors.

In any field, certain styles have definite advantages. For example, certain styles of music or art have wider appeal than others. There are definite styles of writing that can communicate ideas effectively; other styles per-

(By permission of King
Features Syndicate, Inc.)

haps are better at communicating technical detail. Certain styles of architecture are better suited than others for particular climatic conditions.

Style has important consequences in programming, too, some of which we will demonstrate in this chapter. Quite apart from the question of professional standards, considerations of style can actually lead to improved quality of programs. For example, research has shown that certain stylistic practices can assist in reducing the number of errors made during the development of a program. At the same time, the program itself can be more easily read and understood by other programmers, who may at some time be called upon to make modifications to it. Program maintenance, that is, the "tuning" of existing programs to meet ever-changing requirements, consumes a large portion of the professional programmer's working day; it is not uncommon, in fact, for more time to be spent on the maintenance of a program than on its original development. Not surprisingly, there is considerable interest shown, by both programmers and their managers, in suggestions that claim to make the maintenance activity less time consuming, and thus free the programmer for more original development work.

Good programming style can make an important contribution to a programmer's success. The rewards are very tangible, but a conscious effort on the part of the programmer is required. In this chapter we discuss a number of issues that relate to this matter of programming style. We try to provide appropriate motivation by considering in some detail the programming process itself. By examining what a programmer actually does, we can see possible problems and make suggestions as to how these might be alleviated.

7-2   PROGRAM QUALITY

Before we can proceed much further in our consideration of methods of improving the quality of computer programs, perhaps we ought to attempt to define what it is we are looking for. The concept is elusive. If you were to survey programmers for their answers to the question, "What are the characteristics of a good program?", you would probably receive a wide variety of responses, depending on the personal tastes and experiences of the individuals surveyed. A number of thoughts, however, might occur with greater frequency. In this section some common responses are analyzed.

The program works

It should never be forgotten that the single most important characteristic of a program is that it works. This seems obvious, yet is difficult to ascertain for programs of any significant size. The media delight in stories of catastrophic and expensive computer errors. Such incidents not only reflect badly on the industry as a whole, but also contribute in no small measure to the general public's suspicion and mistrust of computers.

Programmers must be very careful that the program eventually implemented is the program that was, in fact, required. It is all too easy to become immersed in details and, as a result, to lose sight of original specifications. Solving *almost* the required problem results in unsatisfied customers and/or unfortunate incidents. It is important that specifications (which may, in fact, change with time) be reviewed continually throughout the design and implementation of the program. The specifications themselves might be erroneous or incomplete, or simply misunderstood. Be careful not to embellish the implementation with features not specifically requested (but fun to do). This introduces even further chance of error.

The program has no bugs

Too many programmers accept bugs (or errors) in programs as a natural consequence of the human condition, and view debugging as an eternal fact of life. There is no particular reason why this should be so. Visions of evil gremlins secretly putting errors into programs spring to mind, when, in fact, the programmers themselves are usually responsible through carelessness, lack of understanding of specifications or language features, or failure to anticipate a particular situation to which the program was applied. Programmers do not do this deliberately (nobody enjoys debugging), and, in fact, can avoid making most of these errors. Yourdan [1975] talks of a style of program development that he calls "antibugging" in which the philosophy is to avoid errors from the outset.

However it might be done, it is clearly the responsibility of the programmer to ensure that his program is error-free. A good deal of research in computer science has been directed toward formal, mathematical proofs of program correctness. It has been established that such proofs are possible; however, the proof procedures are lengthy and difficult—seldom practical for real applications. This being the case, the programmer must resort to other methods, such as testing, to establish the correctness of a program. More will be said about this in the next section.

The program is well documented

It is important that computer programs be well documented. Documentation exists to assist in the understanding or use of a program. This is not only important to those charged with maintaining or modifying a program, but can also be of great value to the programmer himself/herself. Most programmers are forced to share their attentions among several things simultaneously, be they different programs, different parts of one program, or even different tasks of their job. Details of particular programs, or particular pieces of programs, are easily forgotten or confused without suitable documentation.

Documentation comes in two forms: *external documentation*, which includes things like reference manuals, algorithm descriptions, flowcharts, project workbooks, and so on; and *internal documentation*, which appears within the program listing itself (essentially, program code plus comments). The value of internal documentation cannot be overemphasized. For any program, the listing itself constitutes the front line of documentation. For this reason we are emphasizing the importance of highly readable program listings. External documentation is often aimed more at users of the program, who neither need to nor wish to plow through the program code itself, but still wish to understand what it does and how it works. External documentation provides an important complementary description of the program.

The program is efficient

The question of efficiency is a thorny one. In the early days of computing, machines were slow and small by today's standards. Programs had to be carefully designed to make maximum use of scarce resources such as execution time and storage. Programmers would spend hours trying to chop seconds off the execution times of their programs, or trying to squeeze programs into a small space in memory. A program's efficiency, most often measured by its "space-time product," was its primary figure of merit.

Today the situation has changed dramatically. Hardware costs have fallen sharply while human costs have soared. Execution time and memory space are no longer as scarce as they once were. Whereas one should always be watchful for large savings that might result from the incorporation of a different solution technique, such as a choice to replace a linear search technique by the much more efficient binary search, it no longer makes economic sense for the normal programmer to try to squeeze every last drop of efficiency out of his/her programs. Although this might be required in a few sensitive cases, for most programs the effort is not justified.

Nevertheless, there are still many programmers at the altar of the space-time product, with the result of producing unnecessarily difficult program code. A program that does not work or is difficult to maintain because of such contorted code, is clearly of low quality regardless of its space-time product.

As you can see, there are many facets of program quality. Clearly, it is important that a program work correctly and reliably, that is, that *all* the requirements for it have been met and unexpected errors are unlikely to occur; however, the matter does not end there. The evolution of programs, as studied by Lehman and Parr [1976], appears to be a real phenomenon. Programs appear to require a continuing process of maintenance and modification to keep pace with changing requirements and implementation technologies. Maintainability and modifiability are essential characteristics of real programs. A program's ability to be read and understood is an important prerequisite to its maintainability and modifiability. To summarize, we want programs that are correct, reliable, maintainable, modifiable, readable, and understandable.

In the remainder of this chapter our intention is to offer suggestions on the production of "good" programs. We begin by looking at the various phases of the programming process. Some remarks are made on potential difficulties at each stage. We then proceed to a more detailed examination of specific tasks that the programmer must perform. The emphasis throughout is on programming style and its impact on the quality of the product.

7-3   PHASES OF THE PROGRAMMING PROCESS

The term "programmer" means different things to different people. A salary survey published by the Canadian Information Processing Society defines the job "programmer" thus:

> *Analyses problems outlined by systems analysts in terms of detailed equipment requirements and capabilities. Designs detailed machine logic flowcharting. Codes program instructions. Verifies program logic by preparing test data for trial runs. Tests and debugs programs. Prepares instruction sheets to guide computer operators during production runs. Evaluates and modifies existing programs to take into account changes in system requirements or equipment configurations.*

Any consideration of the programming process itself must begin by isolating its individual component phases. One study (Cooke and Bunt [1975a]) identifies the following five phases:

1. Problem analysis
2. Solution development
3. Solution implementation as a program
4. Testing
5. Maintenance

In this section we take this list as our basis, and consider the requirements of each of these stages.

Problem analysis refers to that stage of the process in which the programmer acquires an understanding of the problem before proceeding to devise a solution. It is an "internalization" process, largely cognitive in nature and difficult to describe. Too many programmers move too rapidly through this phase; as a result, the specifications might be misunderstood or misinterpreted. Some programmers like to "echo" the problem specifications back to the analyst or designer, to reduce the chance of misunderstanding. Errors made at this stage are often difficult to detect and very time consuming to remedy later in the project.

The second stage, *solution development*, is predominantly creative. Throughout this book we have stressed the separation of this stage from that of implementation, emphasizing the design of algorithms much more than their coding in any particular programming language. Although some would argue that the ability to solve problems is an inherent gift, that it is difficult to train or to improve creativity, there is enough evidence of the merits of a systematic approach (such as the top-down method described in Sec. 7-4) to suggest that it is not a hopeless task. There is an unfortunate tendency on the part of many programmers to succumb to the lure of the machine by moving to the implementation phase before the problem has been completely solved. This can lead to major problems later in the project.

The third stage identified is that of *implementing the solution* developed in stage 2 *as an actual program* (or *coding*). Provided that the solution has been well defined, this process is largely mechanical—a relatively straightforward mental process. Various rules of the programming language used must be recalled from memory or retrieved from a manual, and the program itself assembled according to required standards of style and structure. Style and structure must be seen as an aid to producing a correct program rather than as an afterthought to be added to an already working program.

The fourth phase is concerned with demonstrating the correctness of the implemented program. Inevitably, some *testing* is performed as part of stages 2 and 3 as well. Any experienced programmer mentally tests each line as it is produced, and mentally simulates the execution of any module prior to any formal testing stage. Testing is never easy. Dijkstra has written that while testing effectively shows the *presence* of errors, it can never show their *absence*. A "successful" test run means only that no errors were uncovered with the particular circumstances tested; it says nothing about other circumstances. In theory, the only way that testing can show that a program is correct is if *all* possible cases are tried (known as an *exhaustive* test), a situation technically impossible for even the simplest programs. Suppose, for example, that we have written a program to compute the average grade on an examination. An exhaustive test would require all possible combinations of marks and class sizes; it would take many years to complete the test.

Does this mean that testing is pointless? Definitely not. The programmer can do much to reduce the number of test cases to be used from the number required by an exhaustive test. With care and effort applied to the design of test cases, many superfluous cases can be eliminated and a reasonable test can possibly be made on a relatively small number of cases.

Testing a program is every bit as much an art as creating it and must be approached with the same diligence and enthusiasm. Certain principles of testing seem clear. Try to approach the testing of a program with a sabotage mentality, the type that delights in forcing an error to reveal itself. Be suspicious of everything. Test cases should be designed from the original program specifications rather than from the program itself; if done from the program, some aspect of the problem overlooked in the implementation is likely to be overlooked in the testing as well. To reduce the chances of this sort of thing happening in professional programming shops, many managers insist that persons other than the original programmer design the test cases for a program. Purchasers of programs often have their own independently developed test data ready when the program is delivered to them. Appreciate that managers view with considerable disfavor programmers whose programs fail a customer acceptance test, since this reflects badly on the entire organization and can affect its reputation in the marketplace. However it is done, a thorough test is an essential part of any programming project.

Student programmers, unfortunately, rarely become involved in the fifth stage of the programming process, *program maintenance*. Its importance in the real world, however, cannot be overemphasized. F. P. Brooks [1975] speculates that the "cost of maintaining a widely-used program is typically 40 percent or more of the cost of developing it." Unlike hardware maintenance, program maintenance deals not with repair of deteriorated components but with repair of design defects, which may include the provision of added functions to meet new needs. The ability of programmers to produce *new* programs is clearly affected by the amount of time they spend maintaining *old* ones. The inevitability of maintenance must be recognized and steps taken to reduce its time consumption.

In the remaining sections of this chapter we concentrate largely on issues of program design and program implementation. We cannot, however, overlook the other components of the task as listed in this section. In practice these are often strongly interconnected; as a result, it is difficult, and perhaps even unwise, to attempt to treat them in isolation. At times, in fact, the most important effects of a decision at some stage may well be felt most strongly in later phases of a project.

7-4 TOP-DOWN DESIGN OF PROGRAMS

Programming is undoubtedly a complex activity—one that combines many mental processes. Many different factors must be brought together in the production of the final program. Perhaps the task is not unlike that of a juggler; if he/she tries to keep too many balls in the air at once, more than likely they will come crashing down (see Fig. 7-1).

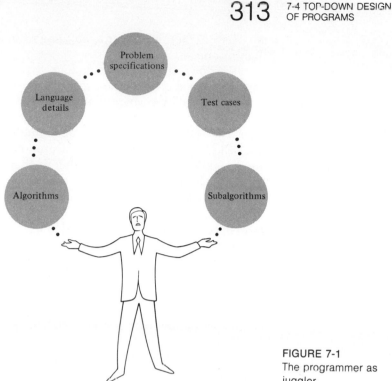

FIGURE 7-1
The programmer as
juggler.

Earlier we spoke of the importance of the "divide and conquer" approach
to the practice of programming. To continue our juggler analogy, we try to
keep only a manageable number of balls in the air at any one time. One step
in this direction has been our separation of the problem-solving task from the
programming task. In this section we consider further divisions of the
problem-solving task, through an approach known as top-down design, whose
goal is to provide some structure to the entire process.

The solution to any problem can exist in several forms or, as we will call
them, *levels of abstraction*. To quote from Niklaus Wirth [1974]: "Our most
important mental tool for coping with complexity is abstraction. Therefore,
a complex problem should not be regarded immediately in terms of computer instructions . . . but rather in terms and entities natural to the problem
itself, abstracted in some suitable sense." In Chap. 2 we spoke of general
algorithms and more detailed or more specific algorithms. We began the
solution process with a very general or abstract statement of the problem
solution, expressed in the terms of the problem itself. We then proceeded to
refine this solution by elaborating on details previously ignored, resulting in
a new solution that was considerably less abstract. This process continued
through a number of refinement stages until the appropriate level of detail
was realized. This is the essence of "top-down" design. We work from a very
abstract solution (the top level) down to an implementation, through a series
of successive refinements. The approach is independent of any programming

language; in fact, we are programming *into* a programming language rather than *in* one.

Top-down design is a technique that has been (perhaps unknowingly) applied by good programmers for years. Only recently has it even been given a name (popularized by, among others, Harlan Mills [1971]). In fact, the same idea has been called different things by different people: "stepwise refinement" (Wirth [1971]), "iterative multilevel modeling" (Zurcher and Randell [1968]), and "hierarchical programming" (Dijkstra [1968]) are examples. The approach is appealing in that it appears to define a structure for the unstructured process of program development. It focuses attention on *design* before details of implementation, and thus cuts down on the number of balls we are forced to juggle at one time. As with any tool, it is used most effectively by someone skilled in its use. Common sense, intuition, and creativity remain valuable progammer attributes.

In this section we illustrate the application of top-down design to the solution of two different problems. The first is a simple problem, intended to focus attention on the solution technique and issues of representation. The second example is more involved, and here the value of the technique in coping with complexity is more evident.

The first problem is to read a number N and list the perfect squares between 1 and N inclusive (assume all values to be positive integers). For example, if N is 30, we would list the following: 1, 4, 9, 16, 25.

We begin with a very abstract solution of this problem, which can be stated simply as:

Print a list of perfect squares between 1 and N

This describes *what* it is we want to do, but does not describe *how* we intend to do it. At this level of the design we are not concerned with this detail—it will come as we refine this basic solution, which we now proceed to do.

As our first refinement, we break this solution into two steps, or *modules*:

Read a number N

Print the perfect squares between 1 and N

To emphasize the hierarchical relationship between these two modules and the original module, we might choose to represent them as shown in Fig. 7-2. Each "level" of the figure represents one of our levels of abstraction, with the most abstract level at the top. The lines shown emphasize the refinement that has been made—the top module has been refined into the two modules at the next level.

We now proceed to refine the individual modules at the second level of abstraction, again by specifying *how* each will be done. The first (or in the diagram, the leftmost) module is already sufficiently detailed that it can be pro-

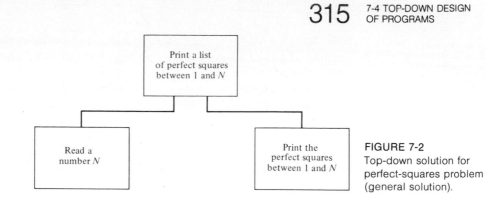

FIGURE 7-2
Top-down solution for
perfect-squares problem
(general solution).

grammed immediately [as **Read(N)** in our algorithmic language], so no
further refinement is necessary. We therefore turn our attention to the
second (or, rightmost) module, which we will break down into the following:

Compile a vector of the perfect squares between 1 and N

Print the vector

Diagrammatically, this yields Fig. 7-3.

This last refinement may not appear to have added much to the solution, but it has, in fact, introduced a vector for representing the list of
perfect squares. This is an implementation detail that was not part of more
abstract solutions. Before we turn our design into an actual algorithm, we
introduce two additional modules of the solution that will form part of the
compilation of the vector of perfect squares. These are:

Is the current number a perfect square?

Insert in next available vector position

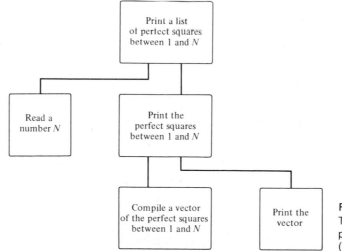

FIGURE 7-3
Top-down solution for
perfect-squares problem
(refined).

This gives us a design which we can then proceed to implement as an algorithm. The final form is shown in Fig. 7-4. Our implementation is guided throughout by this diagram. The four boxes from which no lower lines emanate (sometimes called *terminals*) can be viewed as workers; these are likely to correspond most closely to actual steps of the algorithm. The other boxes (sometimes called *nonterminals*) might define higher-level decision structures that describe how the work is to be carried out. These might appear in the final algorithm in the form of control structures, or perhaps just as comments. The algorithm PERFECT_SQUARES results.

Algorithm PERFECT_SQUARES. This algorithm reads a positive integer N and prints a list of those integers from 1 to N that are perfect squares. A vector SQUARES (indexed by VEC_PTR) is used to hold the squares for printing. T is an integer variable.

1. [Input]
   Read(N)
2. [Compile vector of perfect squares]
   VEC_PTR ← 0
   Repeat thru step 4 for I = 1, 2, . . ., N
3. [Compute truncated square root of current number]
   T ← TRUNC(SQRT(I))   (T is an integer variable)
4. [Is I a perfect square?]
   If T * T = I
   then VEC_PTR ← VEC_PTR + 1
        SQUARES(VEC_PTR) ← I
5. [Print out vector contents]
   Repeat for I = 1, 2, . . ., VEC_PTR
     Write(SQUARES(VEC_PTR))
6. [Finished]
   Exit   □

This example illustrates the application of the technique of top-down design. The solution of the problem has been derived by means of a systematic decomposition into simpler subproblems (again, the divide-and-conquer approach). At each level of abstraction, attention is focused on what it is that you want to do; you then proceed to define the modules that will cause this to happen. This collection of modules defines the next level of abstraction. This process is repeated until you have a set of modules that can be coded with relative ease.

This approach has several other advantages. The individual modules are "small" enough (in terms of function) to be easily understood. The danger of complication by outside effects has been reduced. This ought to lessen the chance of error. An organized testing pattern is suggested by the final structure diagram (Fig. 7-4). Since the purpose of each module is clearly stated, the modules can be separately tested. The testing of a module ought to be

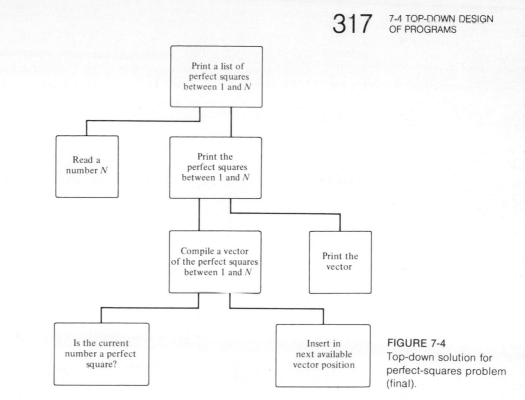

FIGURE 7-4
Top-down solution for
perfect-squares problem
(final).

simpler than the testing of a complete program. The *interfaces*, or relationships between the individual modules, are also defined by the structure diagram, and can be tested after each module has been thoroughly tested.
Perhaps an additional advantage of the top-down approach would be that a
comment structure is suggested, again by the structure diagram. This can
lead to enhanced overall readability of the resulting program.

We now turn our attention to a more complex problem, the "authorship" problem. Simply stated, the problem is one of deciding which of two
authors reputed to have written a given piece of text is the true author.
Using a technique like that described in Sec. 5-4.1, frequencies of use of
certain key discriminator words are gathered for each author based on large
volumes of text known to be written by each of them. These word frequencies are compared to word frequencies in the disputed text and from
this comparison, a conclusion can be made as to the actual author. The
authorship of a number of "Federalists Papers," a series of essays written by
Alexander Hamilton, James Madison, and John Jay in support of the adop-
tion of the Constitution of the United States, has reportedly been decided in
this manner (see Mosteller and Wallace [1963]).

Two authors, Tremblay and Bunt, both claim to have written a given
piece of text. Samples of work from each have been analyzed; the results are
given in Table 7-1. By counting the frequencies of the discriminator words in
the disputed article, and by using the method of *least-squares fit*, we will

attempt to decide which author wrote the text in question. To illustrate the method of least-squares fit, if the word 'ON' is used at a frequency of 22 times for every 1,000 words in the disputed text, then the least-squares difference is

$$(19.3 - 22)^2 \text{ for Tremblay}$$
$$\text{and} \quad (27.4 - 22)^2 \text{ for Bunt}$$

(Notice that the least-squares fit is an attempt to arrive at a measure of "closeness" that is independent of direction or sign and gives added weight to the magnitude of the difference.)

The first data card gives the number of discriminator words. This is followed by the names of the two possible authors (here, Tremblay and Bunt), followed by a deck of discriminator words with the associated authors' rates, one word per card. This is then followed by the disputed text. This is assumed to be in the same form as that in Sec. 5-4.1; words are separated by one blank ('□') and no punctuation is present.

We proceed with the design of the solution as before, beginning with a very abstract solution.

TABLE 7-1 Known Word Frequencies (per 1,000 Words) for Tremblay and Bunt

Discriminator Word	Tremblay Rate	Bunt Rate
'A'	73.2	60.4
'AND'	17.7	18.5
'BE'	18.0	16.0
'DATA'	46.9	32.3
'EFFICIENCY'	4.6	13.8
'EXAMPLE'	10.3	8.0
'FOR'	9.7	3.2
'IN'	62.1	41.0
'IS'	23.6	31.8
'IT'	17.3	5.2
'MEMORY'	8.2	12.0
'OF'	38.7	56.9
'ON'	19.3	27.4
'PRACTICAL'	2.7	15.7
'PROGRAMMER'	9.1	11.1
'QUALITY'	3.1	24.2
'THAT'	1.0	11.4
'THEN'	2.7	4.6
'TO'	22.0	19.7
'WE'	8.3	20.6
'WHEN'	6.0	12.9
'WHICH'	27.2	35.2
'WITH'	3.1	19.3

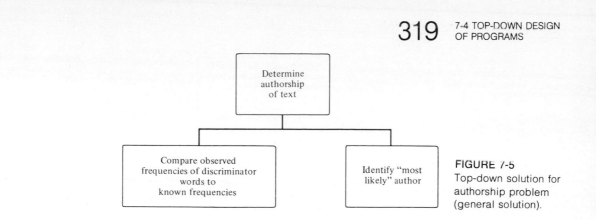

FIGURE 7-5
Top-down solution for
authorship problem
(general solution).

Determine authorship of text

This, we refine immediately into two simpler modules:

Compare observed frequencies of discriminator words to known
frequencies

Identify "most likely" author

The hierarchical organization of these three modules is shown in Fig. 7-5.
Notice that the two modules at level 2 supply details as to *how* the authorship
is to be determined, although these modules themselves require further refinement describing how each of *them* is to be done.

The refinement of the leftmost box at level 2 of our hierarchy is going
to take some doing. Here, we will be introducing details of the data structures to be used, as well as the major computations required. We begin by
breaking the module:

Compare observed frequencies of discriminator words to known
frequencies

into three simpler modules describing how this is to be done:

Obtain known frequencies as a separate vector for each author

Derive frequency vector for disputed text

Compute least-squares fit, observed versus known frequencies, for
each author

Again, the hierarchical arrangement of this portion of the solution is shown
diagrammatically, here in Fig. 7-6.

We will defer further refinement of the last (or, in the diagram, the
rightmost) of these modules at level 3 until actual implementation, since
its coding is reasonably straightforward. Further refinement of the other

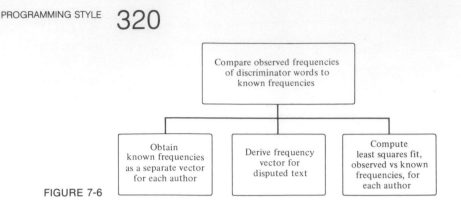

FIGURE 7-6

modules at this level, however, is required. This involves details of the input (which are given in the problem description), details of the data organization (that is, the organization of the respective vectors), and details of the computations required. Beginning with the module,

> Obtain known frequencies as a separate vector for each author

we see that this requires some input data to be read. We therefore introduce the following submodule:

> Read known discriminator word frequencies

Here is a clear instance where an upper-level module describes a controlling action, and a lower-level module does the work, supplying the results needed by the upper-level module. In particular, the upper-level module requires frequency data *for each author*; the lower-level module must supply this in each case. The upper-level module is not concerned with details of input format (this is deferred to the "worker" module), only with the requirement of obtaining separate vectors for *its* upper-level module. This illustrates an important principle of a top-down hierarchy: control information flows *downward*, results flow *upward*. Figure 7-7 shows this portion of the hierarchy.

We turn now to the second level 3 module:

> Derive frequency vector for disputed text

Working from the disputed text (which must, therefore, be read), individual discriminator words must be selected, their occurrences counted, and frequencies (per 1,000 words) computed. Thus, we define the following four submodules:

> Read disputed text
>
> Select next discriminator word

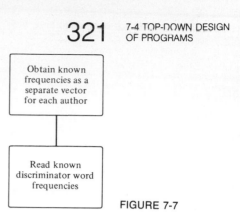

FIGURE 7-7

Count occurrences in text

Compute occurrences per 1,000 words

This portion of the hierarchy is shown in Fig. 7-8.
For our final refinement we return to the module

Identify "most likely" author

which we break down into the following simpler modules (see Fig. 7-9):

Determine each author's total least-squares difference

Select author with lowest total

This completes our design of a solution to the authorship problem. The complete solution is given in Fig. 7-10. Clearly, this solution is not unique—there could easily be others, equally valid. We have, however, by means of a reasonably straightforward process, arrived at one solution. We now proceed to implement this solution as the following algorithm, AUTHORSHIP.

Algorithm AUTHORSHIP. This algorithm tries to resolve the question of the authorship of a passage of disputed text. Data are assumed to be in the forms specified in the problem statement. The word-frequency tables are

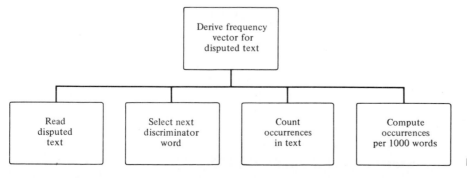

FIGURE 7-8

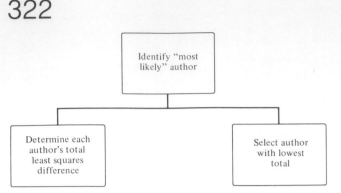

FIGURE 7-9

implemented as parallel vectors: a vector WORDS of type string holds the discriminator words, a real vector A1_RATE holds the recorded frequencies for author number 1, a real vector A2_RATE holds the recorded frequencies for author number 2, and a real vector OBS_RATE holds the frequencies computed from the suspect text. The lengths of the vectors is given by the integer variable WORD_TOTAL. The author determined to have written the passage of text in question is the author whose frequency vector fits more closely (by the method of least-squares fit) with the observed frequency vector. For simplicity, the entire text is assumed to fit into the memory area denoted by the single string variable TEXT. The string variables UNSCANNED and NEXT_WORD, and the integer variable P, will be used for text isolation. The integer variable count serves as a word counter.

1. [Read input vectors]
    Read(WORD_TOTAL)   (number of discriminator words)
    Read(AUTHOR1, AUTHOR2)   (names of authors)
    Repeat for I = 1, 2, . . ., WORD_TOTAL
        Read(WORDS[I], A1_RATE[I], A2_RATE[I])

2. [Derive frequency vector for disputed text]
    Read(TEXT)
    UNSCANNED ← TEXT   (portion as yet unexamined)
    COUNT ← 0   (word counter)
    P ← INDEX(UNSCANNED, '□')   (find first blank)
    Repeat while P ≠ 0   (for each word in text)
        NEXT_WORD ← SUB(UNSCANNED, 1, P − 1)   (pick off next word)
        COUNT ← COUNT + 1   (count total words in text)
        UNSCANNED ← SUB(UNSCANNED, P + 1)
        Repeat for I = 1, 2, . . ., WORD_TOTAL
          (compare text word selected with each discriminator word)
          If WORDS[I] = NEXT_WORD
          then OBS_RATE[I] ← OBS_RATE[I] + 1   (count occurrences)
        P ← INDEX(UNSCANNED, '□')   (find next blank)
    Repeat for I = 1, 2, . . ., WORD_TOTAL

(convert raw counts to counts per 1000 words)
OBS_RATE[I] ← (OBS_RATE[I] / COUNT) ∗ 1000

3. [Compute least-squares fit for each author]
A1_FIT ← 0
A2_FIT ← 0
Repeat for I = 1, 2, . . ., WORD_TOTAL
   A1_FIT ← A1_FIT + (OBS_RATE[I] − A1_RATE[I]) ↑ 2
   A2_FIT ← A2_FIT + (OBS_RATE[I] − A2_RATE[I]) ↑ 2

4. [Print most likely author]
If A1_FIT > A2_FIT
then Write(AUTHOR2, 'IS THE MOST LIKELY AUTHOR.')
else If A2_FIT > A1_FIT
       then Write(AUTHOR1, 'IS THE MOST LIKELY AUTHOR.')
       else Write('AUTHORSHIP UNDECIDABLE.')

5. [Finished]
Exit   □

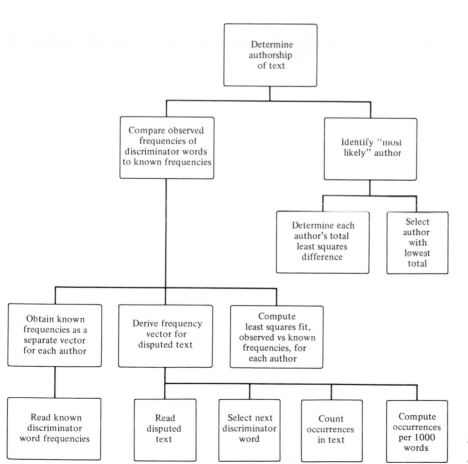

FIGURE 7-10
Top-down solution for authorship problem (final).

This concludes our discussion of top down design. Like any tool, a certain degree of skill is required before it can be used effectively. Do not be afraid to start over at any point, including while you are trying to implement your solution. In actual fact, you may find yourself returning to the "drawing board" several times before you come up with a workable solution. Although the technique requires considerable practice before it is mastered, the evidence is strong that the rewards warrant the effort.

Having considered the design of a solution, we move in the next section to questions pertaining more to implementation of this solution as an actual program.

7-5 ELEMENTS OF PROGRAMMING STYLE

[We] are beginning to see a breakthrough in programming as a mental process. This breakthrough is based more on considerations of style than on detail. It involves taking style seriously, not only in how programs look when they are completed, but in the very mental processes that create them. In programming, it is not enough to be inventive and ingenious. One also needs to be disciplined and controlled in order not to become entangled in one's own complexities.

Harlan Mills, foreword to Programming Proverbs, *by Henry F. Ledgard, Hayden Book Company, 1975.*

Throughout this chaper we have alluded to the importance of good style in programming, not only for reasons of professional standards, but more important, as an aid to the very production of programs. In programming, as in most things, style is largely a matter of personal taste. We do not intend to impose our own style on readers simply for whimsical reasons. Instead, we will present a set of guidelines that we hope will encourage the correct attitude toward the production of programs. Where possible within the limits imposed by the framework of this book, we will try to provide appropriate motivation by showing examples of both good and bad style. In cases where programming language matters are involved (as they very often are, since we are now dealing with actual programs) we can only refer the reader to the appropriate sections in one of the companion language texts.

Recently, several short books have dealt exclusively with programming from the perspective of style, notably *The Elements of Programming Style* (Kernighan and Plauger [1974]) and *Programming Proverbs* (Ledgard [1975]). In addition to the entertaining reading that they provide, these books offer suggestions on how even experienced programmers can write better programs. The material in this section is an attempt to highlight some of the ideas in books such as these.

A major theme in this book is the separation of the problem-solving component of programming from the actual writing of program code. The previous section of this chapter dealt with a structure for the problem-solving process. We turn now to considerations of actual program writing. It is understandably difficult to talk of program writing without referring to the language in which the program is being written. A command of the language being used is certainly an important factor in the production of good programs. There are, however, general issues that are largely independent of language concerns. In this section we concentrate on these general issues of program production.

The importance of two fundamental issues on the production of good-quality computer programs has been stressed throughout this chapter, and in fact throughout much of the book. These are, first, the management of complexity, and, second, readability. Certainly, these are not independent: if a program's complexity is kept under control, it is likely to be more readable than a program in which such care has not been taken; conversely, a readable program is more likely to be easily understood than one that is difficult to read, and thus by definition is less complex. Although they do approach the issue from somewhat different perspectives, they can in practice be combined effectively.

Many have recognized the fundamental importance of the management of complexity in the practice of programming. Kernighan and Plauger [1976] write:

Controlling complexity is the essence of computer programming. We will always be limited by the sheer number of details that we can keep straight in our heads. Much of what we have tried to teach in this book is how to cope with complexity.

Dijkstra [1972a] writes:

We should recognize that . . . the art of programming is the art of organizing complexity, of mastering multitude and avoiding its bastard chaos as effectively as possible.

The importance of readability has been expounded with equal force. While an important function of a program is communication with the computer, it is equally important as an aid to effective communication between human beings. Programs are read more than they are written; while debugging or modifying a program you are required to read and understand what is there before a change can be made. Readability is, as we have said, a key to understanding; programs that cannot be understood can be neither maintained nor modified and are therefore of little value. Even the author of an unreadable program, to whom it may have been clear when it was written originally, will be hard-pressed to remember what is going on after a very short period of time. On the subject of readability, Kernighan and Plauger [1976] write:

> *In our experience, readability is the single best criterion of program quality: if a program is easy to read, it is probably a good program; if it is hard to read, it probably isn't good.*

Spencer, Tremblay, and Sorenson [1977] write:

> *It is vital to distinguish between readability and writeability. It is important to be able to write programs easily. It is necessary to be able to read programs easily.*

In this section we continue to use these as basic themes as we explore the question of programming style. The section is divided into three subsections, dealing with program design issues, program implementation issues, and program presentation issues, respectively. Some of the issues relating to program design have been dealt with in Sec. 7-4. Section 7-4, however, was concerned more with the design of a solution to a problem than with its representation as an actual program. In this section we look at the design of the program itself. Having developed a suitable design, we then consider issues of implementation. Implementation issues are tied closely to the effective use of the programming language in which this implementation is being carried out. Some general principles, however, transcend language. Finally, we consider program presentation issues. The concern here is with the appearance of the program, in particular the program listing. Under each of these subheadings, we offer a number of suggestions, with illustrations where possible, that might lead to improving the quality of your own programs.

7-5.1 PROGRAM DESIGN ISSUES

In one sense this is where the largest gains in program quality can be achieved. Quality is not an "add-on" feature. Many desirable program characteristics (modifiability being a good example) are difficult to add to an existing program; they should be incorporated into the program as it is being designed. The appearance of a program is easy to change at any time; the manner in which it functions is more difficult.

The design phase is considered by many as the appropriate place to face the issue of program bugs. Programs implemented without bugs (that is, the bugs have been "designed out") need no separate debugging phase. Not only does this save valuable programmer time, but bug avoidance is the key to reliable programs.

Cleverness has been the undoing of many a programmer. Perhaps programmers pride themselves on their puzzle-solving abilities and view a program as yet another puzzle. Such an attitude to programming is revealed by excessive use of programming tricks, often tricks taking advantage of quirks in the language or implementation. Such tricks may serve to reduce the space-time product of the program, but at a considerable cost in clarity—a loss in clarity that usually cannot be afforded. As a general rule, never sacrifice clarity for efficiency; never sacrifice clarity for the opportunity to reveal your cleverness.

One of the attractive benefits of the top-down approach to the design of programs is the opportunity for clear separation of function. This is a design decision that can have significant impact on the ease with which the resulting program can be modified should the need arise. Functional separation is based on the premise that the scope of influence of any particular design decision is reasonably small—thus, its impact on the program itself should be localized as much as possible. The top-down design methodology attempts to enforce this sort of discipline by controlling the types of interaction permitted between modules. In this section we offer two small examples illustrating the use and effect of functional separation.

One illustration of the idea of functional separation involves the "magic-number" syndrome. Magic numbers are mysterious numeric constants that appear in program calculations, usually with little or no explanation. Often these are used as loop parameters, to define the size of a vector or array, or perhaps simply as constants in some formula. Although their effect appears harmless, excessive use of magic numbers can seriously compromise the modifiability of the program.

Suppose, for example, that we were to design a modest information-retrieval system to maintain information (or a data base) on the 37 students in a particular class in computer science. For each student we keep the following information: name, number of lab problems completed, lab mark to date. We choose to keep this information in three parallel vectors, each with 37 elements: a string vector **NAMES**, an integer vector **LABS_DONE**, and a real vector **LAB_MARK**. Routines are written to insert new lab marks, to correct marks erroneously recorded, or simply to display selected pieces of information. Each of these operations requires one or more scans through one or more of these vectors, with a loop something like the following:

Repeat for I = 1, 2, . . ., 37

In this particular program the number 37 is clearly a magic number. In fact, the execution of this program is heavily bound to this number. It may appear in more than a dozen different places in the code, ranging from declarations, to computation loops, to print loops. The use of this number ties these various routines closely together, in a way that is not completely obvious.

Suppose that we wish to use this program for a different class, say a larger class with 212 students. First, we must go through this program and change all occurrences of the magic number 37 to the new magic number 212. We must find them all too; any we do not find will eventually (not necessarily immediately) cause an error. In addition, occurrences of 36 or 38 may need to be changed to 211 or 213, and so forth. The use of magic numbers has certainly made this program unnecessarily difficult to modify.

This particular case is quite easy to remedy. Rather than using a magic number for the class size, introduce a variable, say, **CLASS_SIZE**, and use this variable every place the class size is required. In fact, the name chosen adds a measure of internal documentation to the program. Each time the program is used for a different class, all you have to do is give a value to CLASS_SIZE, either by assignment (compile-time) or by input (run-time). This is an example of functional separation—we have designed the routines of this new program to be independent of the actual class size.

Functional separation is often achieved through the use of subprograms; that is, "worker" modules are implemented as subprograms that are called by higher-level modules when their action is required. Functional separation is enhanced if all data values are exchanged through the parameter list rather than through global variables, and if the use of side effects is minimized. In the example of the information-retrieval system, let us suppose that each of the required operations on the data base is encoded as a separate subprogram, and further that all information exchanged between the subprogram and point of call is exchanged through the parameter list. This being the case, the following subalgorithm might represent the retrieval operation.

Procedure RETRIEVE(STUDENT, LABS, MARK, CLASS_SIZE). The purpose of this procedure is to search the student information data base (recorded in global vectors NAMES, LABS_DONE, and LAB_MARK) for the information pertaining to the particular STUDENT name (STUDENT is a string variable). His/her lab mark is to be returned in the real variable MARK, and the number of labs he/she has completed is to be returned in the integer variable LABS. The integer variable CLASS_SIZE gives the number of students in the data base. The integer variable IND is local to the procedure and is used to index into the appropriate vectors.

1. [Initialize]
   IND ← 0
2. [Find the appropriate student (assuming names are unique)]
   Repeat for I = 1, 2, . . ., CLASS_SIZE
      If NAMES[I] = STUDENT
      then IND ← I
3. [Was the correct student found?]
   If IND = 0
   then Write('STUDENT', STUDENT, 'NOT ON FILE')
        LABS ← 0
        MARK ← 0
        Return
   else LABS ← LABS_DONE[IND]
        MARK ← LAB_MARK[IND]
   Return ☐

A call to this procedure might be something like the following:

```
call RETRIEVE('JOHN BROWN',   DONE, GRADE, 37)
Write('JOHN BROWN HAS COMPLETED', DONE, 'LABS FOR A MARK OF',
    GRADE)
```

The procedure **RETRIEVE** performs a simple linear search of the database for the desired information. For larger classes this might get to involve a considerable expense. If this program were to be used routinely for larger classes, the designer might wish to modify the search strategy, say, to a binary search, to cut down on the execution time required. The functional separation of the search activity, in the routine **RETRIEVE**, for example, from the operations of the calling modules has made this a fairly straightforward change, even though its ramifications might be quite profound. This design change requires modifications only to those routines in which searching is an integral component. Routines not intimately connected with the searching operation, requiring only the *results* of the search, need not be concerned with how that search is carried out; therefore, their design should be independent of this decision. This is another example of the value of functional separation.

In this subsection we have been concerned with two issues at the level of program design. We have advocated first that programs always be written in the most straightforward manner possible—that clarity should *never* be compromised for cleverness, and second that the concept of functional separation be an important design consideration. The union of these two ideas can lead to a clean, simple design, the consequences of which will be appreciated throughout the life of the program.

7-5.2 PROGRAM IMPLEMENTATION ISSUES

The actual implementation of a program is felt by many to be the most interesting and the most important phase of the task. Although the problem solution and program design may already be known, the implementation as a program seems never to be as straightforward as believed. The last decade has seen the emergence of an approach to programming known as *structured programming*, naively thought by some to eliminate all implementation problems. Unhappily, this is not the case. As Harlan Mills [1976] observes in an article on the growth of data processing, "there is a great deal of oversell and confusion about structured programming, primarily because an adolescent data processing community is anxious to find simple answers to complex problems."

Structured programming is nothing more than an approach to program implementation in which rigor and structure displace "seat-of-the-pants" programming. Good programmers programmed this way long before it was given a name. The approach employed in this book has been consistent with the philosophy of structured programming.

To a large extent the "structure" of a program is determined by the constructs used to direct the flow of control. It is important to remember that while you read the program listing from top to bottom, execution of the program may proceed in a very different way. One of the main goals of structured programming is to structure the flow of control in such a way that the execution sequence is as close as possible to the reading sequence. This enforces a discipline on the programmer, in terms of control structures that can be used, and further, on the manner in which they can be used. As a first approximation we limit ourselves to the two control structures introduced in Chap. 3—the IF-THEN-ELSE construct and the REPEAT construct—and combinations of these.

According to the strict letter of the law, any program written using only these control structures is, by definition, a structured program. Unfortunately, bad programs can be written using any technique. What is more important is an adherence to the *spirit* of the law or the intent of structured programming.

Consider the algorithm MAXMIN_3, which was given in Sec. 3-2 to illustrate the use of nested IFs.

Algorithm MAXMIN_3. This algorithm reads three numbers, A, B, and C, and prints the largest value and the smallest value. Values are assumed to be distinct.
1. [Input data values]
    Read(A, B, C)
2. [Determine largest and smallest values]
    If A < B
    then If A < C
        then MIN ← A
            If B > C
            then MAX ← B   (A < C < B)
            else MAX ← C   (A < B < C)
        else MIN ← C   (C < A < B)
            MAX ← B
    else If A > C   (A > B at this point)
        then MAX ← A
            If B > C
            then MIN ← C   (A > B > C)
            else MIN ← B   (A > C > B)
        else MAX ← C   (C > A > B)
            MIN ← B
3. [Output results]
    Write('LARGEST VALUE IS', MAX, ', SMALLEST IS', MIN)
4. [Finished]
    Exit   □

By the strict definition of structured programming, this is, in fact, a structured program. Its readability can be improved, however, possibly at a small sacrifice in efficiency, by "unwinding" some of the nesting. The human mind has difficulty comprehending complex nested structures; it requires the retention of several different program states at the same time. Deeply nested structures are highly error-prone and normally can be avoided.

There are several ways of avoiding deeply nested structures. One method, perhaps appropriate in this particular case, is to use compound conditions in an if statement to define the specific alternatives more precisely. You must be aware, however, that the condition itself must be reasonably easy to understand, or else there is little value in the change. A second method is simply to repeat code; for example, a test might be repeated. If to avoid repeating a small section of code requires a contorted structure, dues will be paid later.

The following revision to the algorithm MAXMIN_3 illustrates the application of the second of these methods.

Algorithm MAXMIN_3 (revised). This algorithm reads three numbers, A, B, and C, and prints the largest value and the smallest value. Values are assumed to be distinct.

1. [Input data values]
   Read(A, B, C)
2. [Determine largest value]
   MAX ← A
   If B > MAX then MAX ← B
   If C > MAX then MAX ← C
3. [Determine smallest value]
   MIN ← A
   If B < MIN then MIN ← B
   If C < MIN then MIN ← C
4. [Output results]
   Write('LARGEST VALUE IS', MAX, ', SMALLEST IS', MIN)
5. [Finished]
   Exit   □

Although the revised program is one step longer and may take longer to execute, it is probably easier to understand than the first. Let not the fact that a program *appears* to be structured prevent you from making changes to improve it.

Aside from major issues of program structure, there are other concerns at the implementation stage. Although this may seem like an overconcern with details, as Wirth [1976] observes, "in programming, the devil hides in details."

A number of experiments (see, for example, Weissman [1974]) have shown that the choice of variable names can play a significant role in the understandability of programs. As we have said earlier, the program listing itself serves as the front line of documentation for a program; the use of variable names that clearly describe their purpose in the program can be more valuable in making the program "self-documenting" than can the presence of comments. This is because the variables are a fundamental part of the coding itself, whereas comments are only an appendage to it. Too many programmers overlook this obvious point, writing something like

X ← Y * Z

where, with a little thought, they could have written

FORCE ← MASS * ACCELERATION

Most programming languages allow some expressive capability in the naming of variables. Where this is available, take full advantage of it.

A final note in connection with the use of variables: some programming languages (FORTRAN and PL/I being notable examples) allow the programmer to use variables that have not been *explicitly* defined in the program. The *implicit* declaration feature of some compilers will assign default attributes to an undeclared variable at the time of its first occurrence. As a general rule, it is dangerous to rely on these default declarations. Failing to declare variables is a bad habit to get into, one that can only lead to grief eventually. Consider it a matter of good style when implementing a program always to declare *all* variables.

7-5.3   PROGRAM PRESENTATION ISSUES

The format and appearance of the program listing are not incidental to the quality of the program. More can be done here to improve the readability of a program than at almost any other point. In this subsection we consider two facets of this question: comments and paragraphing.

Comments constitute a major component of a program's internal documentation. They serve to help the reader understand the intent or purpose of portions of code, and can also assist in explaining the logic of difficult sections. Beginning programmers are seldom given any instruction in the writing of comments, yet the writing of good comments is probably as important, and perhaps as difficult to learn, as the writing of good programs. Good comments cannot do much to improve bad code, but bad comments can seriously detract from good code.

One of the most comprehensive discussions of the use of comments is an article by Sachs [1976]. Some good material is also found in Kernighan and Plauger [1974] and Ledgard [1975].

Many programmers fall in one of two extreme categories: those who write few or no comments, and those who overcomment. Each of these extremes detracts from the readability of the program in its own way—undercommenting by failing to provide appropriate supporting information, and overcommenting by increasing the clutter. Comments should not just "parrot" the code, but should explain and support it. This requires, perhaps, that the programmer approach his/her program simultaneously from two points of view: that of a programmer and that of a documentor, fully appreciating the objectives of each.

Much of what you can or cannot do with comments is a function of the programming language being used. Unfortunately, some languages in which comments might be most needed offer little in the way of features to support good comments. Most languages allow the programmer to designate an entire line (or card) as a comment. This feature allows the use of fairly lengthy multiline comments that can be used to explain the purpose of and interface with a particular program component or module. For example, every subprogram should begin with such a comment to explain its purpose, the manner in which it is called (including the meaning of its parameters), and any special situations that might arise. Such a comment might read like the description at the start of each algorithm in this book.

In addition to the separate line facility, some languages (such as PL/I) allow comments to be placed on the same line as program code. This can be very useful in the writing of short comments to explain a single line or single operation, without breaking the overall visual pattern of the program itself. Such comments should be separated as much as possible from the code, perhaps by moving them over to the right-hand side of the listing.

As a final point on comments, always make sure that comments and code agree. If you make a change to the code, be sure that a similar change is made to any comment relating to it. This is often overlooked.

The value of *paragraphing*, or controlled indenting of a program listing, is, once again, in enhancing readability. In any written text paragraphing serves two main purposes: to identify structural units of the text, and to relieve the tedium of the reading process. Both apply to programs as well. The danger of unparagraphed programs is well illustrated with a quote from Robert Frost.

> *The view was all in lines*
> *Straight up and down of tall slim trees*
> *Too much alike to mark or name a place by*
> *So as to say for certain I was here*
> *Or somewhere else;*
>
> Robert Frost, "The Wood-Pile," 1914

Paragraphing can be of great assistance in revealing the logical structure of a program (or algorithm). Throughout this book, for example, we have adopted the convention that, as a rule, a new line is begun for each alternative of an IF construct, and that where alternatives consist of more than a single statement, subsequent statements be indented to the point of the first. This shows the alternatives clearly, and the conditions affecting their choice. Further, if nesting is involved, it, too, is clearly shown. The following example taken from Chap. 4 illustrates this scheme:

```
If SCORE1 > SCORE2
then STATS[ROW1, 2] ← STATS[ROW1, 2] + 1
     STATS[ROW2, 3] ← STATS[ROW2, 3] + 1
else If SCORE2 > SCORE1
```

```
    then STATS[ROW2, 2] ← STATS[ROW2, 2] + 1
         STATS[ROW1, 2] ← STATS[ROW1, 2] + 1
    else STATS[ROW1, 4] ← STATS[ROW1, 4] + 1
         STATS[ROW2, 4] ← STATS[ROW2, 4] + 1
```

Imagine how difficult even this small section of code would be to read without the support provided by the paragraphing. Paragraphing can also be used to delineate the range of a loop if all the statements in the range are indented under the **REPEAT**. Statements following the loop return to the original starting position. Again, a short example from Chap. 4:

```
Repeat for ROW = 1, 2, . . ., 12
  If TEAMS[ROW] = TEAM1
  then ROW1 ← ROW
  else  If TEAMS[ROW] = TEAM2
        then ROW2 ← ROW
STATS[ROW1, 1] ← STATS[ROW1, 1] + 1
STATS[ROW2, 1] ← STATS[ROW2, 1] + 1
```

Issues of program presentation seldom *cause* errors, but they can play a large role in avoiding them. Too many programmers dismiss these matters as simply "window dressing," preferring instead to devote their energies to the more creative aspects of the job. As a key factor in the readability of the final program, its presentation serves to enhance its overall quality, in addition to giving the program a pleasing, professional appearance.

7-5.4 POSTSCRIPT

A shattered flower vase is often cheaper to replace than to repair.
  Henry F. Ledgard, Programming Proverbs, *Hayden Book Company, 1975.*

Paradoxically, it seems to be almost a law of nature that the better the program you produce, the more likely that it will at some time need to be changed. People tend to take good programs and adapt them to their own purposes rather than take the trouble to design their own programs. This is not to imply that this is a bad thing—on the contrary; whenever possible, programmers should be encouraged to adapt, rather than reinvent. Every change, or patch, to a program, however, increases its "disorder." Before long, a program that has undergone many changes becomes a very fragile structure, like a house of cards; if you touch it in the wrong place it all comes tumbling down.

It is a mark of maturity for a programmer to know when it is time to replace a program. When that time comes, do not be afraid to scrap the old program and start over. This is not a defeatist attitude. As Ledgard [1975] points out, "lessons painfully learned on the old program can be applied to the new one to yield a better program in far less time with far less trouble."

Rather than being subjected to the insults and abuses of disgruntled modifiers, old programs that have served well deserve a graceful retirement.

7-6   PROGRAMMING AS A HUMAN ACTIVITY

The successful programmer must learn to master a wide range of different skills, from those of a creative nature such as problem analysis and solution design, through to purely mechanical tasks such as coding. These tasks require different abilities, yet all must be performed correctly if a correct program is to be produced. In this section we put aside technical questions and consider the effect of "the human condition" on the performance of a programmer's activities. Various aspects of the issue are considered, and suggestions are offered as to how the problem might be contained.

Programmers are not machines. They are human beings performing quite a complex task, and for this reason can be expected to commit errors as a result of the basic "human condition." Human beings have real limitations in perceptual capability and performance that vary from one to another. It is important that each programmer recognize and learn to live within his/her own basic limitations.

In this section we speculate on the effects of various factors influencing a programmer writing programs, on the commission of errors. We consider the causes of these errors, their ramifications, and various things that might be done to avoid or contain them. We begin by classifying the causes of error into four broad categories:

1. Information-processing effects
2. Social effects
3. Environmental effects
4. Personality effects

Material for this discussion is drawn primarily from a pair of articles by Cooke and Bunt [1975a, 1975b], and from books by Weinberg [1971] and Brooks [1975].

7-6.1   INFORMATION-PROCESSING EFFECTS

Errors due to information-processing effects result from inherent limitations in the reliability of human perception, memory, and cognition. Programming undeniably has a large information processing component, the information comprising such things as the problem specifications and requirements, the rules and features of the programming language being used, and whatever parts of the program have already been written. Relevant here is the programmer's ability to input information visually, to organize the raw data (characters, spaces, etc.) into meaningful pieces (variable names, keywords, control structures, etc.), and to interact effectively with the various levels of memory. Without going into details, the programmer as a human being is a very imperfect information processor: symbols are confused, language details are forgotten, the execution of routines is misunderstood, and so on. This is a frequent and important cause of error, and one to which a certain amount of attention has recently been devoted (see, for example, studies by Weissman [1973, 1974] and Brooks [1973]).

What can be done to reduce this source of error? The problem can be attacked on several fronts. Language designers can use the results of research in this area in the design of new programming languages better suited to human information-processing limitations (see, for example, Gannon [1975]). This, however, is beyond the scope of this book and certainly beyond the control of most programmers. The individual programmer can do a great deal within the context of his/her own programs by maintaining an appreciation of what the potential difficulties are. For instance, an awareness of the problems of visual input suggests that the layout of a program listing is very important. This is consistent with previous remarks on the importance of program readability. Research in eye movements suggests that effective input takes place only when the eyes fixate on some part of the display. A programmer can enhance desirable eye movements through judicious use of indentation and blank lines ("white space") to mark structural entities of the program (loops, decision alternatives, etc.). This can also assist in the perception and understanding of the structure. Perceptual speed and accuracy are heavily influenced by properties of the display. Errors increase with complexity, unfamiliarity, and clutter. This reinforces the remarks on spacing and indentation to reduce the clutter of the program listing, and discourages experimentation with unfamiliar language features.

Clutter is also increased by injudicious placement of comments. Comments are certainly important in enhancing the understandability of a program, but must be placed so as not to interfere with the perception of its important components. It is far better to have a multiline comment precede a program component rather than have it interleaved within the statements of the component. Another possibility is to take advantage of space to the right of the statements, if the language allows comments to be placed there.

Research on memory suggests definite upper limits on the number of "units" that can be processed effectively, where a unit can be a simple element or a related group of simple elements. For example, we might consider an operator to be a unit, or a statement, or even an entire loop or procedure. This research says something about the size of what we referred to as a "module" in Sec. 7-4. If a module, be it a procedure, a loop, or a decision alternative, is to be fully understood, its size should be limited accordingly. It might also say something about the depth of nesting before confusion sets in. As a programmer, you should try to determine your own personal limits and try not to exceed them.

An additional consequence of memory limits is that a programmer is unlikely to remember all details of a large program while working on one part

of it. This suggests two things: first, that the dependencies on other parts of
a program be kept to a minimum (that is, that interfaces between modules be
as clean and as clear as possible), and second, that ready reference to the
purpose of each module in a program be provided, either in the form of a set
of comments at the head of the module or in the form of external documentation. The method of top-down design tries to enforce this discipline by
limiting the number and direction of intermodule references. In a top-down
hierarchy, modules must be unaware of other modules on the same level
(that is, modules must be "blind" laterally), all references downward in the
hierarchy are *control* references (that is, *do something*), and all references
upward are the return of results. In addition, the top-down structure diagram serves as important external documentation.

In his paper "The Humble Programmer," Dijkstra [1972a] writes of the
importance of keeping programs within the "intellectual grip" of the programmer. This comment clearly relates to the information-processing capabilities of the programmer as a human being. We have suggested a number
of ways in which an appreciation of these factors can reduce the chance of
error. As a general rule, we encourage the practice of "defensive programming," which, like defensive driving, centers on an anticipation of errors,
both your own and those of other people. The most effective programmers
are those who anticipate difficulties and take steps to avoid them.

7-6.2 SOCIAL EFFECTS

*The practice of programming has long since passed the point where it is
entirely a private activity.*

R. Conway and D. Gries, *An Introduction to Programming,*
Winthrop Publishers, Inc. 1973.

Many aspects of programming involve communication between different
people (directly, or indirectly through shared program code). In large programming projects, programmers rarely work on a strictly individual basis.
Group efforts may benefit from cooperation among the members of the
group, or may be hindered by excessive competition and "ego flashing."
There is clearly a connection here with personality effects. Programmers in
a group must be willing to accept suggestions and criticisms made for the
"common good" of the project—errors must not be interpreted as public
advertisements of a programmer's shortcomings.

The structure of a programming group must be designed to help overcome programmer egos. Programmers in a group are in most cases in competition with each other for promotions, raises, and so on, yet all benefit
from successful completion of the project. Brooks [1975] and Weinberg
[1971] write at some length about problems of group work. Weinberg introduces the notion of "egoless" programming that appears to have found
successful application. According to this view, programs, unlike paintings or

sculpture, for example, are not to be viewed as extensions of the creator's ego, and thus can benefit from the suggestions and criticisms of others. The essence of the egoless approach is that each programmer in a group recognize his/her limitations, and calls on other members of the group to read his/her programs for errors and clumsiness. Systematic code perusals form a regular part of teams organized on this basis.

There appear to be definite merits to such an approach. Overall project debugging time appears to be reduced. Programs appear to be more adaptable, since more than one person must be capable of understanding what is going on. Project schedules appear to be less affected by individual programmers missing days for illness, courses, and so on. Finally, each programmer can only improve his/her programming by reading the programs of other people. This leads not only to greater job satisfaction on the part of the group members, but also to an increase in the general level of competence of the group.

7-6.3 ENVIRONMENTAL EFFECTS

The environment in which a programmer works has a clear effect on his/her effectiveness. This ranges from hardware and software support of activities relating to the development of programs to the character of office surroundings. The availability of suitable programming languages and computer systems is a definite factor, but often beyond the control of the individual programmer. Considerable investigation has taken place on the merits of interactive programming as compared to batch programming in terms of programmer effectiveness. Data reported by Brooks [1975] suggest that an interactive facility can double the productivity of programmers in certain applications. This may, however, not be realizable in installations where appropriate languages and support systems (for example, a good system for managing data) are not available interactively. Brooks concludes: "I am convinced that interactive systems will never displace batch systems for many applications."

There may be other valuable assists to the production of programs. For example, a good library of programs and routines can save needless "reinvention of the wheel." Systems are also available to aid in the debugging, testing, performance tuning, and documenting of programs. Such systems, an example of which is the Programmer's Workbench developed at Bell Laboratories (Dolotta and Mashey [1976]), cater specifically to the needs of persons developing computer programs and, as a result, reduce the chance of error in the process.

The environment in which a person works best varies from person to person. Some people demand absolute quiet, others perform best when there is some background noise. Many programmers, by choice, are "night hawks," who relish slaving over a hot computer until the early morning hours. Some like people around at all times, others prefer privacy. It is undoubtedly the task of management to find and set the environment that maximizes the productivity of the programming personnel. Management must be prepared, however, to alter standard practices to accommodate special situations. Both Weinberg [1971] and Brooks [1975] offer amusing anecdotes concerning the discovery of suitable environments.

7-6.4 PERSONALITY EFFECTS

Despite their best intentions, many people may never be able to be effective programmers. In some cases this inability may be due less to information-processing limitations than to individual personality traits. Factors such as carelessness, lack of motivation, lack of organization, inability to take (or give) direction, or inability to work under stress can all lead to errors being made. Since little in the way of useful research has been conducted in this aspect of programming, it is difficult to assess its degree of impact. It is, however, something of which programmers and managers must be aware.

7-6.5 CONCLUDING REMARKS

We conclude our discussion of the human element in programming by observing that recognition of the problem is a large step toward its solution. Individually, each programmer must recognize and learn to live within his/her own human limitations. Management must learn to respect individual differences and recognize the effects of the human element in group activities. At the present time there has been insufficient research in the area to be able to offer concrete suggestions. The need for more activity in this important area is seen by many to be critical.

7-7 SUMMARY

The programming profession is in the midst of a period of critical introspection. The need to improve the quality of programming products is real and immediate. Current thinking is that this might best be done by instilling in programmers a keener appreciation of the importance of programming style.

In this chapter we have tried to give a reasonably comprehensive overview of current thoughts on the practice of programming. We have tried to give a working definition of program quality as well as ideas as to how such quality might be achieved. A formal structure was presented for the problem-solving process. Suggestions were offered on the actual production of programs. Finally, some thoughts were offered on the effect of the human element.

Because of space limitations, we have only been able to scratch the surface of the field. Readers wishing to pursue this subject are encouraged to refer to the list of references at the end of the chapter. In addition to the references cited within the text of the chapter, we recommend a special issue of *Computing Surveys* (ACM [1974]) devoted in its entirety to the subject of programming, and books by Dijkstra [1976] and Jackson [1975].

BIBLIOGRAPHY

Association for Computing Machinery: *Computing Surveys*, special issue on programming, ed. Peter J. Denning, vol. 6, December 1974.

Brooks, Frederick P. Jr.: *The Mythical Man-Month*, Addison-Wesley, Reading, Mass., 1975.

Brooks, R.: "Cognitive Processes in Computer Programming," Psychology Department, Carnegie-Mellon University, 1973.

CIPS: *Computer Magazine*, vol. 4, April 1973.

Cooke, John E., and Bunt, Richard B.: "Human Error in Programming as a Result of Conventional Training Methods," *Proc. IBM Scientific Symposium on Software Engineering Education*, May 1975, pp. 63–69. (a)

———, and Bunt, Richard B.: "Human Error in Programming: The Need to Study the Individual Programmer," *INFOR*, vol. 13, October 1975, p. 296. (b)

Dijkstra, Edsger W.: "Complexity Controlled by Hierarchical Ordering of Function and Variability," in *Software Engineering*, ed. P. Naur and B. Randell, NATO Scientific Affairs Division, 1968, p. 181.

———: "The Humble Programmer," *Communications of the ACM*, vol. 15, October 1972, p. 859. (a)

———: "Notes on Structured Programming," in *Structured Programming*, ed. Dahl, Dijkstra, Hoare, Academic, New York, 1972, p. 1. (b)

———: A *Discipline of Programming*, Prentice-Hall, Englewood Cliffs, N.J., 1976.

Dolotta, T. A., and Mashey, J. R.: "An Introduction to the Programmer's Workbench," *Proc. Second International Conference on Software Engineering*, October 1976, p. 164.

Gannon, John D.: "Language Design to Enhance Programming Reliability," Technical Report CSRG-47, Computer Systems Research Group, University of Toronto, 1975.

Jackson, Michael A.: *Principles of Program Design*, Academic, New York, 1975.

Kernighan, Brian W., and Plauger, P. J.: *The Elements of Programming Style*, McGraw-Hill, New York, 1974.

———, and Plauger, P. J.: *Software Tools*, Addison-Wesley, Reading, Mass., 1976.

Ledgard, Henry F.: *Programming Proverbs*, Hayden, Rochelle Park, N.J., 1975.

Lehman, M. M., and Parr, F. N.: "Program Evolution and Its Impact on Software Engineering," *Proc. Second International Conference on Software Engineering*, October 1976, p. 350.

Mills, Harlan: "Top Down Programming in Large Systems," in *Debugging Techniques in Large Systems*, ed. R. Rustin, Prentice-Hall, Englewood Cliffs, N.J., 1971, p. 41.

————: "Software Development," *Transactions on Software Engineering*, vol. SE-2, 1976, p. 265.

Mosteller, Frederick, and Wallace, David, L.: "Inference in an Authorship Problem, *Journal of the American Statistical Association*, vol. 53, 1963, p. 275.

Sachs, Jon: "Some Comments on Comments," *Systems Documentation Newsletter*, vol. 3, December 1976.

Spencer, Henry A., Tremblay, Jean-Paul, and Sorenson, Paul G.: "Programming Language Design," Department of Computational Science, University of Saskatchewan, Saskatoon, 1977.

Weinberg, Gerald M.: *The Psychology of Computer Programming*, Van Nostrand Reinhold, New York, 1971.

Weissman, L.: "Psychological Complexity of Computer Programs: An Initial Experiment," Technical Report CSRG-26, Computer Systems Research Group, University of Toronto, 1973.

————: "A Methodology for Studying the Psychological Complexity of Computer Programs," Technical Report CSRG-37, Computer Systems Research Group, University of Toronto, 1974.

Wirth, Niklaus: "Program Development by Stepwise Refinement," *Communications of the ACM*, vol. 14, April 1971, p. 221.

————: "On the Composition of Well-Structured Programs," *Computing Surveys*, vol. 6, December 1974, p. 247.

————: *Algorithms + Data Structures = Programs*, Prentice-Hall, Englewood Cliffs, N.J., 1976.

Yourdan, Edward: *Techniques of Program Structure and Design*, Prentice-Hall, Englewood Cliffs, N.J., 1975.

Zurcher, F. W., and Randell, B.: "Iterative Multi-level Modelling—A Methodology for Computer System Design," *Proc. IFIP Congress*, 1968, p. D138.

8

Numerical Computations

One of the earliest and most important applications of computers was to the solution of numerical problems. Vast sums of money are spent each day in the solution of such problems. The study of these problems and techniques for their solution is an important field in both applied mathematics and computer science.

In this chapter we provide a brief introduction to the field. We begin by considering the important question of error in numerical calculations—what causes it, and how to deal with it. We then turn to consider techniques for the solution of a number of important problems in the field of numerical mathematics.

8-1   ERROR

The organization of conventional computer memories imposes certain constraints on the representation of numbers. Fixed word lengths dictate not only the maximum value that these numbers can take, but also the accuracy of the representation. Because the number of decimal places that can be carried is finite, it is not possible to obtain an exact representation of many real numbers. The representation of the number $\frac{1}{3}$, for example, must be terminated after a certain number of decimal digits. In this way, an inaccuracy, or *error*, is introduced.

Any calculation involving real numbers on a computer is subject to several types of error. Although error is always present, the case is not hopeless. Even though the error cannot be eliminated, it is possible to have it contained. There exists an extensive body of research directed toward the problem of error containment in numerical calculations, with and without the use of a computer. In this section three causes of error are discussed. We also consider how errors are accumulated by some basic arithmetic operations. Some suggestions for minimizing the error in a calculation are presented.

8-1.1 THE TYPES OF ERROR

Three types of error are common in numerical calculations. These are inherent error, truncation error, and roundoff error. Many numerical values obtained experimentally necessarily contain *inherent errors* because of the uncertainty of measurements. Generally, it is advisable to state explicitly the error limits of any experimental result. For example, a measurement of $25.4 \pm .05$ degrees Celsius indicates that the temperature is not less than 25.35°C and not more than 25.45°C. Inherent error is also present in finite decimal approximations of numbers such as $\pi, e, \sqrt{2}$, and $\frac{1}{3}$ because they have no exact finite decimal representations. Some numbers, in fact, have a finite representation in one number system but not in another. For example, $\frac{1}{10}$ has a finite decimal representation but not a finite binary representation, and therefore a binary computer will not give exactly 1.0 as the answer to $.1 + .1 + .1 + .1 + .1 + .1 + .1 + .1 + .1 + .1$.

Truncation error occurs when:

1. An infinite mathematical process is approximated by a finite process, or

2. When a finite process is approximated by a smaller number of iterations

An example of (1) is the calculation of $\sin x$ using the infinite Taylor series approximation

$$\sin x = x - \frac{x^3}{3!} + \frac{x^5}{5!} - \frac{x^7}{7!} + \frac{x^9}{9!} - \cdots + \cdots$$

In any practical calculation, the formula must be truncated after a finite number of terms. Thus, the calculated value will be inexact.

As an example of (2), consider the summation of a large number of terms, most of which are very small. For example,

$$\sum_{n=1}^{1,000,000} \frac{1}{n!} = 1 + \frac{1}{2} + \frac{1}{6} + \frac{1}{24} + \frac{1}{120} + \frac{1}{720} + \cdots + \frac{1}{1,000,000!}$$

The terms decrease very rapidly, and such a summation would soon have to be truncated. The error that occurs when digits are dropped from a number without rounding is also called a truncation error. For example, 9.2344778 might be truncated to 9.2344, introducing an error.

Roundoff errors are those errors which result from rounding a number. A decimal number is rounded to n decimal places by adding 5 to the $(n + 1)$th digit to the right of the decimal point and then dropping all digits to the right of the nth digit. Again, since the word size of a computer is finite, roundoff errors occur frequently in computer calculations. Rounding

3.14159 to three decimal places, we get 3.142. The resulting error is (3.14200 − 3.14159), which is .00041. When a number is rounded to n places after the decimal point, the error will always be less than or equal to $5 \times 10^{-n-1}$.

Roundoff error can also result from the shifting of values prior to a computation. Suppose, for example, that we wish to add two real numbers that differ by an order of magnitude or more, say, 999.0 and 1.12954. If we assume floating-point representation to seven decimal places, these are stored as $.9990000 \times 10^3$ and $.1129540 \times 10^1$. Before the calculation is actually performed, the fractional part of the smaller number will be shifted to make its exponent the same as that of the larger number. Thus, $.1129540 \times 10^1$ is changed to $.0011295 \times 10^3$. As a result, some accuracy is lost.

8-1.2 THE EXPRESSION AND PROPAGATION OF ERROR

An error may be expressed in absolute, relative, or percentage terms. The *absolute error* is simply the difference between the exact value of a number and the approximation. The *relative error* is the absolute error divided by the exact value. The *percentage error* is the relative error multiplied by 100%. Let $\bar{x}$ be an approximation of the exact value x. The absolute error of $\bar{x}$ is $x - \bar{x}$, the relative error of $\bar{x}$ is

$$\frac{x - \bar{x}}{x}$$

and the percentage error of $\bar{x}$ is

$$\frac{x - \bar{x}}{x} \times 100\%$$

Since the value x is not known, it is often convenient to define the relative error to be

$$\frac{x - \bar{x}}{\bar{x}}$$

and the percentage error as

$$\frac{x - \bar{x}}{\bar{x}} \times 100\%$$

When arithmetic operations are performed on two approximate numbers, the result is, of course, also an approximation; in fact, a larger error may result. In this manner, errors are said to *accumulate* or *propagate*. If the error bounds of the operands are known, however, then the error of the result can be estimated.

Let $x = \bar{x} + e_x$ and $y = \bar{y} + e_y$, where $\bar{x}$ is an approximation of x, $\bar{y}$ is an approximation of y, and e_x and e_y are the absolute errors of $\bar{x}$ and $\bar{y}$, respectively. Consider the basic arithmetic operations: addition, subtraction, multiplication, and division.

Addition: For the sum of x and y, we have

$$x + y = (\bar{x} + e_x) + (\bar{y} + e_y) = (\bar{x} + \bar{y}) + (e_x + e_y)$$

The absolute error of the sum is

$$e_{x+y} = e_x + e_y$$

and the relative error is

$$r_{x+y} = \frac{e_{x+y}}{\bar{x} + \bar{y}} = \frac{e_x + e_y}{\bar{x} + \bar{y}}$$

Notice that the relative error of the sum is intermediate between the relative errors of the two operands.

Subtraction: The difference between x and y is

$$x - y = (\bar{x} + e_x) - (\bar{y} + e_y) = (\bar{x} - \bar{y}) + (e_x - e_y)$$

Therefore,

$$e_{x-y} = e_x - e_y$$

and

$$r_{x-y} = \frac{e_{x-y}}{\bar{x} - \bar{y}} = \frac{e_x - e_y}{\bar{x} - \bar{y}}$$

Notice that when x and y are nearly equal, the denominator is small, and thus the relative error may become very large.

Multiplication: For the multiplication of x and y,

$$xy = (\bar{x} + e_x)(\bar{y} + e_y) = \bar{x}\bar{y} + \bar{x}e_y + \bar{y}e_x + e_x e_y$$

Assuming that e_x and e_y are much smaller than $\bar{x}$ and $\bar{y}$, we expect the terms involving $\bar{x}$ and $\bar{y}$ to dominate. That is, we expect the term $e_x e_y$ to contribute very little to the magnitude of e_{xy}. Thus, the absolute error of the product

$$e_{xy} \doteq \bar{x}e_y + \bar{y}e_x$$

where the symbol $\doteq$ is read "is approximately equal to." The relative error of the product

$$r_{xy} = \frac{e_{xy}}{\bar{x}\bar{y}} \doteq \frac{xe_y + ye_x}{xy} = \frac{e_x}{\bar{x}} + \frac{e_y}{\bar{y}}$$

is approximately equal to the sum of the relative errors of the operands.

Division: For the division of x and y,

$$\frac{x}{y} = \frac{\bar{x} + e_x}{\bar{y} + e_y}$$

Rationalizing the denominator, we have

$$\frac{x}{y} = \frac{\bar{x}\bar{y} + \bar{y}e_x - \bar{x}e_y - e_x e_y}{\bar{y}^2 - e_y^2}$$

Again, since we expect e_x and e_y to be small, we can neglect the terms involving powers or products of e_x and e_y.

$$\frac{x}{y} \doteq \frac{\bar{x}\bar{y} + \bar{y}e_x - \bar{x}e_y}{\bar{y}^2}$$

Rearranging, the quotient becomes

$$\frac{x}{y} \doteq \frac{\bar{x}}{\bar{y}} + \frac{\bar{x}}{\bar{y}}\left(\frac{e_x}{\bar{x}} - \frac{e_y}{\bar{y}}\right)$$

The absolute error of the quotient is therefore

$$e_{x/y} \doteq \frac{\bar{x}}{\bar{y}}\left(\frac{e_x}{\bar{x}} - \frac{e_y}{\bar{y}}\right)$$

The relative error of the quotient

$$r_{x/y} = \frac{e_{x/y}}{\bar{x}/\bar{y}} \doteq \frac{e_x}{\bar{x}} - \frac{e_y}{\bar{y}}$$

is approximately equal to the difference between the relative errors of the numerator and denominator.

Note that the sign of an error may be either positive or negative. Consequently, the previous discussion in no way implies that the errors incurred by addition and multiplication are greater than those of subtraction and division.

The various types of error discussed combine in any numerical calculation. To illustrate, let us consider once again the Taylor series approximation for $\sin x$:

$$\sin x = x - \frac{x^3}{3!} + \frac{x^5}{5!} - \frac{x^7}{7!} + \frac{x^9}{9!} - \cdots + \cdots$$

which we can also write as

$$\sin x = \sum_{n=1}^{\infty} (-1)^{n-1} \frac{x^{2n-1}}{(2n-1)!}$$

Clearly, the accuracy of this approximation increases as more terms are included, or equivalently as the value of n increases. Thus, the magnitude of the truncation error reduces as n increases.

As the value of n increases, however, the additional terms become increasingly smaller. Before long, roundoff error becomes significant, since it becomes increasingly difficult to represent the values accurately in the word size given. Thus, the magnitude of the roundoff error grows as n increases.

Clearly, the total error in this calculation is the sum of the truncation error and the roundoff error. Figure 8-1 shows the behavior of these various types of error with varying n. The point denoted by $\hat{n}$ represents the value of n for which the total error is minimized. For values of n less than $\hat{n}$, truncation error dominates; for values of n greater than $\hat{n}$, roundoff error dominates.

8-1.3 THE ITERATIVE METHOD

The total error curve in Fig. 8-1 shows that it is a futile exercise to try to compute numerical results that are 100% accurate. Error can never be eliminated totally. This is not to say, however, that the situation is hopeless. In most cases, through the application of some carefully considered techniques, it is possible to contain the error within acceptable limits. Thus, while it may not be possible to get an "exact" result, it is usually possible to obtain a result that is close enough for the purposes required. The definition of "close enough" depends on the particular circumstances of each calculation. The accuracy of the values being used (which may depend, for example, on the accuracy of measuring instruments) is certainly a factor. It is

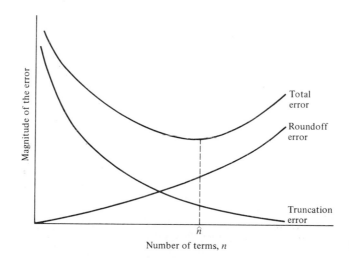

Number of terms, n FIGURE 8-1

unreasonable to expect results that are more accurate than the values used to obtain them.

The study of numerical calculations centers primarily on methods that may be described as *iterative methods*. The method begins with an initial approximation to the solution of a problem. We can denote this by S_0. From this solution, a more accurate solution, say S_1, is derived. This iterative process can continue through any number of subsequent solutions, S_2, S_3, S_4, ... However, as shown in Fig. 8-1, the accuracy can only improve up to a certain point, after which it gets worse. The person desiring a solution must therefore specify what he/she considers to be an acceptable error. This can be expressed in any of the forms described previously. Let us assume, for the sake of simplicity, that an absolute error ϵ has been deemed acceptable. Our iterative process is then terminated when two successive solutions agree within this error tolerance, that is, when

$$\left| S_i - S_{i-1} \right| < \epsilon$$

At this point, our technique is said to have *converged* to the result S_i.

The study of convergence is also important in numerical computations. There may be several reasons why a particular solution technique fails to converge. For example, the specified tolerance may be set too small, or the solution technique itself may be inappropriate to the particular circumstances of the problem posed. It is often good programming practice to impose an additional stopping criterion: to terminate if convergence is not achieved after a given number of iterations.

In this section we have discussed the problem of error in numerical calculations. Although it cannot be eliminated, it can be kept under control, and acceptable answers can be obtained. Also, the iterative method of solution was introduced.

Numerical solution techniques are employed for a number of reasons. It may be, for example, that a solution cannot be obtained by other means, such as analytical techniques. For some categories of problem, analytical solution techniques simply do not exist. An example of such a problem is finding the roots of a polynomial, say, of the tenth degree. In other cases, even though analytical techniques do exist, the solution may be very difficult to derive. For example, the analytical solution of a system of 50 equations in 50 unknowns would be most tedious indeed. Numerical techniques often provide a straightforward method for computing satisfactory solutions. The use of the computer simplifies the process considerably.

In the remaining sections of this chapter, we survey a number of important numerical problems and present several techniques for their solution within the framework just described. These problems have been chosen both to illustrate the type of problem that is relevant and to illustrate the types of solution technique employed.

EXERCISES 8-1

1. Using the Taylor series formula for sin x given in this section, compute by hand the value of sin 1, stopping first after 3 terms, then after 5 terms, then after 10 terms. Assume six decimal digits of accuracy. Comment on the type and magnitude of the error occurring in each case. Observe that x is in radians. (The correct value to six decimals places is .841471.)

2. Assume that you have a computer capable of storing real numbers correct to three decimal places of accuracy. Consider the execution of the following algorithmic statements where x is real.

```
x ← 0.0
Repeat for i = 1, 2, . . ., 10
    x ← x + 1/3
Write(x)
```

Hand-trace this sequence of statements and comment on the value of x that is printed.

3. In conventional number systems, addition is an *associative* operation; that is for numbers, a, b, and c,

$$(a + b) + c = a + (b + c)$$

Unfortunately, the associative property does not always hold on a computer. Explain why this is so, and give an example where addition is not associative.

4. In addition to the associative property described in exercise 3, addition in conventional number systems has two other important properties. First, it is a *commutative* operation; that is,

$$a + b = b + a$$

Also, multiplication *distributes* over addition; that is,

$$a \times (b + c) = a \times b + a \times c$$

Do these properties always hold on a computer? If not, give counter-examples.

5. Calculate the sum of the following floating-point numbers, assuming five decimal digits of accuracy. Add them first in the order given, and then rearrange for maximum accuracy.

$$.24382 \times 10^1$$
$$.85155 \times 10^0$$
$$.79843 \times 10^0$$
$$.62837 \times 10^1$$
$$.48919 \times 10^3$$

8-2 FINDING THE ROOTS OF NONLINEAR FUNCTIONS

The *root* of a function of a single variable is defined to be that value of the variable that results in a value of zero for the function. If we graph the function in the (x,y) plane, as shown in Fig. 8-2, the root is the point (or points) at which the function crosses the x axis.

For linear functions, the roots are easy to find algebraically. For example, the root of the function

$$f(x) = x - 3$$

is easily found by solving for the x value that sets the function to 0.

$$x - 3 = 0$$
$$x = 3$$

For nonlinear functions, that is, functions with powers of x greater than 1, the solution techniques are not so simple. For second-degree polynomial functions of the form

$$f(x) = ax^2 + bx + c$$

the roots are given by the quadratic formula

$$x = \frac{-b \pm \sqrt{b^2 - 4ac}}{2a}$$

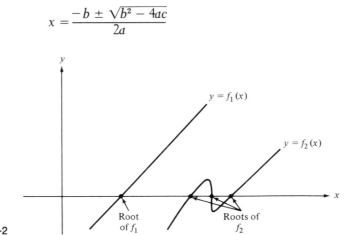

FIGURE 8-2

Solution formulas for third- and fourth-degree polynomials are more complicated still; formulas for higher-degree polynomials do not exist. Suppose, further, that we wish to consider nonpolynomial functions of x, such as

$$\sin x - x + 1 = 0$$

Since there exists no general formula for the roots of this form of equation, we must resort to other solution methods. We might try rearranging the equation to

$$\sin x = x - 1$$

and graphing the two functions of the equality, as shown in Fig. 8-3. The solution is then the point of intersection of the two curves. This method is clearly not acceptable for anything but rough approximations, since the accuracy is very poor. Functions involving trigonometric functions, or logarithms, are transcendental functions, and there are no general formulas for their roots.

We now turn to a discussion of some numerical procedures for finding the roots of functions. These are sufficiently general that they can be used for all higher-degree polynomials and also for transcendental functions.

8-2.1 FIXED-POINT ITERATION

We wish to find all solutions to the equation $f(x) = 0$ for a given function f. That is, we wish to find all values of x which make $f(x) = 0$. One method for doing this is called *fixed-point iteration* and is done as follows.

Fixed-point iteration is a process of making successive approximations which get progressively closer to the root. An initial guess x_0 is used as the starting point for the iteration. Thereafter, each new approximation x_{n+1} is obtained by means of the equation

$$x_{n+1} = g(x_n)$$

The function g must be chosen so that if $x = r$ is a root of $f(x) = 0$, then

$$g(r) = r$$

This ensures that once the approximation becomes exactly the value of the root, it will remain at the root through successive iterations.

Such a function g is obtained by algebraically manipulating the equation

$$f(x) = 0$$

into the form

$$g(x) = x$$

There are typically many ways to do this for a given function f.

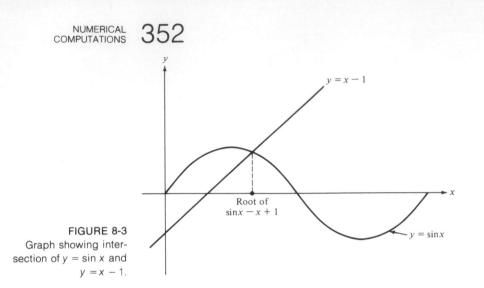

FIGURE 8-3
Graph showing inter-
section of $y = \sin x$ and
$y = x - 1$.

Some of these ways may give a function g which will cause the fixed-point iteration to converge. Other ways, however, may give a function g which will cause the fixed-point iteration to *not* converge (that is, to *diverge*). One condition which is sufficient to guarantee convergence is that

$$\left| g'(x) \right| < 1$$

for all x in the region containing the root r and all the approximations x_0, x_1, x_2, . . . , where $g'(x)$ is the *derivative* of g at x.

Consider the function

$$f(x) = \frac{x + 1}{x - 1}$$

We wish to find a root of the equation

$$\frac{x + 1}{x - 1} = 0$$

We can rewrite this equation as

$$\frac{x + 1}{x - 1} + 1 = 1$$

$$\frac{x + 1}{x - 1} + \frac{x - 1}{x - 1} = 1$$

$$\frac{x + 1 + x - 1}{x - 1} = 1$$

$$\frac{2x}{x-1} = 1$$

$$2x = x - 1$$

$$2x + 1 = x$$

This equation is of the required form, $g(x) = x$. We now examine the derivative of g to determine whether or not the fixed-point iteration will converge.

$$g(x) = 2x + 1$$
$$g'(x) = 2$$

Notice that $\left| g'(x) \right| > 1$ for any x. Therefore, we have no guarantee that the iteration will converge (although it may).

Now let us try the iteration with an initial guess of $x_0 = 0$.

$$x_0 = 0$$
$$x_1 = g(x_0) = 2x_0 + 1 = 0 + 1 = 1$$
$$x_2 = g(x_1) = 2x_1 + 1 = 2 + 1 = 3$$
$$x_3 = g(x_2) = 2x_2 + 1 = 6 + 1 = 7$$
$$\cdot$$
$$\cdot$$
$$\cdot$$

Clearly this iteration is not converging to the root (which is $r = -1$).

Now let us try another formulation of $g(x) = x$, in an attempt to find an iteration which will converge. Recall the original equation:

$$f(x) = \frac{x+1}{x-1} - 0$$

$$\frac{x + 1 + 1 - 1}{x-1} - 0$$

$$\frac{x - 1 + 2}{x-1} = 0$$

$$\frac{x-1}{x-1} + \frac{2}{x-1} = 0$$

$$1 + \frac{2}{x-1} = 0$$

$$\frac{2}{x-1} = -1$$

$$\frac{-2}{x-1} = 1$$

$$-2 = x - 1$$

$$-2 + 1 = x$$

$$-1 = x$$

This equation is again of the form $g(x) = x$, with $g(x) = -1$. The derivative is $g'(x) = 0$. Obviously,

$$|g'(x)| < 1 \qquad \text{for all } x$$

Thus, the iteration is guaranteed to converge. Starting again with $x_0 = 0$, we see a very quick convergence:

$$x_0 = 0$$
$$x_1 = g(x_0) = -1$$
$$x_2 = g(x_1) = -1$$
$$x_3 = g(x_2) = -1$$
$$\cdot$$
$$\cdot$$
$$\cdot$$

In general, however, the convergence of the fixed-point iteration method is very slow and the necessity of taking derivatives to determine convergence poses problems when we are dealing with a function whose derivative is unknown or does not exist. In the following sections we look at root-finding methods which are easier to use and have better convergence properties.

8-2.2 THE METHOD OF SUCCESSIVE BISECTION

Consider the function $f(x)$. We wish to find a value of x for which $f(x) = 0$. We begin by choosing two x values, x_1 and x_2, whose functional values have different signs; that is, one of $f(x_1)$ and $f(x_2)$ is positive and the other is negative; therefore, $f(x_1) \times f(x_2) < 0$. If we assume that $f(x)$ is continuous on (x_1, x_2), there must exist a root between x_1 and x_2 that can be found by the *method of successive bisection*. Clearly, if the function never changes its sign, this method cannot be applied. Such a function never crosses the x axis and thus has no real roots.

The procedure is as follows:

1. Find r, the midpoint of (x_1, x_2), where $f(x_1)$ and $f(x_2)$ are opposite in sign.
2. If $f(r) = 0$, then the root is r. Also, if $|x_1 - x_2|$ is within some previously stated tolerance of 0, then the root is taken to be x_1 or x_2.
3. If $f(r)$ has the same sign as $f(x_1)$, then r must be on the same side of the actual root as x_1. Thus, repeat the procedure with $x_1 = r$.
4. If $f(r)$ has the same sign as $f(x_2)$, then repeat the procedure with $x_2 = r$.

By this method, the search interval (x_1, x_2) always contains the root. The length of the interval is successively halved until the functional value of the midpoint of the interval is sufficiently close to 0, or until x_1 and x_2 are very close together.

This method is illustrated in Fig. 8-4. Initial values x_1 and x_2 are chosen on opposite sides of the root. A preliminary graphing of the function can be used to suggest these initial values. The point x_3 is derived as the midpoint of the interval (x_1, x_2). As shown, $f(x_3)$ is not zero; therefore, we must continue our search. Since $f(x_3)$ has the same sign as $f(x_1)$, our new search interval becomes (x_3, x_2), with x_4 as the midpoint. Since $f(x_4)$ has the same sign as $f(x_2)$, our search interval becomes (x_3, x_4), with x_5 as the midpoint. In this case we are fortunate; since $f(x_5) = 0$, x_5 is the desired root.

The following algorithm formalizes this process.

Algorithm SUCCESSIVE _BISECTION. Given x_1 and x_2, two x values such that $f(x_1) \times f(x_2) < 0$, and ϵ, the desired accuracy, this algorithm finds ROOT, a value of x for which $f(x) = 0$. Assume all variables to be real.

1. [Make at most 30 iterations]
   Repeat thru step 4 for i = 1, 2, . . ., 30
2. [Find the midpoint of the interval]
   $$\text{ROOT} \leftarrow \frac{x_2 + x_1}{2}$$
3. [Root found?]
   If f(ROOT) = 0 or $|x_2 - x_1| < \epsilon$
   then Write(ROOT, f(ROOT))
       Exit
4. [Bisect the interval]
   If f(ROOT) * f(x₁) < 0
   then x₂ ← ROOT
   else x₁ ← ROOT
5. [Print message]
   Write('ROOT NOT FOUND IN 30 INTERATIONS. ROOT SO FAR IS', ROOT)
6. [Finished]
   Exit   □

The error associated with the root found in this manner is bounded by (that is, cannot be greater than) one-half of the length of the last interval used, or the size of the interval which would have been used had another iteration been made.

This method is almost guaranteed to converge to a root if the initial conditions are met, unless the roundoff error becomes too large. However, it converges more slowly (requires more iterations) than most other methods. To illustrate the use of this method on a real problem, we will seek a root of $f(x) = x^3 - x^2 - 2x + 1$ on the interval $(0, 1)$, using the method of successive bisection. We will terminate the process when our search interval is less than .0001. Since $f(0)$ is positive and $f(1)$ is negative, and the function is

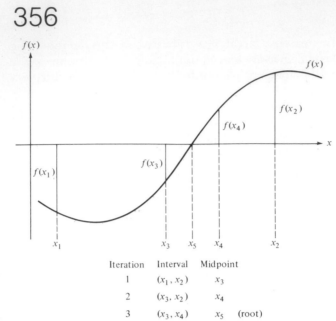

Iteration	Interval	Midpoint	
1	(x_1, x_2)	x_3	
2	(x_3, x_2)	x_4	
3	(x_3, x_4)	x_5	(root)

FIGURE 8-4
Finding the root of $f(x)$ by
the method of successive
bisection.

continuous everywhere, success in finding a root is practically guaranteed.
A trace of the method is given in Table 8-1.

Following the steps given in the table, the size of the search interval is
within our specified tolerance. Thus, we terminate. The resulting root is
then $.445039 \pm .000031$.

An alternative test of convergence for this particular problem is to test
the function value at the computed point [that is, $f(r)$]. If it is within a

TABLE 8-1 Trace of Successive Bisection Method

Interval, (x_1, x_2)	Midpoint, $r = \dfrac{x_1 + x_2}{2}$	$f(r)$
(.000000, 1.000000)	.5	−.125000
(.000000, .500000)	.25	.453125
(.250000, .500000)	.375	.162109
(.375000, .500000)	.4375	.017334
(.437500, .500000)	.46875	−.054230
(.437500, .468750)	.453125	−.018536
(.437500, .453125)	.445313	−.000622
(.437500, .445313)	.441407	.008350
(.441407, .445313)	.443360	.003862
(.443360, .445313)	.444337	.001618
(.444337, .445313)	.444825	.000498
(.444825, .445313)	.445069	−.000062
(.444825, .445069)	.444947	.000218
(.444947, .445069)	.445008	.000078
(.445008, .445069)	.445039	.000007

specified tolerance of 0, the process is said to converge. This particular test may result in faster convergence if the interval endpoints are approximately equally spaced on either side of the root. How would you modify the algorithm SUCCESSIVE_BISECTION for this test for convergence?

8-2.3   THE SECANT METHOD

The *secant method*, sometimes called the *computed line approach*, converges more quickly than the method of successive bisection. The secant method also requires two starting points, x_1 and x_2. The curve of the function between the two points is approximated by a straight line (the secant line) which is then interpolated or extrapolated (depending on whether or not the functional values of the two points are of the same sign) to a third point. The approximation to the root is the point at which the computed line crosses the x axis. The process is repeated, always using the last two points calculated to obtain the next approximation, until the root is obtained within the desired accuracy.

Figure 8-5a shows the situation when $f(x_1) \times f(x_2) < 0$. From the geometry of the diagram, we see that

$$\frac{x_2 - x_3}{x_2 - x_1} = \frac{f(x_2)}{f(x_2) - f(x_1)}$$

and therefore

$$x_3 = x_2 - \frac{x_2 - x_1}{f(x_2) - f(x_1)} \times f(x_2)$$

Figure 8-5b depicts the case where $f(x_1) \times f(x_2) > 0$. Again, there are two similar triangles, and we have

$$\frac{x_2 - x_3}{x_1 - x_2} = \frac{f(x_2)}{f(x_1) - f(x_2)}$$

which can be rearranged to

$$x_3 = x_2 - \frac{x_2 - x_1}{f(x_2) - f(x_1)} \times f(x_2)$$

Both cases result in the same formula, which can be stated generally as

$$x_{n+1} = x_n - \frac{x_n - x_{n-1}}{f(x_n) - f(x_{n-1})} \times f(x_n)$$

Figure 8-6 shows how the process converges to a root.

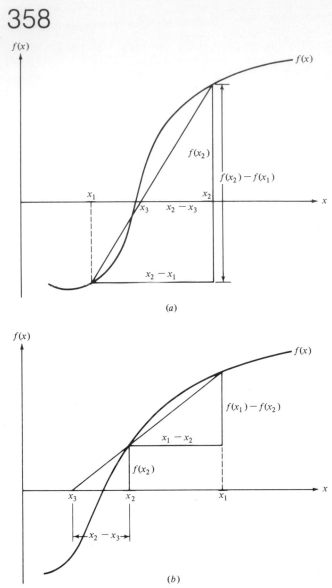

FIGURE 8-5
Finding the root of $f(x) = 0$
by the secant method:
(a) x_1 and x_2 are of
opposite signs; (b) x_1 and
x_2 are of the same sign.

Algorithm SECANT_METHOD. Given x_1 and x_2, the two starting x values, and ϵ, the required accuracy, this algorithm finds ROOT, the value of x for which $f(x) = 0$. Assume all variables to be real.

1. [Initialize]
   xOLD ← x_1
   xNEW ← x_2
2. [Perform at most 30 iterations]
   Repeat thru step 5 for i = 1, 2, . . . , 30
3. [Calculate next estimate]
   ROOT ← xNEW − ((xNEW − xOLD) / (f(xNEW) − f(xOLD))) * f(xNEW)

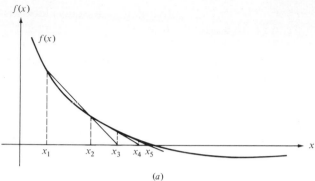

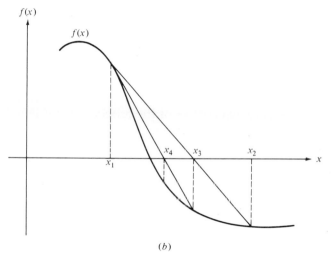

FIGURE 8-6
Finding the root of $f(x) = 0$
using the secant method:
(a) approaching the root
from one side; (b)
approaching the root from
both sides.

4. [Convergence?]
    If $|f(ROOT)| < \epsilon$
    then Write(ROOT, f(ROOT))
        Exit
5. [Shift variables]
    xOLD ← xNEW
    xNEW ← ROOT
6. [Print message]
    Write('ROOT NOT FOUND IN 30 ITERATIONS. ROOT SO FAR IS', ROOT)
7. [Finished]
    Exit   ☐

To illustrate the use of this method we will again find a root of $f(x) = x^3 - x^2 - 2x + 1 = 0$ on the interval $(0,1)$, this time using the computed line approach. Table 8-2 gives the trace. As shown in the table, this method requires considerably fewer iterations than the method of successive bisection. The rate of convergence actually depends upon how similar the function is to a straight line.

The secant method may fail in some cases where the successive bisection method succeeds. In fact, if the starting values are not carefully chosen, the secant method may zero in on a relative minimum instead of a root (see Fig. 8-7a), or it may happen that the extrapolation from two points near a relative extreme of the function will result in a point far from the root (see Fig. 8-7b). Again, a rough graph of the function is a valuable guide to the choice of initial values.

8-2.4 NEWTON'S METHOD

Newton's method was developed by Sir Isaac Newton to assist in hand calculations of roots, but it applies well in a computing environment. In Newton's method, the curve of the function is approximated by the tangent to the curve at a certain value of x. The method requires only one initial guess, x_1. Each successive approximation of the root is the x-axis intersection of the tangent of the curve at the point of previous guess.

Since the derivative of the function at x_1 [denoted by $f'(x_1)$] is equal to the slope of the tangent line at x_1, we see from Fig. 8-8 that

$$f'(x_1) = \frac{f(x_1)}{x_1 - x_2}$$

which, when rearranged, yields

$$x_2 = x_1 - \frac{f(x_1)}{f'(x_1)}$$

or, more generally

$$x_{n+1} = x_n - \frac{f(x_n)}{f'(x_n)}$$

TABLE 8-2 Trace of the Secant Method

n	x_n	$f(x_n)$: $x^3 - x^2 - 2x + 1$	$\dfrac{x_n - x_{n-1}}{f(x_n) - f(x_{n-1})}$
1	0	1	
2	1	-1	$-.5$
3	.5	$-.125$	$-.571429$
4	.428571	.037902	$-.438478$
5	.445190	$-.000340$	$-.434575$
6	.445042	$-.0000003$	$-.435679$

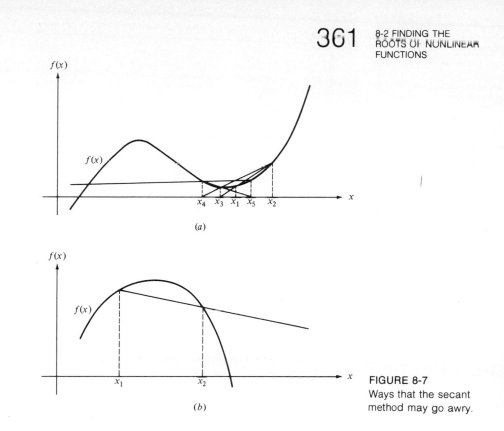

(a)

(b)

FIGURE 8-7
Ways that the secant
method may go awry.

Successive approximations are made until $f(x_n)$ is very small or until x_n and x_{n-1} are very close together (see Fig. 8-9). In the following algorithm we choose the former stopping criterion.

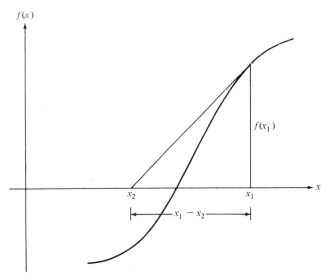

FIGURE 8-8

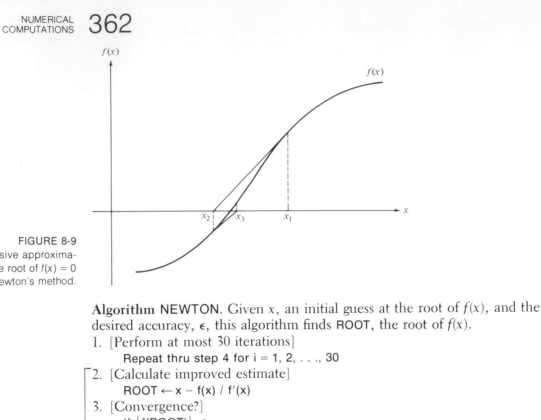

$f(x)$

$f(x)$

x

x_2 x_3 x_1

FIGURE 8-9
Successive approxima-
tions to the root of $f(x) = 0$
by Newton's method.

Algorithm NEWTON. Given x, an initial guess at the root of $f(x)$, and the desired accuracy, ϵ, this algorithm finds ROOT, the root of $f(x)$.

1. [Perform at most 30 iterations]
   Repeat thru step 4 for i = 1, 2, . . ., 30
2. [Calculate improved estimate]
   ROOT ← x − f(x) / f'(x)
3. [Convergence?]
   If $|$f(ROOT)$| < \epsilon$
   then Write(ROOT, f(ROOT))
      Exit
4. [Save new estimate]
   x ← ROOT
5. [Print message]
   Write('ROOT NOT FOUND IN 30 ITERATIONS. ROOT SO FAR', ROOT)
6. [Finished]
   Exit   □

Table 8-3 shows the results of applying Newton's method to our example with the initial approximation $x_1 = 0.5$. The process is to terminate when the functional value of the estimated root is within .00001 of zero.

Notice that Newton's method is much more efficient than the two methods previously discussed. However, it does have certain disadvantages: it requires the derivative of the function and, like the computed line approach, it can go wrong easily (see Fig. 8-10).

Newton's method requires some preliminary setup of the problem to render it amenable for solution. To illustrate, we will develop a formula for x_{n+1} that could be used with Newton's method to find the square root of a nonnegative number, N.

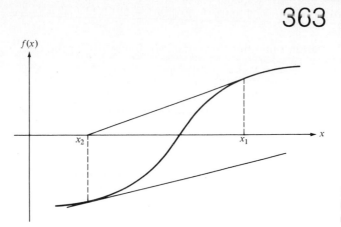

FIGURE 8-10
Example of how Newton's
method can fail.

$$x = \sqrt{N}$$

can be written as

$$x^2 - N = 0$$

Let

$$f(x) = x^2 - N$$

The root of $f(x) = 0$ will be the square root of N. For any n,

$$f'(x_n) = 2x_n$$

so

$$x_{n+1} = x_n - \frac{x_n^2 - N}{2x_n}$$

$$= \frac{x_n^2 + N}{2x_n} = \frac{1}{2}\left(x_n + \frac{N}{x_n}\right)$$

is the formula to be used. This is, in fact, the formula used on many electronic pocket calculators that supply a built-in square-root function. In a

TABLE 8-3 Trace of Newton's Method

n	x_n	$f(x_n)$: $x^3 - x^2 - 2x + 1$	$f'(x_n)$: $3x^2 - 2x - 2$	$\dfrac{f(x_n)}{f'(x_n)}$
1	.5	−.125	−2.25	.055556
2	.444444	.001373	−2.296297	−.000598
3	.445042	−.0000003	−2.295897	.0000001

similar manner, formulas can be found for the cube roots, fourth roots, and so on.

Four methods of finding roots of nonlinear equations have been presented in this section. Newton's method converges to the root most quickly followed, in performance, by the secant method. Finally, the slowest methods to converge are the methods of successive bisection and fixed-point iteration. The safest method to use, and the easiest to calculate an error bound for, is the method of successive bisection. When we have enough knowledge to make a good first guess and to calculate the derivative at any point, Newton's method is the best. If we are not sure of how the function behaves, the method of successive bisection is preferable. The secant method can be used in applications somewhat in between—for example, when we are familiar with the function but not its derivative. In numerical computations, trade-offs such as these must always be considered, and the method chosen is largely dependent on the particular application.

EXERCISES 8-2

1. Consider the function $f(x) = 2x^2 - x$. We wish to find the values of x such that $f(x) = 0$ (there are two such values). We rewrite the equation as $2x^2 = x$. Therefore, the function g required for fixed-point iteration is $g(x) = 2x^2$. Perform a fixed-point iteration beginning with $x_0 = -\frac{1}{8}$. Try again with each of $x_0 = \frac{1}{8}$, $x_0 = \frac{1}{4}$, $x_0 = \frac{1}{2}$, and $x_0 = 1$. What are the roots? Explain the varying convergence properties in terms of the value of the derivative $g'(x) = 4x$ in the intervals $(-\frac{1}{4}, \frac{1}{2})$, $(\frac{1}{2}, \infty)$, and $(-\infty, -\frac{1}{4})$ and also at the points $-\frac{1}{4}$ and $\frac{1}{2}$.

2. Using the method of successive bisection, devise an algorithm to compute a root of each of the following functions to an accuracy of .0001.
   (a) $f(x) = x^2 - 5x + 6$
   (b) $f(x) = x^3 - 4x^2 - 7x + 10$
   (c) $f(x) = xe^x - 1$

3. Using the computed line method, devise an algorithm to compute a root of each of the following functions to an accuracy of .02%.
   (a) $f(x) = x^2 - 3x - 18$
   (b) $f(x) = x^3 + 2x^2 - 29x + 42$
   (c) $f(x) = e^x \cos(x) - 3$

4. Using Newton's method, devise an algorithm to compute a root of each of the following functions to an accuracy of .0005.
   (a) $f(x) = x^2 + x - 90$
   (b) $f(x) = x^3 + 4x^2 - 17x - 60$
   (c) $f(x) = \sqrt{x} - 4\tan(x/2)$

5. Construct an algorithm to compute the cube root of a number N to an

accuracy of six decimal places using Newton's method. Use $x_0 = N$ as the original estimate.

6. Let $f_1(x)$ and $f_2(x)$ denote two functions in x. One method of obtaining their solution is by writing a third function $g(x)$ as

$$g(x) = f_1(x) - f_2(x)$$

and solving for a root of $g(x)$. Using any of the methods of this section, construct an algorithm to solve these equations in this manner.

8-3 NUMERICAL INTEGRATION

Integration is a standard mathematical technique for computing the area of a closed figure. Figure 8-11 shows the area given by the expression

$$\int_a^b f(x)\, dx$$

where the function f (which is continuous) yields the curve shown, and a and b are two points on the x axis.

There are a number of instances where numerical integration (also known as *quadrature*) must be used instead of classical analytical techniques: (1) some functional values are known, but not the function itself; (2) the function would be difficult or impossible to integrate analytically; or (3) a computer is to be used to evaluate the integral.

Several numerical techniques for integration have been formulated. In this section we will present four popular methods of numerical integration, in order of increasing rate of convergence, but also of increasing complexity.

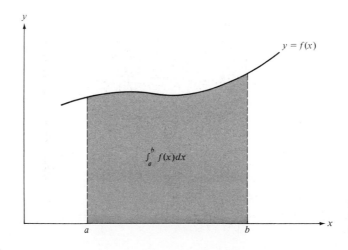

FIGURE 8-11
$\int_a^b f(x)\, dx$ can be thought of as the area under the curve, $f(x)$, between a and b.

8-3.1 THE RECTANGLE RULE

The *rectangle rule* for numerical integration is based on the assumption that the area under the curve can be approximated by summing the areas of a finite number of rectangles, as shown in Fig. 8-12. This assumption gives rise to the following approximation:

$$\int_a^b f(x)\ dx \doteq \sum_{i=0}^{n-1} f(x_i) \cdot h = h[f(x_0) + f(x_0 + h) + \cdots + f(x_0 + (n-1)h)]$$

where

$$h = \frac{b-a}{n}$$

and
$$x_{i+1} = x_i + h$$
$$x_0 = a$$

Clearly, the accuracy of this approximation improves as n gets larger (that is, the size of each rectangle gets smaller). In fact, the limit as $n \to \infty$ constitutes the formal definition of the integral. The rectangle rule is usually used in an iterative procedure to achieve some prescribed accuracy, say, ϵ. The integral is evaluated repeatedly for progressively larger values of n (or, equivalently, smaller values of h).

How do we know when a prescribed accuracy has been reached? Two criteria are popular: the absolute criterion and the relative criterion. Let us first examine the absolute criterion. Let I_n be the last approximation made and I_{n-1}, the second last. The absolute criterion is met when

$$\left| I_n - I_{n-1} \right| < \epsilon$$

This criterion is satisfactory for many functions, but when the true value is very small, the condition may be satisfied even when our answer, I_n, is 100% in error. A more reliable indication is the relative criterion,

$$\left| \frac{I_n - I_{n-1}}{I_n} \right| < \epsilon$$

We will use this criterion throughout the remainder of the chapter. The following algorithm gives a method for numerical integration using the rectangle rule.

Algorithm RECTANGLE. Given a and b, the endpoints of the interval of integration, and ϵ, the prescribed accuracy, this algorithm calculates REC, the rectangle rule approximation of $\int_a^b f(x)\ dx$, where f is assumed to be

known to the algorithm. If the accuracy is not met after 20 iterations, a
message is printed and the procedure is halted. Assume all variables to be
real.

1. [Initialize]
    REC ← 0
    h ← b − a
2. [Perform at most 20 iterations]
    Repeat thru step 8 for i = 1, 2, . . . , 20
3. [Save previous approximation]
    LAST ← REC
4. [Calculate number of subintervals to be used]
    n ← (b − a) / h
5. [Accumulate rectangle sum]
    SUM ← 0
    Repeat for j = 0, 1, . . . , n − 1
        SUM ← SUM + f(a + j ∗ h)
6. [Calculate rectangle rule approximation]
    REC ← h ∗ SUM
7. [Test accuracy]
    If |(REC − LAST) / REC| < ϵ
    then Write('ANSWER IS', REC, 'TOLERANCE MET IN', i,
            'ITERATIONS')
        Exit
8. [Halve the size of the subintervals for next iteration]
    h ← h / 2
9. [No convergence]
    Write('TOLERANCE WAS NOT MET AFTER 20 ITERATIONS',
        'CURRENT ANSWER IS', REC)
10. [Finished]
    Exit   □

This process can be quite expensive. Evaluation of the function f may
require extensive computations, so it may be worthwhile saving functional

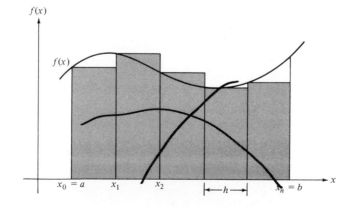

FIGURE 8-12
Rectangle rule approxi-
mation of $\int_a^b f(x)\ dx$.

values in a vector as they are calculated, if they will be used later. The trade-off between storage space and computation time would have to be investigated. However, an even better optimization can be made.

Let h_i and n_i denote the length of the subinterval and the number of subintervals used for the ith iteration, respectively. Also let SUM_i denote the sum of functional values obtained during the ith iteration. Recall that $h_i = h_{i-1}/2$ and $n_i = 2 \times n_{i-1}$. Notice that

$$
\begin{aligned}
\mathsf{SUM}_i &= f(a) + f(a + h_i) + f(a + 2 \times h_i) + \cdots + f(a + (n_i - 1) \times h_i) \\
&= f(a) + f(a + h_i) + f(a + h_{i-1}) + \cdots + f(a + (n_{i-1} - 1) \times h_{i-1}) \\
&\quad + f(a + (n_i - 1) \times h_i) \\
&= \mathsf{SUM}_{i-1} + f(a + h_i) + f(a + 3 \times h_i) + \cdots + f(a + (n_i - 1) \times h_i)
\end{aligned}
$$

A considerable saving in computation can be accomplished by using the already calculated value of SUM_{i-1} in the calculation of SUM_i. This is actually quite a simple change. To incorporate this into algorithm RECTANGLE, only steps 1 and 5 would be changed, as follows:

1. [Initialize]
    REC ← 0
    h ← b − a
    SUM ← f(a)
5. [Accumulate rectangle sum]
    Repeat for j = 1, 3, 5, . . . , n − 1
        SUM ← SUM + f(a + j ∗ h)

To illustrate the use of the rectangle rule (with a single iteration), we will use this method to find the approximation of the integral of the function e^x over the interval $(1.6, 3.2)$, with $h = 0.2$. The integral is

$$
\int_{1.6}^{3.2} f(x)\,dx \doteq 0.2(4.953 + 6.050 + 7.389 + 9.025 + 11.023 + 13.464 \\
+ 16.445 + 20.086) = 17.687
$$

The true value of the integral is $e^{3.2} - e^{1.6} = 19.580$.

The error of this method is very large, but as n increases (and h decreases), the rectangles become narrower, and a better approximation is obtained. The iterative method discussed incorporates this notion; however, convergence is slow.

An additional variation of the rule would be to use $f(x_{i+1})$, the value of the function at the higher endpoint of the interval, or

$$
f\left(\frac{x_{i+1} + x_i}{2}\right)
$$

the value of the function at the midpoint of the interval, as the height of
the approximating rectangle. Also, it is possible to use the rectangle rule
with unevenly spaced x values, often necessary when function values are
derived experimentally. In this case h is different for each component of the
sum because the intervals are of varying lengths. These extensions are not
dealt with in this book.

8-3.2 THE TRAPEZOID RULE

The *trapezoid rule* provides a more accurate approximation to the integral
than does the rectangle rule, although the method is very similar. Instead
of using rectangles to approximate the area, we use the trapezoids formed
by the secant line of the curve on each subinterval, that is, the line formed
by joining the two endpoints of the curve on the subinterval, as shown in
Fig. 8-13. This is an attempt to follow the actual curvature of the function
more closely than the rectangular approximation. The area of a trapezoid
(a four-sided figure with two parallel sides) is given by the average length
of the parallel sides, multiplied by the distance between them. Thus, the area
of one trapezoid, say between x_i and x_{i+1}, is given by the formula

$$\frac{f(x_i) + f(x_{i+1})}{2} \times (x_{i+1} - x_i) = \frac{h}{2} [f(x_i) + f(x_{i+1})]$$

and the total area, that is, the integral, is given by

$$\int_a^b f(x) \, dx \doteq \sum_{i=0}^{n-1} \frac{h}{2} [f(x_i) + f(x_{i+1})]$$

$$= \frac{h}{2} [f(x_0) + f(x_1) + f(x_1) + f(x_2) + \cdots + f(x_{n-1}) + f(x_n)]$$

$$= \frac{h}{2} [f(x_0) + 2f(x_1) + 2f(x_2) + \cdots + 2f(x_{n-1}) + f(x_n)]$$

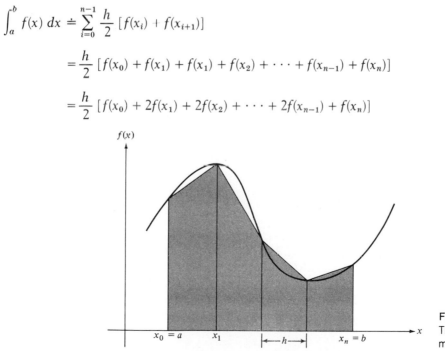

FIGURE 8-13
Trapezoid rule approxi-
mation of $\int_a^b f(x) \, dx$.

To illustrate this technique, we rework the example of the previous section, again with a single iteration.

$$\int_{1.6}^{3.2} f(x)\, dx \doteq \frac{0.2}{2} [4.953 + 2(6.050) + 2(7.389) + 2(9.025) + 2(11.023)$$

$$+ 2(13.464) + 2(16.445) + 2(20.086) + 24.533] = 19.645$$

This is much closer than the rectangle rule approximation to the true value of 19.580, but there is still an error of .065. The trapezoid rule, in fact, has an error of $O(h^2)$, read "error of order h squared" and meaning that as h approaches zero, the error becomes small at a rate proportional to h^2.

The algorithm for the iterative procedure using the trapezoid rule is very similar to algorithm RECTANGLE.

Algorithm TRAPEZOID. Given a, b, and ϵ, the prescribed accuracy, this algorithm calculates TRAPEZOID, the trapezoid rule approximation of $\int_{a}^{b} f(x)\, dx$, where f is assumed to be known to the algorithm. Assume all variables to be real.

1. [Initialize]
      TRAPEZOID ← 0
      h ← b − a
2. [Perform at most 20 iterations]
      Repeat thru step 8 for i = 1, 2, . . . , 20
3. [Save previous approximation]
      LAST ← TRAPEZOID
4. [Calculate number of subintervals to be used]
      n ← (b − a) / h
5. [Accumulate trapezoid sum]
      SUM ← 0
      Repeat for j = 0, 1, . . . , n − 1
         SUM ← SUM + f(a + j ∗ h) + f(a + (j + 1) ∗ h)
6. [Calculate trapezoid rule approximation]
      TRAPEZOID ← $\frac{h}{2}$ ∗ SUM
7. [Test accuracy]
      If |(TRAPEZOID − LAST) / TRAPEZOID| < ϵ
      then Write('ANSWER IS', TRAPEZOID, 'TOLERANCE MET IN', i,
                  'ITERATIONS')
            Exit
8. [Halve the interval size for next iteration]
      h ← h / 2

9. [No convergence]
    Write('TOLERANCE WAS NOT MET AFTER 20 ITERATIONS',
        'CURRENT ANSWER IS', TRAPEZOID)
10. [Finished]
    Exit

☐

As with the rectangle rule, it is possible to optimize the trapezoid rule by using the results of previous calculations in subsequent iterations. It is also possible to eliminate many multiplications and divisions by using a rearranged formula for the integral approximation.

$$\int_a^b f(x)\, dx \doteq \frac{h}{2}\, [f(x_0) + 2f(x_1) + 2f(x_2) + \cdots + 2f(x_{n-1}) + f(x_n)]$$

$$= h[f(x_0)/2 + f(x_1) + f(x_2) + \cdots + f(x_{n-1}) + f(x_n)/2]$$

These extensions are left as exercises. Finally, the trapezoid rule can also be used with unequally spaced x values, making it convenient for use with experimental results as well.

8-3.3 SIMPSON'S RULE

In *Simpson's rule*, second-degree polynomials are fitted to the curve, one polynomial for each pair of subintervals, and the areas under these quadratics are calculated and added together to approximate the integral, as shown in Fig. 8-14. In the rectangle rule the value of the function in the interval (x_i, x_{i+1}) is assumed to be either $f(x_i)$ or $f(x_{i+1})$. In the trapezoid rule the function in the interval is approximated by the secant line. For Simpson's rule, the function in the interval is approximated by a polynomial of the second degree, which coincides with the values of the function at three consecutive points.

To approximate the integral $\int_a^b f(x)\, dx$, the interval (a,b) is divided into n equal subintervals (x_0, x_1), (x_1, x_2), . . . , (x_{n-1}, x_n), where n must be an even number. Now

$$\int_{a=x_0}^{b=x_n} f(x)\, dx = \int_{x_0}^{x_2} f(x)\, dx + \int_{x_2}^{x_4} f(x)\, dx + \cdots + \int_{x_{n-2}}^{x_n} f(x)\, dx$$

First consider the Simpson's rule approximation of $\int^{x_2} f(x)\, dx$. We wish to find a quadratic of the form

$$g(x) = c_0 + c_1 x + c_2 x^2$$

such that

$$g(x_0) = f(x_0)$$
$$g(x_1) = f(x_1)$$
$$g(x_2) = f(x_2))$$

We will then use the function $g(x)$ to approximate $f(x)$ in the interval (x_0,x_2):

$$\int_{x_0}^{x_2} f(x)\,dx \doteq \int_{x_0}^{x_2} g(x)\,dx$$

By solving for c_0, c_1, and c_2 in the equations

$$f(x_0) = c_0 + c_1 x_0 + c_2 x_0^2$$
$$f(x_1) = c_0 + c_1 x_1 + c_2 x_1^2$$
$$f(x_2) = c_0 + c_1 x_2 + c_2 x_2^2$$

we will obtain the approximating function

$$g(x) = c_0 + c_1 x + c_2 x^2$$

Letting $h = x_i - x_{i-1}$, we find

$$c_0 = f(x_0)$$

$$c_1 = \frac{1}{h}\left[-\frac{3}{2}f(x_0) + 2f(x_1) - \frac{1}{2}f(x_2)\right]$$

$$c_2 = \frac{1}{h^2}\left[\frac{1}{2}f(x_0) - f(x_1) + \frac{1}{2}f(x_2)\right]$$

Now,

$$g(x) = f(x_0) + \frac{x}{h}\left[-\frac{3}{2}f(x_0) + 2f(x_1) - \frac{1}{2}f(x_2)\right] + \frac{x^2}{h^2}\left[\frac{1}{2}f(x_0) - f(x_1) + \frac{1}{2}f(x_2)\right]$$

and

$$\int_{x_0}^{x_2} f(x)\,dx \doteq \int_{x_0}^{x_2} g(x)\,dx = \int_{x_0}^{x_2}\left(xc_0 + \frac{x^2}{2}c_1 + \frac{x^3}{3}c_2\right)$$

$$= (x_2 - x_0)c_0 + \frac{(x_2 - x_0)^2}{2}c_1 + \frac{(x_2 - x_0)^3}{3}c^2$$

$$= 2hc_0 + 2h^2c_1 + \frac{8}{3}h^3c_2$$

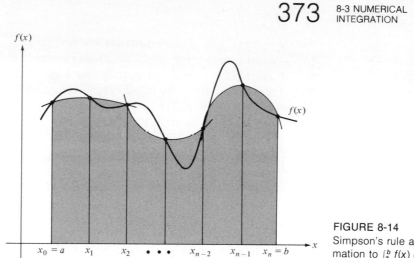

FIGURE 8-14
Simpson's rule approxi-
mation to $\int_a^b f(x)\, dx$.

Substituting the values obtained for the coefficients c_0, c_1, and c_2,

$$2hc_0 + 2h^2c_1 + \frac{8}{3}\,h^3c_2 = 2hf(x_0) + 2h\left[-\frac{3}{2}\,f(x_0) + 2f(x_1) - \frac{1}{2}f(x_2)\right]$$

$$+ \frac{8}{3}h\left[\frac{1}{2}\,f(x_0) - f(x_1) + \frac{1}{2}\,f(x_2)\right]$$

$$= h\left[2f(x_0) - 3f(x_0) + 4f('x_1) - f(x_2)\right.$$

$$\left. + \frac{4}{3}f(x_0) - \frac{8}{3}f(x_1) + \frac{4}{3}\,f(x_2)\right]$$

$$= \frac{h}{3}\left[f(x_0) + 4f(x_1) + f(x_2)\right]$$

Generalizing to the whole interval (a,b)

$$\int_a^b f(x)\, dx \doteq \frac{h}{3}\left[f(x_0) + 4f(x_1) + f(x_2)\right] + \frac{h}{3}\left[f(x_2) + 4f(x_3) + f(x_4)\right]$$

$$+ \cdots + \frac{h}{3}[f(x_{n-2}) + 4f(x_{n-1}) + f(x_n)]$$

$$= \frac{h}{3}[f(x_0) + 4f(x_1) + 2f(x_2) + 4f(x_3) + \cdots + 2f(x_{n-2})$$

$$+ 4f(x_{n-1}) + f(x_n)]$$

This formula constitutes the definition of Simpson's rule. Remember that n must always be an even number.

Simpson's rule is very popular for several reasons. First, the formula turns out to be exact for all polynomials of degree three and lower, because the areas in error cancel out exactly (see Fig. 8-15). Another reason for the wide use of the rule is its small error—only $O(h^4)$. To illustrate the use of Simpson's rule, we will find the integral of the third-degree polynomial $x^3 + 2x^2 + 4$ on the interval $(0,4)$ with $h = .5$. The result

$$\int_0^4 f(x) \, dx \doteq \frac{.5}{3} [4 + 4(4.625) + 2(7) + 4(11.875) + 2(20)$$

$$+ 4(32.125) + 2(49) + 4(71.375) + 100] = 122.6666 \ldots$$

is exactly equal to the analytical answer

$$\frac{4^4}{4} + \frac{2(4)^3}{3} + 4(4) = 122.6666 \ldots$$

As another example, we return to the problem given in the previous two sections, that is, the integral e^x over the interval $(1.6, 3.2)$ with $h = .2$. Using Simpson's rule,

$$\int_{1.6}^{3.2} f(x) \, dx \doteq \frac{.2}{3} [4.953 + 4(6.050) + 2(7.389) + 4(9.025)$$

$$+ 2(11.023) + 4(13.464) + 2(16.445) + (20.086) + 24.533]$$

$$= 19.580$$

Here again, the answer is correct up to three decimal places.

As a final example we calculate $\int_1^2 (6x^5 - 3x^2) \, dx$ using Simpson's rule with $h = 0.1$.

$$\int_1^2 f(x) \, dx \doteq \frac{.1}{3} [3 + 4(6.033) + 2(10.610) + 4(17.208) + 2(26.389)$$

$$+ 4(38.813) + 2(55.235) + 4(76.521) + 2(103.654)$$

$$+ 4(137.736) + 180] = 56.001$$

The exact answer is $2^6 - 2^3 - 0 = 56.000$, so the error of the Simpson's rule approximation is .001.

The following algorithm describes Simpson's rule more precisely. Again, the iterative method is used.

Algorithm SIMPSON. Given a, b, and ϵ, this algorithm calculates SIMP, the Simpson's rule approximation of $\int_a^b f(x) \, dx$, to an accuracy of ϵ by a series of iterations. Assume all variables to be of type real.

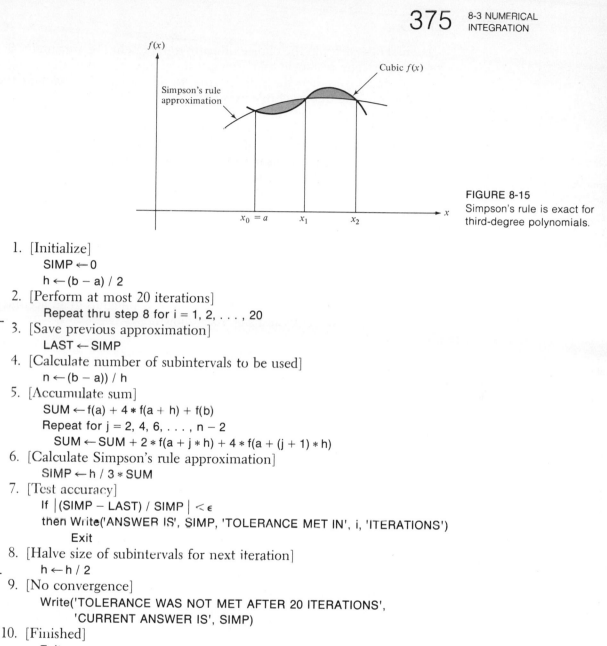

FIGURE 8-15
Simpson's rule is exact for
third-degree polynomials.

1. [Initialize]
   SIMP ← 0
   h ← (b − a) / 2
2. [Perform at most 20 iterations]
   Repeat thru step 8 for i = 1, 2, . . . , 20
3. [Save previous approximation]
   LAST ← SIMP
4. [Calculate number of subintervals to be used]
   n ← (b − a)) / h
5. [Accumulate sum]
   SUM ← f(a) + 4 ∗ f(a + h) + f(b)
   Repeat for j = 2, 4, 6, . . . , n − 2
      SUM ← SUM + 2 ∗ f(a + j ∗ h) + 4 ∗ f(a + (j + 1) ∗ h)
6. [Calculate Simpson's rule approximation]
   SIMP ← h / 3 ∗ SUM
7. [Test accuracy]
   If |(SIMP − LAST) / SIMP | < ε
   then Write('ANSWER IS', SIMP, 'TOLERANCE MET IN', i, 'ITERATIONS')
      Exit
8. [Halve size of subintervals for next iteration]
   h ← h / 2
9. [No convergence]
   Write('TOLERANCE WAS NOT MET AFTER 20 ITERATIONS',
         'CURRENT ANSWER IS', SIMP)
10. [Finished]
    Exit   □

Our earlier optimization of the SUM calculation is not directly applicable in Simpson's rule, but some modifications would yield a working technique. For instance, we might divide the interval size by six, or subtract
some components from the SUM as well as adding to it, on each iteration.
These extentions are left as exercises.

Three numerical integration techniques have been presented in this section. These differ in the choice of figure to represent the area under the given curve. The rectangle rule attempts to approximate this area using a series of rectangles; it is the simplest rule to derive but also contains the largest error. The trapezoid rule uses trapezoids and is a better approximation of the integral, especially when many subintervals are used. By using parabolas, Simpson's rule attempts to follow the curvature more closely than is possible with the straight-line approximations of the other two methods. The error of Simpson's rule is generally smaller, and this, along with the simplicity of the formula, once derived, makes Simpson's rule a popular integration method. When choosing a method for any given application, the trade-off between computation time and resulting accuracy must be considered.

EXERCISES 8-3

1. Devise algorithms to approximate the following integrals to an accuracy of .0001 using the rules and step sizes shown.

(a) $\int_0^2 x^2 e^x \, dx$, using the rectangle rule with $h = .25$.

(b) $\int_0^{\pi/2} \sin^2 x \, dx$, using the trapezoid rule with $h = 0.10$.

(c) $\int_1^{\pi} \frac{1 - \cos x}{x} \, dx$ using Simpson's rule with $h = \pi/9$.

2. Construct an algorithm to compute the area in the first quadrant under the curve $y = x^2$ and inside the circle with unit radius. Use Simpson's rule with a step size of $h = .05$.

3. A method was given in this section for improving the efficiency of the rectangle rule algorithm by saving old values. This method can also be applied to the other rules as well. Using this technique, give improved algorithms for the trapezoid rule and Simpson's rule.

4. Formulate an algorithm to compute the area of a circle with a radius of 2 units, whose center has coordinates (4,4). Use Simpson's rule with a step size of .01.

8-4   SIMULTANEOUS LINEAR EQUATIONS

As a third problem in the field of numerical computations, we introduce, in this section, various methods for solving sets of simultaneous linear equations. This is a frequent problem in many fields of science and engineering, and very often, considerable amounts of data are involved.

The first two methods presented are straightforward and direct; the others are iterative methods. The discussion will be restricted to systems having the same number of equations as unknowns.

Consider the following system of two equations in two unknowns:

$$2x + 3y = 7$$
$$3x + 5y = 11$$

These equations can be solved for x and y in the following manner. First, we must eliminate one variable. To eliminate x from the second equation, we divide the first equation by 2 and then subtract 3 times the first equation from the second.

$$x + \tfrac{3}{2}y = \tfrac{7}{2}$$
$$\tfrac{1}{2}y = \tfrac{1}{2}$$

Dividing the second equation by $\tfrac{1}{2}$, we obtain

$$y = 1$$

and substituting this y value into the first equation, we see that

$$x = 2$$

This system of two equations in two unknowns has been solved by a technique known as *Gaussian elimination*. We will now formalize and generalize this process, and consider some variations.

A set of linear equations

$$a_{11}x_1 + a_{12}x_2 + \cdots + a_{1n}x_n = b_1$$
$$a_{21}x_1 + a_{22}x_2 + \cdots + a_{2n}x_n = b_2$$
$$\vdots \qquad\qquad\qquad\qquad \vdots$$
$$a_{n1}x_1 + a_{n2}x_2 + \cdots + a_{nn}x_n = b_n$$

can be represented in array notation as

$$Ax = b$$

where

$$
A = \begin{bmatrix} a_{11} & a_{12} & \cdots & a_{1n} \\ a_{21} & a_{22} & \cdots & a_{2n} \\ \cdot & \cdot & & \cdot \\ \cdot & \cdot & & \cdot \\ \cdot & \cdot & & \cdot \\ a_{n1} & a_{n2} & \cdots & a_{nn} \end{bmatrix}, \qquad x = \begin{bmatrix} x_1 \\ x_2 \\ \cdot \\ \cdot \\ \cdot \\ x_n \end{bmatrix}, \qquad b = \begin{bmatrix} b_1 \\ b_2 \\ \cdot \\ \cdot \\ \cdot \\ b_n \end{bmatrix}
$$

or simply as an augmented array

$$
\begin{bmatrix} a_{11} & a_{12} & \cdots & a_{1n} & b_1 \\ a_{21} & a_{22} & \cdots & a_{2n} & b_2 \\ \cdot & \cdot & & \cdot & \cdot \\ \cdot & \cdot & & \cdot & \cdot \\ \cdot & \cdot & & \cdot & \cdot \\ a_{n1} & a_{n2} & \cdots & a_{nn} & b_n \end{bmatrix}
$$

with the x's "understood."

The process of Gaussian elimination consists of the following steps:

1. *Forward elimination*: starting from the top row, successively divide the row by its diagonal element, called the *pivot*, leaving a 1 on the diagonal, and then subtract a multiple of the row from each row below it, leaving zeros in the column below the diagonal 1
2. *Back substitution*: subtract multiples of lower rows from the higher rows, leaving zeros in all positions but those on the diagonal, which contain 1's, and the rightmost column

After these two steps have been completed, the solution, that is,

$$
x = \begin{bmatrix} x_1 \\ x_2 \\ \cdot \\ \cdot \\ \cdot \\ x_n \end{bmatrix}
$$

can be read directly from the rightmost column of the remaining array.

We illustrate these steps on our earlier example, which we first represent in array notation (with $n = 2$ in this case) as

$$
\begin{bmatrix} 2 & 3 & 7 \\ 3 & 5 & 11 \end{bmatrix}
$$

We now carry out the operations indicated. Pivots are circled, and the operations are performed as indicated (R1 denotes row 1, R2 denotes row 2) with the result of each step indicated by the arrow.

$$R1 \div 2 \qquad \rightarrow \begin{bmatrix} \textcircled{1} & 3/2 & \vdots & 7/2 \\ 3 & 5 & \vdots & 11 \end{bmatrix}$$

$$R2 - 3 * R1 \qquad \rightarrow \begin{bmatrix} 1 & 3/2 & \vdots & 7/2 \\ 0 & 1/2 & \vdots & 1/2 \end{bmatrix} \Big\} \text{forward elimination}$$

$$R2 \div 1/2 \qquad \rightarrow \begin{bmatrix} 1 & 3/2 & \vdots & 7/2 \\ 0 & \textcircled{1} & \vdots & 1 \end{bmatrix}$$

$$R1 - 3/2 * R2 \qquad \rightarrow \begin{bmatrix} 1 & 0 & \vdots & 2 \\ 0 & 1 & \vdots & 1 \end{bmatrix} \Big\} \text{back substitution}$$

The first three steps shown comprise the forward elimination, and the last step is back substitution. After the final step, the third column contains the solution.

Algorithm GAUSSIAN_ELIMINATION formalizes this procedure.

Algorithm GAUSSIAN_ELIMINATION. Given n, $m = n + 1$, and an augmented array of the form

$$\begin{bmatrix} a_{11} & a_{12} & \cdots & a_{1n} & \vdots & b_1 \\ a_{21} & a_{22} & \cdots & a_{2n} & \vdots & b_2 \\ \cdot & \cdot & & \cdot & \vdots & \cdot \\ \cdot & \cdot & & \cdot & \vdots & \cdot \\ \cdot & \cdot & & \cdot & \vdots & \cdot \\ a_{n1} & a_{n2} & \cdots & a_{nn} & \vdots & b_n \end{bmatrix}$$

this algorithm finds the corresponding solution array,

$$\begin{bmatrix} 1 & 0 & \cdots & 0 & \vdots & x_1 \\ 0 & 1 & \cdots & 0 & \vdots & x_2 \\ \cdot & \cdot & & \cdot & \vdots & \cdot \\ \cdot & \cdot & & \cdot & \vdots & \cdot \\ \cdot & \cdot & & \cdot & \vdots & \cdot \\ 0 & 0 & \cdots & 1 & \vdots & x_n \end{bmatrix}$$

where $Ax = b$. Variables i, j, and k are used as subscripts to the arrays.
1. [Forward elimination]
    Repeat thru step 3 for i = 1, 2, . . . , n
2. [Divide each element in the row by the pivot]
    Repeat for j = i, i + 1, . . . , m
        $a_{ij} \leftarrow a_{ij} / a_{ii}$
3. [Subtract a multiple of the row from each lower row]

Repeat for k = i + 1, i + 2, . . . , n
 Repeat for j = i, i + 1, . . . , m
  $a_{kj} \leftarrow a_{kj} - a_{kj} * a_{ij}$

4. [Back substitution]
 Repeat for i = n, n − 1, . . . , 2
  Repeat for k = 1, 2, . . . , i − 1
   $a_{km} \leftarrow a_{km} - a_{ki} * a_{im}$
   $a_{ki} \leftarrow 0$

5. [Finished]
 Exit ☐

Several variations of the Gaussian scheme are possible. Rows may be divided by their diagonal elements after, instead of during, the forward elimination step, or even after the back substitution.

An important variation is the implementation of pivoting. Roundoff errors can be reduced by interchanging rows, or rows and columns, of the array, so the largest number in magnitude is the next pivot. When we interchange both rows and columns, we are said to be using elimination with pivoting, or *full pivoting*. When we interchange only rows, however, we are using elimination with *partial pivoting*. An interchange of rows corresponds to a simple change of the order of the equations. On the other hand, a column interchange is equivalent to a change in the order of variables within the equations. Therefore, we must keep track of the column interchanges to know in what order the values of the variables, $x_1, x_2, \ldots, x_n$, will appear in the final rightmost column. If the ith and jth columns are interchanged, then x_i and x_j will switch places in the result.

To illustrate these techniques, we will now solve the following system of equations using Gaussian elimination without pivoting, with partial pivoting, and will full pivoting.

$$x_1 + 3x_2 - 2x_3 = 7$$
$$4x_1 - x_2 + 3x_3 = 10$$
$$-5x_1 + 2x_2 + 3x_3 = 7$$

Without pivoting:

$$\begin{bmatrix} 1 & 3 & -2 & \vdots & 7 \\ 4 & -1 & 3 & \vdots & 10 \\ -5 & 2 & 3 & \vdots & 7 \end{bmatrix}$$

$$\begin{array}{c} R2 - 4 * R1 \\ R3 + 5 * R1 \end{array} \rightarrow \begin{bmatrix} 1 & 3 & -2 & \vdots & 7 \\ 0 & -13 & 11 & \vdots & -18 \\ 0 & 17 & -7 & \vdots & 42 \end{bmatrix}$$

$$R2 \div -13 \quad \rightarrow \begin{bmatrix} 1 & 3 & -2 & \vdots & 7 \\ 0 & 1 & -.8462 & \vdots & 1.3846 \\ 0 & 17 & -7 & \vdots & 42 \end{bmatrix}$$

R3 − 17 ∗ R2 → $\begin{bmatrix} 1 & 3 & -2 & | & 7 \\ 0 & 1 & -.8462 & | & 1.3846 \\ 0 & 0 & 7.3854 & | & 18.4618 \end{bmatrix}$

R3 ÷ 7.3854 → $\begin{bmatrix} 1 & 3 & -2 & | & 7 \\ 0 & 1 & -.8462 & | & 1.3846 \\ 0 & 0 & 1 & | & 2.4998 \end{bmatrix}$

R1 + 2 ∗ R3
R2 + .8462 ∗ R3 → $\begin{bmatrix} 1 & 3 & 0 & | & 11.9996 \\ 0 & 1 & 0 & | & 3.4999 \\ 0 & 0 & 1 & | & 2.4998 \end{bmatrix}$

R1 − 3 ∗ R2 → $\begin{bmatrix} 1 & 0 & 0 & | & 1.4999 \\ 0 & 1 & 0 & | & 3.4999 \\ 0 & 0 & 1 & | & 2.4998 \end{bmatrix}$

With partial pivoting:

$\begin{bmatrix} 1 & 3 & -2 & | & 7 \\ 4 & -1 & 3 & | & 10 \\ -5 & 2 & 3 & | & 7 \end{bmatrix}$

R1 → R3 → $\begin{bmatrix} -5 & 2 & 3 & | & 7 \\ 4 & -1 & 3 & | & 10 \\ 1 & 3 & -2 & | & 7 \end{bmatrix}$

R1 ÷ −5 → $\begin{bmatrix} 1 & -.4 & -.6 & | & -1.4 \\ 4 & -1 & 3 & | & 10 \\ 1 & 3 & -2 & | & 7 \end{bmatrix}$

R2 − 4 ∗ R1
R3 − 1 ∗ R1 → $\begin{bmatrix} 1 & -.4 & -.6 & | & -1.4 \\ 0 & .6 & 5.4 & | & 15.6 \\ 0 & 3.4 & -1.4 & | & 8.4 \end{bmatrix}$

R2 ↔ R3 → $\begin{bmatrix} 1 & -.4 & -.6 & | & -1.4 \\ 0 & 3.4 & -1.4 & | & 8.4 \\ 0 & .6 & 5.4 & | & 15.6 \end{bmatrix}$

R2 ÷ 3.4 → $\begin{bmatrix} 1 & -.4 & -.6 & | & -1.4 \\ 0 & 1 & -.4118 & | & 2.4706 \\ 0 & .6 & 5.4 & | & 15.6 \end{bmatrix}$

R3 − 0.6 ∗ R2 → $\begin{bmatrix} 1 & -.4 & -.6 & | & -1.4 \\ 0 & 1 & -.4118 & | & 2.4706 \\ 0 & 0 & 5.6471 & | & 14.1176 \end{bmatrix}$

R3 ÷ 5.6471 → $\begin{bmatrix} 1 & -.4 & -.6 & | & -1.4 \\ 0 & 1 & -.4118 & | & 2.4706 \\ 0 & 0 & 1 & | & 2.5000 \end{bmatrix}$

$$\begin{array}{l} R1 + .6*R3 \\ R2 + .4118*R3 \end{array} \rightarrow \left[\begin{array}{ccc|c} 1 & -.4 & 0 & .1000 \\ 0 & 1 & 0 & 3.5001 \\ 0 & 0 & 1 & 2.500 \end{array}\right]$$

$$R1 + .4*R2 \rightarrow \left[\begin{array}{ccc|c} 1 & 0 & 0 & 1.5000 \\ 0 & 1 & 0 & 3.5001 \\ 0 & 0 & 1 & 2.5000 \end{array}\right]$$

With full pivoting:

$$\left[\begin{array}{ccc|c} 1 & 3 & -2 & 7 \\ 4 & -1 & 3 & 10 \\ -5 & 2 & 3 & 7 \end{array}\right]$$

$$R1 \leftrightarrow R3 \rightarrow \left[\begin{array}{ccc|c} -5 & 2 & 3 & 7 \\ 4 & -1 & 3 & 10 \\ 1 & 3 & -2 & 7 \end{array}\right]$$

$$R1 \div -5 \rightarrow \left[\begin{array}{ccc|c} 1 & -.4 & -.6 & -1.4 \\ 4 & -1 & 3 & 10 \\ 1 & 3 & -2 & 7 \end{array}\right]$$

$$\begin{array}{l} R2 - 4*R1 \\ R3 - 1*R1 \end{array} \rightarrow \left[\begin{array}{ccc|c} 1 & -.4 & -.6 & -1.4 \\ 0 & .6 & 5.4 & 15.6 \\ 0 & 3.4 & -1.4 & 8.4 \end{array}\right]$$

$$\underset{\text{column}}{C2 \leftrightarrow C3} \rightarrow \left[\begin{array}{ccc|c} 1 & -.6 & -.4 & -1.4 \\ 0 & 5.4 & .6 & 15.6 \\ 0 & -1.4 & 3.4 & 8.4 \end{array}\right] \begin{array}{l} \text{value of } x_3 \\ \text{will appear here} \\ \text{value of } x_2 \\ \text{will appear here} \end{array}$$

$$R2 \div 5.4 \rightarrow \left[\begin{array}{ccc|c} 1 & -.6 & -.4 & -1.4 \\ 0 & 1 & .1111 & 2.8889 \\ 0 & -1.4 & 3.4 & 8.4 \end{array}\right]$$

$$R3 + 1.4*R2 \rightarrow \left[\begin{array}{ccc|c} 1 & -.6 & -.4 & -1.4 \\ 0 & 1 & .1111 & 2.8889 \\ 0 & 0 & 3.5555 & 12.4445 \end{array}\right]$$

$$R3 \div 3.5555 \rightarrow \left[\begin{array}{ccc|c} 1 & -.6 & -.4 & -1.4 \\ 0 & 1 & .1111 & 2.8889 \\ 0 & 0 & 1 & 3.5001 \end{array}\right]$$

$$\begin{array}{l} R1 + .4*R3 \\ R2 - .1111*R3 \end{array} \rightarrow \left[\begin{array}{ccc|c} 1 & -.6 & 0 & 0 \\ 0 & 1 & 0 & 2.5000 \\ 0 & 0 & 1 & 3.5001 \end{array}\right]$$

$$R1 + .6*R2 \rightarrow \left[\begin{array}{ccc|c} 1 & 0 & 0 & 1.5000 \\ 0 & 1 & 0 & 2.5000 \\ 0 & 0 & 1 & 3.5001 \end{array}\right] \begin{array}{l} x_1 \\ x_3 \\ x_2 \end{array}$$

The exact answers are $x_1 = 1.5$, $x_2 = 3.5$, and $x_3 = 2.5$. The total absolute errors for the Gaussian elimination without pivoting, with partial pivoting, and with full pivoting are .0004, .0001, and .0001, respectively. Pivoting did reduce the error.

Gauss-Jordan elimination is another method for solving systems of equations. It differs from the Gaussian scheme only in that elements above the pivot are eliminated at the same time as those below, thus making back substitution unnecessary. The algorithm is simpler but less efficient—it requires more arithmetic operations. As in Gaussian elimination, pivoting can sometimes be used in the Gauss-Jordan method.

Algorithm GAUSS_JORDAN_ELIMINATION. This algorithm is given the same information as algorithm GAUSSIAN_ELIMINATION and accomplishes the same result (without pivoting).

1. [Do once for each row]
   Repeat thru step 3 for i = 1, 2, . . . , n
2. [Divide each element in the row by the pivot]
   Repeat for j = i, i + 1, . . . , m
   $a_{ij} \leftarrow a_{ij} / a_{ii}$
3. [Subtract a multiple of the row from each other row]
   Repeat for k = 1, 2, . . . , i − 2, i − 1, i + 1, i + 2, . . . , n
   Repeat for j = i, i + 1, . . . , m
   $a_{kj} \leftarrow a_{kj} - a_{ki} * a_{ij}$
4. [Finished]
   Exit   ☐

Notice that if one of the diagonal elements of the matrix becomes zero, algorithms GAUSSIAN_ELIMINATION and GAUSS_JORDAN_ELIMINATION will certainly fail, since the pivot is used as a divisor in step 2. One way to rectify this shortcoming of both algorithms would be to incorporate in step 2 a check for zero and try the next row (same column) if a zero is encountered. The original row would then have to supply a pivot in a different column. The problem of zero pivots can be completely circumvented by using elimination with pivoting, provided that the system is indeed solvable.

To illustrate the technique of Gauss-Jordan elimination, we return to our earlier example. We will examine two solutions: one without pivoting, and one with partial pivoting.

Without pivoting:

$$\begin{bmatrix} 1 & 3 & -2 & 7 \\ 4 & -1 & 3 & 10 \\ -5 & 2 & 3 & 7 \end{bmatrix}$$

$$\begin{matrix} R3 + 5 * R1 \\ R2 - 4 * R1 \end{matrix} \rightarrow \begin{bmatrix} 1 & 3 & -2 & 7 \\ 0 & -13 & 11 & -18 \\ 0 & 17 & -7 & 42 \end{bmatrix}$$

R2 ÷ −13 →
$$\left[\begin{array}{ccc|c} 1 & 3 & -2 & 7 \\ 0 & 1 & -.8462 & 1.3846 \\ 0 & 17 & -7 & 42 \end{array}\right]$$

R1 − 3 ∗ R2
R3 − 17 ∗ R2 →
$$\left[\begin{array}{ccc|c} 1 & 0 & .5386 & 2.8462 \\ 0 & 1 & -.8462 & 1.3846 \\ 0 & 0 & 7.3854 & 18.4618 \end{array}\right]$$

R3 ÷ 7.3854 →
$$\left[\begin{array}{ccc|c} 1 & 0 & .5386 & 2.8462 \\ 0 & 1 & -.8462 & 1.3846 \\ 0 & 0 & 1 & 2.4998 \end{array}\right]$$

R1 − .5386 ∗ R3
R2 + .8462 ∗ R3 →
$$\left[\begin{array}{ccc|c} 1 & 0 & 0 & 1.4998 \\ 0 & 1 & 0 & 3.4999 \\ 0 & 0 & 1 & 2.4998 \end{array}\right]$$

With partial pivoting:

$$\left[\begin{array}{ccc|c} 1 & 3 & -2 & 7 \\ 4 & -1 & 3 & 10 \\ -5 & 2 & 3 & 7 \end{array}\right]$$

R1 ↔ R3 →
$$\left[\begin{array}{ccc|c} -5 & 2 & 3 & 7 \\ 4 & -1 & 3 & 10 \\ 1 & 3 & -2 & 7 \end{array}\right]$$

R1 ÷ −5 →
$$\left[\begin{array}{ccc|c} 1 & -.4 & -.6 & -1.4 \\ 4 & -1 & 3 & 10 \\ 1 & 3 & -2 & 7 \end{array}\right]$$

R2 − 4 ∗ R1
R3 − 1 ∗ R1 →
$$\left[\begin{array}{ccc|c} 1 & -.4 & -.6 & -1.4 \\ 0 & .6 & 5.4 & 15.6 \\ 0 & 3.4 & -1.4 & 8.4 \end{array}\right]$$

R2 ↔ R3 →
$$\left[\begin{array}{ccc|c} 1 & -.4 & -.6 & -1.4 \\ 0 & 3.4 & -1.4 & 8.4 \\ 0 & .6 & 5.4 & 15.6 \end{array}\right]$$

R2 ÷ 3.4 →
$$\left[\begin{array}{ccc|c} 1 & -.4 & -.6 & -1.4 \\ 0 & 1 & -.4118 & 2.4706 \\ 0 & .6 & 5.4 & 15.6 \end{array}\right]$$

R1 + .4 ∗ R2
R3 − .6 ∗ R2 →
$$\left[\begin{array}{ccc|c} 1 & 0 & -.7647 & -.4118 \\ 0 & 1 & -.4118 & 2.4706 \\ 0 & 0 & 5.6471 & 14.1176 \end{array}\right]$$

R3 ÷ 5.6471 →
$$\left[\begin{array}{ccc|c} 1 & 0 & -.7647 & -.4118 \\ 0 & 1 & -.4118 & 2.4706 \\ 0 & 0 & 1 & 2.5 \end{array}\right]$$

R1 + .7647 ∗ R3
R2 + .4118 ∗ R3 →
$$\left[\begin{array}{ccc|c} 1 & 0 & 0 & 1.5 \\ 0 & 1 & 0 & 3.5001 \\ 0 & 0 & 1 & 2.5 \end{array}\right]$$

Again, we see a decrease in the error when partial pivoting is used.

We now examine two iterative procedures for solving sets of equations. Iterative methods have certain advantages over the direct methods previously discussed. They are useful when the coefficient matrix involved has many zeros, that is, when the matrix is sparse. In general, their computer implementation is easy.

In the *Jacobi* method, the first step is to solve for one of the variables in each equation. Whenever possible, the variable with the largest coefficient should be solved for in each equation. For example, the set of equations

$$5x_1 + x_2 + 3x_3 = 10 \tag{1}$$
$$x_1 + x_2 + 5x_3 = 8 \tag{2}$$
$$2x_1 + 4x_2 + x_3 = 11 \tag{3}$$

might be transformed to

$$x_1 = 2.00 - .20x_2 - .60x_3 \tag{1}$$
$$x_2 = 2.75 - .50x_1 - .25x_3 \tag{3}$$
$$x_3 = 1.60 - .20x_1 - .20x_2 \tag{2}$$

We then make initial guesses at the values of the variables (all zeros will do if no better values are known). Using these guesses on the right-hand sides of the equations, we next solve these equations for better estimates of the variables. We then repeat the procedure, always using the most recently calculated set of approximations to obtain better approximations. When two sets of estimates are close enough together, the procedure is stopped and the last set is taken as the solution.

One stopping criterion might be when the sum of the squares of the differences is less than a prescribed tolerance. That is, if x_j^i is the ith estimation of the variable x_j and there are n variables to be solved for, we would take $x_1^i, x_2^i, \ldots, x_n^i$ to be the solution if

$$\sum_{j=1}^{n} (x_j^i - x_j^{i-1})^2 < \epsilon$$

where ϵ is the specified tolerance. This iterative process is formalized in the following algorithm.

Algorithm JACOBI. Given n, the number of unknowns, $x_1, x_2, \ldots, x_n$, an array of initial guesses, and $f_1, f_2, \ldots, f_n$, the equations to be used to solve for $x_1, x_2, \ldots, x_n$, respectively, this algorithm solves the system of equations

$$x_1 = f_1(x_2, x_3, \ldots, x_n)$$

$$x_2 = f_2(x_1, x_3, x_4, \ldots, x_n)$$

.
.
.

$$x_n = f_n(x_1, x_2, \ldots, x_{n-1})$$

by the Jacobi method. The approximations are stored in the real vector
NEWx. The process stops when the sum of the squares of the differences is
less than some given ϵ.

1. [Limit to 30 iterations]
   Repeat thru step 4 for i = 1, 2, . . . , 30
2. [Calculate the new set of approximations]
   Repeat for j = 1, 2, . . . , n
   $\text{NEWx}_j \leftarrow f_j(x_1, x_2, \ldots, x_{j-1}, x_{j+1}, \ldots, x_n)$
3. [Test accuracy]
   SUM $\leftarrow 0$
   Repeat for j = 1, 2, . . . , n
   SUM $\leftarrow$ SUM + $(\text{NEWx}_j - x_j)^2$
   If SUM $< \epsilon$
   then Write('ANSWER IS', NEWx_1, NEWx_2, . . ., NEWx_n)
   Exit
4. [Use new set of values in next approximation]
   Repeat for j = 1, 2, . . . , n
   $x_j \leftarrow \text{NEWx}_j$
5. [No convergence]
   Write('TOLERANCE NOT MET IN 30 ITERATIONS',
   'ANSWER SO FAR IS', x_1, x_2, . . ., x_n)
6. [Finished]
   Exit   □

The Jacobi procedure applied to the previous example would yield the
sequence of values shown in Table 8-4 (numbers are truncated to four
decimal places). Twenty-one iterations were required for four decimal places
of accuracy.

Notice that improved values for $x_1, x_2, \ldots, x_{j-1}$ are available when the
improved x_j is calculated. It might speed the process to use these new values
in the calculation of x_j. When the most recently calculated individual
values rather than the most recently calculated set of values are used, the
method is called *Gauss-Seidel iteration*. This method is guaranteed to
succeed regardless of the starting values if the coefficient of the variable
solved for in each equation is larger in magnitude than the sum of the
absolute values of the other coefficients in that equation. This condition is
very often satisfied in engineering problems involving the solutions of
simultaneous equations. The algorithm for this modified iterative scheme
follows.

TABLE 8-4 Trace of the Jacobi Iterative Method

x_1	x_2	x_3
0	0	0
2.0	2.75	1.6
.49	1.35	.65
1.34	2.3425	1.232
.7923	1.7720	.8635
1.1275	2.1379	1.0871
.9201	1.9144	.9469
1.0490	2.0532	1.0331
.9695	1.9672	.9796
1.0188	2.0203	1.0126
.9883	1.9874	.9921
1.0072	2.0078	1.0048
.9955	1.9952	.9970
1.0027	2.0030	1.0018
.9983	1.9982	.9988
1.0010	2.0011	1.0007
.9993	1.9993	.9995
1.0004	2.0004	1.0002
.9998	1.9997	.9998
1.0001	2.0001	1.0001
.9999	1.9999	.9999
1.0000	2.0000	1.0000

Algorithm GAUSS_SEIDEL. Given n, $x_1, x_2, \ldots, x_n$, $f_1, f_2, \ldots, f_n$, and ϵ, as in algorithm JACOBI, this algorithm solves the set of equations using Gauss-Seidel iteration. Notice that the array NEWx is not required.

1. [Limit to 30 iterations]
   Repeat thru step 3 for i = 1, 2, . . . , 30
2. [Calculate new set of approximations]
   SUM ← 0
   Repeat for j = 1, 2, . . . , n
   OLD ← x_j
   x_j ← $f_j(x_1, x_2, \ldots, x_{j-1}, x_{j+1}, \ldots, x_n)$
   SUM ← SUM + $(x_j - OLD)^2$
3. [Test accuracy]
   If SUM < ϵ
   then Write('ANSWER IS', $x_1, x_2, \ldots, x_n$)
       Exit
4. [No convergence]
   Write('TOLERANCE NOT MET IN 30 ITERATIONS',
       'ANSWER SO FAR IS', $x_1, x_2, \ldots x_n$)
5. [Finished]
   Exit

□

As an example, Table 8-5 shows the sequence of values obtained when Gauss-Seidel iteration is used to solve the following system of equations.

$$x_1 = -1.4000 + .4000x_2 + .6000x_3$$
$$x_2 = 2.3333 - .3333x_1 + .6666x_3$$
$$x_3 = 3.3333 - 1.3333x_1 + .3333x_2$$

Notice that only six iterations were required for four decimal places of accuracy when the Gauss-Seidel method was used.

In this section we have presented four methods for solution of sets of simultaneous linear equations; of these, the Gaussian elimination method with or without pivoting and the Gauss-Seidel method are most often used.

EXERCISES 8-4

1. By hand, solve the following system of equations using Gaussian elimination without pivoting, with partial pivoting, and with full pivoting. Show the steps of each solution in detail.

$$2x_1 + 3x_2 + 5x_3 = 5$$
$$3x_1 + 4x_2 + 7x_3 = 6$$
$$x_1 + 3x_2 + 2x_3 = 5$$

2. Trace the execution of the algorithm GAUSS_JORDAN_ELIMINATION on the following system of equations:

$$x + 2y + 3z = 5$$
$$2x - y + z = 6$$
$$x + 3y - 5z = 2$$

3. Trace the execution of the algorithm JACOBI on the system of equations in exercise 2. Use $x = y = z = 0$ as the initial guesses and assume that $\epsilon = 0.005$.

4. Repeat exercise 3 using the algorithm GAUSS_SEIDEL.

TABLE 8-5   Trace of Gauss-Seidel Iterative Method

x_1	x_2	x_3
0	0	0
2.0	1.75	.85
1.14	1.9675	.9785
1.0194	1.9956	.9970
1.0026	1.9994	.9996
1.0003	1.9999	.9999
1.0000	2.0000	1.0000

8-5   CURVE FITTING BY LEAST-SQUARES APPROXIMATION

Experimental results are almost always in error to some degree. This is due both to experimental conditions and to the inaccuracy of measuring devices. Consider the problem of finding a curve to "smooth" the data obtained from an experiment. This is a useful way of representing the functional depend-ency on the observed values, and can be used to predict future results. To do this we require a mathematical formula which represents the relationships between the variables involved, as shown in Fig. 8-16. The process should produce the "best" single curve that represents these results.

There are several ways in which the concept of "best" can be defined. One alternative would be to minimize the sum of the differences between the points and the curve, or in other words, the approximation errors. However, this is unsatisfactory, as seen in Figure 8-17a, where line A would be as good an approximation as line B. If the sum of the absolute values of the errors is minimized, ambiguity results. In Fig. 8-17b lines A, B, and C are equally good approximations for the three points, using this criterion.

The method that is commonly used and generally thought to be best is the minimization of the sum of the squares of the errors—the *least-squares technique*. This attempts to concentrate on the magnitude of the difference rather than the direction. Also, it places importance on large errors, while small errors are more easily tolerated.

In this section we develop a procedure to fit a curve to a set of points which we suspect can be approximated by a polynomial of degree n. The n that we choose depends on the data. For a linear relationship, n will be 1. If we choose n to be equal to the number of points minus 1, the curve will fit the points exactly; however, when n is large, the resulting polynomials are awkward to find and they oscillate violently.

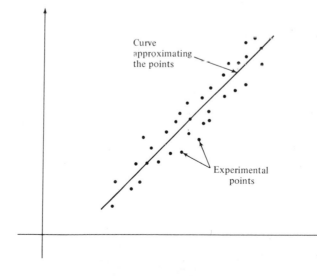

Curve approximating the points

Experimental points

FIGURE 8-16
Least-squares curve fit.

FIGURE 8-17
Two alternatives for
minimizing errors:
(a) minimize the sum of
the errors; (b) minimize the
sum of the absolute values
of the errors.

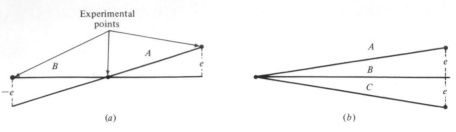

We want to approximate the data by a function of the form

$$f(x) = a_0 + a_1 x + a_2 x^2 + \cdots + a_n x^n$$

Let (x_i, y_i) denote the ith data point and let N be the number of points to which we wish to fit the curve. The error of the ith point is given by the difference between the actual value (y_i) and the estimated value $(f(x_i))$. Thus,

$$e_i = y_i - f(x_i) = y_i - a_0 - a_1 x_i - a_2 x_i^2 - \cdots - a_n x_i^n$$

and the sum of the squares of the errors is therefore

$$S = \sum_{i=1}^{N} (y_i - a_0 - a_1 x_i - a_2 x_i^2 - \cdots - a_n x_i^n)^2$$

This function will be a minimum when its partial derivatives with respect to the a's are all equal to zero, that is, when

$$\frac{\partial s}{\partial a_0} = \sum_{i=1}^{N} 2(y_i - a_0 - a_i x_i - a_2 x_i^2 - \cdots - a_n x_i^n)(-1) = 0$$

$$\frac{\partial s}{\partial a_1} = \sum_{i=1}^{N} 2(y_i - a_0 - a_1 x_i - a_1 x_i^2 - \cdots - a_n x_i^n)(-x_i) = 0$$

$$\frac{\partial s}{\partial a_2} = \sum_{i=1}^{N} 2(y_i - a_0 - a_1 x_i - a_2 x_i^2 - \cdots - a_n x_i^n)(-x_i^2) = 0$$

$$\cdot \qquad \cdot$$
$$\cdot \qquad \cdot$$
$$\cdot \qquad \cdot$$

$$\frac{\partial s}{\partial a_n} = \sum_{i=1}^{N} 2(y_i - a_0 - a_1 x_i - a_2 x_i^2 - \cdots - a_n x_i^n)(-x_i^n) = 0$$

Rearranging, we get $n + 1$ equations in $n + 1$ unknowns $(a_0, a_1, \ldots, a_n)$, which can be solved simultaneously to obtain the coefficients of the approximating polynomial.

$$a_0 N + a_1 \sum_{i=1}^{N} x_i + a_2 \sum_{i=1}^{N} x_i^2 + \cdots + a_n \sum_{i=1}^{N} x_i^n = \sum_{i=1}^{N} y_i$$

$$a_0 \sum_{i=1}^{N} x_i + a_1 \sum_{i=1}^{N} x_i^2 + a_2 \sum_{i=1}^{N} x_i^3 + \cdots + a_n \sum_{i=1}^{N} x_i^{n+1} = \sum_{i=1}^{N} x_i y_i$$

$$a_0 \sum_{i=1}^{N} x_i^2 + a_1 \sum_{i=1}^{N} x_i^3 + a_2 \sum_{i=1}^{N} x_i^4 + \cdots + a_n \sum_{i=1}^{N} x_i^{n+2} = \sum_{i=1}^{N} x_i^2 y_i$$

$$\cdots$$

$$a_0 \sum_{i=1}^{N} x_i^n + a_1 \sum_{i=1}^{N} x_i^{n+1} + a_2 \sum_{i=1}^{N} x_i^{n+2} + \cdots + a_n \sum_{i=1}^{N} x_i^{2n} = \sum_{i=1}^{N} x_i^n y_i$$

These equations are called *normal equations*.

To illustrate the technique we will fit a polynomial of degree 2 to the data of Table 8-6. In this case $N = 7$. The required sums, rounded to four decimal places, are

$$\sum_{i=1}^{N} x_i = 15.2620$$

$$\sum_{i=1}^{N} x_i^2 = 43.1754$$

$$\sum_{i=1}^{N} x_i^3 = 136.4352$$

$$\sum_{i=1}^{N} x_i^4 = 454.1167$$

$$\sum_{i=1}^{N} y_i = 13.0643$$

$$\sum_{i=1}^{N} x_i y_i = 29.4168$$

$$\sum_{i=1}^{N} x_i^2 y_i = 90.7620$$

The formulas to use are

$$a_0 N + a_1 \sum_{i=1}^{N} x_i + a_2 \sum_{i=1}^{N} x_i^2 = \sum_{i=1}^{N} y_i$$

$$a_0 \sum_{i=1}^{N} x_i + a_1 \sum_{i=1}^{N} x_i^2 + a_2 \sum_{i=1}^{N} x_i^3 = \sum_{i=1}^{N} x_i y_i$$

$$a_0 \sum_{i=1}^{N} x_i^2 + a_1 \sum_{i=1}^{N} x_i^3 + a_2 \sum_{i=1}^{N} x_i^4 = \sum_{i=1}^{N} x_i^2 y_i$$

Substitution yields the following system to be solved (presumably, by one of the methods of Sec. 8-4):

$$7.0000a_0 + 15.2620a_1 + 43.1754a_2 = 13.0643$$
$$15.2620a_0 + 43.1754a_1 + 136.4352a_2 = 29.4168$$
$$43.1754a_0 + 136.4352a_1 + 454.1167a_2 = 90.7620$$

Without going into the details, the solution is

$$a_0 = 4.4190$$
$$a_1 = -3.6492$$
$$a_2 = .8761$$

so the curve to fit the data is

$$y = 4.4190 - 3.6492x + .8761x^2$$

From this curve it is possible to predict further y values. Suppose, for example, that the data of Table 8-6 represent the results from an experiment, where x is the controlled variable and y the measured result. The calculated polynomial suggests the functional relationship that appears to exist in these points. Thus, if we wish to predict the experimental results for additional values of the controlled variable (such as, say, $x = 3.0$), these are easily obtained by substitution (yielding in this case $y = 1.3563$).

The formulas just given can be adapted to linear approximations, since a straight line is simply a polynomial of degree 1. This is a special case of this problem, which is so commonly used as to be given a special name—*linear regression*. Suppose, for example, that we wish to fit a straight line to the data of Table 8-7. In this case $N = 8$. We have the following values:

$$\sum_{i=1}^{N} x_i = 28.0 \qquad \sum_{i=1}^{N} y_i = -17.1$$

$$\sum_{i=1}^{N} x_i^2 = 140.0 \qquad \sum_{i=1}^{N} x_i y_i = -21.1$$

Substituting these values into the equations

$$a_0 N + a_1 \sum_{i=1}^{N} x_i = \sum_{i=1}^{N} y_i$$

$$a_0 \sum_{i=1}^{N} x_i + a_1 \sum_{i=1}^{N} x_i^2 = \sum_{i=1}^{N} x_i y_i$$

TABLE 8-6

x	y
.4501	3.0509
.9802	1.5078
1.3376	1.0999
2.3999	.9212
2.7936	1.0012
3.5805	2.2333
3.7201	3.2500

we get

$$8.0a_0 + 28.0a_1 = -17.1$$
$$28.0a_0 + 140.0a_1 = -21.1$$

By solving this pair of equations, we get the line

$$y = -5.36667 + .92262x$$

which is known as the *regression line*. The values -5.36667 and $.92262$ are known as the *regression coefficients*.

The method of curve fitting described in this section is based on a polynomial approximation. This is the most popular technique, but it is not the only technique. Other methods are based on nonpolynomial functions—exponential or logarithmic functions, for example. These are not discussed in this book.

EXERCISES 8-5

1. Compute the regression line for the data of Table 8-6.

2. Formulate an algorithm to read in a set of N (x, y) values and compute the regression coefficients.

TABLE 8-7

x	y
.0	−5.0
1.0	−4.5
2.0	−3.5
3.0	−2.9
4.0	−2.0
5.0	−.8
6.0	.1
7.0	1.5

3. The midterm averages for a selected sample of freshman computer science students were recorded and tabulated along with their final high school averages. The results are given in Table 8-8.

TABLE 8-8

Student	Midterm Average	High School Average
ADAMS	73.5	78.2
BARNES	66.9	71.5
CAMPBELL	83.8	81.7
DOLBY	58.1	77.3
FRIESEN	77.1	85.6
HOOPER	44.3	65.8
JONES	35.8	70.3
MARSHALL	66.2	60.5
REMPEL	72.5	75.3
SCOTT	60.6	66.6
TURNER	85.8	85.2
WALSH	87.9	86.3

Compute and plot the regression line for these data. How closely does it appear to represent the original data? What does this one experiment tell you about the predictability of freshman performance from high school averages?

4. Collect data on the heights and weights of 15 of your classmates. Compute the regression line for these data. Divide the data into male and female categories and compute separate regression lines for each. Compare their slopes.

5. Using the data of exercise 3, obtain the best-fit second-degree polynomial.

8-6 SUMMARY AND REVIEW

In this chapter five aspects of numerical analysis have been presented. This section contains a brief list of important points.

1. *Errors*
   (a) *Classification: Inherent errors* are due to the approximate nature of measurement. *Truncation errors* result from the truncation of a process or a number. *Roundoff errors* occur when a number is rounded. An error may be expressed in *absolute*, *relative*, or *percentage* form.

(b) *Propagation:*

$$e_{x+y} = e_x + e_y \qquad\qquad r_{x+y} = \frac{e_x + e_y}{\bar{x} + \bar{y}}$$

$$e_{x-y} = e_x - e_y \qquad\qquad r_{x-y} = \frac{e_x - e_y}{\bar{x} - \bar{y}}$$

$$e_{xy} \doteq \bar{x}e_y + \bar{y}e_x \qquad\qquad r_{xy} \doteq \frac{e_x}{\bar{x}} + \frac{e_y}{\bar{y}}$$

$$e_{x/y} \doteq \frac{\bar{x}}{\bar{y}}\left(\frac{e_x}{\bar{x}} - \frac{e_y}{\bar{y}}\right) \qquad r_{x/y} \doteq \frac{e_x}{\bar{x}} - \frac{e_y}{\bar{y}}$$

2. Root finding
   (a) *The problem:*   find a value of x for which $f(x) = 0$.
   (b) *The fixed-point iteration approach:*   choose a function g. Generate approximations to the root iteratively by

$$x_{n+1} = g(x_n)$$

   (c) *The method of successive bisection:*   find an interval (x_1, x_2) such that $f(x_1) \times f(x_2) < 0$. If the midpoint,

$$\frac{x_1 + x_2}{2}$$

   is not a root, then bisect the interval, choose the half which contains the root, and repeat the procedure with this new interval.
   (d) *The secant method (also called the computed line approach):* choose two points, x_1 and x_2, and calculate more x values until the root is obtained. Subsequent x values come from the formula

$$x_{n+1} = x_n - \frac{x_n - x_{n-1}}{f(x_n) - f(x_{n-1})} \times f(x_n)$$

   (e) *Newton's method:*   choose one point, x_1, which you suspect is close to the root, and calculate successive x values until the root is found. Subsequent x values come from the formula

$$x_{n+1} = x_n - \frac{f(x_n)}{f'(x_n)}$$

3. *Numerical integration*
   (a) *The problem:*   the integral of a function can be interpreted as the area under the curve on the graph of the function. Numerical integration methods compute approximations to this area.

(b) *The rectangle rule:*

$$\int_a^b f(x)\,dx \doteq \sum_{i=1}^{n-1} f(x_i) \times h = h[f(x_0) + f(x_0 + h) + f(x_0 + 2h)$$
$$+ \cdots + f(x_0 + (n-1)h)]$$

(c) *The trapezoid rule*

$$\int_a^b f(x)\,dx \doteq \sum_{i=0}^{n-1} \frac{h}{2}[f(x_i) + f(x_{i+1})]$$
$$= \frac{h}{2}[f(x_0) + 2f(x_1) + 2f(x_2) + \cdots + 2f(x_{n-1}) + f(x_n)]$$

(d) *Simpson's rule*

$$\int_a^b f(x)\,d_x \doteq \frac{h}{3}[f(x_0) + 4f(x_1) + 2f(x_2) + 4f(x_3)$$
$$+ \cdots + 2f(x_{n-2}) + 4f(x_{n-1}) + f(x_n)]$$

4. *Simultaneous linear equations*
   (a) *Given:*

$$a_{11}x_1 + a_{12}x_2 + \cdots + a_{1n}x_n = b_1$$
$$a_{21}x_1 + a_{22}x_2 + \cdots + a_{2n}x_n = b_2$$
$$\cdot$$
$$\cdot$$
$$\cdot$$
$$a_{n1}x_1 + a_{n2}x_2 + \cdots + a_{nn}x_n = b_n$$

where the a's and b's are known, we wish to find values for x_1, $x_2, \ldots, x_n$ such that the equalities are all true.
   (b) *Gaussian elimination:* forward elimination reduces all elements below the diagonal to zero and all elements on the diagonal to 1. Back substitution then reduces elements above the diagonal to zero.
   (c) *Gauss-Jordan elimination:* elements above and below the diagonal are reduced to zero at the same time. No back substitution is required.
   (d) *Pivoting:* roundoff errors are reduced by full or partial pivoting.

Full pivoting	Interchanging rows and columns so the next pivot is the largest number in magnitude

Partial pivoting   Interchanging rows only so the next
pivot is the largest possible number in
magnitude

(e) *Iterative methods:*   solve for one of the variables in each equation. Make an initial guess at the solution and substitute into the right-hand side of the equations, obtaining a better approximation of the solution. Continue the equation with newer x values.

Jacobi iteration   Use the most recent set of values

Guess-Seidel iteration   Use the most recent individual
values

5. *Curve fitting by least-squares approximation*
   (a) *The problem:*   we wish to "smooth" some experimental data by approximating them with a curve.
   (b) *The method:*   where the experimental data are (x_1, y_1), (x_2, y_2), $\ldots$, (x_N, y_N), solve the following set of simultaneous linear equations:

$$a_0 N + a_1 \sum_{i=1}^{N} x_i + a_2 \sum_{i=1}^{N} x_i^2 + \cdots + a_n \sum_{i=1}^{N} x_i^n = \sum_{i=1}^{N} y_i$$

$$a_0 \sum_{i=1}^{N} x_i + a_1 \sum_{i=1}^{N} x_i^2 + a_2 \sum_{i=1}^{N} x_i^3 + \cdots + a_n \sum_{i=1}^{N} x_i^{n+1} = \sum_{i=1}^{N} x_1 y_i$$

$$a_0 \sum_{i=1}^{N} x_i^2 + a_i \sum_{i=1}^{N} x_i^3 + a_2 \sum_{i=1}^{N} x_i^4 + \cdots + a_n \sum_{i=1}^{N} x_i^{n+2} = \sum_{i=1}^{N} x_i^2 y_i$$

$$\vdots$$

$$a_0 \sum_{i=1}^{N} x_i^n + a_1 \sum_{i=1}^{N} x_i^{n+1} + a_2 \sum_{i=1}^{N} x_i^{n+2} + \cdots + a_n \sum_{i=1}^{N} x_i^{?n} = \sum_{i=1}^{N} x_i^n y_l$$

(c) *Linear regresion:*   an important special case in which a straight line is to be fitted to the given data.

9

Advanced String Processing

The basic notions of string processing were introduced in Chap. 5. In this chapter the notions of pattern matching introduced earlier are extended. Several string applications which involve lexical analysis, text editing, KWIC indexing, and the use of bit strings in information organization and retrieval are described.

9-1 BASIC FUNCTIONS

In Chap. 5 we discussed in detail certain primitive string-handling functions such as concatenation, LENGTH, and SUB. These functions were considered to be primitive in the sense that most other string-handling functions can be obtained in a modular manner from these primitives. In this section we concentrate our efforts on the development of three additional functions—MATCH, SPAN, and BREAK—which are nonprimitive, yet basic, to string-handling problems. These functions are useful in pattern matching. Although several other useful functions could have been introduced, we have chosen these functions on the basis of their popularity in the string-manipulation language SNOBOL. This language is one of the most powerful languages of its kind.

Recall from Sec. 5-3 that pattern matching is the process of searching a certain string (called the *subject*) for the occurrence of a certain substring (called a *pattern*). This searching of the subject string is usually done in a left-to-right manner. For example, the function reference

    INDEX(SUBJECT, PATTERN)

requires that the leftmost occurrence of PATTERN in SUBJECT be located. If the pattern-matching process is successful, then the position of the first character of **PATTERN** in **SUBJECT** is returned; otherwise, the matching process fails and a zero value is returned by the function.

The search for an occurrence of a particular substring in a subject string is performed in a left-to-right manner on a character-by-character basis. It is convenient to associate a *cursor* with the scanning process. The purpose of the cursor is to keep track of where we are in the pattern match. For example, in the reference

INDEX('TREMBLAY□J□P', '□')

we associate a cursor with the left-to-right search of the subject string 'TREMBLAY□J□P' for the leftmost blank. Initially, the cursor denotes the position of the leftmost character in the subject (that is, 'T'). Since this character is not a blank, the cursor is advanced to position 2. At this point it gives the position of the second leftmost character (that is, 'R'). Again, this character is not a blank and therefore the cursor is advanced to the third position, and so on. The process terminates successfully when the cursor refers to the ninth character (that is, a blank). It is important to note that although the presence of a cursor mechanism is of no concern to the programmer using the **INDEX** function (other than giving him/her insight into the pattern-matching process), such a mechanism does, nevertheless, exist. Furthermore, when the search for the occurrence of a pattern string in a subject string does not begin at its leftmost character, the cursor mechanism often becomes the center of attention.

In the development of the algorithms for the string-handling functions **MATCH**, **SPAN**, and **BREAK**, which we now give, particular attention is given to the cursor mechanism just discussed. The position of a character being examined by the mechanism will be denoted by the integer variable **CURSOR**.

Since the three functions to be discussed are all of a pattern-matching nature, we are able to characterize them by using the same set of arguments throughout. In particular, each pattern-matching function has the following six arguments:

SUBJECT	Subject string which is to be examined.
PATTERN	Pattern string which is sought within the subject string.
CURSOR	Character position (integer) in **SUBJECT** at which the pattern-matching process is to begin. Consequently, the subject string is searched from the **CURSOR** position to the last character in the subject string.
MATCH_STR	String variable which contains the desired substring found in **SUBJECT** in the case of a success-

ful pattern match. In the case of an unsuccessful pattern match, the contents of the variable are unaltered.

REPLACE_FLAG — Logical variable which indicates whether or not the matched substring should be replaced. If the value of the flag is *true*, then a replacement is required; otherwise, no replacement is desired.

REPLACE_STR — String which replaces the matched substring when a replacement of the matched substring is specified (that is, when REPLACE_FLAG is *true*).

The three pattern-matching functions are truth-valued functions (that is, each function has a value of *true* or *false*) in the following algorithms. In each case, a value of *true* is returned when a successful pattern match occurs; otherwise, a value of *false* is returned.

Function MATCH(SUBJECT, PATTERN, CURSOR, MATCH_STR, REPLACE_FLAG, REPLACE_STR). Given the six arguments described earlier, this function returns a value of *true* if the pattern string (PATTERN) is found starting at the character position specified by CURSOR in the string SUBJECT; otherwise, the function returns a value of *false*. When a successful pattern match occurs, MATCH_STR is set to the value of PATTERN and if REPLACE_FLAG is *true*, the substring matched in SUBJECT is replaced by the value in REPLACE_STR.

1. [Does the pattern fit within the search bounds of the subject string?]
    If CURSOR + LENGTH(PATTERN) > LENGTH(SUBJECT) + 1
    then Return(false)
2. [Perform pattern match]
    If SUB(SUBJECT, CURSOR, LENGTH(PATTERN)) ≠ PATTERN
    then Return(false)
3. [Set MATCH_STR and perform indicated replacement]
    MATCH_STR ← PATTERN
    If REPLACE_FLAG
    then SUB(SUBJECT, CURSOR, LENGTH(PATTERN)) ← REPLACE_STR
        CURSOR ← CURSOR + LENGTH(REPLACE_STR)
    else CURSOR ← CURSOR + LENGTH(PATTERN)
4. [Successful return]
    Return(true)   ☐

The first step of this algorithm determines whether or not the pattern fits the search bounds of the subject string. If it does not, then an unsuccessful return results. Step 2 examines the subject string from the cursor character position and attempts to match the pattern string. If this pattern-matching process fails, then an unsuccessful return is made; otherwise, the pattern has been found and control passes to step 3. Here, the pattern string is copied into MATCH_STR, and if a replacement is indicated, the contents of REPLACE_STR replace the matched substring in SUBJECT. In either case, CURSOR is appropriately updated.

As an example of how the MATCH function behaves, consider the following parameter values:

```
SUBJECT = 'STRING MANIPULATION IS FUN!'
PATTERN = 'IS'
CURSOR = 21
REPLACE_FLAG = true
REPLACE_STR = 'WAS'
```

In step 1, CURSOR + LENGTH(PATTERN) (equalling 23) is less than LENGTH(SUBJECT) + 1 (equalling 28), and therefore in step 2 we attempt a pattern match using the SUB function. A successful pattern match occurs since the pattern string 'IS' is found in positions 21 and 22 of the subject string. Consequently, we proceed to step 3. In this step MATCH_STR receives the value 'IS', and since a replacement is indicated, the substring 'IS' is replaced by 'WAS', giving a new subject string value of

'STRING MANIPULATION WAS FUN!'

Also, CURSOR's new value is 24.

As another example of using the function MATCH, let the parameter values be the following:

```
SUBJECT = 'STRING MANIPULATION IS FUN'
PATTERN = 'FUN'
CURSOR = 24
REPLACE_FLAG = true
REPLACE_STR = 'FUN□AND□GAMES'
```

In this case a successful pattern match results, with the empty string being matched at the end of SUBJECT. [Recall that SUB(SUBJECT, 27, 0) returns the empty string.] Since a replacement is indicated, the new value of SUBJECT becomes

'STRING MANIPULATION IS FUN AND GAMES'

Finally, CURSOR receives a value of 37 (that is, one more than the length of the new subject string), and the value of MATCH_STR is the empty string.

The second basic pattern-matching function, SPAN, matches all consecutive characters in the subject string from the indicated cursor position and ending at the first character not in the pattern string. For example, the function invocation

SPAN('□□□COMPUTER□SCIENCE', '□', 1, MATCH_STR, true, '')

should match and delete the leftmost three blanks in the subject string, thus yielding the string 'COMPUTER□SCIENCE'.

Function SPAN(SUBJECT, PATTERN, CURSOR, MATCH_STR, REPLACE_FLAG, REPLACE_STR). Given the six parameters described earlier, this algorithm returns *true* if the character denoted by CURSOR matches any of the characters in PATTERN. In the case of a successful pattern match, the value of MATCH_STR becomes a sequence of characters. This sequence of characters contains (as its first character) the character at the position specified by CURSOR and all other characters which are contained in PATTERN. The pattern-matching process terminates on encountering a character not in PATTERN or the end of the subject string. If a replacement operation is specified, this sequence of characters is replaced by the value of REPLACE_STR. I is an integer variable.

1. [Does the pattern fit within the bounds of the subject string?]
   If CURSOR > LENGTH(SUBJECT)
   then Return(false)
2. [Initialize pattern match]
   I ← CURSOR
3. [Is character I in the pattern string?]
   Repeat while I ≤ LENGTH(SUBJECT) and
            INDEX(PATTERN, SUB(SUBJECT, I, 1)) ≠ 0
     I ← I + 1
4. [Unsuccessful pattern match]
   If I = CURSOR
   then Return(false)
5. [Set MATCH_STR and perform indicated replacement]
   MATCH_STR ← SUB(SUBJECT, CURSOR, I − CURSOR)
   If REPLACE_FLAG
   then SUB(SUBJECT, CURSOR, I − CURSOR) ← REPLACE_STR
       CURSOR ← CURSOR + LENGTH(REPLACE_STR)
   else CURSOR ← I
6. [Successful return]
   Return(true)   □

Step 1 determines whether or not CURSOR refers to a character position within the subject string. If it does not, the pattern match fails. The second step initializes the variable I to the position of the first character which is to be examined in the matching process (that is, the value of CURSOR). In step 3 a search in PATTERN is made for the character at position I in SUBJECT. If this search is successful, then I is incremented by one and the search is repeated, using the next character in SUBJECT. Encountering the end of the subject string or failing to find the specified character (of SUBJECT) in PATTERN terminates the third step. Step 4 determines whether or not any character was matched. The failure of the pattern-matching process results in an unsuccessful return from the function; otherwise, control passes to step 5, where the matched substring is copied into MATCH_STR, the indicated replacement (if any is to be performed) is executed, and CURSOR is updated appropriately. Finally, step 6 contains a successful return from the function.

As an example of the behavior of this function, consider the subject string '□□□COMPUTER□SCIENCE'. Assume that we want to delete any leading blanks from this string. We can accomplish this task by invoking the function SPAN with the following assignment of parameters:

PATTERN = '□', CURSOR = 1, REPLACE_STR = '', and
REPLACE_FLAG = true.

Since the value of CURSOR is less than the length of the subject string, I is set to the cursor position 1 in step 3. The condition in the repeat statement in this step succeeds three times (that is, for I = 1, 2, and 3). Therefore, the matched substring is '□□□'. The condition fails, however, when I = 4, since SUB(SUBJECT, 4, 1) is equal to 'C'. Consequently, the INDEX function returns a value of 0. The test in step 4 fails since the current value of I (that is, 4) is not equal to that of CURSOR. We then proceed to step 5, where MATCH_STR receives the value '□□□'. Also, in this step, the empty string replaces the matched substring in SUBJECT and the value of CURSOR remains at 1 because the length of the empty string is zero. Finally, the function returns the logical value *true* in step 6.

Additional examples that exhibit the behavior of the function SPAN follow. In every case SUBJECT contains the value

'THEREFORE,□COMPUTER□SCIENCE□IS□EXCITING.'

SPAN(SUBJECT, 'EFHORT', 1, MATCH_STR, false, REPLACE_STR) succeeds with the substring 'THEREFORE' being matched, with MATCH_STR receiving this value. Since REPLACE_FLAG is false, no replacement is made. The updated value of CURSOR is 10.

SPAN(SUBJECT, ',', 1, MATCH_STR, true, REPLACE_STR) fails because the leftmost character in SUBJECT is a letter, not a comma.

SPAN(SUBJECT, ',□', 10, MATCH_STR, true, '□') succeeds with the substring ',□' being matched. MATCH_STR is set to this substring, whose occurrence in SUBJECT is replaced by a blank. The value of CURSOR becomes 11.

The next pattern-matching function we discuss is called BREAK. This function scans the subject string on a character-by-character basis, starting at the given cursor position and proceeding to the first instance of a character which is also a character in the pattern string. If such a character is found, then the matched string is the substring of the subject string from the initial cursor position up to, but excluding, the character found. For example, the function reference

BREAK('JOHN□DOE', '□', 1, MATCH_STR, false, REPLACE_STR)

matches the substring 'JOHN' in the subject string.

Function BREAK(SUBJECT, PATTERN, CURSOR, MATCH_STR, REPLACE_FLAG, REPLACE_STR). Given the six pattern-matching parameters introduced earlier, this function returns a value of *true* if it encounters, in a character-by-character scan of the subject string from the cursor position onward, a character which is also in PATTERN; otherwise, the function returns a value of *false*. When a successful pattern match occurs, MATCH_STR is set to the substring of the subject string from the initial cursor position up to, but not including, the character found to be also in PATTERN. If REPLACE_FLAG is *true*, the matched substring in SUBJECT is replaced by the value of REPLACE_STR. Finally, CURSOR is updated, accordingly. I is an integer variable.

1. [Does the pattern fit within the search bounds of the subject string?]
   If CURSOR > LENGTH(SUBJECT)
   then Return(false)
2. [Initialize pattern match]
   I ← CURSOR
3. [Is character I in the pattern string?]
   Repeat while I ≤ LENGTH(SUBJECT)
           and INDEX(PATTERN, SUB(SUBJECT, I, 1)) = 0
      I ← I + 1
4. [Successful pattern match?]
   If I = LENGTH(SUBJECT) + 1
   then Return(false)
5. [Set MATCH_STR and perform indicated replacement]
   MATCH_STR ← SUB(SUBJECT, CURSOR, I – CURSOR)
   If REPLACE_FLAG
   then SUB(SUBJECT, CURSOR, I – CURSOR) ← REPLACE_STR
        CURSOR ← CURSOR + LENGTH(REPLACE_STR)
   else CURSOR ← I
6. [Successful return]
   Return(true)   ☐

This algorithm is very similar to the previous function, SPAN. Note, however, that in step 3 we are searching for the first character (from the cursor position onward in SUBJECT) which is also in PATTERN. Also, observe that the test in step 4 for an unsuccessful pattern match involves checking I for being out of the subject string bounds.

As an example of the use of this function, consider the case in which SUBJECT = 'PAUL□TREMBLAY'. Let us assume that we want first to obtain the first name (that is, 'PAUL') in the subject string and then delete it from SUBJECT. We can accomplish this task by invoking the function BREAK with the following assignment of parameters:

PATTERN = '□', CURSOR = 1, REPLACE_FLAG = true,
REPLACE_STR = ''.

Since the value of CURSOR is less than the length of SUBJECT, I is initialized to the value of CURSOR (that is, 1) in step 2. The condition in the repeat

statement of step 3 succeeds four times (that is, for I = 1, 2, 3, and 4). Therefore, the matched substring is 'PAUL'. The condition fails when I = 5, since
SUB(SUBJECT, 5, 1) is a blank. Consequently, the INDEX function returns a
value of 1. The test in step 4 fails because the current value of I (that is, 5) is
not equal to 14 (that is, LENGTH(SUBJECT) + 1). Therefore, we proceed to
step 5, where MATCH_STR receives the value 'PAUL'. In this step the empty
string replaces the matched substring in SUBJECT and the value of CURSOR
remains at 1 because the length of the empty string is zero. Finally, the
function returns the logical value *true* in step 6.

The following are additional examples which illustrate the use of the
BREAK function. In every case, SUBJECT has the value

'COMPUTERS,□HOWEVER,□HAVE□GREAT□SPEED.'

SPAN(SUBJECT, ',' 1, MATCH_STR, false, REPLACE_STR) succeeds with
the substring 'COMPUTERS' being matched and MATCH_STR receiving this
value. Since REPLACE_FLAG is false, no replacement is made. The updated
value of CURSOR is 10.

SPAN(SUBJECT, ';!:', 1, MATCH_STR, true, REPLACE_STR) fails since the
subject string does not contain the symbols ';', '!', or ':'.

SPAN(SUBJECT, '.', 1, MATCH_STR, false, REPLACE_STR) succeeds with
all the subject string except the period being matched. This matched substring
is assigned to MATCH_STR. The updated value of CURSOR is 37.

This concludes the subsection on basic string-manipulation functions.
In the next subsection we apply these functions to several applications.

EXERCISES 9-1

1. Design a function REPEAT for replicating a given character string a specified number of times. For example, REPEAT('HA!', 3) should generate
   the string 'HA!HA!HA!'.
2. Design a function TRIM which deletes all trailing blanks of a given character string. For example, given the argument

   'IT IS A NICE DAY!□□□□□',
   TRIM should return the string
   'IT IS A NICE DAY!'.

9-2 STRING MANIPULATION APPLICATIONS

This section contains four applications illustrating how the concepts and,
specifically, the string-handling functions discussed in this chapter can be
applied. The first and second applications deal with lexical analysis and

KWIC index generation. Both of these applications are interesting and practical. The third application is concerned with an information-retrieval problem which is solved using bit strings. Since bit strings were not discussed earlier in the book, this third application introduces them and their associated operations, as well as showing how such strings can be applied. In the fourth application we introduce a simple text editor and describe the text-handling operations that it provides. It is an excellent application because it involves a wide variety of character string-manipulation problems.

9-2.1 LEXICAL ANALYSIS

In Sec. 2-1 we briefly described the problem of translating a higher-level language (such as the programming language used in this book) to a lower-level target language (such as machine language). In this subsection we examine certain aspects of the translation process more closely.

In an attempt to describe the translation process, let us approach the problem by first analyzing a similar task in the translation of natural languages: that is, the translation from one natural language (for example, French) into another (for example, English). It is to be stressed, however, that the analogy between programming language translation and natural language translation is valid only in an informal sense.

Consider the French-to-English translation, by a novice translator, of a string of characters that represent French text. Such a translation would probably consist of the following steps:

1. Recognize the French words in the given text. This step involves performing a lexical scan of the text in order to classify the words into grammatical categories (noun, adjective, verb, etc.).
2. Verify the grammatical structure of each sentence in the text (that is, parse each sentence).
3. Using a French-to-English dictionary, look up the translation of the French words found in step 1 to their English equivalent. Then, according to certain rules, transform the grammatical structure obtained in step 2 into its equivalent English structure. From the latter structure, generate the translated sentences which represent the output string of English text.

Equivalently, a compiler for a certain language can be viewed as consisting of three separate tasks, as shown in Fig. 9-1. The source program is input to a *scanner* whose purpose is to separate the incoming text into words (or *tokens*) such as constants, variable names, keywords (such as **THEN**, **ELSE**, and **REPEAT**), operators, special symbols [such as (and)], and so on. The scanner supplies these tokens to the syntax analyzer, whose task is essentially to construct a parse of the given sentence. The output of the syntax analyzer is passed on to the code generator. Here, the parse and other things, such as symbol tables for variables and constants, are used to generate object code for that sentence.

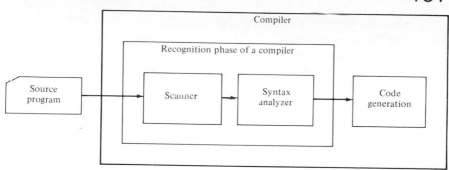

FIGURE 9-1
Model of a compiler.

In this subsection we are concerned with the lexical analysis of source statements from a subset of our algorithmic programming language. Nevertheless, many of the techniques presented are applicable to the lexical analysis phase of natural language machine translation. A discussion of the syntax analysis and code-generation phases is given in Sec. 10-5.2.

In the discussion to follow, we consider the design of a scanner algorithm for the assignment statement of our algorithmic notation. Actually, we concern ourselves with a retricted form of this statement in which only integer arithmetic expressions are permitted. We also ignore subscripted variables and functional references. The words or tokens for such a language are identifiers, integers, the arithmetic operators (namely, the addition, subtraction, multiplication, division, and exponentiation operators), and the assignment operator ($\leftarrow$).

The words or tokens of this simple language are

Identifier	String of alphanumeric characters starting with a letter and whose remaining characters can be a letter, a digit, or an underscore ($\_$)
Integer	String of digits
Addition and subtraction operators	+ and $-$
Multiplication and division operators	* and /
Exponentiation operator	$\uparrow$
Assignment operator	$\leftarrow$
Left parenthesis	(
Right parenthesis	)

In addition to these tokens, there are certain symbols which must be handled by the scanner, and yet their presence is not passed on to the syntactic analyzer. In our simple language the blank character is such a symbol (that is, $A\square\leftarrow\square B\square+\square C$ is syntactically equivalent to $A\leftarrow B+C$).

Before formulating a scanning algorithm, several observations concerning the previous classification of tokens can be made. In particular, the definition of an identifier does not make any distinction among the letters A through Z. Therefore, it is not necessary to distinguish among any of these characters; consequently, they can be treated as a single group whose generic name is "letter." Regardless of the algorithm which we may devise, it will not be required to perform different actions for different identifiers. A similar observation can be made about the digits 0 through 9.

In most instances it is inefficient to pass a class name, such as "integer" or "identifier," to the syntactic analyzer. Instead, we associate a unique representation number with each class, and it is this number along with the source form (for example, the integer) which is given to the syntactic analyzer. For our example language, we adopt the following representation number assignments: "identifier" is 1, "integer" is 2, "addition/subtraction operator" is 3, "multiplication/division operator" is 4, "exponentiation operator" is 5, "assignment operator" is 6, "left parenthesis" is 7, and "right parenthesis" is 8.

We are now prepared to give an algorithm for scanning a source statement from the simple programming language just described.

Algorithm SCAN. Given a source statement SOURCE (of type string), this algorithm separates this statement's tokens with their corresponding representation numbers, as described earlier. CHAR represents the current character being examined in SOURCE. TOKEN contains the present token being isolated and REP_NO contains the representation number of that token. LETTERS and DIGITS are character variables which contain the values 'ABCDEFGHIJKLMNOPQRSTUVWXYZ' and '0123456789', respectively. The integer variable CURSOR denotes the position of the character being examined in SOURCE. DUMMY is a logical variable which is used to hold the logical value returned by the function SPAN. F and T are temporary integer and string variables, respectively. ALPHA is a string variable which contains all letters and digits.

1. [Initialize]
    CURSOR ← 1
    ALPHA ← LETTERS ∘ DIGITS
2. [Output source statement]
    Write(SOURCE)
3. [Scan source statement]
    Repeat step 4 while CURSOR ≤ LENGTH(SOURCE)
4. [Obtain next token]
    If ¬ SPAN(SOURCE, '□', CURSOR, T, false, ")
                                                (check for nonblank symbol)
    then CHAR ← SUB(SOURCE, CURSOR, 1)   (isolate next character)
        If INDEX(LETTERS, CHAR) ≠ 0   (check for identifier)
        then DUMMY ← SPAN(SOURCE, ALPHA, CURSOR, TOKEN, FALSE,")
            Write(1, TOKEN)   (output identifier)

```
        else If INDEX(DIGITS, CHAR) ≠ 0        (check for integer)
            then DUMMY ← SPAN(SOURCE, DIGITS, CURSOR, TOKEN,
                FALSE,")
                Write(2, TOKEN)                 (output integer)
        else F ← INDEX('+*↑←()', CHAR)
                                                (check for +, *, ↑, ←, (, or ))
            If F ≠ 0
            then REP_NO ← F + 2
                Write(REP_NO, CHAR)
                                                (output operator)
            else F ← INDEX('s/', CHAR)
                                                (check for − or /)
                If F ≠ 0
                then REP_NO ← F + 2
                    Write(REP_NO, CHAR)
                                                (output operator)
                else Write('ILLEGAL CHARACTER', CHAR)
            CURSOR ← CURSOR + 1
5. [Finished]
    Exit                                                                    □
```

The first step of the algorithm initializes CURSOR to a character position of 1. Also, ALPHA is set to the string of alphanumeric characters which contains all letters of the English alphabet followed by 10 decimal digits. Step 2 outputs the original source statement which is to be scanned. The third step controls the scanning of the source statement. Step 4 determines the next token (along with its representation number) to be output. Observe that blank characters are ignored. The first character of a potential token is copied into CHAR. This character is first checked to determine if it is a letter. If it is, then the SPAN function returns the identifier found in TOKEN, and the token representation number (that is, 1) and TOKEN are written. If, however, CHAR is a digit, the SPAN function places a digit string into TOKEN. The representation number of the digit string (that is, 2) and TOKEN are then written. If the next token is neither an identifier nor a number, then an attempt is made to classify CHAR as an operator or a parenthesis. The INDEX function is again used for this purpose. A check is first made for +, *, ↑, ←, (, or). If this check succeeds, the appropriate representation number of the character is computed (by the statement REP_NO ← F + 2) and this value and CHAR are written. In case of a failure, the character is checked against − and /. If this last check fails, however, the current character is illegal. Note that if an identifier, a number, or a blank is not found, then CURSOR must be explicitly incremented to the next character position.

The output from this algorithm for the source statement

'□□A1□←□A□+□5' is

```
     □□A1□←□A□+□5
  1        A1
  6        ←
  1        A
  3        +
  2        5
```

The output from the algorithm for the source statement

'75□←□X□Y□−□↑□Z' is

```
     75□←□X□Y□−□↑□Z
  2        75
  6        ←
  1        X
  1        Y
  3        −
  5        ↑
  1        Z
```

In the last example, notice that the scanner is not responsible for the checking of proper syntax at the statement or sentence level. Clearly, the given statement is syntactically incorrect. The problem of parsing sentences from a programming language will be discussed in more detail in Sec. 10-5.2.

In certain compilers or translators a token and its associated representation number are often passed to a symbol-table routine. This routine generates a number of tables, such as an identifier table, a statement-label table, a programmer-defined function table, and so on. The type of table organization that can be used depends on a number of factors, such as table size. A simple linear table such as that discussed in Sec. 6-5.1 may be used. For large symbol tables, however, more sophisticated table organization techniques must be used. Such techniques are discussed in Sec. 10-10.2.

EXERCISES 9-2.1

1. Revise the algorithm SCAN so that it can handle the sample language which contains real numbers. A real number is defined to be a sequence of digits (possibly empty) followed by a decimal point (.) followed by a (nonempty) sequence of digits. For example, 7.25 and .50 are examples of valid real numbers.

2. Alter the algorithm obtained in exercise 1 such that comments can be handled. These comments should be printed by the scanner with the assignment statements, but no indication of the presence of comments should be passed on to the syntactic analyzer. Therefore, the statement

    [Update A] A□←□B□+□C

should only generate the following:

1	A
6	←
1	B
3	+
1	C

3. Modify the algorithm SCAN in the text such that a table which contains all tokens is maintained. Use a sequential unordered table for this purpose.

4. As an extension of exercise 2, formulate a scanner for the algorithmic language introduced so far except for subscripted variables and subalgorithm references.

9-2.2 KEYWORD-IN-CONTEXT (KWIC) INDEXING

An important area within computer science is information retrieval. In information-retrieval applications, a data base (that is, a large amount of information encompassing many aspects of a particular application) is maintained on an on-line system. The data base is often kept in large random-access (for example, disk) files. For example, a literature search of a certain topic area in a library system or the number of graduate students who are married and in the College of Engineering in a student record application may be required. This type of application occurs in so many different situations that a special class of languages known as *query languages* has been created. The data base is searched for requested information based on index items generated by a user query. An important problem associated with information-retrieval systems (and, in particular, in automatic library systems) is that of creating a good *indexing scheme*.

A popular method of indexing that is used in library systems is the permuted or KWIC (keyword-in-context) indexing method. This method produces a listing of all the lines in a document that contain any of a list of prespecified keywords. If a particular line contains several keywords, then this line appears several times in the listing, once for each keyword it contains. A KWIC index provides the context surrounding every occurrence of each keyword. In practice, KWIC indexing is often applied to phrases, especially titles, selected from a set of documents of interest. For example, the data base to be searched might contain the titles of research papers in the area of computer science. An individual expressing interest in a specialized research area (such as programming languages or compiler writing) could then query this data base for the titles dealing with this area. While KWIC indexing allows a user to determine the role of a word quickly, this method of indexing requires a large amount of memory because of the amount of contextual information that must be stored.

To illustrate this storage requirement, consider, as an example, the title of the following book:

An Introduction to Computer Science an Algorithmic Approach

In a KWIC indexing scheme, each word in the phrase, or title in this case, is scanned for keywords and reproduced once in a permuted manner for each keyword. In the current example, the list of permuted indices is:

> *Introduction to Computer Science an Algorithmic Approach // An*
> *Computer Science an Algorithmic Approach // An Introduction to*
> *Science an Algorithmic Approach // An Introduction to Computer*
> *Algorithmic Approach // An Introduction to Computer Science an*
> *Approach // An Introduction to Computer Science an Algorithmic*

The underscored first word in each index line is called an *index word*. The set of index words generated from a title is the set of keywords for that title (for example, Introduction, Computer, Science, Algorithmic, and Approach are the keywords in the current example). Keywords are considered to be those words which have meaning as to the nature of the document. Ordinary words such as "a," "to," "and," "not," "an," and so on, reveal little about the contents of a document. In general, prepositions, conjunctions, pronouns, and, in many instances, adverbs, adjectives, and verbs are not considered to be keywords.

Let us now proceed to design a KWIC index generator. We shall assume that the information necessary to generate the index is stored in four vectors. The first string vector, ORD_WORDS, holds an ordered list of ordinary words for the index-generation system—one word per vector element. In this subsection we assume that the vector ORD_WORDS contains seven words: 'a', 'an', 'and', 'its', 'the', 'to', and 'with'. The second vector (also of type string), TITLE, contains the titles of the documents with one element having, as its value, the title of one document. A third vector, KEYWORD, contains an ordered list of the keywords that are present in the current list of titles. Associated with these keywords, and stored in a fourth vector, T_INDEX, is a list of vector indices which refer to the titles stored in TITLE. If a given keyword occupies the Ith element in KEYWORD, then T_INDEX[I] contains a character string which holds the positions in TITLE of all titles which contain the given keyword. The position indices are separated by blank characters.

As an example of this organization, let us assume that TITLE[1], TITLE[2], ..., TITLE[5] have, as values, the following titles:

    'AN INTRODUCTION TO COMPUTER SCIENCE AN ALGORITHMIC
    APPROACH //'
    'STRUCTURED PL/I PROGRAMMING //'
    'STRUCTURED WATFIV-S PROGRAMMING //'
    'PL/I PROGRAMMING WITH APPLICATIONS //'
    'AN INTRODUCTION TO PASCAL PROGRAMMING //'

The KEYWORD and T_INDEX vectors for this set of five documents are given in Table 9-1. The keyword STRUCTURED, for example, appears in TITLE[2] and TITLE[3].

TABLE 9-1   Keywords and Title Indices for KWIC Example

I	KEYWORD	T_INDEX
1	'ALGORITHMIC'	'1'
2	'APPLICATIONS'	'4'
3	'APPROACH'	'1'
4	'COMPUTER'	'1'
5	'INTRODUCTION'	'1□5'
6	'PASCAL'	'5'
7	'PL/I'	'2□4'
8	'PROGRAMMING'	'2□3□4□5'
9	'SCIENCE'	'1'
10	'STRUCTURED'	'2□3'
11	'WATFIV-S'	'3'

The following algorithm analyzes the input phrases or titles and constructs the vectors KEYWORD, TITLE, and T_INDEX. Each title input is assumed to consist of a single card. Each word in the title is separated by a blank character, and the last word of the title is followed by a single blank. Other punctuation symbols are not allowed.

Algorithm KWIC_CREATE. Given an input sequence of phrases (that is, titles) in the form of character strings and the ordered list of ordinary words, ORD_WORDS, as described earlier, this algorithm constructs the vectors TITLE, KEYWORD, and T_INDEX. PHRASE is a string variable which contains the current title being input. ORD_SEARCH is a function which searches the vector ORD_WORDS to determine whether or not a word in a particular title is an ordinary word. If the word in question is ordinary, then its vector index is returned; otherwise, a value of zero is returned. Similarly, KEY_SEARCH is a function which searches the vector KEYWORD to determine whether or not a keyword has previously been encountered in the titles. If it has, the vector index of that keyword is returned; otherwise, the new keyword is inserted appropriately in KEYWORD (recall that KEYWORD is ordered alphabetically) and the index associated with the position occupied by this keyword is returned. The integer variable LAST_TITLE denotes the index of the last title to be added to the vector TITLE. WORD is as intermediate string variable which contains the current word being processed in a title. KEYIND is an integer variable which is used as an index for vector KEYWORD. BLANKS and DUMMY are temporary string and logical variables, respectively. These variables do not contain any useful data.

1. [Initialize]
   LAST_TITLE ← 0
2. [Process all titles]
   Repeat thru step 8 while there is still an input title

3. [Read a document title]
   Read(PHRASE)
4. [Remove any leading blanks and append end markers // to title]
   DUMMY ← SPAN(PHRASE, '□', 1, BLANKS, true, '')
   PHRASE ← PHRASE ∘ '//'
5. [Store current document title]
   LAST_TITLE ← LAST_TITLE + 1
   TITLE[LAST_TITLE] ← PHRASE
6. [Process all words in current title]
   Repeat thru step 8 while SUB(PHRASE, 1, 2) ≠ '//'
7. [Scan and remove next word from current title]
   WORD ← SUB(PHRASE, 1, INDEX(PHRASE, '□') − 1)
   PHRASE ← SUB(PHRASE, INDEX(PHRASE, '□') + 1)
8. [Is word a keyword?]
   If ORD_SEARCH(WORD) = 0
   then KEYIND ← KEY_SEARCH(WORD)
      T_INDEX[KEYIND] ← T_INDEX[KEYIND] ∘ '□' ∘ LAST_TITLE
      If SUB(T_INDEX[KEYIND], 1, 1) + '□'
      then SUB(T_INDEX[KEYIND], 1, 1) = ''
9. [Finished]
   Exit   □

Step 1 of the algorithm initializes the counter LAST_TITLE to zero. The second step controls the input of all phrases or titles while step 3 reads a title. Step 4 deletes any leading blanks in the title and then appends the two marker symbols //. The fifth step first updates the LAST_TITLE counter and then stores the title in the vector TITLE. Step 6 controls the processing of each word in the title. In step 7 we isolate the next word (in a left-to-right scan) in the title. We also delete this word (and the following blank character) from PHRASE. If the word is a keyword, then a search of the vector KEYWORD for that keyword is made. Otherwise, the word is ignored, If the keyword is not found (that is, this is its first occurrence in the input), then we proceed to perform a lexically ordered insertion into vector KEYWORD. If, on the other hand, the keyword is not new, then we return its position in KEYWORD. The functions ORD_SEARCH and KEY_SEARCH are left as exercises.

With the construction of the pertinent vectors out of the way, we can now proceed to formulate an algorithm to generate a KWIC index.

Algorithm KWIC_GEN. Given four string vectors—ORD_WORDS, TITLE, and KEYWORD, whose number of elements is specified by LAST_KEY, and T_INDEX as discussed earlier—this algorithm generates a KWIC index ordered lexically by keywords. KEYSTRING is a string variable which contains the string of TITLE indices (for a particular keyword) which are stored as an element of T_INDEX. INDEX_NO is an integer variable which contains a particular TITLE vector index. KWIC_LINE is a temporary string variable which contains a line of output. Finally, I is a counter variable and POS is a temporary integer variable.

1. [Process each keyword in KEYWORD]
    Repeat thru step 5 for I = 1, 2, . . ., LAST_KEY
2. [Obtain index list for current keyword]
    KEYSTRING ← T_INDEX[I] ∘ '□'
3. [Process all title indices in KEYSTRING]
    Repeat thru step 5 while LENGTH(KEYSTRING) > 1
4. [Obtain and delete next title index]
    INDEX_NO ← SUB(KEYSTRING, 1, INDEX(KEYSTRING, '□') − 1)
    KEYSTRING ← SUB(KEYSTRING, INDEX(KEYSTRING, '□') + 1)
5. [Obtain and output KWIC line]
    KWIC_LINE ← TITLE[INDEX_NO]
    POS ← INDEX(KWIC_LINE, KEYWORD[I])
    If POS ≠ 0
    then KWIC_LINE ← KEYWORD[I] ∘ SUB(KWIC_LINE, POS +
        LENGTH(KEYWORD[I])) ∘ '□' ∘ SUB(KWIC_LINE, 1, POS − 1)
        Write(KWIC_LINE)
    else Write('ERROR-KEYWORD NOT FOUND IN TITLE')
6. [Finished]
    Exit   □

The first step of the algorithm establishes a loop which examines all the keywords which are presently in the system. Step 2 copies the string which contains the title locations of a particular keyword. The third step controls the processing of each title given by the title locations for the current keyword. Step 4 obtains, and deletes, the next title location from KEYSTRING. In step 5 a search of the appropriate title is made for the keyword in question. If the search is successful, then the permuted index line is output; otherwise, an error message results.

As an example, let us examine the behavior of this algorithm, using, as data, Table 9-1 and the five titles given earlier. Step 1 sets up a loop which is repeated 11 times. The first time step 2 is executed, KEYSTRING has the value '1□'. In step 3 a check is made to determine whether or not all title indices corresponding to a given keyword have been processed. The result of the test indicates that there are other title indices remaining to be processed. Step 4 sets INDEX_NO to 1 and reduces KEYSTRING to the empty string. In step 5 KWIC_LINE is initially set to

    'AN INTRODUCTION TO COMPUTER SCIENCE AN ALGORITHMIC
    APPROACH //'

and then reset to

    'ALGORITHMIC APPROACH // AN INTRODUCTION TO COMPUTER
    SCIENCE AN□'

The while condition in the repeat loop of step 3 is reexamined and the test fails, with the control being returned to step 1. The process is now repeated for the second keyword, that is, 'APPLICATIONS'.

If a complete trace of the output produced by the algorithm were given, then the next three lines and the last three lines printed for the given data would be

```
APPLICATIONS // PL/I PROGRAMMING WITH
APPROACH // AN INTRODUCTION TO COMPUTER SCIENCE AN
ALGORITHMIC
COMPUTER SCIENCE AN ALGORITHMIC APPROACH // AN
INTRODUCTION TO


     .

     .

     .

STRUCTURED PL/I PROGRAMMING //
STRUCTURED WATFIV-S PROGRAMMING //
WATFIV-S PROGRAMMING // STRUCTURED
```

EXERCISES 9-2.2

1. Given a word as an input parameter assigned to the variable WORD, construct a function subalgorithm ORD_SEARCH which searches the ORD_WORDS vector for the word. If the word is present, its index location in ORD_WORDS is returned; otherwise, a value of zero is returned.

2. Given as an input parameter a keyword assigned to the variable WORD, contruct algorithm KEY_SEARCH, which searches the KEYWORD array looking for the keyword. If the keyword is present, its index location in KEYWORD is returned; otherwise, the keyword is inserted in the KEYWORD vector at the appropriate location as determined by the lexical ordering of keywords. Note that space must also be left at the corresponding location in the T_INDEX to hold the string of TITLE vector indexes for the keyword.

3. Algorithm KWIC_GEN does not handle the situation in which a keyword appears more than once in a title. For example, the title

```
'TECHNIQUES FOR SYSTEM DESIGN AND SYSTEM
IMPLEMENTATION□'
```

should be printed twice:

```
SYSTEM DESIGN AND SYSTEM IMPLEMENTATION // TECHNIQUES
FOR
SYSTEM IMPLEMENTATION // TECHNIQUES FOR SYSTEM DESIGN
AND
```

Alter KWIC_GEN so that it provides this facility.

4. As more and more documents are added to a KWIC-index-generating system, it becomes less attractive to receive a KWIC-index printout for the complete list of documents. Instead, it is more desirable to receive a KWIC-index listing for those documents which contain in their title (or some other representable phrase) certain select index words. For example, we might pose a command of the form

LIST KWIC FOR "index word expression"

where "index word expression" is defined to be:

1. An index word
2. A sequence of index words delimited by parentheses and the logical operators OR, AND, and NOT such that the sequence forms a valid expression

Example commands are

LIST KWIC FOR PROGRAMMING
LIST KWIC FOR (APPLICATIONS AND PROGRAMMING)
LIST KWIC FOR (NOT(WATFIV-S) OR PL/1)

Construct an algorithm which interprets the "LIST KWIC" commands and outputs only the KWIC-index terms, as specified in the index-term expression for the command.

9-2.3 THE APPLICATION OF BIT STRINGS TO INFORMATION RETRIEVAL

So far in this book the strings on which we have focused have been character strings. There is another type of string, however, which is provided by some programming languages. This type of string is called a *bit string*. A bit string is a string which contains only characters from the alphabet $\{0, 1\}$. Examples of bit strings are the following:

'1'
'0'
'1010'
'000101'

The operations available for the manipulation of bit strings are the same as those discussed for character strings (that is, concatenation, SUB, INDEX, etc.). In addition, however, bit strings can be "ANDed," "ORed," and "NOTed" (or negated). These logical operations are performed bit by bit on a bit string. A '0' bit is interpreted as the logical value *false* while the '1' bit is taken to mean *true*. For example, the result of "ORing" the bit strings '1001' and '0100' yields the following:

'1001' OR '0100' = '1101'

Similarly, the negated value of the bit string '1001' is '0110'. Logical expressions of bit strings are easily formed as in the following:

('1001' OR '0010') AND ('1011')
NOT ('0110') AND ('1111')

Recall that the "NOT" operator has precedence over the "AND" operator, which in turn has precedence over the "OR" operator.

In this subsection we consider an information-retrieval application which provides us with an opportunity to exhibit a solution which makes use of bit strings. It should be noted that not all information-retrieval problems lend themselves to solutions involving bit strings. Even for those that do, such a solution may not be the best solution. We will not pursue, at this time, other methods of information organization which are alternatives to the use of bit strings.

We first give a number of definitions which are used in dealing with information, and then proceed to consider as a concrete example a simple student record application.

A collection of information concerning a particular item or individual is called a *record*. Each portion of this record which gives one specific attribute of the item or individual is a *field*. A set of records is a *file*. If a file contains information on a complete set, for example, all books in a library or all employees working for a corporation, then the file is a *master file*. A subset of a file is a *subfile*. In information retrieval, records having specified values in particular fields must be searched for. A *query* is used to describe formally the subfile of records being requested.

Consider a file of student records which have the following five fields:

Student's number (NUMBER)	Integer
Student's name (NAME)	Character string
Student's sex (SEX)	One-digit field coded as 1 for male and 2 for female
Student's college (COLLEGE)	One-digit field coded as follows:
	1 Science
	2 Commerce
	3 Engineering
	4 Graduate Studies
	5 Home Economics
	6 Agriculture
Student's marital status (MARITAL_STATUS)	One-digit field coded as follows:
	1 Single
	2 Married
	3 Other

If a request is made for information pertaining to students in the College of Engineering, then every record must be scanned to determine which records are required. Even greater difficulty would be encountered in searching for information on married students in engineering, as both the college and marital status fields must be considered. Is there an easier way to find the records containing these data?

If we assume that the master file is ordered, that is, we can identify a first record, second record, and so on, then we can specify beforehand a number of subsets in the file. This can be done compactly by using bit-represented sets. Obviously, many bit strings would be required to specify directly all possible subfiles which may be requested. To resolve this problem, only one bit string should be established for each specific value a file may have. Most information requests will be satisfied by the specified subsets or by logical operations performed on their bit string representations, as previously described.

The purpose of this approach is to avoid the complex and lengthy search often necessary for the retrieval of information. Bit-string-represented subsets sacrifice a minimum of storage space to decrease search time. Note, however, that they should not be used to specify records having a certain value in a file which has many possible values. For example, a student number field will have a different value for each record, and it is therefore not practical to allocate a bit string for each student number. When used in conjunction with a field such as marital status, only several bit strings are necessary to specify each possible status subset. For example, the bit string 10010110 . . . could be used to indicate that the first, fourth, sixth, and so on, records pertain to married students.

As a concrete example, the following list of student records will be used throughout the discussion in this subsection. For clarity, blanks have been inserted between the fields and titles have been added.

Number	Name	Sex	College	Marital Status
596426	LARRY R BROWN	1	2	2
600868	ROY B ANDERSON	1	4	2
621656	DAVID N PARKER	1	3	2
640621	JOHN M BROWN	1	1	2
652079	PATRICIA L FOX	2	1	2
672915	JOE E WALL	1	3	3
672919	LINDA R GARDNER	2	2	1
683369	SUSAN C FROST	2	1	3
690528	SUSAN L WONG	2	5	2
703062	JAKE L FARMER	1	6	1

To solve this information-retrieval problem, we make use of 11 bit strings. The length of each bit string is equal to the total number of students currently in the file. For example, one string represents male students; another, female students; a third, students in the College of Science; and so on.

The 11-bit strings correspond to the 11 values which are possible for SEX(2 values), COLLEGE(6 values), and MARITAL_STATUS(3 values). A '1' in the ith position of the bit string for female students indicates that the ith student is a female. A '0' in the ith position of the bit string for Commerce designates that the ith student is not in Commerce. For our sample file of 10 records, the bit strings are as follows:

Keyword	Bit String	Keyword	Bit String
Male	'1111010001'	Home Economics	'0000000010'
Female	'0000101110'	Agriculture	'0000000001'
Science	'0001100100'	Single	'0000001001'
Commerce	'1000001000'	Married	'1111100010'
Engineering	'0010010000'	Other status	'0000010100'
Graduate Studies	'0100000000'		

For convenience, we shall represent the two bit strings associated with SEX by a two-element bit-string vector, SEX_FILE. Each bit-string element in this vector can be indexed by the code for SEX. That is, SEX_FILE[1] and SEX_FILE[2] indicate the subfiles of records to male and female students, respectively. Similarly, the bit strings associated with COLLEGE and MARITAL_STATUS fields are represented by a six-element and three-element bit-string vectors COLLEGE_FILE and STATUS_FILE, respectively. COLLEGE_FILE[2] then represents the subfile of Commerce students. We can find the bit string which corresponds to the request of all male Commerce students simply by ANDing SEX_FILE[1] and COLLEGE_FILE[2]. This is done by performing the logical operations

    SEX_FILE[1] AND COLLEGE_FILE[2]

yielding the value

    '1111010001' AND '1000001000' = '1000000000'

which indicates that the first student in the file (that is, Larry R. Brown) is a male Commerce student.

Observe that since, for efficiency reasons, we have chosen to encode the SEX, COLLEGE, and MARITAL_STATUS fields associated with the student record applications, we require the keywords, such as Male, Female, Science, Commerce, and so on, to be maintained by the system. An obvious approach which satisfies this requirement is to have the keywords stored in three separate vectors, as follows:

SEX_WORD[1]   = 'MALE'   COLLEGE_WORD[1] = 'SCIENCE'
SEX_WORD[2]   = 'FEMALE'   COLLEGE_WORD[2] = 'COMMERCE'
STATUS_WORD[1] = 'SINGLE'   COLLEGE_WORD[3] = 'ENGINEERING'
STATUS_WORD[2] = 'MARRIED'   COLLEGE_WORD[4] = 'GRADUATE STUDIES'
STATUS_WORD[3] = 'OTHER'   COLLEGE_WORD[5] = 'HOME ECONOMICS'
                              COLLEGE_WORD[6] = 'AGRICULTURE'

We now proceed to formulate an algorithm which builds the student master file and the 11-bit strings associated with the student records application.

Procedure BUILD. Given the number of student records, N(an integer), followed by the student records (one per card), this algorithm constructs the master file consisting of the four vectors NUMBER, NAME, SEX, COLLEGE, and MARITAL_STATUS as described earlier. The algorithm also creates the bit-string vectors SEX_FILE, COLLEGE_FILE, and STATUS_FILE. The function REPEAT replicates a given string a specified number of times. I is a temporary index variable.

1. [Input number of student records]
   Read(N)
2. [Initialize]
   Repeat for I = 1, 2
     SEX_FILE[I] ← REPEAT('0', N)
   Repeat for I = 1, 2, . . ., 6
     COLLEGE_FILE[I] ← REPEAT('0', N)
   Repeat for I = 1, 2, 3
     STATUS_FILE[I] ← REPEAT('0', N)
3. [Process the student records]
   Repeat thru step 5 for I = 1, 2, . . ., N
4. [Input a student record]
   Read(NUMBER[I], NAME[I], SEX[I], COLLEGE[I], MARITAL_STATUS[I])
5. [Update bit-string vectors]
   SUB(SEX_FILE[SEX[I]], I, 1) ←'1'
   SUB(COLLEGE_FILE[COLLEGE[I]], I, 1) ←'1'
   SUB(STATUS_FILE[MARITAL_STATUS[I]], I, 1) ←'1'
6. [Finished]
   Return

The first step of the procedure initializes each of the 11 required bit strings to a sequence of N zero bits. The REPEAT function, which was introduced in exercise 1 of Exercises 9-1, performs this task. Step 2 inputs the value of N. The third step controls the processing of all student records. Step 4 inputs the five fields associated with a particular student, while step 5 uses this student's data to update the 11-bit strings. The next subalgorithm outputs a list of all students who satisfy a particular query.

Procedure OUTPUT(TITLE, STRING). Given a TITLE (a character string) associated with a particular query and STRING (a bit string), which describes (in an encoded manner) those students who satisfy the query, this procedure generates a report listing all student information pertaining to that query. The vectors NUMBER, NAME, SEX, COLLEGE, and MARITAL_STATUS are assumed to be global. The vectors SEX_WORD, COLLEGE_WORD, and STATUS_WORD, which are also global, are used to replace the codes for sex, college, and marital status with their English equivalents. I is an index variable. Note that STRING is assumed to be passed by value.

1. [Output title of report]
    Write(TITLE)
2. [Generate the details of the report]
    Repeat thru step 5 while INDEX(STRING, '1') $\neq$ 0
3. [Obtain position of next student to be output]
    I ← INDEX(STRING, '1')
4. [Change the corresponding bit in STRING to '0']
    SUB(STRING, I, 1) ← '0'
5. [Output details of next student]
    Write(NUMBER[I], NAME[I], SEX_WORD[SEX[I]],
        COLLEGE_WORD[COLLEGE[I]],
        STATUS_WORD[MARITAL_STATUS[I]])
6. [Finished]
    Return

Step 1 prints the given title. The second step controls the output of a detail line in the report for every student that satisfies a particular query. Step 3 scans the bit string for its leftmost '1' bit. The position of this bit specifies the position in the master file where the corresponding student can be found. Step 4 changes the value of the bit in that position to zero. The fifth step outputs the detail line for the student in question. Note that the English equivalent of the codes are output.

The previous subalgorithms can be invoked as part of an algorithm which also specifies queries concerning the master file. For example, the algorithm segment

```
Call BUILD
Call OUTPUT('ALL STUDENTS', SEX_FILE[1] AND SEX_FILE[2])
Call OUTPUT('FEMALE STUDENTS', SEX_FILE[2])
Call OUTPUT('SCIENCE STUDENTS', COLLEGE_FILE[1])
Call OUTPUT('UNMARRIED STUDENTS IN SCIENCE', COLLEGE_FILE[1]
    AND NOT STATUS_FILE[2])
```

produces the reports consisting of all students, all female students, all science students, and all unmarried science students.

Clearly, this approach of specifying a query leaves something to be desired. An alternative approach which would provide an easy mode of system interaction for the student-record application would be to create a command language which is English-like in nature. For example, commands of the form

```
LIST STUDENTS
LIST STUDENTS FEMALE
LIST STUDENTS SCIENCE
LIST STUDENTS UNMARRIED AND SCIENCE
```

could specify the four reports just mentioned. This simple command language, however, would require a compiler. The pursuit of creating such a compiler at this point would be premature. This enhancement of the system is therefore left as an exercise for Chap. 10, where compilation techniques are discussed (Sec. 10-5.2).

EXERCISES 9-2.3

1. Given the sample file of 10 records as presented in this subsection, state the output for the following statements:
   (*a*) Call OUTPUT('ALL SCIENCE STUDENTS', COLLEGE_FILE[1])
   (*b*) Call OUTPUT('FEMALE STUDENTS IN COMMERCE', SEX_FILE[2] AND COLLEGE_FILE[2])
   (*c*) Call OUTPUT('ALL STUDENTS NOT IN SCIENCE', NOT COLLEGE_FILE[1])

2. A transportation company operating in the mountain national parks of Alberta is computerizing its payroll and accounting procedures. After some initial planning and problem analysis, it is decided that an employee file should be created containing the following information:

   1. Employee's social insurance number (SIN)—a nine-digit field.
   2. Employee's name (NAME)—a 25-character field.
   3. Employee's sex (SEX)—a one-character field coded as 'M' for male and 'F' for female.
   4. Employee's type of work (TYPE)—encoded as a one-character field with 'A' denoting agent, 'D' denoting driver, 'H' denoting driver's helper, and 'P' denoting payroll.
   5. Employee's wage (WAGE)—encoded as a one-digit field where 1 denotes $4.00/hour, 2 denotes $4.80/hour, 3 denotes $6.00/hour, and 4 denotes $7.00/hour.
   6. Employee's location (LOCATION)—encoded as a one-character field where 'B' denotes Banff, 'J' denotes Jasper, and 'L' denotes Lake Louise.

   The values that the fields can assume are given.
   The following list of employee records is to be used. For clarity, blanks have been inserted between the fields and titles have been added.

Soc. Ins. No.	Name	Sex	Type	Wage	Location
693121053	LEW ARCHER	M	D	2	J
686725001	NANCY DREW	F	D	2	B
591146235	ARCHIE GOODWIN	M	H	1	B
661301964	PHILIP MARLOWE	M	A	3	J
529270792	JANE MARPLE	F	P	2	B
637263675	TRAVIS MCGEE	M	P	3	J

When computing the bimonthly paychecks, the company feels it is desirable to extract specific information from this file. For example, they may want a listing of all employees earning $6.00 per hour, a listing of all drivers in Lake Louise, a listing of all female employees in Jasper, etc. Construct the required number of bit strings (there are 13 of them) for this application.

3. Formulate a procedure (similar to the algorithm BUILD) to build a master file and the 13 bit strings obtained in exercise 2.

4. Construct a procedure (similar to the procedure TITLE) to output a list of employees who satisfy a particular query.

9-2.4 TEXT EDITING

In Sec. 5-4.1 the basic notions of text editing were introduced. In particular, an algorithm was developed for the right justification of textual material. In this subsection we continue the discussion of computer-generated text. We do so by introducing a particular example of a text editor. This editor is capable of performing the basic operations required of a text editor and does so by using a list of commands. Each command starts in character position 1 of a line of input with the two-character sequence ## so as to clearly distinguish a command from text material (that is, we assume that a line of text never begins with ##). Also, blank lines (that is, blank cards) of input text are not allowed. Each command must appear with its arguments on a separate line of input. The text editor commands are summarized in Table 9-2. The code column in the table will be discussed shortly.

Let us now formulate a general solution to the text-editing problem. One possible solution is the following:

1. Input the desired line width
2. Initialize variables
3. Repeat thru step 7 while there still remains input data
4. Input a line of text
5. Repeat thru step 7 while there still remains an item to be processed
6. Scan the next item (that is, word or command) from the current input line and obtain its type
7. If the next item is a word,
   then add this word to the current output line being prepared;
      if there are enough words to output the line,
         then print an edited line and start accumulating the next output line;
   else execute the action specified by the command.

The code of each command is given in Table 9-2. If the next item is not a command, however, then its code (or type) is assumed to be '1'. We assume that it is illegal to have another command immediately follow a ##NP com-

TABLE 9-2   Summary of Commands in a Sample Text Editor

Command	Code	Description
##JS	2	Right-justify the text; i.e., add extra blanks such that the printed form of the text is aligned in in the right margin
##NJ	3	Do not right-justify text
##NP	4	Start a new paragraph
##CN/.../	5	Center the text between the two slashes
##BL□n	6	Add n blank lines
##ED	7	End of text

mand. Since a solution to this problem is somewhat lengthy, it is convenient to separate some of the tasks in the preceding general algorithm into sub-algorithms. Figure 9-2 exhibits such an approach, where the main algorithm is given the name TEXT_EDIT. This algorithm, in turn, invokes a subalgorithm, FORMATOR. This subalgorithm is responsible for the production of the edited version of all input text. FORMATOR uses subalgorithms SCANNER, PRINT_JUST, and PRINT_UNJUST. SCANNER inputs, when necessary, a new input line and isolates the next item (that is, word or command) from the current input line. The scanner also determines the type of this item and returns it and the item to FORMATOR. The subalgorithm PRINT_JUST outputs a right-justified edited line of text. It determines and inserts the number of extra blanks required between the words which form this line of output. PRINT_UNJUST prints the current accumulated line of output without any right justification.

Let us now formulate a subalgorithm for the scanning module.

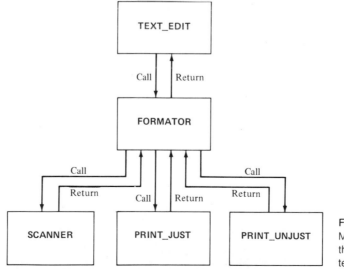

FIGURE 9-2
Modular description of the solution to a sample text editor.

Procedure SCANNER(INPUT_LINE, CURSOR, TYPE, ITEM). Given a character string (INPUT_LINE), which represents a line of input text, and CURSOR (of type integer), which denotes the current character position being examined, this procedure isolates the next word or command ITEM in INPUT_LINE and determines its type (TYPE). DUMMY is a temporary logical variable and BLANKS is a string variable which holds any leading blanks that are scanned from INPUT_LINE.

1. [Scan any leading blank characters]
    DUMMY ← SPAN(INPUT_LINE, '□', CURSOR, BLANKS, false, '')

2. [Has the current line been entirely scanned?]
    If CURSOR > LENGTH(INPUT_LINE)
    then Read(INPUT_LINE)
        INPUT_LINE ← INPUT_LINE ∘ '□'
        CURSOR ← 1
        DUMMY ← SPAN(INPUT_LINE, '□', CURSOR, BLANKS, false, '')
        DUMMY ← BREAK(INPUT_LINE, '□', CURSOR, ITEM, false, '')
        If SUB(INPUT_LINE, 1, 4) = '##JS'
        then TYPE ← 2
        else If SUB(INPUT_LINE, 1, 4) = '##NJ'
            then TYPE ← 3
            else If SUB(INPUT_LINE, 1, 4) = '##NP'
                then TYPE ← 4
                else If SUB(INPUT_LINE, 1, 4) = '##CN'
                    then TYPE ← 5
                        ITEM ← SUB(INPUT_LINE, 6,
                            INDEX(SUB(INPUT_LINE, 6), '/'))
                        CURSOR ← LENGTH(INPUT_LINE) + 1
                else If SUB(INPUT_LINE, 1, 4) = '##BL'
                    then TYPE ← 6
                        ITEM ← SUB(INPUT_LINE, 6)
                        CURSOR ← LENGTH(INPUT_LINE) + 1
                  else If SUB(INPUT_LINE, 1, 4) = '##ED'
                    then TYPE ← 7
                    else TYPE ← 1
    Return

3. [Obtain ordinary word of text]
    DUMMY ← BREAK(INPUT_LINE, '□', CURSOR, ITEM, false, '')
    TYPE ← 1
    Return   □

The first step of this procedure moves the CURSOR past the positions of any leading blanks which precede the next item in the input line.

Step 2 first determines whether or not the current input line has been entirely scanned. If it has, then a new input line of text is read and CURSOR is set to '1'. Note that the input line is always padded on the right end with an additional blank. The next task which this step performs is to determine whether or not the next item is a command or a word. A nested conditional statement accomplishes this task and a return to the invoking algorithm results. If, on the other hand, the current input line still contains unscanned items, then control passes on to step 3.

The final step of the procedure breaks on the next blank of the current input line and places the next word in ITEM. Also, the TYPE of this word is set to '1'.

The next procedure generates an output line of text in unjustified form.

Procedure PRINT_UNJUST(WORDS, NO_WORDS). Given a string vector WORDS which contains the words of an output line and NO_WORDS, the number of words in this line, this procedure outputs the next line without any right justification. LINE is a string variable that represents the edited line to be output. I is an index variable.
1. [Obtain the current line in unjustified form]
      LINE ← WORDS[1]
      Repeat for I = 2, 3, . . . , NO_WORDS
         LINE ← LINE ∘ '□' ∘ WORDS[I]
2. [Output current line]
      Write(LINE)
3. [Finished]
      Return   □

This procedure is very simple and therefore requires no further comment.

The following procedure is similar to PRINT_UNJUST except that it right-justifies the output line.

Procedure PRINT_JUST(WORDS, NO_WORDS, LINE_LENGTH, WIDTH). Given the string vector WORDS which contains the words to be output, NO_WORDS (an integer), which denotes the number of words, LINE_LENGTH (of type string), which represents the length of an unedited line, and WIDTH (an integer), which specifies the number of characters in an edited line, this procedure produces an output line of the specified width by inserting extra blanks between the words. TOTAL_PAD contains the number of additional blanks required. AV_PAD specifies the number of blanks that must be inserted between each pair of words. EXTRA_BLANKS contains the number of extra blanks that must be distributed among some of the words in the output line. The string variable LINE contains the edited line to be output and I is an index variable. TRUNC, MOD, and REPEAT are functions.
1. [Compute total number of blanks to be inserted]
      TOTAL_PAD ← WIDTH − LINE_LENGTH
2. [Compute average number of blanks to be padded between each word]
      AV_PAD ← TRUNC(TOTAL_PAD / (NO_WORDS − 1))

3. [Compute extra blanks to be distributed between certain words]
   EXTRA_BLANKS ← MOD(TOTAL_PAD, NO_WORDS − 1)
4. [Obtain output line]
   LINE ← WORD[1]
   Repeat for I = 2, 3, . . . , NO_WORDS − EXTRA_BLANKS
      LINE ← LINE ∘ REPEAT('□', AV_PAD + 1) ∘ WORDS[I]
   Repeat for I = NO_WORDS − EXTRA_BLANKS + 1, NO_WORDS −
      NO_WORDS − EXTRA_BLANKS + 2, . . . , NO_WORDS
      LINE ← LINE ∘ REPEAT('□', AV_PAD + 2) ∘ WORDS[I]
5. [Output line]
   Write(LINE)
6. [Finished]
   Return   □

The first step of the procedure computes the number of blanks which are to be inserted in order to obtain a right-justified output line. Step 2 determines the number of blanks (AV_PAD) which are to be inserted between each pair of words in the edited output line. The third step determines how many additional blanks are to be distributed between some of the words in the output line. Step 4 generates the desired line with all padding blanks inserted. Step 5 outputs the right-justified line, with a return being made in step 6.

The following procedure controls the generation of output from the text editor.

Procedure FORMATOR(WIDTH). Given the line width (WIDTH) of an edited line, this procedure controls the input of a passage of text (words and commands) in the text-editing language described earlier and produces the desired edited version of this input text. INPUT_LINE is a string variable which contains the current input line being processed, and CURSOR specifies the current character positions being examined. The string vector WORDS is used to represent the words to be output in the next edited line. NO_WORDS accumulates the number of words in this line while LINE_LENGTH contains the length of the unedited line at any given time. ITEM and TYPE represent the current item (that is, word or command) and its type, respectively, being isolated by procedure SCANNER. BLANK_LINE is a string variable which contains a sequence of blank characters. NO_LINES specifies the number of blank lines that is to be output for a ##BL command. JUSTIFY is a logical variable that denotes whether or not the current output line is to be right-justified. I is an index variable. Procedures PRINT_JUST and PRINT_UNJUST control the generation of a right-justified and an unjustified output line, respectively.

1. [Initialize]
   INPUT_LINE ←"   (force SCANNER to read first line)
   CURSOR ← 1

```
        JUSTIFY ← true
        LINE_LENGTH ← NO_WORDS ← 0
 2. [Edit the given text]
    Repeat thru step 4
 3. [Obtain next word or command]
    Call SCANNER(INPUT_LINE, CURSOR, TYPE, ITEM)
 4. [Process current word or command]
    If TYPE = 1
    then If LINE_LENGTH + LENGTH(ITEM) + 1 ≤ WIDTH
        then   (add the word to the current line)
              NO_WORDS ← NO_WORDS + 1
              WORDS(NO_WORDS) ← ITEM
              LINE_LENGTH ← LINE_LENGTH + LENGTH(ITEM) + 1
        else   (print previous line and use new item to start new line)
              If JUSTIFY
              then Call PRINT_JUST(WORDS, NO_WORDS, LINE_LENGTH,
                   WIDTH)
              else Call PRINT_UNJUST(WORDS, NO_WORDS)
              WORDS[1] ← ITEM
              NO_WORDS ← 1
              LINE_LENGTH ← LENGTH(ITEM)
    else If TYPE = 2
        then     (right-justify text)
              JUSTIFY ← true
        else  If TYPE = 3
              then   (do not right-justify text)
                  JUSTIFY ← false
              else  If TYPE = 4
                  then If LINE_LENGTH ≠ 0
                      then     (print the previous line and a blank line
                               and then indent the first word on the next
                               line)
                        Call PRINT_UNJUST(WORDS, NO_WORDS)
                      Write(BLANK_LINE)
                      Call SCANNER(INPUT_LINE, CURSOR, TYPE, ITEM)
                      If TYPE ≠ 1
                      then Write('ERROR IN INPUT')
                      else NO_WORDS ← 1
                          WORDS[1] ← REPEAT('□', 5) ○ ITEM
                          LINE_LENGTH ← LENGTH(WORDS[1])
                  else If TYPE = 5
                      then If LINE_LENGTH ≠ 0     (center the item)
                          then Call PRINT_UNJUST(WORDS, NO_WORDS)
                          NO_BLANKS ← (WIDTH − LENGTH(ITEM)) / 2
                          Write(REPEAT('□', NO_BLANKS) ○ ITEM)
```

```
                    LINE_LENGTH ← 0
                    NO_WORDS ← 0
           else If TYPE = 6
              then   (add a number of blank lines specified by
                         ITEM)
                    If LINE_LENGTH ≠ 0
                    then CALL PRINT_UNJUST(WORDS,
                         NO_WORDS)
                    NO_LINES ← ITEM
                    Repeat for I = 1, 2, . . ., NO_LINES
                       Write(BLANK_LINE)
                    LINE_LENGTH ← 0
                    NO_WORDS ← 0
           else   (end of text)
                    If LINE_LENGTH ≠ 0
                    then Call PRINT_UNJUST(WORDS,
                         NO_WORDS)
                    Return                              □
```

Although this procedure is somewhat lengthy, it is reasonably straight-forward. The first step of this procedure initializes several variables. First, INPUT_LINE and CURSOR are set to be the empty string and 1, respectively. In so doing, this guarantees that an input line will be read in when the procedure SCANNER is first invoked. Second, the logical variable JUSTIFY is set to *true*. The default option is then to right-justify an output line. Finally, the line length and number of words in the current line being prepared for output are set to zero. Observe that each word in this line is placed in the vector WORD, and NO_WORDS monitors the number of words in that array at any given time. Step 2 controls the output of all the input text. Step 3 invokes the scanner to supply it with the next item in the input line. The fourth step processes the current word or command. This step involves the checking of TYPE supplied by the scanner as to whether or not ITEM is a word or command. If ITEM contains a word, then it is either added to the current line, or the previous line is output and a new line is started. If, on the other hand, TYPE denotes a command, then the various commands are handled in the nested if statement.

Finally, the main algorithm which controls the four previous procedures follows.

Algorithm TEXT_EDIT. Given the desired edited line width (WIDTH) for each line, this algorithm invokes procedure FORMATOR to process the given input and produce its edited version.

1. [Input line width]
    Read(WIDTH)

2. [Invoke **FORMATOR** routine]
   Call FORMATOR(WIDTH)
3. [Finished]
   Exit   □

   The tracing of this algorithm for some sample input data is left as an exercise.

EXERCISE 9-2.4

1. Trace the algorithm **PRINT_JUST** for the following argument values:
   NO_WORDS = 8
   LINE_LENGTH = 59
   WIDTH = 65
   WORDS[1] = 'AN□'
   WORDS[2] = 'INTRODUCTION□'
   WORDS[3] = 'TO□'
   WORDS[4] = 'COMPUTER□'
   WORDS[5] = 'SCIENCE□'
   WORDS[6] = 'AN□'
   WORDS[7] = 'ALGORITHMIC□'
   WORDS[8] = 'APPROACH'

10

Linear Data Structures and Their Applications

The basic notions of data structures were introduced in earlier chapters. In this chapter we extend these notions to encompass the set of linear lists. All data structures introduced thus far are special subcases of a linear list. There are two ways of storing data structures in the computer's memory. The first of these storage-allocation methods, which takes advantage of the one-dimensional property of the computer's memory, is called *sequential allocation*. The second allocation method, which is based on the storage of the address or location of each element in the list, is known as *linked allocation*. Both methods of allocation are discussed in detail in this chapter. We shall see that each method of allocation has merits.

Several subclasses of linear lists can be defined. The most important of these subclasses are called *stacks* and *queues*. These structures, along with several associated applications, are discussed at length in this chapter. In particular, certain classical applications, such as recursion and simulation, are described. Applications based on the linked storage representation of linear lists are also included in the discussion. The most important of these is the application dealing with hash-table methods.

10-1 LINEAR LISTS

Chapter 2 introduced several primitive data structures, such as real numbers, integers, and logical values. These structures were called "primitive" because most computers have machine language instructions which manipulate them directly. Basic string-processing notions were introduced in Chap. 5. Most computers also have a variety of machine language instructions which manipulate strings. Chapter 6 used another primitive structure, called an address (or pointer), to explain the argument-parameter correspondence in subalgorithms. Finally, Chap. 4 introduced the first nonprimitive data structure in this book—the array.

We now define the most important class of linear data structures. A *list* is an ordered set consisting of a variable number of elements to which additions and deletions can be made. A list which displays the relationship of physical adjacency is called a *linear list*. Note that no deletion or addition operations can be performed on arrays. At best, we can change the value of an array element to denote an element which is to be subsequently ignored. The setting of an array element to zero to denote deletion is an example of this approach. Also, once the size of an array is chosen, it cannot be changed without being redeclared. These differences clearly distinguish a linear list from an array.

Operations performed on linear lists include those which have been introduced earlier for arrays. However, since the size of a list can vary, insertions and deletions can be performed on a linear list. Such an insertion or deletion operation can be specified by position. As an example, we may want to delete the ith element of a list or insert a new element before or after the ith existing element. Often, it may be required to insert or delete an element whose position in a linear list depends on the values of other elements in the list (as in sorting). In general, it may be required to insert or delete a given element to or from a list, respectively. This element can precede or follow an element having a specified value or satisfying a particular relationship.

Of course, we are interested in performing other important operations besides insertion and deletion. Each element in a linear list is assumed to contain one or more *fields*. For our purposes, we consider a field to be a primitive, such as a string, an integer, or a pointer. Important operations on linear lists include the following:

1. Determine the size or number of elements in a list
2. Search a list for a particular element which contains a field having a certain value
3. Copy a list
4. Sort the elements of a list into ascending or descending order, depending on the values of one or more fields within each element
5. Combine two or more lists to form a new list
6. Divide or split a list into several sublists

As just mentioned, an element of a linear list contains one or more reals, integers, strings, and so on. As an example, recall from Sec. 4-1 the discussion of a complex number. Each complex number contains an ordered pair of real numbers. Each number in this ordered pair is handled differently. Let us now consider a linear list of complex numbers. The element or *record* structure of each item in the list can be represented as

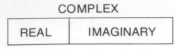

where the name of the entire record is **COMPLEX**. The record contains two fields, **REAL** and **IMAGINARY**. Both of these fields contain real numbers. Although the record structure for a complex number contains two fields of the same type, this need not always be the case. For example, the record structure

EMPLOYEE

NAME	ADDRESS	RATE_OF_PAY	DEPENDENTS

describes an element (**EMPLOYEE**) which contains four fields. The first two (**NAME** and **ADDRESS**) are of type string, the third is of type real, and **DEPENDENTS** is of type integer. Throughout the remainder of this book we shall define the structure of an element or record in this way.

Throughout the remainder of the chapter we will formulate algorithms for most of these operations. Such algorithms, however, depend on how a linear list is stored or organized in memory (that is, its storage structure). The next section deals with this extremely important point.

10-2 STORAGE STRUCTURE CONCEPTS

A *pointer* (or *link*) is an address or reference to a data structure. We illustrate the use of a pointer by considering a problem which arises in many compilers during the translation of source programs. Assume that a certain source program contains five occurrences of the real constant 2.7128. Clearly, five copies of 2.7128 could be created during the translation of this program. From a storage efficiency point of view, it is better to create only one copy of 2.7128 and use four pointers to reference this single copy. On most computers, less memory space is required to represent a pointer than a real number. The value of a pointer is an integer. This integer value represents the location or address of a word or half-word of memory in a computer (recall Sec. 2-1). For example, if we assume that in the current example the number 2.7128 is stored in location (or word number) 1000, then the four pointers in question would all have as a value the address 1000.

An important property of a pointer is its size. All possible values (or addresses) that a pointer can have are, for a particular computer system, fixed-sized integers. This uniformity permits the referencing of any data structure, regardless of its complexity, to be made in a uniform or consistent manner. In the previous example the method of referencing would be exactly the same if the four pointers referred to a logical constant or an integer constant rather than a real constant.

There are essentially two ways that can be used to obtain the address of an element in a data structure. The first method involves using the description of the data being sought. This method of obtaining an address is called the *computed address method* and is often used in many compilers to compute the address of an element of an array. We will pursue the details of this example in the next section. The second method of obtaining an address is to store it somewhere in the computer's memory. This method of access is called a *link* or *pointer address method*. To access an element of a particular structure, we load a pointer value. In certain programming languages the address of each argument of a procedure is stored in memory (recall the *pass as variable* concept from Sec. 6-4). Also, the return address used by a procedure to return to the calling program is stored and not computed. Several data structures require a combination of computed and link addresses.

In the first part of this chapter we discuss storage structures for linear lists which are based on the computed-address method. The second part of the chapter describes storage structures based on the link-address technique.

In discussing storage structures, we are concerned with the main memory of the conventional digital computer. As mentioned in Chap. 2, main memory is organized as an ordered sequence of words.

For efficiency reasons, it is desirable to organize data in the computer's memory so that a particular element of these data can be referenced by computing its address rather than performing a search. Many data structures (such as the array) permit the referencing of any element through its position in the structure. This referencing is accomplished through the use of an *addressing function*. Computationally simple addressing functions are of particular interest. Several examples of these functions will be given for arrays in the next section.

10-3 SEQUENTIAL STORAGE STRUCTURES FOR ARRAYS

One of the simplest data structures which makes use of the computed address method to locate an element is a vector. Usually, several contiguous memory locations are sequentially allocated to the vector. For an n-element vector, each element of which occupies one word of memory, n consecutive words of memory are used. Since a vector contains a fixed number of elements, the number of memory locations required is also fixed. More generally, the sequential storage representation of a vector A with a subscript lower bound of 1 can be viewed as in Fig. 10-1, where L_0 denotes the address of the first word allocated to the first element of A, and c specifies the number of words occupied by each element. The position of the ith element of A (that is, A_i) is given by

$$loc(A_i) = L_0 + c * (i - 1)$$

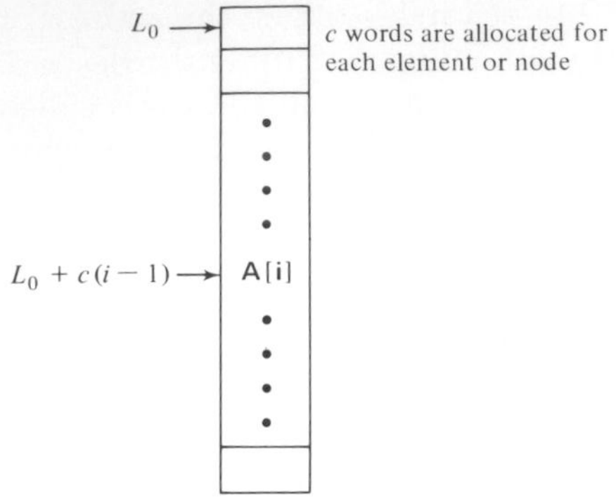
L_0
c words are allocated for each element or node
$L_0 + c(i - 1)$
A[i]

FIGURE 10-1
Sequential storage representation of a vector.

In certain programming languages such as FORTRAN, vectors are allocated memory at compile time since the size of each vector is declared to be a constant. In other programming languages, such as PL/I, the size of a vector can be specified during program execution. Actually, the size of a vector can be specified by a computation. Such programs are said to be able to allocate memory *dynamically*.

It is possible to represent a higher-dimensional array by an equivalent vector. For example, in PL/I a two-dimensional array, A, consisting of three rows and four columns is stored sequentially by rows as

A[1,1] A[1,2] A[1,3] A[1,4] A[2,1] A[2,2] A[2,3] A[2,4] A[3,1] A[3,2] A[3,3] A[3,4]
↑
L_0
1st row — 2nd row — 3rd row

The location of element A[i,j] is obtained from the equation

$$\text{loc}(A[i,j]) = L_0 + (i - 1) * 4 + j - 1$$

As an example, the address of element A[3,2] is given as

$$\text{loc}(A[3,2]) = L_0 + 2 * 4 + 2 - 1 = L_0 + 9$$

More generally, a two-dimensional array of n rows and m columns (with a subscript lower bound of 1) which is stored row by row (that is, in *row major order*) has the location of its element A[i,j] given by the equation

$$\text{loc}(A[i,j]) = L_0 + (i - 1) * m + (j - 1)$$

In a few programming languages, arrays are stored column by column in memory (that is, in *column major order*) rather than row by row. The address of element A[i,j] in an array with n rows and m columns stored in column-major order is given by

$$\text{loc}(A[i,j]) = L_0 + (j - 1) * n + (i - 1)$$

The representation of a two-dimensional array can be further generalized to handle arbitrary lower and upper bounds on its subscripts. Let us assume that the subscripts fall within the following ranges:

$$b_1 \leq i \leq u_1 \quad \text{and} \quad b_2 \leq j \leq u_2$$

The address of element A[i,j], assuming a row major storage order, becomes

$$loc(A[i,j]) = L_0 + (i - b_1) * (u_2 - b_2 + 1) + (j - b_2)$$

where each row in the array has $u_2 - b_2 + 1$ elements. For example, the address of A[−1,2], when $b_1 = -3$, $b_2 = 0$, and $u_2 = 2$ is

$$loc(A[-1,2]) = L_0 + (-1-(-3)) * (2 - 0 + 1) + (2 - 0)$$
$$= L_0 + 8$$

These notions can be extended easily to higher-dimensional arrays. As an example, consider a three-dimensional array A whose representative element is denoted by A[i,j,k] and whose subscript limits are $1 \leq i \leq 3$, $1 \leq j \leq 2$, and $1 \leq k \leq 4$. The row-major-order storage of this array is

A[1,1,1] A[1,1,2] A[1,1,3] A[1,1,4] A[1,2,1] A[1,2,2] A[1,2,3] A[1,2,4]
↑
L_0
A[2,1,1] A[2,1,2] A[2,1,3] A[2,1,4] A[2,2,1] A[2,2,2] A[2,2,3] A[2,2,4]
A[3,1,1] A[3,1,2] A[3,1,3] A[3,1,4] A[3,2,1] A[3,2,2] A[3,2,3] A[3,2,4]

A pictorial representation of this array is given in Fig. 10-2 by a cube consisting of three planes with each plane having eight points. The address of element A[i,j,k] is

$$loc(A[i,j,k]) = L_0 + (i - 1) * 8 + (j - 1) * 4 + k - 1$$

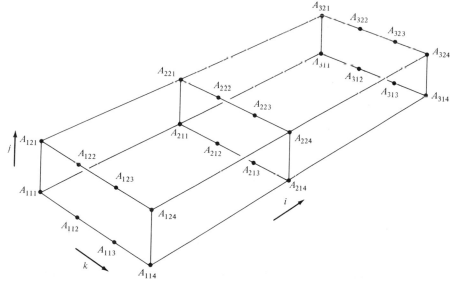

FIGURE 10-2
Pictorial representation of a three-dimensional array, A.

We can now generalize the sequential storage concept to an n-dimensional array with general element $A[s_1, s_2, \ldots, s_n]$ and subscript limits given by $1 \le s_1 \le u_1$, $1 \le s_2 \le u_2$, ..., $1 \le s_n \le u_n$. The storage representation of this array in row major order has the form

$$A[1,1,\ldots,1,1] \qquad A[1,1,\ldots,1,2] \ldots \qquad A[1,1,\ldots,1,u_n]$$
$$A[1,1,\ldots,2,1] \qquad A[1,1,\ldots,2,2] \ldots \qquad A[1,1,\ldots,2,u_n]$$
$$. \quad . \quad . \quad . \quad . \quad . \quad . \quad . \quad . \quad . \quad . \quad .$$
$$A[u_1,u_2,\ldots,u_{n-1},1] \quad A[u_1,u_2,\ldots,u_{n-1},2] \ldots A[u_1,u_2,\ldots,u_{n-1},u_n]$$

The address of element $A[s_1,s_2,\ldots,s_n]$ is given by

$$loc(A[s_1,\ldots,s_n]) = L_0 + u_2 u_3 \cdots u_n (s_1 - 1) + u_2 u_3 \cdots u_n (s_2 - 1)$$
$$+ \cdots + u_n(s_{n-1} - 1) + (s_n - 1)$$

It is more convenient to rewrite this equation as

$$loc(A[s_1,s_2,\ldots,s_n]) = L_0 + \sum_{1 \le i \le n} p_i(s_1 - 1)$$

where $p_i = \prod_{i < r \le n} u_r$ and is a constant and Σ and Π denote summation and product, respectively. For arbitrary lower and upper subscript bounds $b_i \le s_i \le u_i$ for $1 \le i \le n$, the location of element $A[s_1,s_2,\ldots,s_n]$ becomes

$$loc(A[s_1,s_2,\ldots,s_n]) = L_0 + \sum_{1 \le i \le n} p_1(s_i - b_i)$$

where

$$p_i = \prod_{i < j \le n} (u_j - b_j + 1)$$

The approach just used can also be used to store *n*-dimensional arrays in column major order. We leave this alternative representation, however, as an exercise. It should be noted that certain compilers allocate storage in a sequentially decreasing manner from the "high" end of the computer's memory. This is opposite to what we have assumed throughout our previous discussion. Clearly, both approaches are equivalent, with the exception that we subtract from the base address L_0 rather than add as we have done.

In many applications involving arrays, there are situations where only a portion of the array need be stored. An example of such a situation involves the solution of the following specialized system of equations:

$$
\begin{aligned}
A_{11}X_1 &= b_1 \\
A_{21}X_1 + A_{22}X_2 &= b_2 \\
A_{31}X_1 + A_{32}X_2 + A_{33}X_3 &= b_3 \\
. \quad . \quad . \quad . \quad . \quad . \quad . \quad . \quad . \quad . \quad . \quad . \quad . \quad . \\
A_{n1}X_1 + A_{n2}X_2 + \cdots\cdots\cdots\cdots\cdots\cdots\cdots + A_{nn}X_n &= b_n
\end{aligned}
$$

A two-dimensional array containing n^2 elements can be used to represent the matrix A and the system of equations can be solved using the numerical techniques discussed in Sec. 8-4. However, if we solve these equations by those conventional methods, nearly half of the elements in A are not used. We can economize on storage by representing the "triangular" array A by an equivalent vector. In so doing, we can solve a larger system of equations than is possible using the previous approach. The system under consideration has $[n(n + 1)]/2$ nonzero elements. Consequently, we can represent this array with a vector having the same number of elements.

The elements of A can be stored as a vector in the following row-by-row order:

$$A_{11}A_{21}A_{22}A_{31}A_{32}A_{33} \cdots A_{nn}$$

Using this representation, the address of A_{ij} is given by (assuming that A_{11} is stored at position 1)

$$\frac{(i - 1) * i}{2} + j$$

As an example, the position of element A_{42} is $[(4 - 1) * (4)]/2 + 2 = 8$. *Symmetric* arrays that have

$$A_{ij} = A_{ji} \qquad \text{for all } 1 \leqslant i \leqslant n \text{ and } 1 \leqslant j \leqslant m$$

can also be represented in this manner. We leave, as an exercise, the formulation of an algorithm for the given system of equations.

We have not to this point seen a general example of a linear list. Arrays are linear lists of fixed size. In the next section we examine a list whose size is variable.

EXERCISES 10-3

1. A two-dimensional array is stored column by column in memory. The row and column subscripts have the following bounds:

$$3 \leqslant i \leqslant 8 \qquad \text{and} \qquad -5 \leqslant j \leqslant 0$$

Obtain an equation for locating the storage position of element A_{ij} assuming that the first element of the first row is stored at position L_0.

2. Obtain the location of $A[s_1, s_2, \ldots, s_n]$ for an n-dimensional array with subscript bounds $b_i \leqslant s_i \leqslant u_i$ for $1 \leqslant i \leqslant n$, assuming a column major storage order.

$$A = \begin{bmatrix} a_{11} & a_{12} & 0 & \cdots & \cdots & \cdots & 0 \\ a_{21} & a_{22} & a_{23} & 0 & \cdots & \cdots & 0 \\ 0 & a_{32} & a_{33} & a_{34} & \cdots & \cdots & 0 \\ \cdot & \cdot & \cdot & & & & \\ \cdot & \cdot & \cdot & & & & \\ 0 & 0 & 0 & \cdots & \cdots & a_{n,n-1} & a_{nn} \end{bmatrix}$$

FIGURE 10-3
Band matrix.

3. Formulate a detailed algorithm for the triangular system of equations based on the storage-representation approach given in the text.

4. In certain numerical applications a special kind of two-dimensional matrix consisting of a band of nonzero elements occurs. All other elements in the array are zero. Such an array, A, is shown in Fig. 10-3, where the elements on the leading diagonal (that is, where the row and column subscripts are equal) are nonzero. The diagonal elements immediately above and below this leading diagonal are also nonzero. If the elements in the non-zero-element band are stored as a vector B in a row major order, then obtain an equation to determine the location of element A_{ij}.

10-4 STACKS

One of the most important subclasses of linear lists is the family of stack structures. In this section we first introduce the concepts associated with this subclass of linear structures. Next, several important algorithms, such as insertion and deletion, for a stack structure are given. In particular, we describe these algorithms for a stack which is represented by a vector.

As mentioned earlier, the most general form of a linear list permits the insertion and deletion of an element at any position in the list. If we restrict occurrence of insertions and deletions to one end of a linear list, then a member of the resulting subclass of linear lists thus obtained is called a *stack*. Using stack terminology, insert and delete operations are commonly referred to as "push" and "pop" operations, respectively. The only directly accessible element of a stack is its *top* element. The least accessible element is its *bottom* element. Since insert and delete operations are performed at the same end of the stack (that is, its top end), elements can only be removed in the opposite order from that in which they were inserted onto the stack. This interesting LIFO (last-in, first-out) phenomenon will be observed in the applications to be discussed in the next section.

A familiar example of a stack, which permits the selection of only its top element, is a pile of trays in a cafeteria. Such an arrangement of trays is supported by a spring so that a person desiring a tray finds that only one is available to him/her at the surface of the tray counter. Such an arrangement of trays is shown in Fig. 10-4. The removal of the top tray from such a

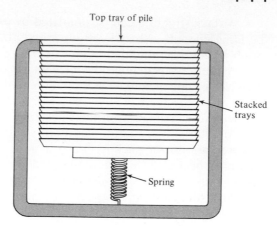

Top tray of pile

Stacked trays

Spring

FIGURE 10-4
Cafeteria-tray holder.

system causes the weight on the spring to lighten and the next tray on the pile to appear at the surface of the counter. Conversely, a tray which is deposited on the top of the pile causes the entire pile to be pushed down and the new tray to appear above the tray counter.

Another common example of a stack is a railway system for the shunting of railway cars. In this system, which is shown in Fig. 10-5, the last railway car to be placed on the stack is the first car to leave. The repeated use of the insert and delete operations permits various arrangements of cars to be realized on the output railway line.

The four common operations which are associated with a stack S are:

PUSH(S, NEW)	Inserts element NEW on top of stack S and returns the new stack
POP(S)	Removes the top element from the stack S and returns the updated stack
TOP(S)	Returns the top element of stack S
EMPTY(S)	Returns true if S contains no elements or otherwise returns false

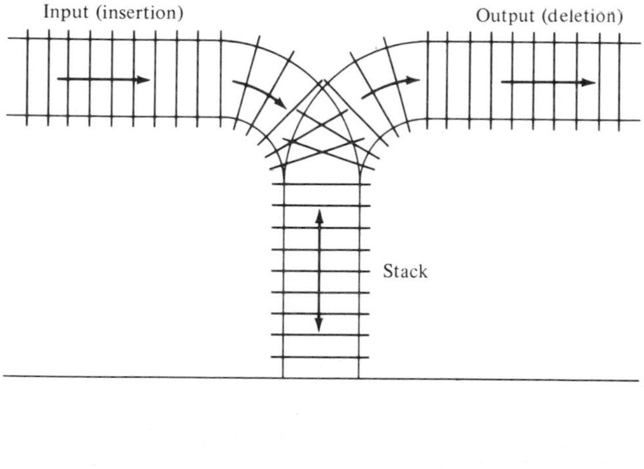

Input (insertion)   Output (deletion)

Stack

FIGURE 10-5
Railway shunting system
representation of a stack.

A stack structure can be simulated by using a vector whose size should be large enough to handle all insertions that could be made to the stack. Such a vector representation of a stack is given in Fig. 10-6.

A pointer TOP denotes the top element of the stack. For an empty stack, TOP has a value of zero. TOP is incremented by "one" prior to placing a new element on the stack. Conversely, TOP is decremented by "one" each time an element is deleted from the stack.

An alternative, and for algorithm-tracing purposes, a more convenient representation of a stack is shown in Fig. 10-7. In this representation the rightmost occupied element of the stack represents its top element. The leftmost element of the stack denotes its bottom element.

The following are implementations of the previously defined stack operations based on a vector implementation.

Procedure PUSH(S, TOP, X). Given a vector S (consisting of N elements) representing a sequentially allocated stack and a pointer TOP denoting the top element in the stack, this procedure inserts element X on the stack.

1. [Overflow?]
   If TOP $\geq$ N
   then Call STACK_OVERFLOW
2. [Increment TOP]
   TOP $\leftarrow$ TOP + 1
3. [Insert new element]
   S[TOP] $\leftarrow$ X
   Return □

The first step of the algorithm checks for an overflow situation. If this situation arises, then the procedure STACK_OVERFLOW is invoked. Although this procedure is application-dependent, usually its invocation gener-

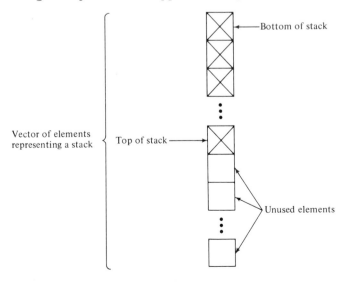

FIGURE 10-6
Representation of a stack
by a vector.

ates a signal which indicates that more storage is required for the stack and that the program must be rerun.

Function POP(S, TOP). Given a vector S (consisting of N elements) representing a sequentially allocated stack and a pointer TOP denoting the top element of the stack, this function returns the top element in the stack. TEMP is a temporary variable.

1. [Underflow?]
   If TOP = 0
   then Call STACK_UNDERFLOW
2. [Unstack element]
   TEMP ← S[TOP]
3. [Decrement pointer]
   TOP ← TOP − 1
4. [Finished]
   Return(TEMP)   □

An underflow situation is checked for in the first step of the algorithm. If this situation arises, then procedure STACK_UNDERFLOW is invoked. Although this procedure is also application-dependent, an empty stack is valid in many applications.

Function EMPTY(S, TOP). Given a vector S representing a sequentially allocated stack, this function determines whether or not this stack is empty.
1. [Empty stack?]
   If TOP = 0
   then Return (true)
   else Return (false)   □

Function TOPV(S, TOP). Given a vector S representing a sequentially allocated stack, this function returns the top element of the stack.
1. [Empty stack?]
   If TOP = 0 then Call ERROR_ROUTINE
2. [Return top element]
   Return (S[TOP])   □

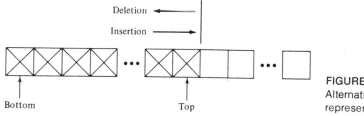

FIGURE 10-7
Alternative vector representation of a stack.

The ERROR_ROUTINE in the last algorithm depends on the application under consideration. Since the last two algorithms are so simple, there is little advantage to be gained by having them as separate subalgorithms. In the stack applications throughout the remainder of the text we shall use them in-line rather than invoke the corresponding subalgorithms.

Thus far in this section we have illustrated the operations on a stack by using a vector representation. We now examine briefly an alternative and "pure" way of representing a stack. Many higher-level programming languages allow explicit control, through special instructions, of when and how much storage is to be allocated for a certain stack. We shall refer to this type of storage as *controlled* storage. Whenever an additional instance (or element) of a stack element is required, it can be created by a special statement. The new element created by such a statement can then be assigned a value. The only element which is accessible in such systems is the most recently created one. Consequently, the programmer is not burdened with the chore of maintaining a separate index to the top element in the stack. The element referred to, by default, is the top element. In essence, then, the creation of a new element and its subsequent assignment of a value, is equivalent (without the variable TOP) to the subalgorithm PUSH introduced earlier. Alternatively, when the latest copy of a variable is no longer required (that is, its stack top), it can be deleted from the stack by executing another special statement. Again, the top element index of the stack is handled automatically by the compiler of the language. Hence, the operations of PUSH and POP do not require programmer knowledge of the index TOP when the implementation language used has programmer-controlled storage.

Although in the remainder of this text we use subalgorithms PUSH and POP as described for the vector representation of a stack, it should be remembered that programmer-controlled storage provides the same insert and delete capabilities without the necessity of updating the index TOP.

Applications often require more than one stack. In such applications, one stack may overflow while others are far from being full. Instead of imposing an individual maximum size for each stack, it becomes desirable to allocate one block of memory which all stacks can use.

Figure 10-8 shows such a memory layout for the case of two stacks. The first and second stacks expand to the right and left, respectively. In such a configuration an overflow condition arises only when the combined size of the two stacks exceeds the allocated memory space. Note that the bottom element of each stack is in a fixed position within the allocated memory space.

FIGURE 10-8
Sequential allocation
scheme for two stacks.

This memory-sharing configuration cannot be extended to more than two stacks while maintaining the common overflow property and the property of each stack having a fixed bottom element. The overflow property can only be maintained at the expense of the other. The amount of overhead incurred in keeping track of the bottom and top elements of several sequentially allocated stacks sharing a common piece of memory is very substantial. At times, entire stacks must be moved to preserve the physical adjacency relation between the elements in each stack. A more efficient storage-allocation scheme for such a situation is discussed later in the chapter.

Now that the basic notions of a stack have been introduced, the natural question which arises is: What is it used for? The next section examines in detail three applications that use stack structures.

EXERCISES 10-4

1. Consider the infinite set of strings

    'c', 'aca', 'bcb', 'abcba', 'bacab', 'abbcbba', 'abacaba', 'aabcbaa', . . .

   and so on. A typical string in this set can be specified as wcw^R, where w contains a sequence of a's and b's and w^R is the reverse of w. For example, if $w = $ 'ab', then $w^R = $ 'ba'. Given an input string x, formulate an algorithm which uses a stack to determine whether or not x belongs to the set of strings as described by wcw^n.

2. Using a vector representation of a stack, formulate an algorithm to obtain its ith element without deleting it.

3. Using a vector representation of a stack, construct an algorithm which changes the ith element of the stack to the value contained in X.

4. Figure 10-8 gave a storage representation of two stacks sharing a maximum of m elements. Assuming a vector representation S, formulate algorithms PUSH(i, X, TOP_i) and DELETE(i, TOP_i) to add element X and to delete the top element from stack i, $1 \le i \le 2$, respectively.

10-5 APPLICATIONS OF STACKS

This section contains three well-known applications of stacks. The first (and perhaps most important) application describes recursion. Recursion is an important facility in several programming languages, and there are many problems whose solutions are best described in a recursive manner. Instances of such problems pervade the remainder of this text. The second application of stacks given deals with the conversion of symbolic infix expressions to machine code. This application of a stack structure is classical in the sense that it was one of the earliest nonnumeric applications to be attempted on computers. The third and final application of stacks involves sorting. Although we do not pursue the topic here, stack structures have influenced the design of computers. In particular, several computers can perform stack operations (such as insertion and deletion) at the hardware or machine level.

10-5.1 RECURSION

In Chap. 6 we saw an instance of a procedure calling another procedure. This is a very useful concept and one that presents no particular problems for the programmer. In this subsection we consider a special case of this facility, in which a procedure is allowed to call itself. Such a procedure is said to be recursive. We will introduce the concept of recursion through several examples. Although the technique can be applied to both procedures and functions, it is somewhat simpler to illustrate with functions. All our examples in this section will be recursive functions. Chapter 11, however, contains several examples of both types of subalgorithms. The remainder of the subsection examines how recursion is actually done by a computer.

An algorithm was developed in earlier chapters to compute the factorial of a number n. You will recall that the factorial of n, written $n!$, was defined as $n(n-1)(n-2)\cdots(1)$. The algorithm calculated each successive multiplier by subtracting 1 from the value used previously (starting with n), computing products until the multiplier became 1. This approach is said to be iterative since it iterates over successive values of the multiplier.

Consider the following alternative definition of factorial:

$$n! = \begin{cases} 1 & \text{if } n = 0 \\ n(n-1)! & \text{if } n > 0 \end{cases}$$

In this formulation, factorial is defined in terms of itself. On the surface, this may appear to be a case of circular definition, but a quick check shows that this is not so, because of the special case when $n = 0$. Let us consider a simple example, $3!$. Since $n > 0$, we apply the first line of the definition, to get

$$3! = 3 \times 2!$$

The right-hand side, however, is not completely processed yet. Continuing, we get

$$
\begin{aligned}
3! &= 3 \times 2! && \text{(use line 2 of definition)} \\
&= 3 \times 2 \times 1! && \text{(use line 2 of definition)} \\
&= 3 \times 2 \times 1 \times 0! && \text{(use line 2 of definition)} \\
&= 3 \times 2 \times 1 \times 1 && \text{(use line 1 of definition)} \\
&= 6
\end{aligned}
$$

Observe that the calculation is complete when we have processed the case of 0! using line 1 of the definition. Thus, although the definition may appear to be circular, we can see that it is not, since it does terminate with the correct value of the factorial. This is, in fact, a recursive approach to the calculation of factorial.

Factorial is perhaps the most widely used example of a problem that is amenable to recursive formulation. Clearly, a special type of problem is needed. There are other good examples, however, for which a recursive formulation is appropriate. Before considering these, let us first consider the algorithmic specification of factorial defined recursively. The function FACT_RECURSIVE follows almost immediately from the definition.

Function FACT_RECURSIVE(N). This function computes N! recursively. N is assumed to be integer and nonnegative.
1. [Apply recursive definition]
   If $N = 0$
   then Return(1)
   else Return($N *$ FACT_RECURSIVE($N - 1$))   ☐

As you can see, this algorithm is very concise. Compare it with the iterative algorithm that requires more steps. There are evidently clear advantages for the programmer if a recursive formulation can be found for a problem. This is not to say, however, that the recursive algorithm will execute any faster. The same computation is required, as the example above shows; it is simply expressed in a different way. Figure 10-9 shows the function calls required to compute FACT_RECURSIVE(4) and the manner in which the results are returned as the computation proceeds. Notice that with the exception of the bottom line (FACT_RECURSIVE(0)), the value of no line can be obtained until the result of the line below is received. Thus, we work down a chain of calls to FACT_RECURSIVE(0), which we call the *base value*, at which point no further calls are required, and then back up the chain of returns. The *depth of recursion* refers to the number of times the function or procedure is called recursively in the process of evaluating a given argument or set of arguments. For FACT_RECURSIVE(4), the depth of recursion is seen to be 4.

As mentioned, factorial is only one example of a problem that can be formulated recursively; there are many other interesting examples. We now consider two of them: the computation of Fibonacci numbers, and the binary search.

The *Fibonacci series* was originally conceived in the thirteenth century by Leonardo of Pisa, nicknamed Fibonacci, as a model for the breeding of rabbits. This particular series turns up again and again, not only in branches of mathematics and computer science, but also in various biological phenomena.

The series itself is very simple:

0, 1, 1, 2, 3, 5, 8, 13, 21, . . .

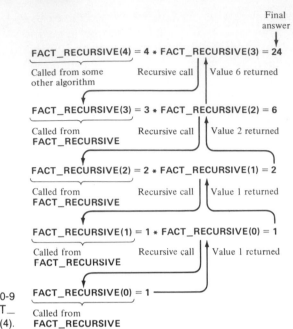

FIGURE 10-9
Trace of FACT—
RECURSIVE (4).

The first two terms are 0 and 1; each subsequent term is calculated as the sum of the two immediately preceding it. In general, $t_i = t_{i-1} + t_{i-2}$.

The problem of finding the nth Fibonacci number has a clear recursive formulation, which is

$$
FIBONACCI(N) = \begin{cases}
FIBONACCI(N-1) + FIBONACCI(N-2) & \text{if } N > 2 \\
1 & \text{if } N = 2 \\
0 & \text{if } N = 1
\end{cases}
$$

The recursive function FIBONACCI follows directly from this definition.

Function FIBONACCI(N). This function computes the nth Fibonacci number recursively. N is assumed to be integer and nonnegative.
1. [Apply recursive definition]
   If N = 1
   then Return(0)
   else If N = 2
         then Return(1)
         else Return(FIBONACCI(N − 1) + FIBONACCI(N − 2))   □

Since the function contains possibly two recursive calls, the execution trace is more complex than that for the recursive factorial. A value is obtained for the first recursive call by the same technique of following a chain of calls down to a base value and then coming back up with returned values.

The process then repeats for the second recursive call. We leave this trace as an exercise.

The technique of binary search was introduced in Chap. 4 as a reasonably efficient means of searching an ordered list of items. The algorithm BINARY_SEARCH (see Sec. 4-3.2) uses an iterative method to accomplish the task.

The binary search problem can also be formulated recursively. Essentially, it amounts to dividing the list of items in half recursively, and performing a binary search on the "new list." The process continues until the desired element is found, or until the "list" comprises a single element. Unlike the factorial and Fibonacci problems, there exists no recursive formula from which to work. We shall proceed, instead, from a general description of the approach. Before reading further, you would be wise to review the algorithm BINARY_SEARCH in Sec. 4-3.2.

We assume, as before, that the list of items has been sorted into ascending order. A general formulation of the recursive solution is as follows:

1. Obtain the position of the midpoint element in the current search interval.
2. If the values of the desired and midpoint elements are equal, then the search is successful.
   If the value of the desired element is less than the value of the midpoint element,
   then perform a binary search on upper half of current search interval (recursive call);
   else perform a binary search on lower half of current search interval (recursive call).

This algorithm will find the desired element if it is in the list, but this may not be the case. We can allow for this possibility by testing first to see if the current search interval comprises a single element that is not the desired element. If this is so, then the desired element cannot be in the list. We therefore rewrite step 1 as

1. If current search interval is a single element whose value is not equal to the value of the desired element,
   then search is unsuccessful;
   else obtain the position of the midpoint of the current search interval.

Step 2 remains as before.

This formulation is clearly recursive, with two recursive calls in step 2. The following function results from formalizing these steps.

Function BIN_SEARCH_RECURSIVE (K, X, TOP, BOTTOM). Given a vector K whose elements are in ascending order, this function searches the vector for a given element whose value is given by X and returns its position in the vector to the calling program. The function operates recursively. TOP and BOTTOM are parameters defining the current search interval. MIDDLE denotes the midpoint of the interval. All values are assumed to be integer.

1. [Is desired element absent?]
        If TOP = BOTTOM and K[TOP] ≠ X
        then Write('SEARCH IS UNSUCCESSFUL')
            Return(0)
2. [Obtain position of midpoint of interval]
        MIDDLE ← TRUNC((TOP + BOTTOM) / 2)
3. [Compare]
        If K[MIDDLE] = X
        then Write('SEARCH IS SUCCESSFUL')
            Return(MIDDLE)
        else If K[MIDDLE] < X
            then Return(BIN_SEARCH_RECURSIVE(K, X, MIDDLE+1, BOTTOM))
            else Return(BIN_SEARCH_RECURSIVE(K, X, TOP, MIDDLE−1))   □

As an example, if we wished to search the vector V, comprising elements 61, 147, 197, 217, 309, 448, and 503, for 197, we would say simply BIN_SEARCH_RECURSIVE(V, 197, 1, 7). As shown in the trace in Table 10-1, the result returned would be 3.

This function executes somewhat differently from the two previous examples. It moves down through the chain of recursive calls, dividing the search interval in half each time, until a "base condition" is reached—either the element is found, or it can be determined that it is not present. At this point, the function terminates. There is no chain backup of returned values, since the recursive function calls are not part of expressions.

From an implementation point of view, there are special problems associated with a recursive function that do not occur in a nonrecursive function. First, since a recursive function can be invoked either externally

TABLE 10-1   Trace of Recursive Binary Search

Function Call	Parameter		Values		Local Variable
	K	X	TOP	BOTTOM	MIDDLE
1. B_S_R(V, 197, 1, 7)	61	197	1	7	4
	147				
	197				
	217				
	309				
	448				
	503				
2. B_S_R(, 197, 1, 3)	Same	Same	1	3	2
3. B_S_R(, 197, 3, 3)	Same	Same	3	3	3 success

(that is, from outside the definition) or internally (that is, recursively), this implies that such a function (if it is to work properly) has to save, in some order, the different return addresses so that a return to the proper location will occur when a return to a calling statement is made. Second, the recursive function must also save the values of all parameters and local variables upon entry and restore these parameters and variables at completion.

It is to be emphasized here that in programming languages in which recursion is allowed and in our algorithmic notation, these problems are handled by the associated compilers. The programmer is, therefore, relieved of these bookkeeping tasks. It is instructive, however, to examine how the compiler for a recursive language might handle these problems. Using similar techniques, a recursive environment can be simulated in a programming language which does not allow recursion.

A general algorithmic simulation model for a recursive subalgorithm contains the following steps:

1. [Prologue]
   Save the values of the parameters and local variables
   Save the return address
2. [Body]
   If the base criterion has been reached,
   then perform the final computation and proceed to step 3
   else perform a partial computation and proceed to step 1
          (initiate a recursive call)
3. [Epilogue]
   Restore the values of the most recently saved parameters and local variables
   Restore the most recently saved return address
   Proceed to the statement which has this address

The model consists of a prologue, a body, and an epilogue. The prologue is responsible for saving the values of the parameters and local variables. In addition, it also saves the return address. Conversely, the epilogue restores these values and the return address. Observe that the parameter values, local variable values, and return address saved are the first to be restored; that is, the process operates in a last-in first-out manner. The body of the subalgorithm contains an invocation to itself. In general, a recursive subalgorithm can contain several recursive calls to itself. Note that in the body we have distinguished between a partial computation and a final computation. The final computation gives the explicit or base definition of the process for some value(s) of the argument(s). The base criterion is a test which determines whether or not the argument value(s) is that for which an explicit definition of the process is given.

The last-in first-out characteristics of a recursive subalgorithm imply that a stack is the most obvious data structure to use to implement steps 1 and 3 of the preceding general model. At each invocation of the sub-algorithm (or level of recursion), the necessary values and return address are pushed onto the stack; upon exit from that level, the stack is popped to restore the saved values of the preceding (or calling) level.

The previous recursive mechanism is best explained by a concrete example. Consider an algorithm to calculate FACTORIAL(N) recursively, which explicitly shows the recursive framework through our simulation model.

Algorithm FACTORIAL. Given an integer N, this algorithm computes N!. The stack S is used to store an activation record associated with each recursive call. Each activation record contains the current value of N and the current return address RET_ADDR. TEMP_REC is also a record which contains two variables (PARM and ADDRESS). This temporary record is required to simulate the proper transfer of control from one activation of the algorithm FACTORIAL to another. Whenever a TEMP_REC is placed on the stack S, copies of PARM and ADDRESS are pushed onto S and assigned to N and RET_ADDR, respectively. TOP points to the top element of S and its value is initially zero. Initially, the return address is set to the main calling address (that is, ADDRESS ← main address). PARM is set to the initial value of N.

1. [Save N and return address]
   Call PUSH(S, TOP, TEMP_REC)
2. [Is the base criterion found?]
   If N = 0
   then FACTORIAL ← 1
       Go to step 4
   else PARM ← N − 1
       ADDRESS ← step 3
       Go to step 1
3. [Calculate N!]
   FACTORIAL ← N ∗ FACTORIAL   (the factorial of N − 1)
4. [Restore previous N and return address]
   TEMP_REC ← POP(S, TOP) (that is, PARM ← N,
       ADDRESS ← RET_ADDR, and pop stack)
   Go to ADDRESS   □

In this algorithm steps 1 and 4 denote its prologue and epilogue, respectively. Steps 2 and 3 represent the body. The base criterion and final computation are contained in step 2. The partial computation and the recursive call are contained in step 3. Note the use of such statements as Go to step 1 and Go to step 4. These statements are called *unconditional transfer statements*. The location of the next statement to be executed after encountering such a go-to statement is specified by the label following the "go to." For example, the execution of the statement "Go to step 1" causes the statement whose label is step 1 to be executed next. A trace of this algorithm with N = 3 is shown in Fig. 10-10.

The application of a stack in this subsection has been concerned with the storage of values for parameters and local variables in the evaluation of recursive functions. In the next subsection we examine the use of a stack in the compilation of arithmetic expressions.

Level number	Description	Stack contents
Enter level 1 (main call)	Step 1: PUSH(S, 0, (3, main address)) Step 2: N ≠ 0 PARM ← 2, ADDRESS ← Step 3	3 / main address ↑ TOP
Enter level 2 (first recursive call)	Step 1: PUSH(S, 1, (2, Step 3)) Step 2: N ≠ 0 PARM ← 1, ADDRESS ← Step 3	3, 2 / main address, Step 3 ↑ TOP
Enter level 3 (second recursive call)	Step 1: PUSH(S, 2, (1, Step 3)) Step 2: N ≠ 0 PARM ← 0, ADDRESS ← Step 3	3, 2, 1 / main address, Step 3, Step 3 ↑ TOP
Enter level 4 (third recursive call)	Step 1: PUSH(S, 3, (0, Step 3)) Step 2: N − 0 FACTORIAL ← 1	3, 2, 1, 0 / main address, Step 3, Step 3, Step 3 ↑ TOP
	Step 4: POP(S, 4) go to Step 3	3, 2, 1 / main address, Step 3, Step 3 ↑ TOP
Return to level 3	Step 3: FACTORIAL ← 1 * 1 Step 4: POP(S, 3)	2 / main address, Step 3 ↑ TOP
Return to level 2	Step 3: FACTORIAL ← 2 * 1 Step 4: POP(S, 2) go to Step 3	3 / main address ↑ TOP
Return to level 1	Step 3: FACTORIAL ← 3 * 2 = 6 Step 4: POP(S, 1) go to main address	(empty) ↑ TOP

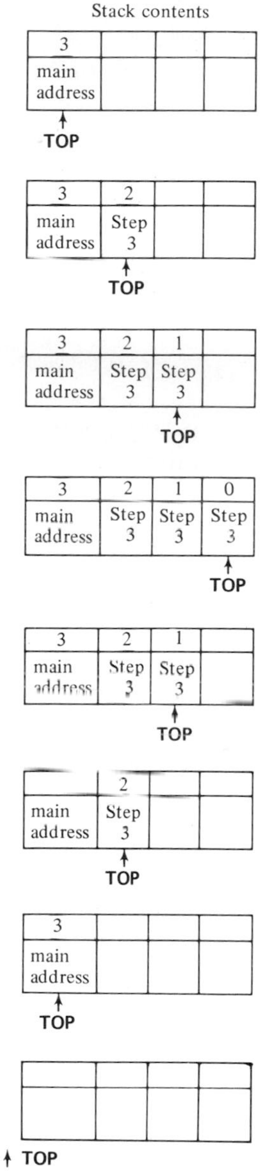

FIGURE 10-10
Trace of algorithm
FACTORIAL with N = 3.

EXERCISES 10-5.1

1. Trace the function FIBONACCI (as in Fig. 10-9) for N = 4.

2. A well-known algorithm for finding the greatest common divisor of two integers is *Euclid's algorithm*. The greatest common divisor function is defined by the following:

$$GCD(M, N) = \begin{cases} GCD(N, M) & \text{if } N > M \\ M & \text{if } N = 0 \\ GCD(N, MOD(M, N)) & \text{if } N > 0 \end{cases}$$

   where MOD(M, N) is M modulo N, the remainder on dividing M by N. Construct a recursive function subalgorithm for this problem.

3. Obtain a trace (as in Fig. 10-9) for GCD(20, 6).

4. The usual method used in evaluating a polynomial of the form

$$p_n(x) = a_0x^n + a_1x^{n-1} + a_2x^{n-2} + \cdots + a_{n-1}x + a_n$$

   is by using the technique known as *nesting*, or *Horner's rule*. This is an interative method which is described as follows:

$$b_0 = a_0$$
$$b_{i+1} = x \cdot b_i + a_{i+1} \qquad i = 0, 1, \ldots, n - 1$$

   from which one can obtain $b_n = p_n(x)$.

   An alternative solution to the problem is to write

$$p_n(x) = x \cdot p_{n-1}(x) + a_n$$

   where

$$p_{n-1}(x) = a_0x^{n-1} + a_1x^{n-2} + \cdots + a_{n-2}x + a_{n-1}$$

   which is a recursive formulation of the problem. Formulate a recursive function subalgorithm for this problem.

5. Trace the algorithm obtained in exercise 4 for the data

$$n = 3, \quad a_0 = 1, \quad a_1 = 3, \quad a_2 = 3, \quad a_3 = 1, \quad \text{and} \quad x = 3.$$

6. Consider the set of all valid, completely parenthesized, infix arithmetic expressions consisting of single-letter variable names, a digit, and the four operators +, −, *, and /. The following recursive definition specifies the set of valid expressions:

1. Any single-letter variable (A – Z) or a digit is a valid infix expression
2. If α and β are valid infix expressions, then $(\alpha + \beta)$, $(\alpha - \beta)$, $(\alpha * \beta)$, and (α / β) are valid infix expressions
3. The only valid infix expressions are those defined by steps 1 and 2

Formulate a recursive function subalgorithm that will input a string of symbols and output either **VALID EXPRESSION** for a valid infix expression or **INVALID EXPRESSION** for an invalid expression.

7. Write a recursive function to compute the square root of a number. Read in triples of numbers N, A, and E, where N is the number for which the square root is to be found, A is an approximation of the square root, and E is the allowable error in the result. Use as your function

$$
ROOT(N, A, E) = \begin{cases} A & \text{if } |A^2 - N| < E \\ ROOT(N, \dfrac{A^2 + N}{2A}, E) & \text{otherwise} \end{cases}
$$

8. The Tower of Hanoi is a game with an alleged historical basis in a ritual practiced by Brahman priests to predict the end of the world. The game begins with a series of gold rings of decreasing size stacked on a needle (the Brahman priests used 64 rings). The object is to stack all the rings onto a second needle in decreasing order of size. Before this is done, the end of the world will be upon us. A third needle is available for use as intermediate storage. The scenario is shown in Fig. 10-11.

The movement of the rings is restricted by the following three rules:

1. Only one ring may be moved at a time
2. A ring may be moved from any needle to any other
3. At no time may a larger ring rest on a smaller ring

Design a recursive *procedure* to solve the Tower of Hanoi problem.

9. Simulate the recursive operation (as in Fig. 10-10) of the procedure produced for exercise 8 for N = 2.

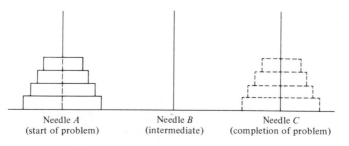

Needle *A*
(start of problem)

Needle *B*
(intermediate)

Needle *C*
(completion of problem)

FIGURE 10-11
Tower of Hanoi.

10. An important theoretical function, known as *Ackermann's function*, is defined as

$$A(M, N) = \begin{cases} N + 1 & \text{if } M = 0 \\ A(M - 1, 1) & \text{if } N = 0 \\ A(M - 1, A(M, N - 1)) & \text{otherwise} \end{cases}$$

Obtain a recursive function subalgorithm for this function.

11. Evaluate the function of exercise 10 for $M = 2$ and $N = 2$.

12. Recursion can be used to generate all possible permutations of a set of symbols. For example, there are six permutations on the set of symbols A, B, and C; namely, ABC, ACB, BAC, BCA, CBA, and CAB. The set of permutations of N symbols is generated by taking each symbol in turn and prefixing it to all the permutations which result from the remaining $N - 1$ symbols. Consequently, permutations on a set of symbols can be specified in terms of permutation on a smaller set of symbols. Formulate a recursive procedure for this problem.

13. Certain applications require a knowledge of the number of different partitions of a given integer N, that is, how many different ways N can be expressed as a sum of integer summands. For example, $N = 5$ yields the partitions $1 + 1 + 1 + 1 + 1$, 5, $1 + 2 + 2$, $3 + 1 + 1$, $2 + 3$, $1 + 4$, and $1 + 1 + 1 + 2$. If we denote by Q_{MN} the number of ways in which an integer M can be expressed as a sum, each summand of which is no larger than N, then the number of partitions of N is given by Q_{NN}. The function Q_{MN} is defined recursively as

$$Q_{MN} = \begin{cases} 1 & \text{if } M = 1 \text{ and for all } N \\ 1 & \text{if } N = 1 \text{ and for all } M \\ Q_{MN} & \text{if } M < N \\ 1 + Q_{M,M-1} & \text{if } M = N \\ Q_{M,N-1} + Q_{M-N,N} & \text{if } M > N \end{cases}$$

Formulate a recursive function subalgorithm for Q_{NN}.

14. Evaluate the function obtained in exercise 13 for $N = 3$ and $N = 5$.

10-5.2 POLISH EXPRESSIONS AND THEIR COMPILATION

As a second classical application of a stack structure, this section examines the evaluation or compilation of infix expressions. It is found to be more efficient to evaluate an infix expression by first converting it to an alternative form and then evaluating the given expression in this latter form. Such an approach eliminates the need to scan the original infix expression repeatedly to obtain its value. Throughout this section we use examples of expressions found in higher-level programming languages. It is to be emphasized, however, that the theory developed in this section applies to any type of expression.

This section is initially concerned with introducing Polish notation. This introduction also examines a method which permits the detection of errors (that is, invalid expressions). The translation of infix expressions into Polish notation is then examined in detail. Finally, the generation of assembly language instructions for Polish expressions is discussed at some length. Certain properties of the arithmetic operators can be used in obtaining some degree of code optimization. An optimized program is one which usually contains fewer programming instructions than its corresponding unoptimized version.

10-5.2.1 Polish notation

We introduce in this subsection *Polish notation*. This notation, which is due to the Polish logician Jan Łukasiewicz, offers certain computational advantages over the more traditional infix notation. The transformation of infix notation to Polish notation is informal. Formal algorithms for such conversions are given in the next subsection. The evaluation of Polish notation is also briefly mentioned. Finally, the notion of error detection as it pertains to Polish notation is introduced. This idea permits the identification of invalid expressions. Usually, the detection of an invalid expression is to be reported to the programmer.

First, let us consider the evaluation of unparenthesized arithmetic expressions consisting of single-letter variable names, nonnegative integers, and the four arithmetic operators $+$, $-$, $*$, and $/$. Typically, a scanner, such as that discussed in Chap. 9, has the task of isolating the symbols in the expression. The precedences of the operators $*$ and $/$ are considered to be equal and of higher value than that of $+$ and $-$ (which are also considered to have equal precedence). As an example, consider the unparenthesized expression

$$
\underbrace{\underbrace{a - \underbrace{b / c}_{1} + \underbrace{d * e}_{2}}_{3}}_{4}
$$

To evaluate this expression in the conventional manner, as discussed in Chap. 2, it must be scanned repeatedly from left to right. The numbers associated with the subexpressions denote the steps of such an evaluation. Clearly, this process is inefficient because of the repeated scanning that must be done.

The presence of parentheses in an expression alters the order of evaluation of its subexpressions. For example, in the expression $(a + b) * (c + d)$ we first evaluate $a + b$, then $c + d$, and finally $(a + b) * (c + d)$. Parentheses can be used in writing expressions such that the order of evaluation of subexpressions is independent of the precedence of the operators. This is accomplished by parenthesizing subexpressions in such a way that, corresponding to each operator, there is a pair of parentheses. This pair of parentheses encloses its operator and operands. Such an expression is said to be *fully parenthesized*. A recursive definition of fully parenthesized infix arithmetic expressions which contain single-letter names, nonnegative integers, and the four arithmetic operators follows:

1. Any letter (a–z) or nonnegative integer is a fully parenthesized infix expression
2. If α and β are fully parenthesized infix expressions, then $(\alpha + \beta)$, $(\alpha - \beta)$, $(\alpha * \beta)$, and (α / β) are fully parenthesized infix expressions
3. The only fully parenthesized infix expressions are those defined by steps 1 and 2

According to this definition, expressions such as ((a + b) * c) and ((a + b) * (c + d)) are considered to be fully parenthesized, while expressions such as x – 5 and (a – b / c) are considered to be invalid.

For a fully parenthesized expression, we can define the parenthetical level of an operator as the total number of pairs of parentheses that surround it. A pair of parentheses has the same parenthetical level as that of the operator to which it corresponds, that is, of the operator which is immediately enclosed by this pair. For example, in the fully parenthesized expression

(a – ((b / 2) * (c + d)))
  1   3 2 3

the integers below the operators specify the parenthetical level of each operator. In the evaluation of such an expression, the subexpression which contains the operator with the highest parenthetical level is evaluated first. Operators with the same parenthetical level (as in the current example) are evaluated from left to right. After the subexpressions containing operators at the highest parenthetical level have been evaluated, the remaining subexpressions which contain operators at the next highest parenthetical level are then evaluated in the same manner. Therefore, in the current example, the subexpressions are evaluated in the following order:

(b / 2), (c + d), ((b / 2) * (c + d)), (a – ((b / 2) * (c + d)))

We mention again that fully parenthesized expressions require no convention regarding operator precedence.

From the preceding discussion it is clear that for a fully parenthesized expression, a repeated scanning of the expression is still required in order to evaluate the expression. This phenomenon is due to the fact that operators appear with the operands inside the expression. The notation used so far in writing the operator between its operands is called *infix notation*. This repeated scanning of an infix expression is avoided if it is converted first to an equivalent parenthesis-free or Polish expression in which the subexpressions have the form

operand   operand   operator   (known as *suffix Polish*)
or, operator   operand   operand   (known as *prefix Polish*)

in place of an infix form where we have

operand   operator   operand

We can directly translate a fully parenthesized infix expression to suffix or prefix notation by beginning with the conversion of the innermost parenthesized subexpression and then proceeding toward the outside of the expression. For example, to convert the fully parenthesized expression

(a − ((b + 5) / c))

1   3   2

to suffix Polish, the innermost parenthesized subexpression at level 3 is

(b + 5)

and is converted to b5+. This suffix subexpression becomes the first operand of the operator / at level 2. Consequently, the subexpression b5+/c at level 2 is then converted to the suffix expression b5+c/. Finally, the expression a − b5+c/ is converted to the suffix expression ab5+c/−.

Table 10-2 gives the equivalent forms of several fully parenthesized expressions. Note that in both the prefix and suffix equivalents of such an infix expression, the variable names are all in the same relative position.

Computer programmers, of course, rarely write expressions in fully parenthesized form. It is to be noted, however, that certain compilers convert partially parenthesized expressions to their fully parenthesized equivalent forms before ultimately performing a conversion to suffix or prefix form. It should also be noted that certain pocket calculators require Polish expressions.

Let us now examine the conversion of a parenthesis-free arithmetic expression into suffix form. As was just mentioned, only the operators need be rearranged to generate the suffix equivalent form. In a left-to-right scan, the leftmost operator having the highest precedence is the first operator

TABLE 10-2   Equivalent Infix, Prefix, and Suffix Forms of Expressions

Infix	Prefix	Suffix
((a * b) + c)	+*abc	ab*c+
((a + b) * (c + d))	*+ab+cd	ab+cd+*
((a + b * c) − (d / f + e))	−+a*bc+/dfe	abc*+df/e+−

encountered in the suffix string. If in an infix expression we do not specify that a leftmost operator has precedence over other operators of equal precedence, then the suffix form of this expression is not unique. As an example, the expression a * b * c could be converted to either ab*c* or abc** if no mention was made that the leftmost operator * in the infix string has precedence over the remaining operator. From this observation it is clear that on scanning a suffix expression from left to right, the operators thus encountered are in the same order as the evaluation of the corresponding infix expression using the operator precedence convention. For example, the suffix equivalent of a−b / c is abc/−. In a left-to-right scan of this suffix form we encounter the operator / before the operator −, indicating that division is to be evaluated before subtraction. An algorithm for converting partially parenthesized infix expressions to their suffix forms will be given in the next subsection.

Let us now examine the evaluation of a suffix expression. This can be accomplished easily by scanning the expression once from left to right. For example, to evaluate the suffix expression ab−c*, we scan this string from left to right until we encounter −. The two operands a and b, which appear to the immediate left of this operator, are its operands. The subexpression ab− is then replaced by its value. Assign this value to the temporary variable T_1. The original suffix string now reduces to T_1c*. On continuing the scanning process, we encounter the next operator, *. The operands of this operator are T_1 and c. The evaluation of the remaining subexpression yields a value which we denote by T_2.

This method of evaluation is summarized in the following general algorithm:

1. Repeat thru step 5 while there is an operator remaining
2. Find the leftmost operator in the expression
3. Obtain the two operands which immediately precede the operator just found
4. Perform the indicated operation
5. Replace the operator and operands with the result

As a further example, the suffix expression abc+de+*−, which corresponds to the infix expression a − (b + c) * (d + e), is evaluated as follows:

Suffix Form	Current Operator	Current Operands
abc+de+*−	+	b, c
aT_1de+*−	+	d, e
a$T_1$$T_2$*−	*	T_1, T_2
aT_3−	−	a, T_3
T_4		

For the values $a = 1$, $b = 2$, $c = 3$, $d = 4$, and $e = 5$ in this example, $T_1 = 5$, $T_2 = 9$, $T_3 = 45$, and $T_4 = -44$.

As mentioned earlier, certain compilers convert infix arithmetic expressions into Polish notation. In these compilers it is important to detect invalid Polish expressions. These invalid Polish expressions correspond to invalid infix arithmetic expressions. We now describe a method that detects such invalid expressions. Although the discussion applies to suffix expressions, an analogous technique can also be developed for prefix expressions.

Suffix expressions contain symbols such as variable names and constants, and operators such as the arithmetic operators. Let these symbols and operators be denoted by the sets $S = \{s_1, s_2, \ldots, s_q\}$ and $O = \{o_1, o_2, \ldots, o_m\}$, respectively. The *degree* of an operator is defined as the number of operands which that operator has. For example, the multiplication operator has a degree of 2. The set of all suffix expressions is defined as follows:

1. Any symbol s_i is a suffix expression
2. If $y_1, y_2, \ldots, y_n$ are suffix expressions and o_i is an operator of degree n, then $y_1 y_2 \ldots y_n o_i$ is a suffix expression
3. The only valid suffix expressions are those obtained by steps 1 and 2

For example, if the set S contains the single-letter variables (a–z) and the operator set O consists of the standard four arithmetic operators, then the following are all valid expressions according to the previous definition:

ab+

abc*+

ab+cd−*

ab*c ∣ dc/+f−

Examples of invalid expressions are:

ab * +	(missing variable name)
abc+	(missing operator)
abd/+−	(missing variable name)
ab+cd+	(missing operator)

In order to detect invalid expressions, it is useful to associate a *rank* with each expression, which is defined as follows:

1. The rank of a symbol s_i is 1
2. The rank of an operator symbol o_j of degree n is $1 - n$
3. The rank of a sequence of symbols and operators is the arithmetic sum of the ranks of the individual symbols and operators

As an example, if the set of symbols is the English alphabet and the operator set is the four arithmetic operators, then the rank function r is defined as

$$r(s_i) = 1 \quad \text{for } 1 \leqslant i \leqslant 26$$
$$r(+) = r(-) = r(*) = r(/) = -1$$

The rank of the formula ab+cd+/ is obtained by adding the ranks of each individual symbol:

$$r(ab+cd+/) = r(a) + r(b) + r(+) + r(c) + r(d) + r(+) + r(/) = 1$$

An observation to follow is very important since its use can detect invalid expressions. Before stating this observation, some basic terminology must be described. If $z = x \circ y$ is a string of symbols, then x is a *head* of z. Also, x is a *proper head* if y is not the empty string. For example, if z = 'aabcd', then the strings 'a', 'aa', 'aab', 'aabc' are all its proper heads. The following statement is central to the detection of invalid Polish expressions:

> A suffix (prefix) expression is valid if and only if the rank of the expression is 1 and the rank of any proper head of the expression is greater than (less than) or equal to 1.

This observation is very important in the compilation of infix expressions since it permits us to detect an invalid Polish expression and, consequently, its invalid infix counterpart. Table 10-3 contains a number of valid and invalid expressions.

In this subsection we have introduced the notions of prefix and suffix notation. We have not, however, formulated any algorithms which convert infix expressions to their prefix and suffix forms. The next subsection is concerned with this task.

EXERCISES 10-5.2.1

1. Using the recursive definition for fully parenthesized infix expressions given at the beginning of this section, establish the validity or invalidity of the following expressions:
   (a) (((a * b) − c) + d)
   (b) (((a / b) * (c / d)) + e)
   (c) ((a + b * c) * d)
   (d) ((((a * x) + b) * x) + c)
   (e) (a − b) * c

2. Express each of the valid expressions in exercise 1 as suffix Polish.

3. Express each of the valid expressions in exercise 1 as prefix Polish.

TABLE 10-3   Valid and Invalid Polish Expressions

Infix	Suffix Polish	Validity Check		Prefix Polish	Validity Check		Valid or Invalid
		Head	Rank		Head	Rank	
a − b * c	abc*−	a	1	−a*bc	−	−1	Valid
		ab	2		−a	0	
		abc	3		a*	−1	
		abc*	2		−a*b	0	
		abc*−	1		−a*bc	1	
a + * b	ab*+	a	1	+*ab	+	−1	Invalid
		ab	2		+*	−2	
		ab*	1		+*a	−1	
		ab*+	0←		+*ab	0←	
(a + b) * (c + d)	ab+cd+*	a	1	*+ab+cd	*	−1	Valid
		ab	2		*+	−2	
		ab+	1		*+a	−1	
		ab+c	2		*+ab	0	
		ab+cd	3		*+ab+	−1	
		ab+cd+	2		*+ab+c	0	
		ab+cd+/	1		*+ab+cd	1	
a * + bc	a*bc+	a	1				Invalid
		a*	0←				
		a*b	1				
		a*bc	2				
		a*bc+	1				

4. Using the recursive definition for suffix expressions given in this section, establish the validity or invalidity of the following:
   (a) ab * c+
   (b) ab−cd+*
   (c) ab+*cd−
   (d) abc* ı dcf*+*

5. Give a recursive definition for valid prefix expressions.

6. Evaluate the valid suffix expressions of exercise 4. Assume that a = 1, b = 2, c = 3, d = 4, e = 5, and f = 6.

7. Give a general algorithm for the evaluation of prefix expressions.

8. Convert the following infix expressions to their prefix and suffix forms and in each case give a trace of the rank computation.
   (a) a − b * c + d
   (b) a * (b + c − d)
   (c) a − b * (e + d)

9. Formulate a recursive algorithm which will recognize whether or not a given suffix expression is well formed. Assume that the suffix expression

contains sungle-letter variable names and the four basic arithmetic operators.

10-5.2.2 Conversion of infix expressions to Polish notation

This subsection formalizes the process of converting infix expressions into their equivalent prefix and suffix forms. The discussion is concerned primarily with the conversion of arithmetic expressions to suffix form. Initially, an algorithm is formulated for the conversion of unparenthesized expressions to suffix notation. This algorithm is then generalized to handle partially parenthesized infix expressions. Both algorithms are able to detect invalid expressions. A brief discussion of the conversion of infix expressions to their prefix equivalent is then given.

Unparenthesized infix expressions are easily converted into suffix notation. Such a conversion is based on the precedence of the arithmetic operators. Also, as we shall see, a stack is required. We represent the desired output (that is, suffix expression) by a string which will be used later in the generation of object code. Recall from the discussion of the previous subsection that variables and constants remain in the same relative position to each other when performing the desired conversion. The operators, however, are reordered in the desired output string to reflect their relative precedence. This reordering process requires the use of a stack.

Consider the infix expressions which contain the four arithmetic operators whose precedence values are given in Table 10-4. Multiplication and division have priority over addition and subtraction. This table also contains a precedence value for single-letter variables. The reason for assigning a precedence value to variable-name symbols will be given shortly. Note that the table contains the rank of the symbols which can occur in infix expressions. The notion of rank was discussed in the previous subsection.

Let us first assume that the stack contains a special symbol (# in Table 10-4) whose precedence value is less than that of all other symbols in the table. The purpose of this special symbol is to prevent the stack from being empty during the conversion process. To compensate for this special symbol, the given infix expression is padded on the right with the symbol #.

A general algorithm for the conversion process might take the following form:

1. Initialize stack contents to the special symbol #
2. Scan the leftmost symbol in the given infix expression and denote it as the current input symbol
3. Repeat thru step 6 while the current input symbol is not #
4. Remove and output all stack symbols whose precedence values are greater than or equal to the precedence of the current input symbol
5. Push the current input symbol onto the stack
6. Scan the next leftmost symbol in the infix expression and let it be the current input symbol

TABLE 10-4

Symbol	Precedence Function, f	Rank Function, r
+, −	1	−1
*, /	2	−1
a, b, c, . . .	3	1
#	0	

7. Remove and output all remaining symbols (except #) from the stack

From this general algorithm we can, in a straightforward manner, formulate a detailed algorithm.

Algorithm UNPARENTHESIZED_SUFFIX. Given an input string INFIX representing an infix expression whose single character symbols have precedence values and ranks as given in Table 10-4, a vector S representing a stack, and a string function NEXTCHAR which, when invoked, returns the next character of the input string, this algorithm converts the string INFIX to its suffix string equivalent, POLISH. RANK contains the value of each head of the suffix string. CURRENT contains the symbol being examined and TEMP is a temporary variable which contains the unstacked element. We assume that the given input string is padded on the right with the special symbol #. As usual, TOP denotes the stack top.

1. [Initialize stack]
   TOP ← 1
   Call PUSH(S, TOP, '#')
2. [Initialize output string and rank count]
   POLISH ←"   (empty string)
   RANK ← 0
3. [Get first input symbol]
   CURRENT ← NEXTCHAR(INFIX)
4. [Translate the infix expression]
   Repeat thru step 6 while CURRENT ≠ '#'
5. [Remove symbols with greater or equal precedence from stack]
   If TOP < 1
   then Write('INVALID')
       Exit
   Repeat while f(CURRENT) ≤ f(S[TOP])
     TEMP ← POP(S, TOP)   (this copies the stack contents into TEMP)
     POLISH ← POLISH ○ TEMP
     RANK ← RANK + r(TEMP)
     If RANK < 1
     then Write('INVALID')
         Exit

6. [Push current symbol onto stack and obtain next input symbol]
   Call PUSH(S, TOP, CURRENT)
   CURRENT ← NEXTCHAR(INFIX)
7. [Remove remaining elements from stack]
   Repeat while S[TOP] ≠ '#'
     TEMP ← POP(S, TOP)
     POLISH ← POLISH ○ TEMP
     RANK ← RANK + r(TEMP)
     If RANK < 1
     then Write('INVALID')
          Exit
8. [Is the expression valid?]
   If RANK = 1 then Write('VALID') else Write('INVALID')
   Exit ☐

The operation of this algorithm is simple. Initially, the marker symbol # is stacked. Step 2 initializes the output string POLISH to the empty string and assigns a zero value to RANK. The third step of the algorithm scans the leftmost symbol of the infix expression and assigns this symbol to CURRENT. Step 5 compares the precedence value of the input symbol CURRENT to that of the top element of the stack. If the precedence value of CURRENT is less than or equal to that of the stack top, then the latter is removed from the stack and placed in string POLISH. The precedence values of CURRENT and the new element at the top of the stack are then compared. This process continues until the test fails. The rank counter is updated as each symbol is removed from the stack. If, on the other hand, the precedence value of CURRENT is greater than that of the stack top, then the current input is stacked in step 6.

Since a variable has the highest precedence value, it is pushed onto the stack. When scanning the very next input symbol, however, such a variable on top of the stack will be deleted and written out into POLISH. This is always the case because, in valid infix expressions, consecutive variable symbols cannot occur. The algorithm could be changed easily so that the precedence value of CURRENT is tested against 3. A successful test would indicate that CURRENT represents a variable and, therefore, it could be written into POLISH without being stacked. We do not, however, pursue this approach for reasons of generality which become more important as infix expressions are permitted to be more complex.

As mentioned earlier, a current symbol whose precedence value is greater than that of the stack element will result in the stacking of the former. Such an action indicates that the operation which corresponds to the incoming operator is to be executed before all operations which correspond to the operators on the stack. The "last-in, first-out" property of a stack will indeed guarantee this behavior. Notice that when both the input and stack symbols have equal precedence, then the stack symbol is written out into POLISH. This behavior ensures that in an expression which contains operators with equal precedence, the leftmost operator is the one which is executed. Consequently, the current algorithm converts a * b * c to ab∗c∗ and not to abc∗∗. The suffix string ab∗c∗ corresponds to the infix expression (a∗b)∗c and abc∗∗ is equivalent to the expression a∗(b∗c). A sample trace of this algorithm for the expression a−b∗c/d+e/f# is given in Table 10-5.

We now turn to the problem of converting parenthesized infix expressions to suffix notation. Programmers do not usually write such expressions in a completely parenthesized form. Intuitively, when the current symbol is a left parenthesis, it should be pushed onto the stack. This should be done regardless of the stack contents. When a left parenthesis is on the stack, however, it should remain there until a matching right parenthesis is encountered in the input expression. At this point, the left parenthesis should be removed from the stack and the matching right parenthesis in the input should be ignored. We can force an incoming left parenthesis on the stack by simply forcing its *input precedence value* to be greater than that of any other symbol. Once on the stack, however, this same left parenthesis must have another precedence value (called its *stack precedence*), which is smaller than that of any other symbol. A stacked left parenthesis is discarded on encountering a matching right parenthesis in the infix expression. Note that this right parenthesis need never be stacked. Our previous algorithm for

TABLE 10-5   Translation of Infix Expression a−b∗c/d+e/f# to Suffix

Character Scanned	Contents of Stack (Rightmost Symbol Is Top of Stack)	Suffix Expression	Rank
	#		
a	#a		
−	#−	a	1
b	# b	a	1
*	#−∗	ab	2
c	#−∗c	ab	2
/	#−∗	abc	3
	#−/	abc∗	2
d	#−/d	abc∗	2
+	#−/	abc∗d	3
	#−	abc∗d/	2
	#+	abc∗d/−	1
e	#+e	abc∗d/−	1
/	#+/	abc∗d/−e	2
f	#+/f	abc∗d/−e	2
#	#+/	abc∗d/−ef	3
	#+	abc∗d/−ef/	2
	#	abc∗d/−ef/+	1

unparenthesized expressions can be modified easily in such a manner that the parenthesis can perform the same function as the special symbol # used earlier. Table 10-4 is easily changed to contain both an input and stack-precedence value for each symbol. In addition, we can get rid of the special symbol #. In so doing, the algorithm becomes more general in the sense that its complexity does not grow significantly as additional operators, such as relational, logical, and unary operators, are added to the system. Such a revised table is given in Table 10-6. Note that this table contains parentheses. Also, each symbol has both an input- and a stack-precedence value, except for a right parenthesis, which has no stack-precedence value since this symbol is never stacked. The infix expressions under consideration have been expanded to contain the exponentiation operator, ↑. The input precedence of each arithmetic operator (except exponentiation) is less than its corresponding stack precedence. Such a relationship preserves the desired left-to-right processing property of equal-precedence operators. The exponentiation operator, however, is right-associative. That is, the expression $2 \uparrow 2 \uparrow 4$ is equivalent to the parenthesized expression $2 \uparrow (2 \uparrow 4)$ and not to the expression $(2 \uparrow 2) \uparrow 4$. This right-associative property is automatically enforced by having the input precedence of the exponentiation operator greater than its stack precedence.

The conversion of a parenthesized infix expression into suffix notation is very similar in operation to the previous algorithm. Initially, we place a left parenthesis on the stack and pad the given infix expression with a matching right parenthesis. The revised algorithm is formulated as follows:

Algorithm SUFFIX. Given an input string INFIX containing an infix expression which has been padded on the right with a right parenthesis and whose symbols have precedence values given by Table 10-6, a vector S, used as a stack, and a function NEXTCHAR, which when invoked returns the next character of its argument, this algorithm converts INFIX into suffix notation and places the result in the string POLISH. The integer variable TOP denotes the stack top. The algorithm uses the stack subalgorithms PUSH

TABLE 10-6 Input- and Stack-Precedence Values for Arithmetic Expressions

| Symbol | Precedence | | Rank |
	Input Function (f)	Stack Function (g)	Rank Function (r)
+, −	1	2	−1
*, /	3	4	−1
↑	6	5	−1
Single-letter variables	7	8	1
(	9	0	−
)	0	−	−

and POP introduced earlier in the chapter. The integer variable RANK accumulates the rank of the suffix expression. Finally, the string variable TEMP is used for temporary storage purposes.

1. [Initialize stack]
    TOP ← 1
    Call PUSH(S, TOP, '(')
2. [Initialize output string and rank count]
    POLISH ← "
    RANK ← 0
3. [Get first input symbol]
    CURRENT ← NEXTCHAR(INFIX)
4. [Translate the infix expression]
    Repeat thru step 7 while CURRENT ≠ "
5. [Remove symbols with greater precedence from stack]
    If TOP < 1 then Write('INVALID')
            Exit
    Repeat while f(CURRENT) < g(S[TOP])
      TEMP ← POP(S, TOP)
      POLISH ← POLISH ○ TEMP
      RANK ← RANK + r(TEMP)
      If RANK < 1 then Write('INVALID')
            Exit
6. [Are there matching parentheses?]
    If f(CURRENT) ≠ g(S[TOP])
    then Call PUSH(S, TOP, CURRENT)
    else POP(S, TOP)
7. [Get next input symbol]
    CURRENT ← NEXTCHAR(INFIX)
8. [Is the expression valid?]
    If TOP ≠ 0 or RANK ≠ 1 then Write('INVALID') else Write('VALID')
    Exit   □

Step 1 places a left parenthesis on the stack. The second step initializes the desired suffix string POLISH (to empty) and RANK (to zero). Step 3 scans the leftmost character in the given infix expression and places it in CURRENT. Steps 4 to 7 perform the desired translation of the given expression. The fifth step compares the current input symbol with the top element of the stack. If the precedence of the stack symbol is greater than that of the current input symbol, then we remove the former from the stack and write it into the suffix string in this step. A rank value of less than 1 implies an invalid expression and causes an exit from the algorithm. Once we remove a symbol from the stack, the entire process is repeated with the new symbol on top of the stack. When all stack symbols whose precedence values are greater than that of the current input symbol have been written out into POLISH, control passes to step 6. In this step a check is made for a left parenthesis on top of the stack and a matching right parenthesis in the input. If this check fails, then the current input symbol is placed on the stack; otherwise, we ignore the right parenthesis in the input and delete the corresponding left parenthesis from the stack. The final step of the algorithm checks the validity of the expression. If the accumulated rank value of the expression is 1 and the stack is empty, then the expression is valid; otherwise, it is invalid.

A trace of the stack contents and the suffix string for the padded infix expression

$$(a + b * c) * (d + e \uparrow f \uparrow a))$$

is given in Table 10-7. We encourage the reader to trace the algorithm for the padded invalid expression $((a + b / d)c + e))$.

TABLE 10-7   Translation of Padded Infix Expression
$(a + b * c) * (d + e \uparrow f \uparrow a))$ to Suffix

Character Scanned	Contents of Stack (Rightmost Symbol Is Top of Stack)	Suffix Expression	Rank
	(		
(	((		
a	((a		
+	((+	a	1
b	((+b	a	1
*	((+*	ab	2
c	((+*c	ab	2
)	((+*	abc	3
	((+	abc*	2
	((	abc*+	1
	(	abc*+	1
*	(*	abc*+	1
(	(*(	abc*+	1
d	(*(d	abc*+	1
+	(*(+	abc*+d	2
e	(*(+e	abc*+d	2
↑	(*(+↑	abc*+de	3
f	(*(+↑f	abc*+de	3
↑	(*(+↑↑	abc*+def	4
a	(*(+↑↑a	abc*+def	4
)	(*(+↑↑	abc*+defa	5
	(*(+↑	abc*+defa↑	4
	(*(+	abc*+defa↑↑	3
	(*(	abc*+defa↑↑+	2
	(*	abc*+defa↑↑+	2
)	(*	abc*+defa↑↑+	2
	(	abc*+defa↑↑+*	1
		abc*+defa↑↑+*	1

So far, we have developed algorithms for converting infix expressions to suffix notation. We now turn to the problem of converting infix expressions to prefix notation. We do not, however, formulate a detailed algorithm for this process. Only the general approach is discussed here.

A simple algorithm based on the scanning of the infix expression from right to left can be obtained in a straightforward manner. In many cases the entire infix expression is not available, but it is obtained one symbol at a time in a left-to-right manner. Consequently, a practical algorithm for the desired conversion should be based on a left-to-right scan of the infix string.

To facilitate such an algorithm, however, it is desirable to use two stacks instead of the one stack used in the conversion of infix to suffix. More specifically, we find it convenient to use an operator stack and an operand stack. The purpose of the latter stack is to store temporarily the intermediate operands. Recall from Sec. 10-5.2.1 that all variables and constants retain their relative order when an infix expression is converted to prefix form. The operators, however, are reordered according to their relative precedence, and the operator stack is used in this reordering. The operand stack, on the other hand, is used for temporary storage of intermediate operands so that finally, when the associated operator is found to be applicable, it can be placed in front of the concatenated operands.

A general algorithm based on this approach follows:

1. Put a left parenthesis on the operator stack
2. Scan the first input symbol
3. Repeat thru step 6 while there is still some input
4. If the current input symbol is a variable or a constant,
    then push this variable or constant on the operand list;
    else Repeat while the precedence of the current input is less than
            that of the operator stack top
        Replace the two topmost operands on the operand stack
            by the operator stack concatenated with these operands
5. If the current symbol is not a right parenthesis or the symbol on the
        operator stack is not a left parenthesis,
    then push the current operator on the operator stack;
    else delete the left parenthesis from the operator stack
6. Obtain the next input symbol

A trace of this general algorithm for the padded input string (a+b) ∗ (c + d)) is given in Table 10-8. To avoid confusion, the operands on the operand stack have been delimited by quote symbols. When the algorithm terminates, the desired prefix expression is the top (and only) element on the operand stack. We leave the detailed algorithm to the exercises.

The notions discussed in this subsection can be extended easily to handle relational operators, logical operators, conditional statements, and

TABLE 10-8   Translation of (a + b) * (c + d)) to Prefix Notation

Character Scanned	Operator Stack (Rightmost Symbol Is Top of Stack)	Operand Stack (Rightmost String Is Top of Stack)
	(	
(	((	
a	((	'a'
+	((+	'a'
b	((+	'a' 'b'
)	((	'+ab'
	(	'+ab'
*	(*	'+ab'
(	(*(	'+ab'
c	(*(	'+ab' 'c'
+	(*(+	'+ab' 'c'
d	(*(+	'+ab' 'c' 'd'
)	(*(	'+ab' '+cd'
)	(*	'*+ab+cd'
	(	'*+ab+cd'
		'*+ab+cd'

many other features found in modern programming languages. A number of exercises at the end of this subsection deal with these extensions.

We have been concerned until now with the conversion of infix expressions to their prefix and suffix counterparts. The motivation behind this conversion is that the latter forms can be converted into object code by a single linear scan of the prefix or suffix expression. The next subsection deals with this problem.

EXERCISES 10-5.2.2

1. Trace (as in Table 10-5) the algorithm UNPARENTHESIZED_SUFFIX for the following input expressions:
   (a) a * b − c * d / e#
   (b) a + * b c#

2. Trace (as in Table 10-7) the algorithm SUFFIX for the following input expressions:
   (a) (a + b) * (c − d + e))
   (b) a * b * (c + d))
   (c) (a + b)(c + d))

3. Consider expressions which contain relational operators. In particular,

obtain the precedence functions which will handle the relational operators

$<, \leq, =, \neq, >,$ and $\geq$

To avoid excessive parenthesization, the relational operators should have a lower priority than the arithmetic operators.

4. As a continuation of exercise 3, consider extending the expressions so that they contain logical operators. In particular, obtain the precedence functions that will handle the logical operators

$\neg$ (not)

&   (and)

| (or)

which are given in decreasing order of priority. These operators have lower priority than the relational and arithmetic operators.

5. Describe how conditional statements (if . . . then . . . else) can be implemented in the suffix Polish framework. In particular extend the precedence functions obtained in exercise 4 so as to incorporate conditional statements.

6. Based on the general algorithm given in the text, formulate a detailed algorithm for converting infix expressions to their equivalent prefix forms.

10-5.2.3   Conversion of Polish expressions to object code

Throughout this discussion we assume that the object code desired is in the form of assembly language instructions. As a matter of convenience, we also assume that the object computer which will execute the object code produced by the translation process is a single-address, single-accumulator machine whose memory is sequentially organized into words. Such a computer was described in Sec. 2-1. In this subsection we assume a simple assembly language representation for the instructions introduced there. A brief description of the symbolic operations is given in Table 10-9. Note that we have added two new instructions: multiplication (MUL) and division (DIV).

An informal algorithm for the evaluation of a suffix string was given in Sec. 10-5.2.1. Let us consider, initially, a "brute-force" algorithm for converting suffix expressions consisting of the four basic arithmetic operators and single-letter variables to assembly language. Finally, assume that the basic arithmetic operators generate the following code:

a + b (ab+)   LOD a
                   ADD b
                   STO T_i

a − b (ab−)	LOD a
	SUB b
	STO T_i
a ∗ b (ab∗)	LOD a
	MUL b
	STO T_i
a / b (ab/)	LOD a
	DIV b
	STO T_i

Note that each operator generates three assembly language instructions. The third instruction in each group is of the form STO T_i, where T_i denotes an address of a location (word) in the computer's memory that is to contain the value of the intermediate result. These addresses are to be created by the desired suffix-to-assembly language algorithm.

A straightforward algorithm involves the use of a stack. This entails the scanning of the suffix expression in a left-to-right manner. Each variable name in the input must be placed on the stack. On encountering an operator, the topmost two operands are unstacked and used (along with the operator) to generate the desired sequence of assembly instructions. The intermediate result corresponding to the operator in question is also placed on the stack. A general algorithm based on this approach follows:

1. Repeat thru step 3 while there still remains an input symbol
2. Obtain the current input symbol

TABLE 10-9 Sample Assembly Language Instruction Set

Operations	Meaning
LOD A	Load: copy the value of the word addressed by A into the accumulator
STO A	Store: copy the value of the accumulator into the word addressed by A
ADD A	Add: replace the present value of the accumulator with the sum of its present value and the value of the word addressed by A
SUB A	Subtract: replace the present value of the accumulator with the result obtained by subtracting from its present value the value of the word addressed by A
MUL A	Multiply: replace the present value of the accumulator with the result obtained by multiplying its present value by the value of the word addressed by A
DIV A	Divide: replace the present value of the accumulator with the result obtained by dividing its present value by the value of the word addressed by A

3. If the current input symbol is a variable,
   then push this variable on the stack;
   else remove the two topmost operands from the stack,
       generate the sequence of assembly language instructions which
           corresponds to the current arithmetic operator,
       stack the intermediate result

A detailed algorithm based on this general approach is now given.

Algorithm ASSEMBLY_CODE. Given a string SUFFIX representing a suffix expression (which contains the four basic operators and single-letter variables) equivalent to a valid infix expression, this algorithm translates the string SUFFIX to assembly language instructions as previously specified. The algorithm uses a stack S with its associated top pointer TOP. The integer variable i is associated with the generation of an intermediate result. The string variable OPCODE contains the operation code which corresponds to the current operator being processed.

1. [Initialize]
   TOP ← i ← 0
2. [Process the suffix expression]
   Repeat thru step 4 for j = 1, 2, . . . , LENGTH(SUFFIX)
3. [Obtain current input symbol]
   CURRENT ← SUB(SUFFIX, j, 1)
4. [Process current input symbol]
   If 'a' ≤ CURRENT and CURRENT ≤ 'z'
   then (push current variable on the stack)
       Call PUSH(S, TOP, CURRENT)
   else (process current operator)
       If CURRENT = '+'
       then OPCODE ← 'ADD□'
       else If CURRENT = '−'
           then OPCODE ← 'SUD□'
           else If CURRENT − '*'
               then OPCODE ← 'MUL□'
               else OPCODE ← 'DIV□'
       RIGHT ← POP(S, TOP)   (unstack two operands)
       LEFT ← POP(S, TOP)
       Write('LOD□' ∘ LEFT)   (output load instruction)
       Write(OPCODE ∘ RIGHT)   (output arithmetic instruction)
       i ← i + 1   (obtain temporary storage index)
       TEMP ← 'T' ∘ i
       Write('STO□' ∘ TEMP)   (output temporary store instruction)
       Call PUSH(S, TOP, TEMP)   (stack intermediate result)
5. [Finished]
   Exit   □

An important point should be made concerning this algorithm. In general, variables and constants (and also operators) can be more than one character in length. Instead of storing the variable names, values of constants, and so on, in the string SUFFIX, integer pointers (giving the index of a variable or constant or operator) to a vector containing these items are used. The label T_i, in practice, would be an index to a vector containing all created variable names used in storing temporary results. This extension is left to the exercises.

The trace of this algorithm for the suffix string ab*c+def/+* is given in Table 10-10. The assembly language program generated by the algorithm is given in the rightmost column of the table. Upon examination of this program, it is clear that the generated code is inefficient. First, in the output code there exist redundant pairs of instructions such as

```
LOD T₁
STO T₁
```

in the output program. Second, there is no advantage taken of the commutative property of the addition and multiplication operators. An example of this occurs in the code generated for the subexpression d + e /f: namely, the sequence

```
LOD e
DIV  f
STO T₃
LOD d
ADD T₃
STO T₄
```

which can be replaced by the equivalent program segment

```
LOD e
DIV  f
ADD d
STO T₄
```

since the right operand (ef/) is already in the accumulator and the values of d + e / f and e / f + d are identical. The last sequence of instructions takes advantage of the commutative property of addition to eliminate two instructions. Finally, there is no effort made to reduce the number of temporary locations which are required to store intermediate results. In particular, the sequence of instructions

```
LOD a
MUL b
STO T₁
LOD T₁
ADD c
STO T₂
```

TABLE 10-10   Sample Code Generated by the Algorithm ASSEMBLY_CODE
for the Suffix String ab*c+def/+*

Character Scanned	Contents of Stack (Rightmost Symbol Is Top of Stack)	Left Operand	Right Operand	Code Generated
a	a			
b	ab			
*	T_1	a	b	LOD a
				MUL b
				STO T_1
c	$T_1 c$			
+	T_2	T_1	c	LOD T_1
				ADD c
				STO T_2
d	$T_2 d$			
e	$T_2 de$			
f	$T_2 def$			
/	$T_2 dT_3$	e	f	LOD e
				DIV f
				STO T_3
+	$T_2 T_4$	d	T_3	LOD d
				ADD T_3
				STO T_4
*	T_5	T_2	T_4	LOD T_2
				MUL T_4
				STO T_5

can be replaced by the equivalent sequence where all instructions are the same, except for the last, which becomes STO T_1. Thus, the temporary variable T_2 becomes unnecessary.

The number of temporary variables required can be reduced easily by making the following simple check. Before generating the assembly instruction for a particular operator, a check of the contents of the left (LEFT) and right (RIGHT) operands associated with the operator in question can be made. For each operand which corresponds to a created variable T_i, the temporary variable counter i is decremented by 1.

The redundant pairs of store and load instructions and the unnecessary temporary storing and subsequent reloading of a right operand for commutative operators is eliminated in the following way. Instead of always storing a partial result in temporary storage, as was done in the statement

Write('STO□' ∘ TEMP)

in step 5 of Algorithm ASSEMBLY_CODE, one can delay the generation of such an instruction until it is deemed absolutely necessary. The previous statement can be altered so as to place an intermediate result marker @ on the stack instead of always generating a store instruction. If such a marker is never pushed down in the stack deeper than the next-to-top position, then an intermediate result need not be saved by the generation of a store instruction. A revised algorithm based on the previous comments is left as an exercise.

Although in this subsection we have only dealt with the generation of code from suffix expressions, similar notions hold for generating code from prefix expressions. We leave these details to the exercises.

EXERCISES 10-5.2.3

1. Trace (as in Table 10-10) the algorithm ASSEMBLY_CODE for the following suffix expressions:
   (a) ab−cd+∗
   (b) ab∗c+de−−
   (c) abc∗+de/f∗−

2. Based on the discussion in the text following the algorithm ASSEMBLY_ CODE, formulate an algorithm which will generate more efficient code when the cummutativity of the operators + and ∗ is taken into consideration.

3. Modify the algorithm obtained in exercise 2 so that the required number of temporary positions is reduced.

4. Modify the algorithm obtained in exercise 3 so as to incorporate the assignment operator.

5. Modify the algorithm obtained in exercise 4 so that it will handle the unary minus operator.

6. Formulate an algorithm similar to the algorithm ASSEMBLY_CODE for generating code for prefix expressions.

7. Modify the algorithm obtained in exercise 6 so that advantage of the commutative operators is taken into consideration.

10-5.3 PARTITION-EXCHANGE SORTING

As the third and final application of a stack structure, we now consider a sorting technique which performs well on large tables. The approach is to place initially a particular record in its final position within the sorted table. Once this is done, all records which precede this record have smaller keys, while all records that follow it have larger keys. This technique partitions the original table into two subtables. The same process is then applied to each of these subtables and repeated until all records are placed in their final positions.

As an example of this approach to sorting, let us consider the placement of 73 in its final position in the following key set:

73   65   52   24   83   17   35   96   41   9

We use two index variables, I and J, with initial values of 2 and 10, respectively. The two keys 73 and K[I] are compared, and if an exchange is required (that is, K[I] < 73), then I is incremented by 1 and the process is repeated. When K[I] ≥ 73, we proceed to compare keys K[J] and 73. If an exchange is required, then J is decremented by 1 and the process is repeated until K[J] ≤ 73. At this point, the keys K[I] and K[J] (that is, 83 and 9) are interchanged. The entire process is then repeated with J fixed and I being incremented once again. When I ≥ J, the desired key is placed in its final position by interchanging keys 73 and K[J].

The sequence of exchanges for placing 73 in its final position is given as follows, where the encircled entries on each line denote the keys being compared:

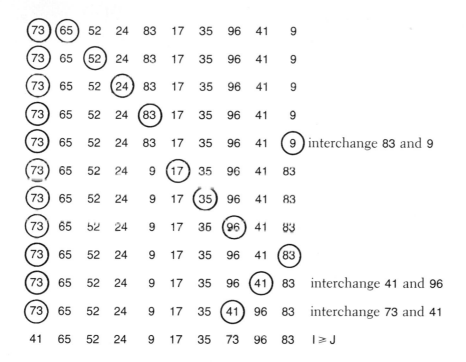

The original key set has been partitioned into two subtables: {41, 65, 52, 24, 9, 17, 35} and {96, 83}. The previous process can be applied to each of these subtables until the original table is sorted. This partition-exchange technique of sorting is sometimes called *quicksort*.

This approach to sorting is formalized in the following procedure.

Procedure QUICK_SORT(K, LB, UB). Given a table K of N records, this recursive procedure sorts the table, as previously described, into ascending order. A dummy record with a key K[N + 1] is assumed where K[I] ≤ K[N + 1] for all $1 \le I \le N$. The integer parameters LB and UB denote the lower and upper bounds of the current subtable being processed. The indices I and J are used to select certain keys during the processing of each subtable. KEY contains the key value which is being placed in its final position within the sorted subtable. FLAG is a logical variable which indicates the end of the process that places a record in its final position. When FLAG becomes false, the input subtable has been partitioned into two disjoint parts.

1. [Initialize]
   FLAG ← true
2. [Perform sort]
   If LB < UB
   then I ← LB
        J ← UB + 1
        KEY ← K[LB]
        Repeat while FLAG
          I ← I + 1
          Repeat while K[I] < KEY   (scan the keys from left to right)
            I ← I + 1
          J ← J − 1
          Repeat while K[J] > KEY   (scan the keys from right to left)
            J ← J − 1
          If I < J
          then K[I] ⇔ K[J]   (interchange records)
          else FLAG ← false
        K[LB] ⇔ K[J]   (interchange records)
        Call QUICK_SORT(K, LB, J − 1)   (sort first subtable)
        Call QUICK_SORT(K, J + 1, UB)   (sort second subtable)
3. [Finished]
   Return   ☐

The behavior of this procedure on the sample key set used earlier is given in Table 10-11.

TABLE 10-11   Behavior of Procedure QUICK_SORT

K_1	K_2	K_3	K_4	K_5	K_6	K_7	K_8	K_9	K_{10}	LB	UB
{76	65	52	24	83	17	35	96	41	9}	1	10
{41	65	52	24	9	17	35}	73	{96	83}	1	7
{9	35	17	24}	41	{52	65}	73	{96	83}	1	4
9	{35	17	24}	41	{52	65}	73	{96	83}	2	4
9	{24	17}	35	41	{52	65}	73	{96	83}	2	3
9	17	24	35	41	{52	65}	73	{96	83}	6	7
9	17	24	35	41	52	65	73	{96	83}	9	10
9	17	24	35	41	52	65	73	83	96		

This algorithm requires, on the average, $O(n \log_2 n)$ comparisons to sort a table of n records. The worst case for this algorithm, however, requires $O(n^2)$ comparisons. This situation occurs when the table is already sorted! The performance of the previous algorithm is no better than a selection sort at this point.

EXERCISES 10-5.3

1. Describe the behavior of the procedure QUICK_SORT (as in Table 10-11) for the sample key set:

    42 23 74 11 65 58 94 36 99 87

2. Formulate an iterative algorithm for the partition-exchange method of sorting.

3. Using the sample key set given in exercise 1, trace the algorithm obtained in exercise 2.

10-6   QUEUES

Another important subclass of linear lists is one which permits deletions to be performed from the beginning (or front) of the list. Insertions, on the other hand, are performed at the opposite end (or rear) of the list. Elements in such an organization are processed in the same order as they were received, that is, on a first-in, first-out (FIFO) or a first-come, first-served (FCFS) basis. A linear list which belongs to this subclass is called a *queue*. Figure 10-12 illustrates a queue structure in which insertions are performed at the rear and elements are deleted from the front of the structure. Perhaps one of the most familiar examples of a queue is a customer checkout line at a supermarket cash register. The first customer in this line is the first to be checked out and a new customer joins the end of the line.

Most computer science students have encountered another familiar example of a queue in a timesharing computer system where many users share the computer system simultaneously. Often, such a computer system has a single central processing unit and one main memory. These resources are shared by allowing one user's program to execute for a short period of time, followed by the execution of a second user's program, and so on, until there is a resumption of execution of the initial user's program. A queue is used to store the user programs which are awaiting execution. Typically, however, this waiting queue does not operate in a strict first-in, first-out manner. Rather, it operates on a complex priority scheme which is based on several factors, such as the execution time required, the number of lines of output, the compiler being used, and the time of day. A queue of this type is some-

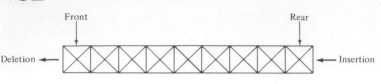

FIGURE 10-12
Representation of a queue.

times called a *priority queue*. More will be said about this type of queue in
Sec. 10-7.

Most people are part of another familiar queue every day. This queue
takes the form of a line of cars waiting to proceed in some direction at a
street intersection. An insertion of a car in the queue involves a car joining
the end of the line of existing cars waiting to proceed through the intersection, while the deletion of a car consists of the front car passing through the
intersection.

As was done in the previous discussion of stacks, we can give a vector
representation of a queue. Again, this vector is assumed to contain a sufficiently large number of elements to handle the variable-length property of a
queue. Another representation, which better reflects the variable-size property of a queue, is given in Sec. 10-9.3. The vector representation of a queue
requires having pointers F and R to denote the positions of its front and rear
elements, respectively. Figure 10-13 illustrates the representation scheme.
Using this representation, we are now in a position to formulate insertion
and deletion algorithms. We first formulate an insertion algorithm.

Procedure QINSERT(Q, F, R, X). Given a vector Q containing N elements
and integer variables F and R, which denote the front and rear elements of a
queue, respectively, it is required to insert X at the rear of the queue. An
empty queue is signaled by F and R both having a value of zero.
1. [Overflow condition?]
     If R ⩾ N then Call QUEUE_OVERFLOW
2. [Increment rear pointer]
     R ← R + 1
3. [Insert new element]
     Q[R] ← X
4. [Is front pointer properly set?]
     If F = 0 then F ← 1
5. [Finished]
     Return   □

The first step of the algorithm checks for an overflow situation. If this
situation arises, then the procedure QUEUE_OVERFLOW is invoked. Although this procedure is application-dependent, usually its invocation generates a signal which indicates that more storage is required for the queue
and that the program must be rerun. Note that in step 4 a check is made for
an insertion into an empty queue. In such an instance the front pointer
must be set to 1.

The corresponding deletion algorithm follows.

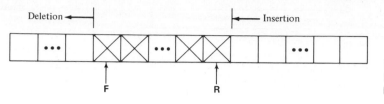

FIGURE 10-13
Representation of a queue
by a vector.

Procedure QDELETE(Q, F, R, X). The details of this procedure are the same as those in procedure QINSERT, except that the deleted element is to be placed in X.

1. [Underflow?]
   If F = 0 then Call QUEUE_UNDERFLOW
2. [Delete front element]
   X ← Q[F]
3. [Is queue now empty?]
   If F = R
   then F ← R ← 0
       Return
4. [Increment front pointer]
   F ← F + 1
   Return   ☐

An underflow situation is checked for in the first step of the algorithm. If this situation arises, then procedure QUEUE_UNDERFLOW is invoked. Although this procedure is also application-dependent, an empty queue is usually a valid state of affairs. Note that if a deletion is performed on a queue which contains a single element prior to deletion time, then the front and rear pointers are reset to zero in step 3.

The previous pair of algorithms can be very wasteful of memory if the front pointer F never catches up to the rear pointer. In such a situation, a very large amount of memory would be required to accommodate the elements of the queue. Therefore, this representation of a queue should only be used when the queue becomes empty at various points in time.

As an example of this representation of a queue, consider a vector which contains five elements. Let us assume that the queue is initially empty. It is required to insert elements 'PAUL', 'RICK', and 'BOB', delete 'PAUL' and 'RICK', and insert 'JOHN', 'GRANT', and 'KEN'. Figure 10-14 gives a trace of the queue contents for this sequence of operations. Note that an overflow occurs on trying to insert 'KEN', even though the first two locations are not being used.

A more suitable method of representing a queue, which gets around this obvious drawback, is to arrange the elements Q[1], Q[2], . . . , Q[N] in a circular fashion, with Q[1] following Q[N]. Figure 10-15 illustrates such a representation. The insertion and deletion subalgorithms for a *circular queue* follow.

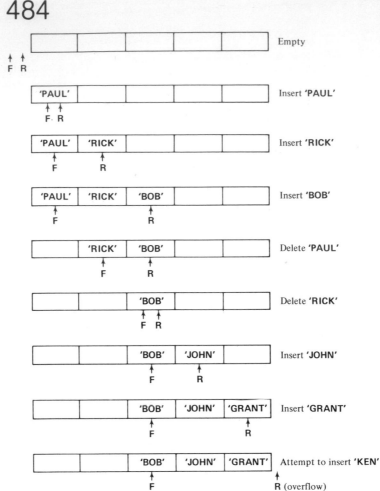

FIGURE 10-14
Trace of operations
on a simple queue.

Procedure CQINSERT(Q, F, R, X). Given values for F and R which denote the front and rear elements of a queue, respectively, a vector Q of N elements which is assumed to be arranged in a similar manner, and a new element X, this procedure performs the required insertion. An empty queue has both of its pointers set to zero.

1. [Reset rear pointer?]
   If R = N
   then R ← 1
   else R ← R + 1
2. [Overflow condition?]
   If F = R then Call QUEUE_OVERFLOW
3. [Insert new element]
   Q[R] ← X
4. [Is front pointer properly set?]
   If F = 0
   then F ← 1

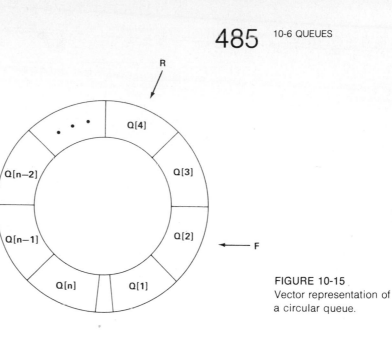

FIGURE 10-15
Vector representation of
a circular queue.

5. [Finished]
    Return   ☐

This algorithm is very similar to the corresponding algorithm for a simple queue except for wraparound detection.

Procedure CQDELETE(Q, F, R, X). The details are the same as that for the procedure CQINSERT, except that the deleted element is copied into X.
1. [Underflow condition?]
     If F = 0 then Call QUEUE_UNDERFLOW
2. [Delete front element]
     X ← Q[F]
3. [Is queue now empty?]
     If F = R
     then F ← R ← 0
         Return
4. [Increment front pointer]
     If F = N
     then F ← 1
     else F ← F + 1
     Return   ☐

An example of a circular queue capable of holding five elements is given in Fig. 10-16, where several operations are performed on an initially empty queue. Note that in this figure the queue is not shown to be circular for convenience purposes.

     The preceding algorithms all deal with a queue which behaves in a first-in, first-out manner. A more general queue structure involves a linear list in which insertions and deletions are made to and from either end of the list. Such a structure is called a *deque* (double-ended queue) and is illus-

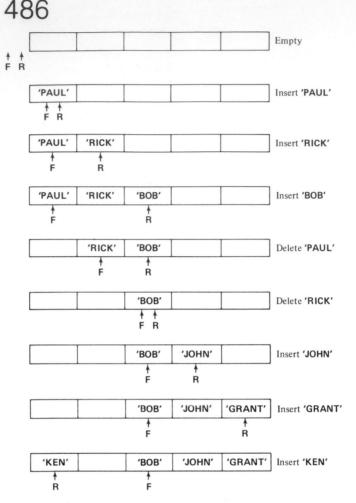

FIGURE 10-16
Trace of operations on a
circular queue.

trated in Fig. 10-17. Observe that a deque is more general than either a stack or a simple queue. There are two variations of a deque: an *input-restricted deque* and an *output-restricted deque*. An input-restricted deque only allows insertions at the rear of the queue. An output-restricted deque, on the other hand, only allows deletions from the front of the queue. We leave the insert and delete algorithms for these structures as exercises.

In this section we have concentrated on the basic notions of queues. In the next sections some of their applications will be examined.

EXERCISES 10-6

1. Construct a procedure for inserting an element into a deque. Assume a vector representation for the queue. Note that you must have a parameter that specifies at which end of the deque the insertion is to be made.

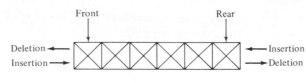

FIGURE 10-17
Deque.

2. Repeat exercise 1 for the deletion operation.

3. Formulate a procedure for performing an insertion into an input-restricted deque.

4. Construct a procedure for performing a deletion from an input-restricted deque. A parameter is required to denote from which end of the deque a deletion is to be made.

5. Repeat exercises 3 and 4 for an output-restricted deque.

10-7 SIMULATION

Simulation has traditionally been one of the most common applications of computers and therefore one with which students of computer science should be acquainted. Simulation involves the construction of models in the form of computer programs—models that are used in the study of some system or phenomenon. Queues often form an integral part of computer simulation models, and that is the reason for the consideration of simulation at this point in the book.

Any model is by nature an abstraction of reality. This abstraction serves two purposes. First, the abstracting process itself increases our understanding of the thing being abstracted, since it requires that we extract only the essential elements. Second, the model usually provides a more convenient experimental subject than the real system, since it is easier for us to control. Both of these purposes are well served by computer simulation models.

In this section we will consider the design of a simulation model of the loading of jobs into main memory by a hypothetical operating system. Through this example we will illustrate the types of problems faced by a designer of simulation models and the types of techniques that can lead to solutions. In fact, the requirements of our solution will force us to introduce a new organization for queues and, accordingly, to design new procedures for handling queues of this type.

We turn now to the particular problem at hand. As you recall from Chap. 2, any job intended for execution in a computer system must first be loaded into main memory by the operating system. This requires that the operating system allocate to the job an area of main memory that is large enough to hold it. For this example we will assume a very straightforward scheme for the allocation of memory; in practice, the schemes are somewhat more elaborate. At any time certain areas of memory are occupied by executing jobs; the unoccupied areas we will refer to as "holes" into which new jobs can be loaded. Clearly, when a job completes and leaves the system, it no longer requires the memory space that it occupied, and a new hole results. As each new job arrives, its size is noted by the operating system. If it will fit into the hole that is currently the largest, then the operating system will load it there. Since it is unlikely that the size of the job will match exactly the size of the hole, a smaller hole is usually created as a side effect of the loading process. The loading process is depicted in Fig. 10-18.

To help to understand this process you might note the similarity between the situation described and that of cars trying to park on an open street. You can view Fig. 10-18 as representing such a parking system, with "cars" taking the place of "jobs," and "parking spaces" taking the place of "holes" in memory. The driver trying to park the car is in a situation analogous to that of the operating system trying to load the job. A parking space sufficiently large to accommodate the car must be found in the same way that an area of memory must be found that is sufficiently large to accommodate the incoming job.

A great many simulation models are organized as systems of queues. Our model of the memory allocation process will be described in terms of three queues. One of these queues is a strict *first-in-first-out* queue as described in the previous section. The other two are examples of *priority queues*, in which the elements in the queue are ordered by some criterion. The element next to be removed is the one with the highest (or lowest) value in terms of this criterion. Arriving elements are placed in the queue in the position dictated by their criterion value on arrival. As you can imagine, priority queues require more sophisticated algorithms for insertion and deletion than those given in Sec. 10-6.

Let us consider, as an example, a queue whose elements consist of a pair of values. The first of these values is a memory address; the second is the size of the "hole" beginning at that address. Thus, the elements in this queue correspond to the holes in our memory. Where there is no hole, there is no entry in the queue. Our memory allocation strategy states explicitly that the operating system tries to fit each arriving job into the currently

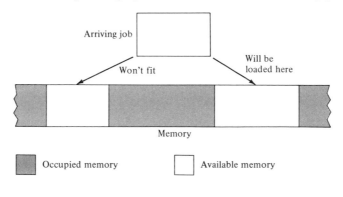

FIGURE 10-18
Loading in a hypothetical
computer memory.

largest hole. To save search time, then, it would be convenient if the queue of holes were ordered so that the largest hole is first. Thus, we will arrange the elements in this queue in decreasing order of hole size. Looking ahead to the exercises at the end of this section, the ordering of the elements in this manner permits the simulation of an interesting alternative strategy for memory allocation.

A second priority queue will be used to simulate the execution of the jobs so that they depart the system in the correct order. Because the execution times of the jobs can vary, they do not depart the system in the same order that they arrive. Suppose, for example, that a job requiring 10 minutes of execution time begins executing. After it has executed for 1 minute, a job requiring only 4 minutes of execution time begins executing as well. Recall that simultaneous execution of several jobs is possible in many computer systems through *multiprogramming*, in which the execution of the jobs is interleaved by the operating system. We will omit the details of how this is done since it is not essential to our discussion. Clearly, the second job will complete before the first job despite the fact that it began to execute 1 minute later. Our second priority queue will be used to ensure the proper completion sequence.

Elements in this second queue must have four components (or, to use the notation of Sec. 10-1, each record must have four fields). The first of these is a job identification number; the second is the address at which this job has been loaded in memory; the third is the amount of execution time remaining for this job; the fourth is the amount of memory occupied by the job. For reasons that will become clear, we will arrange the elements in this queue in increasing order of remaining execution time. For the sake of simplicity, we will assume that all execution times are in multiples of 1 minute.

The final queue needed for this model is a straightforward first-in, first-out queue to represent the arrival of jobs to the system. Elements in this queue consist of three components: a job identification number, a memory request, and a request for a certain amount of execution time. Again, to keep things as simple as possible, we will assume that a new job arrives every minute for 30 minutes.

Figure 10-19 gives a summary of the three queues we have described.

Any computer simulation is the study of the behavior of some system over some period of time. The passage of time is a key ingredient in any simulation. For this example, since we have deliberately kept the units of time very simple, this is quite easy to do. The passage of time is handled simply by updating the state of the system at regular 1-minute intervals, taking care to account for all changes taking place during that period. In our system, all changes occur at time units that are multiples of 1 minute, so no accuracy is lost. In systems where events occur irregularly, more elaborate mechanisms must be defined to model the passage of time.

We are now ready to begin to specify our model. We will adopt a simple

Memory queue

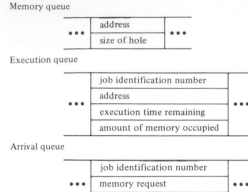

Execution queue

Arrival queue

FIGURE 10-19
The queues of the
simulation model.

naming convention to account for our multiple-component queue entries. Each of the components will, in fact, be a separate vector; this will enable us to take advantage of some of the algorithms developed in Sec. 10-6. The similarities in names will indicate the respective queues. Table 10-12 describes the vectors involved and their respective queue pointers. We will assume that all values are integers.

Our model clearly requires mechanisms for insertion and deletion for priority queues analogous to those given in Sec. 10-6. The deletion mechanism is simply that of algorithm QDELETE, since we will be deleting from the front of the queues in all cases. For example, when the remaining execution time of the element at the front of the execution queue becomes zero, the job corresponding to this element is removed from the system. Insertion is more complicated, however, as illustrated in Fig. 10-20. Here a new job has been loaded whose execution time dictates that it be inserted somewhere in the middle of the execution queue rather than at the rear, as in algorithm QINSERT. Algorithm QFILE accomplishes this task on a single vector by in-

TABLE 10-12   Queue Naming Conventions

Queue	Name of Vector	Queue Pointers	Component
Memory	MEM_Q_ADDR	MF1, MR1	Address of hole
queue	MEM_Q_SIZE	MF2, MR2	Size of hole
Execution	EXEC_Q_ID	EF1, ER1	Job identification number
queue	EXEC_Q_ADDR	EF2, ER2	Address of job (after loading)
	EXEC_Q_TIME	EF3, ER3	Remaining execution time
	EXEC_Q_SIZE	EF4, ER4	Memory occupied by job
Arrival	ARR_Q_ID	AF1, AR1	Job identification number
queue	ARR_Q_MEM	AF2, AR2	Memory request
	ARR_Q_TIME	AF3, AR3	Execution time request

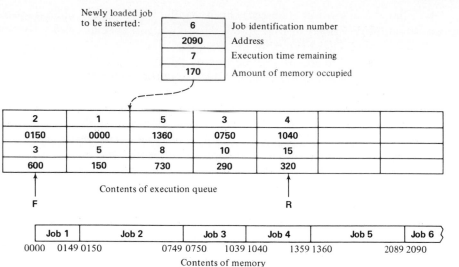

Newly loaded job to be inserted:

6	Job identification number
2090	Address
7	Execution time remaining
170	Amount of memory occupied

2	1	5	3	4		
0150	0000	1360	0750	1040		
3	5	8	10	15		
600	150	730	290	320		

Contents of execution queue

F   R

Job 1	Job 2	Job 3	Job 4	Job 5	Job 6

0000   0149 0150   0749 0750   1039 1040   1359 1360   2089 2090

Contents of memory

FIGURE 10-20
Insertion of an element into the execution queue.

serting the new element in its appropriate place and shifting the other elements, if necessary, to accommodate it.

Procedure QFILE (Q, F, R, X, P). Given a vector Q containing N elements and integer variables F and R, which denote the front and rear elements, respectively, of a priority queue, it is required to insert X in the appropriate place in the queue. P points to the place at which X is inserted. The queue is assumed to be in ascending order from front to rear. An empty queue is signaled by F and R both having a value of zero.

1. [Overflow condition?]
   If R ≥ N then Call QUEUE_OVERFLOW
2. [Increment rear pointer to accommodate insertion of new element]
   R ← R + 1
3. [Find appropriate position for new element]
   P ← F
   Repeat while X > Q[P] and P < R
   P ← P + 1
4. [Shift remaining elements one position]
   Repeat for L = R, R − 1, R − 2, . . ., P + 2, P + 1
   Q[L] ← Q[L − 1]
5. [Insert new element]
   Q[P] ← X
6. [Is front pointer properly set?]
   If F = 0 then F ← 1
7. [Finished]
   Return   ☐

That this algorithm is correct can be verified by tracing its behavior on the situation described in Fig. 10-20. There are two special cases. First, if the element belongs at the front of the queue (that is, X < Q[F]), then P will be set to F and all elements will be moved down one position.

Second, if the element belongs at the rear of the queue (that is, X > Q[R]), then the loop in step 3 will terminate with P = R, the loop in step 4 will be executed zero times (since R < P + 1), and the element will be inserted at the rear of the queue with no elements moved. The rest of the algorithm is identical to algorithm QINSERT given in Sec. 10-6. It is worth noting at this point that the use of a linked representation for the queue, as introduced in Sec. 10-9, makes these operations much more straightforward. We invite you to reconsider these algorithms after you have read Sec. 10-9.

Algorithm QFILE as given only accommodates queues with single-component entries. However, because the position of the inserted elements is returned through parameter P, the algorithm can easily be used for queues with multiple component keys, where one of these components is the ordering criterion. This is done as follows. First, algorithm QFILE is used to insert into the ordering criterion vector; then the returned value of P is used to place the remaining components in parallel positions in their respective vectors by calling the following procedure, QPLACE.

Procedure QPLACE (Q, F, R, X, P). Given parameters identical to those of procedure QFILE, this procedure inserts element X at position P, adjusting other elements accordingly.

1. [Overflow?]
   If R ⩾ N then Call QUEUE_OVERFLOW
2. [Adjust element positions accordingly]
   R ← R + 1
   Repeat for L = R, R − 1, R − 2, . . ., P + 2, P + 1
     Q[L] ← Q[L − 1]
3. [Insert new element at position P]
   Q[P] ← X
4. [Is front pointer properly set?]
   If F = 0 then F ← 1
5. [Finished]
   Return   □

To illustrate, if we wish to file the complete element shown in Fig. 10-20, we would do the following:

Call QFILE(EXEC_Q_TIME, EF3, ER3, 7, P)   (insert criterion component)
Call QPLACE(EXEC_Q_ID, EF1, ER1, 6, P)   (insert others as given by P)
Call QPLACE(EXEC_Q_ADDR, EF2, ER2, 2090, P)
Call QPLACE(EXEC_Q_SIZE, EF4, ER4, 170, P)

We are now ready to give the complete simulation model. We will assume a memory of 10,000 words, which is initially one big hole. This means that the memory queue has one entry with address 0000 (assume four-digit addresses) and size 10,000. The other two queues are empty, initially. We will assume that the simulation is to be run for 30 minutes, with a new job arriving each minute. We will also assume that the characteristics of each job are read from the input stream at the time of its arrival. In the simulation, we will count the number of completed jobs. We will use the time of arrival as the job identification number.

Algorithm MEMORY_SIM. This algorithm is a simulation model of memory management in a hypothetical computer system as described. Three queues are used; these are described in Table 10-12. All variables are assumed to be integers. These include COMPLETIONS (a count of jobs completed) and MINUTE (the current time), as well as other variables whose meanings are clear from their context. Procedures QINSERT, QDELETE, QFILE, and QPLACE are used for queue manipulations.

1. [Initialize counter, all queue pointers, and memory queue]
   COMPLETIONS ← 0
   MF1 ← MR1 ← EF1 ← ER1 ← AF1 ← AR1 ← 0
   MF2 ← MR2 ← EF2 ← ER2 ← AF2 ← AR2 ← 0
   EF3 ← ER3 ← AF3 ← AR3 ← 0
   EF4 ← ER4 ← 0
   Call QFILE(MEM_Q_SIZE, MF2, MR2, 10000, P)
   Call QPLACE(MEM_Q_ADDR, MF1, MR1, 0000, P)
2. [Engage timing loop]
   Repeat thru step 5 for MINUTE = 1, 2, 3, ..., 30
3. [Read information for next arrival and file in arrivals queue]
   Read(MEM_REQUEST, TIME_REQUEST)
   Call QINSERT(ARR_Q_ID, AF1, AR1, MINUTE)   (FIFO queue)
   Call QINSERT(ARR_Q_MEM, AF2, AR2, MEM_REQUEST)
   Call QINSERT(ARR_Q_TIME, AF3, AR3, TIME_REQUEST)
4. [Is enough memory available to satisfy first job in arrivals queue?]
   Call QDELETE(MEM_Q_SIZE, MF2, MR2, HOLE_SIZE)   (largest hole)
   Call QDELETE(MEM_Q_ADDR, MF1, MR1, ADDR)
   MEM_REQUEST ← ARR_Q_MEM[AF2]
   If HOLE_SIZE > MEM_REQUEST
   then (remove first job from arrivals queue)
       Call QDELETE(ARR_Q_ID, AF1, AR1, ID)
       Call QDELETE(ARR_Q_MEM, AF2, AR2, MEM)
       Call QDELETE(ARR_Q_TIME, AF3, AR3, TIME)
          (adjust hole accordingly)
       HOLE_SIZE ← HOLE_SIZE − MEM
       ADDR ← ADDR + MEM
          (file new hole in memory queue and job in execution queue)
       Call QFILE(MEM_Q_SIZE, MF2, MR2, HOLE_SIZE, P)
       Call QPLACE(MEM_Q_ADDR, MF1, MR1, ADDR, P)
       Call QFILE(EXEC_Q_TIME, EF3, ER3, TIME, P)

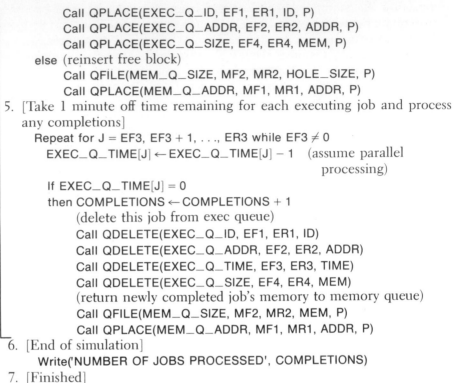

```
            Call QPLACE(EXEC_Q_ID, EF1, ER1, ID, P)
            Call QPLACE(EXEC_Q_ADDR, EF2, ER2, ADDR, P)
            Call QPLACE(EXEC_Q_SIZE, EF4, ER4, MEM, P)
        else (reinsert free block)
            Call QFILE(MEM_Q_SIZE, MF2, MR2, HOLE_SIZE, P)
            Call QPLACE(MEM_Q_ADDR, MF1, MR1, ADDR, P)
```

5. [Take 1 minute off time remaining for each executing job and process any completions]

```
    Repeat for J = EF3, EF3 + 1, . . ., ER3 while EF3 ≠ 0
        EXEC_Q_TIME[J] ← EXEC_Q_TIME[J] – 1   (assume parallel
                                                        processing)

        If EXEC_Q_TIME[J] = 0
        then COMPLETIONS ← COMPLETIONS + 1
            (delete this job from exec queue)
            Call QDELETE(EXEC_Q_ID, EF1, ER1, ID)
            Call QDELETE(EXEC_Q_ADDR, EF2, ER2, ADDR)
            Call QDELETE(EXEC_Q_TIME, EF3, ER3, TIME)
            Call QDELETE(EXEC_Q_SIZE, EF4, ER4, MEM)
            (return newly completed job's memory to memory queue)
            Call QFILE(MEM_Q_SIZE, MF2, MR2, MEM, P)
            Call QPLACE(MEM_Q_ADDR, MF1, MR1, ADDR, P)
```

6. [End of simulation]
   Write('NUMBER OF JOBS PROCESSED', COMPLETIONS)

7. [Finished]
   Exit   □

This model works in the following fashion. First, the count of completed jobs is initialized, as are all 18 queue pointers described in Table 10-12. Then, the memory queue is initialized to contain the entire memory as one hole with address 0000. Note the use of the two procedures QFILE and QPLACE, communicating through a common value of P, to insert the two components of the memory queue entry.

Step 2 is the timing loop for the simulation, which is to run for 30 minutes in steps of 1 minute each.

In step 3, a new job arrives and is inserted into the arrivals queue. The time of arrival provides a unique identification for the job. Step 4 embodies the memory management strategy. If there is a hole in memory sufficiently large to accommodate the first job in the arrivals queue, then that hole is allocated to the job, and its size is adjusted accordingly. The loaded job is then removed from the arrivals queue and inserted in the execution queue.

Step 5 simulates the execution of the jobs in the execution queue by subtracting 1 minute from the time remaining for each of them to account for their execution during the period since the last update. (Note that this assumes that the jobs execute in parallel. This may not be physically possible on many systems, but it is a useful simplification for our purposes.) Any completions are counted, and the memory occupied is once again made available.

When 30 minutes have elapsed, control passes to step 6, and the final result is printed.

This completes our discussion of this particular model. Many other questions remain concerning the use of computer simulation. One of the most important of these is the issue of model validation. For a model to have any value as an experimental tool, it must be demonstrated that it is an adequate representation of the real system, at least for the purpose of the particular experiment. Otherwise, one can have little confidence in any results from the model. Validity can be established in a number of ways. The most straightforward approach is to compare data obtained from the model with actual measurements taken from the real system. If close agreement can be attained, then the model can be used with confidence as a predictive tool. In many cases, however, for example when the system being modeled is only hypothetical, this is not possible and other methods must be employed. We will not go into these here. However it is done, the validation of a simulation model is an essential part of the simulation process.

In this section we have attempted to convey the flavor of computer simulation. It is a very broad subject, one that can be studied in considerable depth, and we have clearly only scratched the surface. Armed with this brief introduction, however, you ought to be prepared for a more complete treatment should the need arise. In this section we have also introduced the important concept of priority queues, along with procedures for manipulating them. Priority queues are very common in simulation applications, but have as well many important applications beyond the realm of simulation.

EXERCISES 10-7

1. The memory management technique implemented in the algorithm MEMORY_SIM makes no attempt to consolidate adjacent holes into one larger hole. This probably means that, over time, memory will be divided into a large number of very small holes. Memory can undoubtedly be allocated more effectively if such consolidation were implemented. Modify the algorithm MEMORY_SIM to perform this consolidation.

2. A strategy for memory management known as the *best-fit* rule tries to fit a job into the *smallest* memory hole that is large enough to hold the job. Construct a simulation model based on the best-fit rule, and design an experiment to measure its effectiveness as compared to the strategy implemented in the algorithm MEMORY_SIM.

3. The concept of "randomness" is the key to many simulation models, as it allows the experimenter to incorporate an element of uncertainty into

his/her models. In computer simulations, randomness is handled through the use of specially designed subprograms which return values that, when taken collectively, satisfy statistical properties of randomness. The design and use of such subprograms is an interesting aspect of simulation.

Let us assume the existence of a *random-number generator* RAND(a, b), where [a, b] is the desired range of random numbers. A call to RAND with a given pair of arguments, such as

$$X \leftarrow RAND(0, 5)$$

returns a number chosen at random from the range indicated. Use the RAND function to design into the algorithm MEMORY_SIM the concept of random memory and time requests by replacing the Read statement in step 3 by two calls to the function RAND. Assume that memory requests are randomly distributed in the range [0, 10000] words, and time requests in the range [1, 9] minutes.

4. A grocery store firm is considering the addition of a new service counter in one of its stores. Currently, the store has three checkouts, but customer volume has increased to the point where a new counter is warranted. To determine if the new counter should be a regular counter or an express counter (that is, eight items or less), a simulation of customer flow through the checkout area is required.

Our initial simulation is of a checkout area consisting of one express counter and three regular checkouts. All customers with eight or fewer items are assumed to proceed to an express counter. Customers with more than eight items go to the standard checkout with the shortest waiting line.

Customers enter the checkout area randomly, with the time of next arrival determined by adding to the present time a random number chosen from the range [0, 360] seconds. (0 is interpreted as a simultaneous arrival of two customers.) The number of items bought by each customer can also be approximated by selecting a random number in the range 1 to 40. The time taken for a customer to proceed through a checkout once the cashier begins "ringing up" his/her groceries can be calculated by using an average rate of 30 seconds per item (ringing plus wrapping time).

In setting up the simulation we should realize that prior to bringing a new customer into the checkout area, we must ensure that all customers who have had their groceries processed are removed from the waiting lines. Assume that no more than 10 customers are waiting in line at any one time for a regular checkout, and no more than 15 are waiting in line at any one time for an express checkout.

You are to formulate an algorithm which simulates the checkout service just described. The desired output should contain the number of customers going through each checkout per hour, the total number of customers handled per hour, the average waiting time at each checkout, the overall average waiting time in minutes, the number of items processed at each checkout per hour, and the total number of items processed per hour. The waiting time is the time a customer spends in the checkout area.

Output having the following format is desirable:

	1	2	3	Express	Total
Number of customers/hour	10	11	14	20	55
Average waiting time	2.80	3.01	2.96	0.22	1.94
Items processed/hour	192	261	210	65	728

Simulate as well the situation in which there are four standard checkouts, using the same method of generating arrival times and number of items purchased. Your algorithm should be designed so that the second simulation can be implemented with very few changes to the first algorithm.

5. A tool frequently used by transportation and city planners is the computer simulation of traffic systems. The systems modeled range from the traffic network of a nation, a city, or area of a city, right down to the traffic flow in one bridge or intersection. The models are used to pinpoint present or future bottlenecks and to suggest and test proposed changes or new systems.

A light-controlled intersection is one example of a traffic system for which the simulation model is relatively simple. Such a model would be used to evaluate intersection performance. The primary quantity measured would be the length of time motorists were stopped at the intersection. The performance of the intersection would be indicated by the average and maximum waiting times experienced by the motorists.

The specific model we will consider consists of an intersection of two two-lane streets, each lane being controlled by a three-color traffic light. The street and lane codes used throughout this section are shown in Fig 10-21.

To simplify the model, we make some assumptions about traffic flow and driver behavior. First, we assume that all traffic entering the intersection proceeds straight ahead; no right or left turns are allowed. Second, we assume that the car and driver response times are the same for all vehicles; that is, given that the path is clear, it takes the same length of time for each car to respond and enter the intersection. The possibilities of stalling cars and accidents are ignored. We assume that all drivers are extremely law abiding and thus stop for both red and amber lights.

At this point we acknowledge the reader's comment that no such intersection exists. That is true; however, for the sake of clarity of description, such a simplistic intersection will be our subject.

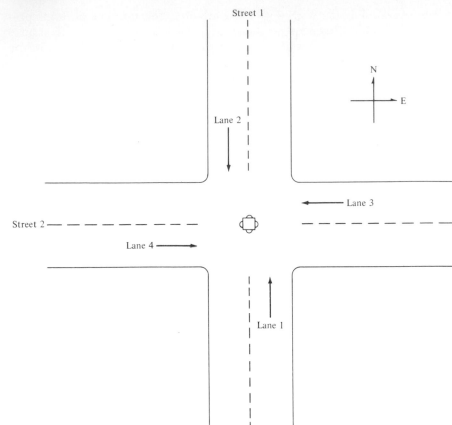

Street 1
Lane 2
N
E
Lane 3
Street 2
Lane 4
Lane 1

FIGURE 10-21
Traffic-light intersection.

The traffic lights are assumed to have the following characteristics:

1. The traffic lights for the two lanes on each street have identical signal timing and are viewed as a set
2. The light cycle times for both sets of lights are equal
3. There are no right- or left-turn arrows and no four-way walk signal
4. There is a short length of time, known as the *delay time*, between one set of lights turning red and the other set turning green
5. The red-light period on each set of lights is longer than the combined periods of the green light, amber light, and delay of the other set of lights

Construct a simulation model of this traffic intersection using queues to represent the oncoming traffic. State any other assumptions that you make.

10-8 LINKED STORAGE ALLOCATION

Thus far in this chapter we have allocated storage for linear lists in a sequential manner. We have relied on the one-dimensional property of a computer's main memory to represent the physical adjacency relationship of the elements in a linear list. Although this method of allocating storage is adequate for many applications, there are many other applications where the sequential-allocation method is inefficient and, therefore, unacceptable. The characteristics that are often found in the latter class of applications are:

1. Storage requirements are unpredictable. The precise amount of data storage in these applications is data-dependent; therefore the storage requirements cannot be easily determined before the associated program is executed.
2. The stored data must be manipulated extensively. These applications are characterized by frequent insertion and deletion activity on the data.

In such application areas, the linked-allocation method of storage can lead to both efficient use of computer storage and computer time. Therefore, the remainder of this chapter examines the concepts of linked allocation and applies them to linear lists. In this section we introduce the basics of linked allocation.

The previous sections have dealt with how the address of an element in a linear list can be obtained through direct computation. Since the data structures were linear and this property was preserved in the corresponding storage structures by using sequential allocation, it was unnecessary for an element to specify the location of the next element in the structure.

An alternative storage-allocation approach is to use pointers or links (see Sec. 10-2) to refer to elements of a linear list. The approach is to store the address of the successor of a particular element in that element. Using this approach, the elements of the linear list need not be physically adjacent in memory. This method of allocating storage is called *linked allocation*. Let us examine how to represent a linear list by this method of allocation.

An obvious way to represent a linear list by linked allocation is to expand the element (or node structure) so as to incorporate an extra field. The field is a pointer or a link which denotes the position of the next element in the list. Such a representation is called a *singly linked linear list* or a *one-way chain*. Figure 10-22a gives a box-and-arrow representation of a linked linear list, where the pointer variable FIRST contains the address or location of the first node in the list. Observe that each element (or node) has two parts. The first part contains the information contents of the element and

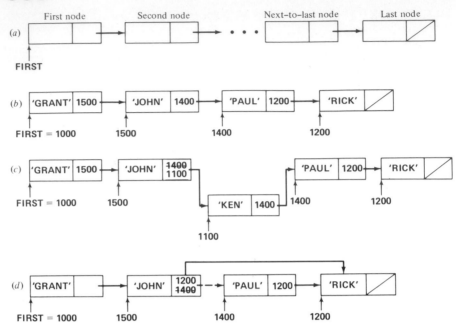

FIGURE 10-22
Linked storage representation of a linear list.

the second part contains the address of the next node in the list. Note that, since the last node in the list does not have a successor, its pointer field does not contain an actual address. A NULL or empty address (denoted by a dash in the diagram) is assigned to the pointer field in this case. Note that NULL is not an actual address but a special value which cannot be mistaken for a real address. The arrow emanating from the link field of a particular node terminates at its successor node. For example, Fig. 10-22b represents a four-node linear list whose elements are stored at memory locations 1000, 1500, 1400, and 1200, respectively. Note that the elements are not stored sequentially within memory. If a list contains no elements (that is, it is empty), this fact is denoted by assigning the value of NULL to FIRST.

Let us now compare sequentially allocated and linked allocated linear lists with respect to several commonly performed operations. In particular, consider the operations of insertion and deletion in a sequentially allocated list. Assume that we have an n-element list and that it is required to insert a new element between the second and third elements. In this case the last $n - 2$ elements of the list must be physically moved to make room for the new element. For large-sized lists which are subjected to many insertions, this insertion approach can be very costly. The same conclusion holds in the case of deletion, since all elements after the deleted element must be moved up so as to use the vacated space caused by the deletion.

An insert operation into a linked list is easy. An insertion involves the interchange of pointers. For example, Fig. 10-22c illustrates the insertion of the new element 'KEN' between the second and third elements of a list of four elements. The new element is assumed to be stored at position 1100. The link field of the second element is changed from 1400 to 1100. Also the link field of the new node is assigned a value of 1400. Similarly, a delete operation involves changing a single pointer. Figure 10-22d illustrates the deletion of the third element, 'PAUL', from the original list. The link value of the second node (1400) is changed to the address of the fourth node (1200). Clearly, both insert and delete operations are more efficiently performed on linked lists than on their sequentially allocated counterparts.

Both allocation methods can be compared with respect to other operations. For a search operation in a linked list, we must follow the links from the first node onward until the desired node is found. This operation is certainly inferior to the computed-address method of locating an element in a sequentially allocated list. If we want to split or join two linked linear lists, then these operations involve the changing of pointer fields without having to move any nodes. Such is not the case for sequentially allocated lists.

Clearly, pointers or links consume additional memory. The cost of this additional memory becomes less important as the information contents of a node require more memory. If the memory requirements for the information contents of each node are small, it may be possible to store the information portion in one part of a memory word and store a pointer in the remaining part. At any rate, a pointer usually takes one-half word of memory on most computers. The extra cost of storing pointers is not too expensive in many applications.

In the previous discussion the actual address of a node was used in the link field for illustration purposes. In practice, however, this address may be of no concern (and indeed unknown) to the programmer. Therefore, in the remainder of our discussion of linked structures, the arrow symbol is used exclusively to denote a successor node.

In this chapter we are interested in the use of pointers only to specify the relationship of logical adjacency among elements of a linear list. However, pointers can be used to specify more complex relationships, such as those found in tree and graph structures (recall Sec. 6-5.3). These relationships are sometimes very difficult to specify using sequential-allocation techniques. Such complex structures are easily represented using linked allocation by placing several pointers in each node. In such instances, it is possible for a node to belong to several different structures. Chapter 11 will examine the linked representation of tree structures in detail.

From our current discussion and some of the applications to follow, it will be evident that in certain instances, linked allocation is more efficient than sequential allocation. In other cases, however, the opposite is true. Yet, in other applications, both types of allocation methods are used.

An important consideration in both methods of allocation is the management of available storage required to perform certain operations, such as insertion and deletion. For linked allocation, a *pool* or list of *free* nodes called the *availability list* must be maintained. On an insert operation, a free node is removed from the availability list and placed in the designated linear list. Conversely, a deleted node from a linear list results in its return to the availability list, where it can be reused for subsequent insertions. This storage management scheme is an obvious advantage—at any particular time, the only space which is actually used is what is really required.

Storage of singly linked lists can be managed in a very straightforward manner. For more complex structures, however, this simplicity vanishes. Since in these structures a node can belong to many different structures, the deletion of a node from one structure does not imply that it can be returned to the availability list. This topic is beyond the scope of this book.

We are now ready to examine several operations which can be performed on linked linear lists. This topic is pursued in the next subsection.

10-9 LINKED LINEAR LISTS

The basic concepts of applying the linked method of storage allocation to represent linear lists were introduced in the previous section. In this section we examine the application of linked allocation to linear lists in greater detail. The first subsection concentrates on the formulation of several algorithms, such as insertion, deletion, traversal, and copying, which deal with singly linked linear lists. This subsection examines the simulation of linked allocation by arrays. The second subsection describes a circularly linked representation of a linear list. Such a representation offers certain computational advantages over singly linked linear lists.

Since the traversal of a singly linked list is performed in one direction, a deletion of an element from such a structure is inefficient. Also, several applications require the traversal of a linear list in both forward and reverse directions. For these reasons we introduce the doubly linked representation of a linear list and its associated operations in the third subsection.

10-9.1 OPERATIONS ON SINGLY LINKED LINEAR LISTS
This section examines in detail the singly linked storage representation of a linear list. Several algorithms based on this representation are also included. Requests and releases of nodes from and to the availability list of storage are also described.

Before proceeding to formulate several algorithms for operations associated with linked linear lists, it is instructive to classify these operations. If the processing of data in a particular application is to be performed by a computer, we must first adequately represent these data in the computer's memory. The difficulty of this task depends to a large extent on the particular programming languages available. Certain programming languages were specifically designed for manipulating linked lists but others were not. Between these two extremes are languages in which common procedures or functions on linked lists have been written as subprograms in a simple "host" language. We do not, however, pursue any of these programming details in this book.

There are three broad classes of operations associated with linked lists. The first class contains operations which are independent of the data values in the list nodes. Operations in this class include creation, insertion, and deletion. These operations are usually provided by those programming languages which support list-processing activities.

A second class of related operations concerns the conversion of raw data from a human-readable form to a corresponding machine form. This class of operations also contains the inverse operation of converting a computer-stored structure into a suitable human-readable form. These are clearly data-dependent operations which depend on the interpretation associated with the structures. Most list-processing languages contain a set of standard basic routines. Any additional routines, however, must be programmed.

The third and final class of operations relate to the data-manipulation requirements which are application-dependent. All of these operations must be programmed. A programmer who has at his/her disposal all three classes of routines will find the task of programming an algorithm much easier.

Throughout this subsection we assume that an element or node of a list contains two fields: an information field (INFO) and a pointer field (LINK). NODE represents the name of an element in the structure. The following box-and-arrow notation illustrates this node structure:

NODE

INFO	LINK

↑
P

where P denotes the address of the node. The selection of a particular field within this node structure is an important matter. Our algorithmic notation must be extended so that this type of selection or referencing can be specified. We let INFO(P) denote the information field of the node with address P. Similarly, the value of the link field is LINK(P). This value represents the address or position of the next node in the linked list. We further assume that the availability list of nodes takes the form of a linked stack as shown in the before part of Fig. 10-23, where the pointer variable AVAIL contains the address of the top node of the availability stack.

We now formulate a request for a node from the availability stack. Let NEW be the pointer variable which is to contain the address of the next available node. Since for practical reasons a stack contains only a finite number of nodes, it must be checked for an underflow condition. An underflow occurs if AVAIL has a value of NULL. If the stack contains an available node, then the new topmost element of the stack after fulfilling the storage request is given by LINK(AVAIL). The fields of the node whose address is given by

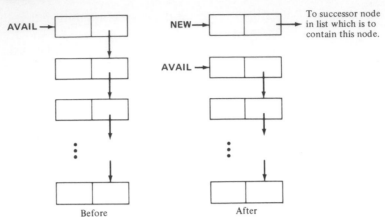

FIGURE 10-23
Availability stack before
and after obtaining a
node from it.

NEW can be filled in with the appropriate values which will insert this node into an existing linked list. The state of the availability stack before and after a free node has been obtained is shown in Fig. 10-23.

Sometimes throughout the remainder of this book we will not be concerned with the availability stack per se. It will be assumed that an available area of storage exists and that we can request a node from this area. In our algorithmic notation, the special assignment statement

$$P \Leftarrow NODE$$

creates a new node consisting of two fields (INFO and LINK), as previously described, with the location of the first of these fields being copied into the pointer variable P. Note the special operator $\Leftarrow$ in the previous statement. At creation time, the fields INFO(P) and LINK(P) have undefined values.

A discarded node is returned to the availability stack in a converse manner. Assuming that FREE denotes the address of the discarded node, then the link field of this node is set to the current value of AVAIL and the value of FREE then becomes the new value of AVAIL. A pictorial representation of this return process is given in Fig. 10-24.

If we are not particularly concerned with the maintenance of the availability list of nodes, then we return it to available storage in a general way. In particular, we specify such a release of storage in our algorithmic notation by the statement

Restore node P to the availability area

After taking this action, the node P is assumed to be inaccessible and P, undefined.

Let us now consider the operation of inserting a node into a linked linear list. Several steps are required in this operation. First, the values for the fields of the new node must be acquired either from an input operation or through computation. Second, we must obtain a node from available stor-

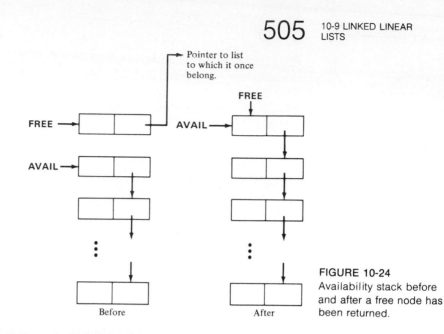

FIGURE 10-24
Availability stack before and after a free node has been returned.

age. Finally, the values of the fields obtained in the first step are copied into the appropriate field positions of the new node, which is then placed in the designated linked list. The linking of a new node to its successor node in the existing linked list is realized by setting the pointer field of the former to a value that gives the address of the latter. We can now proceed to a detailed algorithm.

Function INS_FRONT(FIRST, X) (version 1). Given a linked linear list whose node structure contains INFO and LINK fields, as previously described, and the information contents X of a new node, this function inserts a node at the front of the list. FIRST is a pointer variable which denotes the first node in the list. AVAIL is a pointer to the top of the availability stack. NEW is a temporary pointer variable which denotes the address of the new node. The availability stack and the node structure (NODE) are assumed to be global to this function.

1. [Underflow condition?]
   If AVAIL = NULL then Call STACK_UNDERFLOW
2. [Obtain address of next free node]
   NEW ← AVAIL
3. [Remove free node from availability stack]
   AVAIL ← LINK(AVAIL)
4. [Initialize information and link fields]
   INFO(NEW) ← X
   LINK(NEW) ← FIRST
5. [Return address of new node]
   Return(NEW)   □

This algorithm can be invoked repeatedly as a function resulting in the construction of a linked linear list. Initially, the construction process begins with an empty list. A new node is inserted by each invocation of function INS_FRONT. The following sequence of assignment statements constructs a linked list of four nodes.

```
LIST ← NULL
LIST ← INS_FRONT(LIST, 'RICK')
LIST ← INS_FRONT(LIST, 'PAUL')
LIST ← INS_FRONT(LIST, 'GRANT')
LIST ← INS_FRONT(LIST, 'BOB')
```

The behavior of function INS_FRONT is exhibited in Fig. 10-25. Note that insertions are performed in a stack-like manner.

If we are not concerned with the details of managing the availability stack, we can state the previous algorithm equivalently as follows.

Function INS_FRONT(FIRST, X) (version 2). This function is the same as that given earlier except that in this case the details of the availability stack are ignored.

1. [Create a new node]
   NEW ⇐ NODE
2. [Initialize information and link fields]
   INFO(NEW) ← X
   LINK(NEW) ← FIRST
3. [Return address of new node]
   Return(NEW)   □

This second version of the function is more representative of what a programmer encounters in programming this function in a programming language that supports list-processing activities.

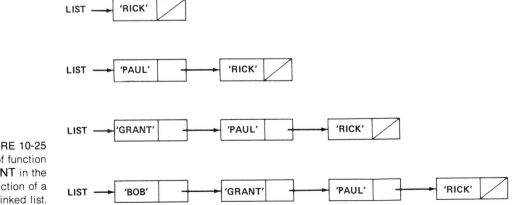

FIGURE 10-25
Behavior of function
INS_FRONT in the
construction of a
four-element linked list.

Rather than inserting a node at the front of an existing list, we can also insert a node at the end of the list. This second approach is more difficult to formulate than function INS_FRONT. We must deal with two situations in inserting a new node at the end of a list. The first case involves an insertion into an empty list. In such an instance the new node merely becomes the front node of the list. In the second case, however, the existing list is not empty. To perform the desired insertion, the end node of the original list must be found. This is accomplished by chaining through the original list until a node with a LINK field of NULL is found. At this point the link field of this node is assigned a value which points to the new node being inserted. The chaining operation and node insertion is accomplished by the following algorithm segment:

```
SAVE ← FIRST
Repeat while LINK(SAVE) ≠ NULL
    SAVE ← LINK(SAVE)
LINK(SAVE) ← NEW
```

We can incorporate this algorithm segment in the following function.

Function INS_END(FIRST, X). Given a linked linear list with a node structure, as previously described, this function inserts a new node (with information contents given by X) at the end of the list whose first node is given by the pointer variable FIRST. NEW and SAVE are temporary pointer variables. AVAIL denotes the top node in the availability stack.

1. [Underflow condition?]
   If AVAIL = NULL then Call STACK_UNDERFLOW
2. [Obtain location of next free node]
   NEW ← AVAIL
3. [Remove node from availability stack]
   AVAIL ← LINK(AVAIL)
4. [Initialize contents of new node]
   INFO(NEW) ← X
   LINK(NEW) ← NULL
5. [Is the original list empty?]
   If FIRST = NULL then Return(NEW)
6. [Initialize search for last node in list]
   SAVE ← FIRST
7. [Search for last node in list]
   Repeat while LINK(SAVE) ≠ NULL
       SAVE ← LINK(SAVE)
8. [Set link field of last node to point to new node]
   LINK(SAVE) ← NEW
9. [Return first node pointer]
   Return(FIRST)   □

Note that in the case of an empty list (step 5), the address of the new node is returned while in the second case the first node in the list remains the same as that before insertion and, consequently, the value of FIRST is returned.

This function can be invoked repeatedly to construct a linked list of four nodes as follows:

```
LIST ← NULL
LIST ← INS_END(LIST, 'BOB')
LIST ← INS_END(LIST, 'GRANT')
LIST ← INS_END(LIST, 'PAUL')
LIST ← INS_END(LIST, 'RICK')
```

The behavior of function INS_END is portrayed in Fig. 10-26.

Observe that the performance of function INS_END degenerates progressively as the number of nodes in a linear list becomes larger. In this situation the entire list must be traversed to perform the indicated insertion. We can avoid such a long search for the end element of a list by simply keeping the address of the last node which was inserted in the list. The next insertion can then be performed without traversing the entire list to find its last element. Rather, the empty link field of the previously inserted node need only be changed to the address of the new node being inserted. A revised function for this modified insertion method follows.

Function INS_LAST(FIRST, LAST, X). This function is the same as function INS_END except for keeping track of the address of the last node inserted. LAST is a pointer variable which contains this address.

1. [Underflow condition?]
   If AVAIL = NULL then Call STACK_UNDERFLOW
2. [Obtain location of next free node]
   NEW ← AVAIL

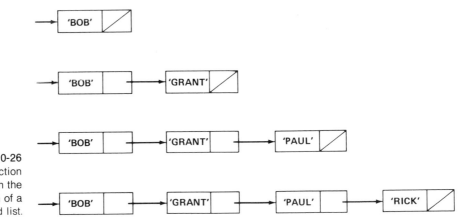

FIGURE 10-26
Behavior of function
INS_END in the
construction of a
four-element linked list.

3. [Remove node from availability stack]
  AVAIL ← LINK(AVAIL)
4. [Initialize contents of new node]
  INFO(NEW) ← X
  LINK(NEW) ← NULL
5. [Is list empty?]
  If FIRST = NULL
  then LAST ← NEW
      Return(NEW)
6. [Insert node at end of nonempty list]
  LINK(LAST) ← NEW
  LAST ← NEW
  Return(FIRST)   □

Again, this function can be invoked repeatedly to construct a linked list of four elements as follows:

LIST ← LAST ← NULL
LIST ← INS_LAST(LIST, LAST, 'BOB')
LIST ← INS_LAST(LIST, LAST, 'GRANT')
LIST ← INS_LAST(LIST, LAST, 'PAUL')
LIST ← INS_LAST(LIST, LAST, 'RICK')

The behavior of function INS_LAST is exhibited in Fig. 10-27.

Certain applications require that a linear list be maintained in order. Such an ordering is in increasing or decreasing order on the information field (INFO). The maintaining of an ordered list frequently results in more efficient processing.

Let us consider the details of such an algorithm. There are essentially three cases to handle. The first case involves the insertion of an element into an empty list. Clearly, no search is required in this case. In the second case

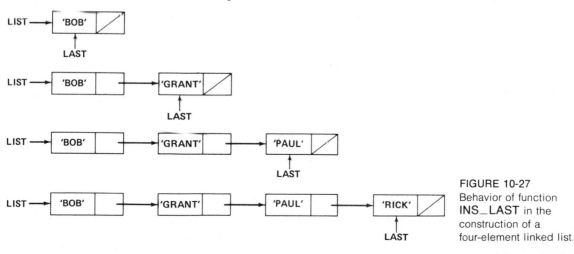

FIGURE 10-27
Behavior of function INS_LAST in the construction of a four-element linked list.

the new element is to precede the first element in the existing list. In such a case the new element becomes the new front element of the updated list and the link field of the new node is assigned the address of the front node (FIRST) before the insertion. In the third and final case a search is carried out to determine where the new node should be inserted. An algorithm segment to perform this search and insertion consists of the following statements:

```
SAVE ← FIRST
Repeat while LINK(SAVE) ≠ NULL and INFO(LINK(SAVE)) ≤ INFO(NEW)
    SAVE ← LINK(SAVE)
LINK(NEW) ← LINK(SAVE)
LINK(SAVE) ← NEW
```

where FIRST contains the address of the front node of the list and NEW is the address of the node being inserted. The insertion process for this third case is shown in Fig. 10-28. We can now proceed to the formulation of a detailed algorithm.

Function INS_ORDER(FIRST, X). Given a linked linear list with a front node of FIRST as previously described, this function inserts a node (with information contents X) into the existing linked list so that the ordering of the nodes in increasing order of their INFO field is maintained. AVAIL denotes the top element in the availability stack. NEW and SAVE are temporary pointer variables.

1. [Underflow condition?]
    If AVAIL = NULL then Call STACK_UNDERFLOW
2. [Obtain location of next free node]
    NEW ← AVAIL
3. [Remove node from availability stack]
    AVAIL ← LINK(AVAIL)
4. [Copy information contents into new node]
    INFO(NEW) ← X
5. [Empty list?]
    If FIRST = NULL
    THEN LINK(NEW) ← NULL
        Return(NEW)

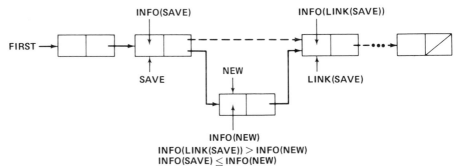

FIGURE 10-28
Insertion in an ordered
linked list.

6. [If new node precedes first node, then insert at front of list]
    If INFO(FIRST) ≥ INFO(NEW)
    then LINK(NEW) ← FIRST
        Return(NEW)
7. [Initialize temporary search pointer]
    SAVE ← FIRST
8. [Search for the predecessor of new node]
    Repeat while LINK(SAVE) ≠ NULL and INFO(LINK(SAVE)) ≤ INFO(NEW)
        SAVE ← LINK(SAVE)
9. [Set link fields of new node and its predecessor]
    LINK(NEW) ← LINK(SAVE)
    LINK(SAVE) ← NEW
10. [Finished]
    Return(FIRST)   □

Steps 1 through 4 require no further comment. The fifth step represents case 1 in our previous discussion. Step 6 represents case 2, where the new element becomes the front element of the updated list. Steps 7, 8, and 9 correspond to the third case. SAVE is first initialized to point to the front element of the list. This temporary variable then points to successive nodes until the value of INFO(LINK(SAVE)) is greater than the information value of the new node, or an attempt is made to "walk off" the end of the list (that is, LINK(SAVE) = NULL). In either case the link fields of the new node and its predecessor are set to their appropriate values.

The following statements construct the linked list of Fig. 10-29:

```
LIST ← NULL
LIST ← INS_ORDER(LIST, 'GRANT')
LIST ← INS_ORDER(LIST, 'BOB')
LIST ← INS_ORDER(LIST, 'RICK')
LIST ← INS_ORDER(LIST, 'PAUL')
```

We have examined in detail the insert operation. An equally important operation involves the deletion of a specified node from a given linked list. There are several ways of specifying which node should be deleted. For example, a node to be deleted can be specified by giving its INFO value. Another approach is to specify the address of the node to be deleted. We formulate a deletion algorithm based on the latter approach.

Procedure DELETE(FIRST, X). Given a linked list (with front node FIRST) whose node structure was previously described and a pointer variable X which denotes the address of the node to be deleted, this procedure performs the indicated deletion. PRED is a temporary pointer variable.
1. [Empty list?]
    If FIRST = NULL
    then Write('NODE NOT FOUND')
        Return

2. [Delete first node?]
    If X = FIRST
    then FIRST ← LINK(FIRST)
        (return node to availability stack)
        LINK(X) ← AVAIL
        AVAIL ← X
        Return
3. [Initialize search for predecessor of X]
    PRED ← FIRST
4. [Perform indicated search for X]
    Repeat while LINK(PRED) ≠ NULL and LINK(PRED) ≠ X
        PRED ← LINK(PRED)
5. [Delete indicated node, if found]
    If LINK(PRED) ≠ NULL
    then LINK(PRED) ← LINK(X)
        (return node to availability stack)
        LINK(X) ← AVAIL
        AVAIL ← X
    else (node not found)
        Write('NODE NOT FOUND')
    Return   □

The first step of the procedure checks for an empty list condition. If this condition is true, then we signal that the indicated node was not found. Step 2 determines whether or not the node to be deleted is the front node of the list. If X denotes the front node, then the second node in the list becomes its new front node. Observe that if the list only contains X, then the updated list becomes empty. The deleted node is also returned to the availability stack in this step.

If X is not the first node of the list, then a search for its immediate predecessor (PRED) is launched. Steps 3 and 4 represent this effort. Step 5 determines whether or not the indicated node was actually found. If the search

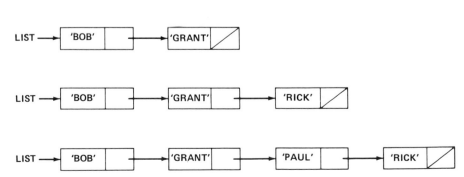

FIGURE 10-29
Behavior of function
INS_ORDER in the
construction of an ordered
linked linear list.

is successful, then the link field of node X is copied into the link field of PRED. The deleted node is then returned to the availability stack. If the search fails then this fact is reported.

Another familiar operation which is performed on a linked list is making a copy of it.

Function COPY(FIRST). Given a linked linear list (with front element FIRST) whose node structure is the same as previously described, this function makes a duplicate copy of it. The address of the front node of the duplicate copy is contained in the pointer variable BEGIN. NEW, SAVE, and PRED are temporary pointer variables.

1. [Empty list?]
      If FIRST = NULL
      then Return(NULL)
2. [Copy first node]
      If AVAIL = NULL
      then Call STACK_UNDERFLOW
      else NEW ← AVAIL
            AVAIL ← LINK(AVAIL)
            INFO(NEW) ← INFO(FIRST)
            BEGIN ← NEW
3. [Initialize traversal]
      SAVE ← FIRST
4. [Perform traversal of remainder of list]
      Repeat thru step 6 while LINK(SAVE) ≠ NULL
5. [Update predecessor and save pointers]
      PRED ← NEW
      SAVE ← LINK(SAVE)
6. [Copy node]
      If AVAIL = NULL
      then Call STACK_UNDERFLOW
      else NEW ← AVAIL
            AVAIL ← LINK(AVAIL)
            INFO(NEW) ← INFO(SAVE)
            LINK(PRED) ← NEW
7. [Set link field of last node]
      LINK(NEW) ← NULL
      Return(BEGIN)   □

The first step of the function checks for an empty list. In such a case a NULL pointer is returned. Step 2 copies the first node of the original list. The remaining steps traverse the rest of the original list. Note that as each new node is created, its address must be copied into the link field of its predecessor node.

Let us now consider the task of implementing or simulating linked linear lists with arrays. As mentioned earlier, several programming languages (for example, FORTRAN) do not permit programmer-defined data structures, since pointer-type variables do not exist in such languages. The concepts presented here are directly applicable to programming languages without pointer data types.

In an array implementation of a linked list, an element or a node consists of a sequence of fields, each of which is an integer, a real number, a string, and so on, except for one field (usually the last field), which represents a pointer to the next node in the linear list. This pointer takes the form of an index or subscript which denotes or references the next node (that is, array element) in the linked linear list. It is easy to understand (and recall Sec. 10-3) how a field name and an index permit the selection or referencing of a field from a particular node.

Consider as an example a linked linear list whose node structure contains INFO and LINK fields. Let us assume that there are at most 100 nodes in such a linear list structure. We can represent these four nodes of the sample linked list of Fig. 10-22b by a pair of vectors INFO and LINK as follows:

$$
\begin{array}{ll}
\text{INFO}[3] = \text{'GRANT'} & \text{LINK}[3] = 7 \\
\text{INFO}[7] = \text{'JOHN'} & \text{LINK}[7] = 10 \\
\text{FIRST} = 15 \rightarrow \text{INFO}[15] = \text{'PAUL'} & \text{LINK}[15] = 3 \\
\text{INFO}[10] = \text{'RICK'} & \text{LINK}[10] = 0
\end{array}
$$

where the first node of the list is denoted by assigning an index value of 15 to the integer variable FIRST. The value of LINK[15] gives the index value of the second name in the list. Knowing that LINK[15] has a value of 3 means that the INFO field of the second node can be referenced by writing INFO[LINK[15]]. The index of the third node in the list is denoted by LINK[3], which contains a value of 7. Once the index of the first node is known, it is an easy matter to follow the link pointers and access all the nodes in a linear list. Note that the last or fourth node (that is, INFO[10], LINK[10]) in the previous example is the last node in the list, consequently its link value is zero. Observe that when arrays are used to represent a linear list, the NULL value becomes the number zero.

As mentioned earlier, the fields within a particular node of a programmer-defined data structure are assumed to be sequentially allocated within that node. However, this is not the case in the previous representation of linear lists by arrays. That is, INFO[3] and LINK[3] are not physically adjacent in memory. This is because of the way that vectors are sequentially allocated in memory.

An alternative representation of a linked list is sometimes possible by using a two-dimensional array. This representation is only possible, however, if all fields in a node are of type integer. Since the LINK field must be an integer, all other entries in the array must be of the same type. If the node structure contains n information fields, then a two-dimensional array with $n + 1$ columns can be used to represent the list. In such a representation each of the n information fields occupies a column in the array. The $(n + 1)$st column can be used to represent the link field. This two-dimensional representation does make the fields of a node adjacent in memory. Note that the two-dimensional array representation offers no programming advantage over the vector approach. Furthermore, the running time of a program using the former approach is increased when compared with the vector approach because of double subscripting.

Individual fields within a certain node can be easily selected. Using the vector-representation approach, we can select any field of a node given the index or subscript P to that node. For example, INFO[P] and LINK[P] in our previous example denote the information and link values, respectively.

For a vector representation of a linear list, we are now in a position to consider the reformulation of the algorithms introduced earlier in the subsection. Recall that each algorithm obtained a free node from the area of available storage. With a vector representation of a linked linear list, however, the available storage area must be maintained by the programmer. The form of this area is usually a linked stack. When a free node is requested, the top node of the availability stack is deleted from the stack and used as the new node. The following algorithm constructs a 100-node availability stack.

Procedure AVAIL_AREA. This procedure constructs an availability stack of 100 nodes. The node structure consists of an information field and a link field. The vectors INFO and LINK are used to represent these fields. I is a temporary integer variable and AVAIL is the index which denotes the top of the stack. All variables except I are assumed to be global to the procedure.

1. [Set the link field of each node to its successor]
   Repeat for I = 1, 2, . . ., 99
     LINK[I] ← I + 1
   LINK[100] ← 0
2. [Set top of stack index]
   AVAIL ← 1
3. [Finished]
   Return   □

This parameterless procedure creates a linked stack with the linked fields ordered in the sense LINK[I] > I. This ordering, however, changes as nodes are removed from and returned to the availability stack. The index AVAIL, which is initially set to 1, gives the subscript corresponding to the top node in the stack.

We can now obtain easily a node from the availability list. We assume that the index of the next available node is to be stored in the integer variable NEW. Since, for practical reasons, the availability list contains only a finite number of nodes (100 in this case), an underflow test must be made. The algorithm segment to obtain a free node from the availability stack is

```
If AVAIL = 0 then Call STACK_UNDERFLOW
NEW ← AVAIL
AVAIL ← LINK[AVAIL]
```

where the procedure STACK_UNDERFLOW denotes the routine which handles an underflow condition. The fields of the node which correspond to NEW can now be filled in.

A similar algorithm segment for the return of a freed node to the availability stack can now be written. Assuming that the variable FREE denotes the index of the discarded node, the following algorithm segment returns this node to the availability area:

```
LINK[FREE] ← AVAIL
AVAIL ← FREE
```

The previous algorithm segments can be incorporated in a function which corresponds to the function INS_FRONT given earlier. In this revised procedure the vectors INFO, LINK, and AVAIL are assumed to be global. P is an integer variable which denotes the index of the requested node.

Function INS_FRONT(FIRST, X) (vector version). This function is a reformulation of the corresponding function given earlier.
1. [Check for stack underflow]
   ```
   If AVAIL = 0 then Call STACK_UNDERFLOW
   ```
2. [Obtain next free node from availability list]
   ```
   P ← AVAIL
   AVAIL ← LINK[AVAIL]
   ```
3. [Initialize information and link fields of new node]
   ```
   INFO[P] ← X
   LINK[P] ← FIRST
   ```
4. [Finished]
   ```
   Return(P)
   ```                                                                    ☐

Assuming that LIST denotes the front node of the linked list, the following algorithm segment constructs the example list of Fig. 10-22*b*:

```
LIST ← 0
LIST ← INS_FRONT(LIST, 'RICK')
LIST ← INS_FRONT(LIST, 'PAUL')
LIST ← INS_FRONT(LIST, 'JOHN')
LIST ← INS_FRONT(LIST, 'GRANT')
```

Note the remarkable similarity between the vector approach to representing linked lists and the programmer-defined approach introduced earlier. In an analogous manner we could reformulate the other algorithms introduced earlier. These reformulations are left as exercises.

On reexamining some of the previous algorithms, such as algorithms DELETE and COPY, we notice that the special case of an empty list must be checked for separately. We can modify the structure of a linked linear list to make this special case disappear. The next subsection explores this possibility.

EXERCISES 10-9.1

1. Given a singly linked linear list whose typical node structure consists of an INFO and LINK field, construct an algorithm which counts the number of nodes in the list.

2. Formulate an algorithm that changes the INFO field of the kth node in a singly linked list to a value given by X.

3. Construct an algorithm which performs an insertion to the immediate left of the kth node in a linked list.

4. Repeat exercise 3 for an insertion to the immediate right of the kth node.

5. Given two linked lists whose front nodes are denoted by the pointers FIRST and SECOND, respectively, formulate an algorithm that will concatenate the two lists. The front node address of the new list is to be stored in THIRD.

6. Obtain an algorithm which will deconcatenate (or split) a given linked list into two separate linked lists. The first node of the original linked list is denoted by the pointer variable FIRST. SPLIT denotes the address of the node which is to become the first node of the second linked list.

7. Assume that you are given an ordered singly linked list whose node structure contains a KEY and LINK field. The list is ordered on the KEY field in increasing order. It is desired to delete a number of consecutive nodes whose KEY values are greater than or equal to KMIN and less than or equal to KMAX. For example, an ordered seven-element linked list containing

'ANDY', 'BOB', 'GRANT', 'KEN', 'PAUL', 'RICK', and 'SAM'

with KMIN = 'BOB' and KMAX = 'KEN', would result in an updated list containing

'ANDY' 'PAUL', 'RICK', and 'SAM'

Assuming that the first node is denoted by the pointer variable FIRST, construct an algorithm which accomplishes this task.

10-9.2 CIRCULARLY LINKED LINEAR LISTS
The previous discussion dealt with linked linear lists in which the link of the last node is NULL. A slight modification of this representation can yield an

improvement in processing. The approach is to replace the NULL pointer in the last node of a list by the address of its first node. Such a modified list is called a *circularly linked linear list*, or simply *circular list*. Figure 10-30 demonstrates the circular structure of this type of list.

Circular linked lists have several advantages over their singly linked list counterparts. First, each node in a circular list is accessible from any given node. That is, from a given node, we can traverse the entire list. In a singly linked list we can only traverse all of it by starting at its first node. An important example of this restriction in a singly linked list is in the deletion algorithm given in the previous subsection. In that algorithm, the address of the first node of the list was required so that the predecessor of the node to be deleted could be found. This predecessor node was obtained by chaining through the list from its first node. Clearly, the address of the first node is not needed in a circular list, since a search for the predecessor of a given node can be initiated from that node. There are certain operations, such as concatenation and splitting, which become more efficient with a circular list structure.

There is, however, a major disadvantage in using a circular list. Care must be exercised to avoid an infinite loop! Such a loop can occur if special care is not taken to detect the end of the list. One way of avoiding this loop conditon is to place a special node which is permanently associated with the existence of a circular list. We call this special node the *list head* of the list. Another important advantage of using a list head is that a list can never be empty. Recall that several algorithms from the previous subsection require testing for an empty list condition. With a circular list representation, the need for testing this condition disappears. A representation of a circular list is given in Fig. 10-31a, where HEAD denotes the head of the list. An empty list is represented by having LINK(HEAD) = HEAD, and this possibility is shown in Fig. 10-31b. Observe that the information field in the list head node is not used. This fact is indicated by shading that field.

An algorithm for inserting a node at the front of a circular list is the following:

```
NEW ⇐ NODE
INFO(NEW) ← X
LINK(NEW) ← LINK(HEAD)
LINK(HEAD) ← NEW
```

Simplification of the previously encountered algorithms is left to the exercises.

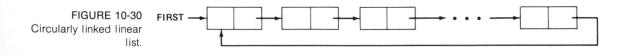

FIGURE 10-30
Circularly linked linear
list.

FIRST →

EXERCISES 10-9.2

1. Construct an algorithm which counts the number of nodes in a circular list with a list head.

2. Formulate an algorithm which inserts a node at the end of a circular list, that is, between the node which points to the list head and the list head.

3. Construct an algorithm which deletes from a circular list a node whose information content is given by the variable X.

4. Obtain an algorithm for the concatenation of two circular lists with list heads (see exercise 5 in Exercises 10-9.1).

5. Construct an algorithm to split a circular list into two circular lists (see exercise 6 in Exercises 10-9.1). Use list heads.

6. Obtain insertion and deletion algorithms for a circular linked list.

10-9.3 DOUBLY LINKED LINEAR LISTS

So far we have traversed linked linear lists in a left-to-right manner. In many applications it is required to traverse a list in both directions. This two-way motion can be realized by maintaining two link fields in each node instead of one. These links are used to denote the address of the predecessor and successor of a given node. The predecessor link is called the *left link* and the successor link is known as the *right link*. A list whose node structure contains two link fields will be called a *doubly linked linear list*, or *two-way chain*. Such a list structure is exhibited in Fig. 10-32, where LEFT and RIGHT are pointer variables that denote the leftmost and rightmost nodes in the list, respectively. Observe that the left link of the leftmost node is NULL. The same is true of the right link of the rightmost node. Finally, the left and right link fields of a field are called LPTR and RPTR, respectively.

Consider the problem of inserting a node into a doubly linked linear list to the right of a specified node whose address is given by the variable M. Several cases can occur. First, the list might be empty. This is indicated by M = NULL. In such a situation LEFT and RIGHT are both NULL. An insertion in this situation involves setting LEFT and RIGHT to the address of the new node. Also, the left and right links of this node are set to NULL.

A second possibility is an insertion within the list; that is, there will be a predecessor and successor to the given node after insertion. Such a case is shown in Fig. 10-33, where NEW denotes the node being inserted.

A third possibility involves the insertion to the right of the rightmost node in the list, thereby requiring the pointer RIGHT to be changed. The state of a list before and after insertion is given in Fig. 10-34.

The following algorithm inserts a node to the immediate right of a specified node.

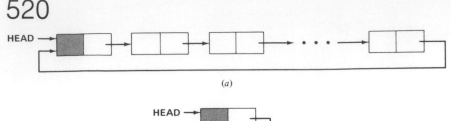

(a)

FIGURE 10-31
Circular linked linear
list with a list head.

(b)

Algorithm DOUBLE_INS(version 1). Given a doubly linked linear list (with node structure LPTR, INFO, and RPTR as just described) whose associated leftmost and rightmost pointers are LEFT and RIGHT, respectively, this algorithm inserts a new node with address NEW to the immediate right of a specified node with address M. The node structure is as previously described. X contains the information contents of the new node.

1. [Obtain new node from availability area]
   NEW ⇐ NODE
2. [Set information contents of new node]
   INFO(NEW) ← X
3. [Insertion in an empty list?]
   If RIGHT = NULL
   then LEFT ← RIGHT ← NEW
       LPTR(NEW) ← RPTR(NEW) ← NULL
       Exit
4. [Rightmost insertion?]
   If M = RIGHT
   then RPTR(NEW) ← NULL
       RIGHT ← NEW
       LPTR(NEW) ← M
       RPTR(M) ← NEW
       Exit
5. [Insert within list]
   LPTR(NEW) ← M
   RPTR(NEW) ← RPTR(M)
   LPTR(RPTR(M)) ← NEW
   RPTR(M) ← NEW
   Exit   □

NODE

LPTR	INFO	RPTR

(a)

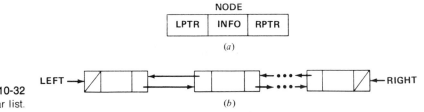

FIGURE 10-32
Doubly linked linear list.

(b)

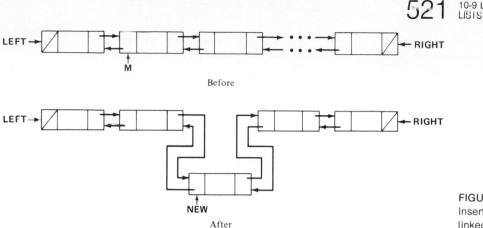

Before

After

FIGURE 10-33
Insertion within a doubly
linked linear list.

In step 5 the term "LPTR(RPTR(M))" represents the left link of the successor node of M.

The deletion of a node from a doubly linked list is a straightforward operation. Several cases are possible. If the list contains a single node, then the leftmost and rightmost pointers associated with the list must be set to NULL. In the case of the rightmost node in the list being marked for deletion, the pointer variable RIGHT must be changed to point to the predecessor of the node being dropped from the list. An analogous situation is possible at the left end of the list. Finally a deletion can occur within the list. A representation of this last case is shown in Fig. 10-35. The right link of the predecessor of OLD must be set to point to the successor of OLD. Also, the left link of the successor node of OLD must be set to point to the predecessor node of OLD. The following pair of statements accomplishes the required task·

RPTR(LPTR(OLD)) ← RPTR(OLD)
LPTR(RPTR(OLD)) ← LPTR(OLD)

A deletion algorithm can now be formulated.

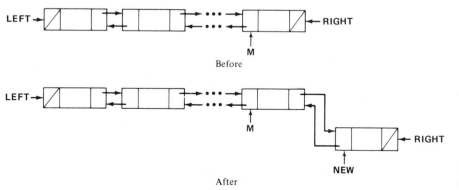

Before

After

FIGURE 10-34
Rightmost insertion in a
doubly linked linear list.

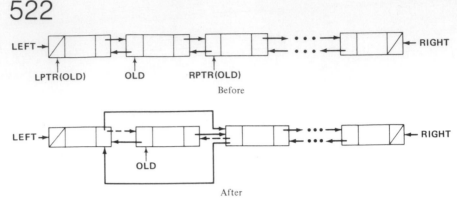

FIGURE 10-35
Deletion of a node from
within a doubly linked
linear list.

Algorithm DOUBLE_DEL (version 1). Given a doubly linked linear list whose node structure is as described in the previous algorithm, this algorithm deletes a node whose address is OLD from the list.

1. [Underflow condition?]
     If LEFT = NULL
     then Call LIST_UNDERFLOW
2. [Single node in list?]
     If LEFT = RIGHT
     then If LEFT = OLD
             then LEFT ← RIGHT ← NULL
                  Return OLD to availability list
                  Exit
             else Write('NODE', INFO(X), 'NOT FOUND')
                  Exit
3. [Is leftmost node being deleted?]
     If LEFT = OLD
     then LEFT ← RPTR(LEFT)
          LPTR(LEFT) ← NULL
          Return OLD to availability list
          Exit
4. [Is rightmost node being deleted?]
     If RIGHT = OLD
     then RIGHT ← LPTR(RIGHT)
          RPTR(RIGHT) ← NULL
          Return OLD to availability list
          Exit
5. [Delete within list]
     RPTR(LPTR(OLD)) ← RPTR(OLD)
     LPTR(RPTR(OLD)) ← LPTR(OLD)
     Return OLD to availability list
     Exit   □

Note that the statement "Return OLD to availability list" means that the discarded node is being returned to the availability area.

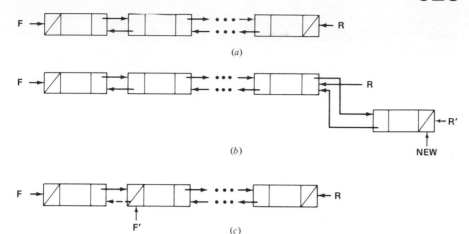

FIGURE 10-36

Observe that the deletion of a node from a doubly linked list is much more efficient, in general, than the deletion of a node from a singly linked list. In particular, in doubly linked list deletion, no search for the predecessor of the node being deleted is necessary. Given the address of the node marked for deletion, the predecessor and successor of this node are immediately known.

A doubly linked list is well suited to represent a queue. Such a representation of a queue is shown in Fig. 10-36a, where R and F are pointer variables which denote the rear and front of the queue, respectively. The insertion of a node with address NEW to the rear of the queue is given in Fig. 10-36b, where R' denotes the rear of the queue after the insertion. An algorithm segment for this insertion is

```
RPTR(R) ← NEW
RPTR(NEW) ← NULL
LPTR(NEW) ← R
R ← NEW
```

In a similar way, a deletion of a node from a doubly linked queue is shown in Fig. 10-36c, where F' denotes the front of the queue after the deletion. The following algorithm steps accomplish this task:

```
F ← RPTR(F)
LPTR(F) ← NULL
```

The algorithms given earlier for inserting a node into and deleting a node from a doubly linked list had several special cases. Let us now consider the possibility of simplifying these algorithms. By first introducing the notion of a list head, as was done in the previous subsection, we can eliminate the case of an empty list. Using this approach, an empty list will always contain the list head node. Second, if we make the list circular, this adds a certain degree of symmetry to the structure. A doubly linked circular list with a list head is shown in Fig. 10-37a, where the pointer variable HEAD denotes the address of the list head. In this representation the right pointer of the rightmost information node points to the head node. Also, the left pointer of the list head points to the rightmost information node. An empty list has the left and right pointers of the list head pointing to the list head. This condition is exemplified in Fig. 10-37b.

Based on this new doubly linked circular list representation, the insertion algorithm introduced earlier becomes much simpler. The following is a revised formulation of this algorithm.

Algorithm DOUBLE_INS (version 2). This is a revised formulation of the first version of the algorithm which was given earlier in this subsection. M and NEW denote the designated and new nodes, respectively.
1. [Insert node at the immediate right of node M]
    LPTR(NEW) ← M
    RPTR(NEW) ← RPTR(M)
    LPTR(RPTR(M)) ← NEW
    RPTR(M) ← NEW
2. [Finished]
    Exit   ☐

This algorithm consists of only one case! The reader should verify that this algorithm works for an insertion into an empty list (that is, into a list which contains only a list head).

In a similar manner, the corresponding deletion algorithm can be reformulated as follows:

Algorithm DOUBLE_DEL (version 2). This is a reformulation of version 1 of the algorithm given earlier. OLD denotes the address of the node marked for deletion.
1. [Delete indicated node]
    RPTR(LPTR(OLD)) ← RPTR(OLD)
    LPTR(RPTR(OLD)) ← LPTR(OLD)

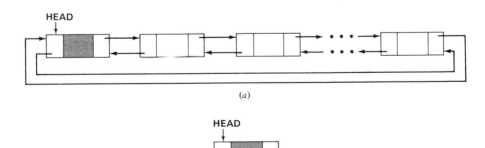

(a)

(b)

FIGURE 10-37
Doubly linked circular list
with a head node.

2. [Return node to availability area]
   Return OLD to availability list
   Exit   □

The doubly linked method of representing list structures is used extensively.
In particular, a variation of this method is used in Chap. 11 to represent
tree structures. In this section we have given the basics of representing linear
lists by linked-allocation techniques. The next section introduces several
applications of linked structures.

EXERCISES 10-9.3

1. Construct an insertion algorithm for a deque which is represented by a
   doubly linked linear list with a list-head node. The information associated with the new node is given by X.

2. Obtain a deletion algorithm for deleting a node from a deque which is
   stored as a doubly linked list with a list head.

3. Repeat exercises 1 and 2 for an input-restricted deque.

4. Repeat exercises 1 and 2 for an output-restricted deque.

10-10 APPLICATIONS OF LINKED LINEAR LISTS

This section discusses three applications of linked linear lists. One of the first
nonnumeric applications to be computerized dealt with the symbolic
manipulation of polynomial expressions. In fact, several special-purpose
programming languages have been designed over the years to manipulate
such expressions. Although many operations are performed on polynomials,
the discussion will be restricted to polynomial addition. Actually, addition
and subtraction are operations that can be implemented much more easily
than other operations such as multiplication and differentiation. Other polynomial operations will be discussed in Chap. 11.

   The second topic deals with the application of hashing functions to the
operations of searching and sorting, and, more generally, to the area of information organization and retrieval. The key value of a record is used in a
computational way to locate that record in memory. The idea of associating
a key with an address in memory is applicable to many situations which arise
in computer science.

   The third and final application is an application of linked queues to
sorting.

10-10.1 POLYNOMIAL MANIPULATION

Consider the familiar symbol-manipulation problem of performing various symbolic operations on polynomials, such as addition, subtraction, multiplication, division, and differentiation. The current discussion will concentrate on the manipulation of polynomials in three variables. For example, the addition of polynomial $2x^2 - xy + x + y^3 + z^2$ to polynomial $x^2 + 2xy + yz - 2z^2$ gives the result $3x^2 + xy + x + y^3 + yz - z^2$. We must first find a suitable representation for polynomials that permits the previous operations to be performed in a reasonably efficient manner. In the manipulation of polynomials, it is clear that individual terms must be selected. In particular, within each term, we must be able to select the coefficient and the exponents.

A node representation for a term of a polynomial in the variables x, y, and z is given in Fig. 10-38a. Such a node consists of four sequentially allocated fields that are collectively referred to as TERM. The first three fields represent the powers of the variables x, y, and z, respectively. The fourth and fifth fields in the node represent the coefficient of the term and the address of the next term, respectively. As an example, the representation of the term yz is shown in Fig. 10-38b.

Given a node with address P, COEFF(P) denotes the coefficient field of the associated term. Similarly, the exponents of x, y, and z are given by POWER_X(P), POWER_Y(P), and POWER_Z(P), respectively. Finally, the address of the next term of the polynomial is given by LINK(P).

Using this representation, the polynomial $x^2 + 2xy - y^3 + yz$ can be represented as a singly linked list. It is convenient to assume that the terms in the list are stored in a certain order. In particular, a term with address P precedes another term with address Q, if POWER_X(P) > POWER_X(Q); or, if these are equal, then we must have POWER_Y(P) > POWER_Y(Q); or, if they are equal, then POWER_Z(P) > POWER_Z(Q). The singly linked list representation of the example polynomial is given in Fig. 10-38c.

The previous ordering of polynomials makes the addition of polynomials an easy matter. In fact, two polynomials ordered in this way can be added by scanning each of their terms only once. For example, a polynomial term $D_1 x^{A_1} y^{B_1} z^{C_1}$ in one polynomial can be added to its corresponding term

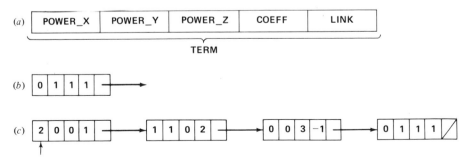

FIGURE 10-38
Singly linked list representation of a polynomial in three variables.

$D_2 x^{A_2} y^{B_2} z^{C_2}$ if $A_1 = A_2$, $B_1 = B_2$, and $C_1 = C_2$. In such a case the coefficient of the sum term is $D_1 + D_2$.

In order to construct an ordered list such as that given in Fig. 10-38c, we must formulate an algorithm which can perform an ordered insertion. The following function inserts a new term into an ordered linked list so that the updated list is still ordered. This function is very similar in structure to the function INS_ORD given in Sec. 10-9.1.

Function POLY_INSERT(NX, NY, NZ, NCOEFF, FIRST). Given an ordered singly linked linear list whose node structure is as just described, this function inserts a new term in the linked list which retains its order. The field names (POWER_X, POWER_Y, POWER_Z, and COEFF) and the node structure name (TERM) are assumed to be global to this procedure. The parameters of the procedure denote the fields of the new term, that is, NX, NY, NZ, and NCOEFF, which correspond to the exponents for x, y, and z and the coefficient value of the term, respectively. FIRST denotes the address of the first term of the polynomial. NEW and SAVE are local pointer variables. A, B, C are local integer variables.

1. [Create a new node]
   NEW ⇐ TERM
2. [Copy information into new node]
   POWER_X(NEW) ← NX
   POWER_Y(NEW) ← NY
   POWER_Z(NEW) ← NZ
   COEFF(NEW) ← NCOEFF
3. [Is the list empty?]
   If FIRST = NULL
   then LINK(NEW) ← NULL
       Return(NEW)
4. [Does new node precede first node of list?]
   A ← POWER_X(FIRST)
   B ← POWER_Y(FIRST)
   C ← POWER_Z(FIRST)
   If (A < NX) or (A = NX and B < NY) or
     (A = NX and B = NY and C < NZ)
   then LINK(NEW) ← FIRST
       Return(NEW)
5. [Initialize temporary pointer]
   SAVE ← FIRST
6. [Search for predecessor and successor of new node]
   Repeat while LINK(SAVE) ≠ NULL
     A ← POWER_X(LINK(SAVE))
     B ← POWER_Y(LINK(SAVE))
     C ← POWER_Z(LINK(SAVE))
     If (A > NX) or (A = NX and B > NY) or
       (A = NX and B = NY and C > NZ)

```
            then SAVE ← LINK(SAVE)
            else  (insert new node)
                     LINK(NEW) ← LINK(SAVE)
                     LINK(SAVE) ← NEW
                     Return(FIRST)
  7. [Insert new node at end of list]
        LINK(NEW) ← NULL
        LINK(SAVE) ← NEW
        Return(FIRST)                                          □
```

Since this function is very similar to the function INS_ORD given earlier, no further comments are necessary. This function can be repeatedly invoked to create an ordered linked list for a polynomial.

The sum polynomial resulting from the addition of two polynomials must also be created. This polynomial can be generated term by term as the two original ordered polynomials are scanned from left to right. Since the resulting terms in the sum polynomial are also ordered, we need only have an algorithm which inserts the next term at the end of the linked list which represents this sum. A general algorithm for doing this task was formulated in Sec. 10-9.1. Two versions of this algorithm were given. The second version was more efficient than the first since in the former we kept track of the last node in the list. On performing the next insertion, we could then directly append the new node at the end of the list. The following function follows this approach.

Function POLY_LAST(NX, NY, NZ, NCOEFF, FIRST). This function contains the same parameters and node structure as in the previous function (POLY_INSERT) and performs an insertion at the end of the list. LAST is a global pointer variable which denotes the address of the last node in the existing list. NEW is a local pointer variable.

```
  1. [Create a new node]
        NEW ⇐ TERM
  2. [Initialize fields of new node]
        POWER_X(NEW) ← NX
        POWER_Y(NEW) ← NY
        POWER_Z(NEW) ← NZ
        COEFF(NEW) ← NCOEFF
        LINK(NEW) ← NULL
  3. [Is the list empty?]
        If FIRST = NULL
        then LAST ← NEW
              Return(NEW)
  4. [Insert new node at end of nonempty list]
        LINK(LAST) ← NEW
        LAST ← NEW
        Return(FIRST)                                          □
```

Again, since this function is very similar to the function INS_LAST of Sec.
10-9.1, no additional comments are given.

The next procedure uses the function POLY_LAST in creating the sum
polynomial, which is derived from adding two given polynomials.

Procedure POLY_ADD(P, Q, R). Given two polynomials that are represented
by linked lists as previously described and whose first terms are denoted by
the pointer variables P and Q, respectively, this procedure symbolically adds
these polynomials and stores the ordered sum as a linked list whose first node
is denoted by the pointer variable R. PSAVE and QSAVE are local pointer
variables. A_1, A_2, B_1, B_2, C_1, C_2, D_1, and D_2 are local numeric variables.

1. [Initialize]
      R ← NULL
      PSAVE ← P
      QSAVE ← Q
2. [End of any polynomial?]
      Repeat thru step 4 while P ≠ NULL and Q ≠ NULL
3. [Obtain field values for each term]
      A_1 ← POWER_X(P)
      A_2 ← POWER_X(Q)
      B_1 ← POWER_Y(P)
      B_2 ← POWER_Y(Q)
      C_1 ← POWER_Z(P)
      C_2 ← POWER_Z(Q)
      D_1 ← COEFF(P)
      D_2 ← COEFF(Q)
4. [Compare terms]
      If $A_1 = A_2$ and $B_1 = B_2$ and $C_1 = C_2$
      then (corresponding terms)
            If COEFF(P) + COEFF(Q) ≠ 0
            then R ← POLY_LAST(A_1, B_1, C_1, $D_1 + D_2$, R)
                  P ← LINK(P)
                  Q ← LINK(Q)
            else P ← LINK(P)
                  Q ← LINK(Q)
      else (terms do not match)
            If $(A_1 > A_2)$ or $(A_1 = A_2$ and $B_1 > B_2)$ or
            $(A_1 = A_2$ and $B_1 = B_2$ and $C_1 > C_2)$
            then (output term from polynomial P)
                  R ← POLY_LAST(A_1, B_1, C_1, D_1, R)
                  P ← LINK(P)
            else (output term from polynomial Q)
                  R ← POLY_LAST(A_2, B_2, C_2, D_2, R)
                  Q ← LINK(Q)
5. [Is polynomial P processed?]
      If P ≠ NULL

```
            then LINK(LAST) ← P
            else  If Q ≠ NULL
                    then LINK(LAST) ← Q
```
6. [Restore initial pointer values for P and Q]
```
        P ← PSAVE
        Q ← QSAVE
```
7. [Finished]
```
    Return
```
☐

Step 1 of the procedure initializes the front pointer of the sum polynomial to NULL. Also, the front pointers of P and Q are copied into PSAVE and QSAVE, respectively.

Step 2 determines whether or not there are unprocessed terms in both polynomials. If there are such terms, then they can be compared. Steps 3 and 4 accomplish this task.

Step 3 obtains the values of A_1, A_2, B_1, B_2, C_1, C_2, D_1, and D_2. The next step compares corresponding terms of the polynomials. If these terms match (that is, $A_1 = A_2$, $B_1 = B_2$, and $C_1 = C_2$), then the sum term is inserted at the end of R, provided that the matching terms do not cancel each other out. Otherwise, if the powers of the P term are greater than those of the Q term, then the former term is inserted at the end of polynomial R, and conversely. Step 6 restores P and Q to their original values.

We are now in a position to formulate a main algorithm which inputs two polynomials and repeatedly invokes the function POLY_INSERT in order to construct the linked lists for these polynomials. We assume that the input consists of a sequence of cards each of which represents a term; that is, each card contains four values. The first three values correspond to the powers of x, y, and z, respectively, for a term. The last value is the coefficient value of that term. A dummy card with a coefficient value of 0 is used to signal the end of a polynomial.

Algorithm POLYNOMIAL. Given the procedures POLY_ADD and POLY_ INSERT, as described earlier, this algorithm inputs two polynomials and obtains their sum. The pointer variables POLY1, POLY2, and POLY3 denote the front terms of the polynomials. X, Y, Z, and C are variables which denote the powers of x, y, z, and the coefficient value of an input term.

1. [Initialize]
```
    POLY1 ← POLY2 ← NULL
```
2. [Input and construct first polynomial]
```
    Read(X, Y, Z, C)
    Repeat while C ≠ 0
      POLY1 ← POLY_INSERT(X, Y, Z, C, POLY1)
      Read(X, Y, Z, C)
```
3. [Input and construct second polynomial]
```
    Read(X, Y, Z, C)
```

```
        Repeat while C ≠ 0
            POLY2 ← POLY_INSERT(X, Y, Z, C, POLY2)
            Read(X, Y, Z, C)
4.  [Add polynomials]
        Call POLY_ADD(POLY1, POLY2, POLY3)
5.  [Finished]
        Exit
```
☐

In terminating this subsection we note that a circular linked representation of a polynomial could have been used throughout the discussion. This choice would have resulted in simpler algorithms. We leave this approach, however, as an exercise.

EXERCISES 10-10.1

1. Construct an algorithm to subtract two polynomials in three variables.

2. Obtain an algorithm for eliminating duplicate terms in an ordered linked list of terms (as described in the subsection). Assume that a term contains three variables.

3. Construct an algorithm to multiply two polynomials in three variables.

4. Formulate an algorithm for evaluating a polynomial of three variables which is represented by a linked list. The values for x, y, and z are given as a, b, and c, respectively.

5. Rewrite the algorithm POLY_ADD in the text for two polynomials which are represented by circular linked lists (with list heads).

10-10.2   HASH-TABLE TECHNIQUES

The best search method introduced so far (binary search) has a search time which is proportional to $\log_2 n$ for a table of n entries. In this section we investigate a class of search techniques whose search times can be independent of the number of entries in a table. To achieve this goal, an entirely new approach to searching must be used. With this approach, the position of a particular entry in the table is determined by the value of the key for that entry. This association is realized through the use of a hashing function.

The notions and concepts of this new approach are first discussed. General simple hashing functions are then introduced. Unfortunately, more than one key can be mapped into the same table position. In such a case, collisions are said to occur. These collisions must be resolved using a collision-resolution technique. Two broad classes of such techniques are examined. Finally, the notions of hashing are applied to the operation of sorting. An algorithm whose performance can be proportional to the size of the table (that is, n) is formulated.

10-10.2.1 Notation and concepts

Data can be represented in many ways. In this section we are given a collection of data entities. Each entity contains, in general, several information fields. We shall represent such an entity by a *record*. The records are combined into a table which represents the information upon which the operations of searching, insertion, and deletion are to be performed. Each field in a record contains, in general, alphanumeric data. The organization of a record is application-dependent and has little bearing on the algorithms discussed later in the section.

A *table* is defined to be an ordered sequence of n records. Each record in a table contains one or more keys. It is with respect to these keys that processing is performed. For example, the key associated with a record could be a variable name in a symbol-table application in a compiler. Equally, it could be an employee number in a payroll application. Each record, for our purpose, is assumed to contain a single key field K_i and additional information which is irrelevant to the present discussion.

There are many applications of tables. For example, an important part of a compiler is the construction and maintenance of a symbol table of variable names for a source program as discussed in Sec. 6-5.1. The importance of a symbol table is best realized when we consider that every occurrence of a variable name in a program requires continual table interaction. Symbol-table access time in a PL/I-like compiler can represent a significant portion of the translation time if improper searching techniques are used. This time can be significantly reduced if a suitable table organization is chosen.

The best search method introduced so far is the binary search technique, which has a search time of $O(\log_2 n)$. The linear search and binary search techniques are both based exclusively on comparing keys. Another approach, however, is to compute the location of the desired record. The nature of this computation depends on the key set (or space) and the memory-space requirements of the particular table. For example, if we have a table of n employee records, each of which is identified by an employee number key whose value lies between 1 and n, the key value of a particular employee, used as a subscript, directly locates the employee in question. Such a convenient key-location relationship rarely exists in real-world applications. This is so because key values are chosen for many reasons, most of which are unrelated to efficient computer processing goals.

For example, in a symbol-table application involving FORTRAN, the key space involves all the valid variable names in that language. This name space has a size of

$$26 + 26 \times 36 + 26 \times 36^2 + \cdots + 26 \times 36^5 \simeq 1.6 \times 10^9$$

that is, 26 names of one letter, 26×36 names of two characters, and so on. For a typical program, this name space is associated with a table space of perhaps 100 or 200 record locations. Consequently, the problem of directly associating a key with the storage location of its associated record in a search is more difficult.

Formally, this *key-to-address transformation* problem is defined as a mapping or a *hashing function* H, which maps the key space (K) into an address space (A). That is, given a key value, a hashing function H produces a table address or location of the corresponding record. The function generates this address by performing some simple arithmetical or logical operations on the key or some part of the key.

Since, as previously indicated, the key space is usually much larger than the address space, many keys will be matched to the same address. Such a many-to-one mapping results in *collisions* between records. A collision-resolution technique is required to resolve these collisions.

In the next subsection we examine some of the most popular hashing functions. The collision-resolution problem is examined in a subsequent subsection. The most important operations which are performed on hash tables are those of access and retrieval.

10-10.2.2   Hashing functions

In this subsection we examine several simple hashing functions. Some of the desirable properties of a hashing function include speed and the generation of addresses uniformly. Before describing these functions, however, we introduce the notion of preconditioning as it relates to the key space K. Each element of K often contains alphanumeric characters. Some of these characters may be arithmetically or logically difficult to manipulate. It is sometimes convenient to convert such keys into a form which can be more easily manipulated by a hashing function. This conversion process is often called *preconditioning*. As an example, let us consider the preconditioning of the key RATE1. One possibility is to encode the letters as the numbers 11, 12, ..., 36 and the set of special symbols (for example, +, −, *, /, . . .) as 37, 38, 39, ... Using this approach, RATE1 is encoded as 2811301501 (that is, the symbols R, A, T, E, and 1 are replaced by the integers 28, 11, 30, 15, 01, respectively).

Preconditioning is most efficiently performed by using the numerically coded internal representation (for example, EBCDIC or ASCII) of each character in the key. For example, in ASCII, the key A1 is binary-encoded as 1000001 0110001. Interpreted as a 14-digit binary number, this has a decimal equivalent of 8,369. Similarly, the EBCDIC representation of A1 is binary-encoded as 11000001 11110001 or 49,649. In general, the preconditioned result of a key may not fit into a word of memory. In such instances we can ignore certain digits of the preconditioned result. Another approach is to use a hashing function which performs a size-reduction transformation. Frequently, one hashing function generates the preconditioned result, and then a second function maps this result into a table location.

We now proceed to describe five simple hashing functions. Other more complex functions exist, but they are not generally used with tables that are stored entirely in the main memory of the computer.

THE DIVISION METHOD

One of the first hashing functions, and perhaps the most widely accepted, is the *division method*, which is defined as

$$H(x) = x \bmod m + 1$$

for some integer divisor m. The operator mod denotes the modulo arithmetic system. In this system the term $x \bmod m$ has a value which is equal to the remainder of dividing x by m. For example, if $x = 35$ and $m = 11$, then

$$H(35) = 35 \bmod 11 + 1 = 2 + 1 = 3$$

The division method yields a "hash value" which belongs to the set $\{1, 2, \ldots, m\}$.

In mapping keys to addresses, the division method preserves, to a certain extent, the uniformity that exists in a key set. Keys which are close to each other or clustered are mapped to unique addresses. For example, for a divisor $m = 31$, the keys 1000, 1001, . . ., and 1010 are mapped to the addresses 9, 10, . . ., and 19. This preservation of uniformity, however, is a disadvantage if two or more clusters are mapped to the same addresses. For example, if another cluster of keys is 2300, 2301, . . ., and 2313, then these keys are mapped to addresses 7, 8, . . ., and 20, and there are many collisions with keys from the cluster starting at 1000. The reason for this phenomenon is that keys in the two clusters yield the same remainder when divided by $m = 31$.

In general, it is uncommon for a number of keys to yield the same remainder when m is a large prime number. In practice it has been found that odd divisors without factors less than 20 are also satisfactory. In particular, divisors which are even numbers are to be avoided since even and odd keys would be mapped to odd and even addresses, respectively (assuming an address space of $\{1, 2, \ldots, m\}$).

THE MIDSQUARE METHOD

Another hashing function which has been widely used in many applications is the *midsquare method*. In this method a key is multiplied by itself and an address is obtained by selecting an appropriate number of bits or digits from the middle of the square. Usually, the number of bits or digits chosen depends on the table size and, consequently, can fit into one computer word of memory. The same positions in the square must be used for all products. As an example, consider a six-digit key, 123456. Squaring this key results in the value 15241383936. If a three-digit address is required, positions 5 to 7 could be chosen, giving address 138. The midsquare method has been criticized by some, but it has given good results when applied to certain key sets.

THE FOLDING METHOD

In the *folding method* a key is partitioned into a number of parts, each of which has the same length as the required address with the possible exception of the last part. The parts are then added together, ignoring the final carry, to form an address. If the keys are in binary form, the "exclusive-or" operation may be substituted for addition. As an example, assume that the key 356942781 is to be transformed into a three-digit address. In the *fold-shifting method*, 356, 942, and 781 are added to yield 079. A variation of the basic method involves the reversal of the digits in the outermost partitions. This variation is called the *fold-boundary method*. In the previous example, 653, 942, and 187 are added together, yielding 782. Folding is a hashing function which is also useful in converting multiword keys into a single word so that other hashing functions can be used.

DIGIT ANALYSIS

A hashing function referred to as *digit analysis* forms addresses by selecting and shifting digits or bits of the original key. For example, a key 7546123 is transformed to the address 2164 by selecting digits in positions 3 to 6 and reversing their order. For a given key set, the same positions in the key and the same rearrangement pattern must be used consistently. Initially, an analysis on a sample of the key set is performed to determine which key positions should be used in forming an address. Digit positions having the most uniform distributions (that is, the smallest peaks and valleys) are selected. As an example, consider the digit analysis of the sample key set shown in Table 10-13. A total of 5,000 ten-digit keys are analyzed to determine which key positions should be used in forming addresses in the address

TABLE 10-13   Digit Analysis of a Sample Set of 10-Digit Part Numbers

	Key Position									
Digit	1	2	3	4	5	6	7	8	9	10
0	5000	531	594	499	590	721	1565	1133	562	2540
1	0	582	568	536	467	905	874	759	612	1581
2	0	571	620	531	563	553	657	606	542	557
3	0	546	565	511	512	277	555	482	522	332
4	0	518	529	495	461	0	284	521	546	0
5	0	503	503	500	463	673	276	469	472	0
6	0	488	456	469	510	629	263	296	426	0
7	0	449	411	500	459	0	212	365	425	0
8	0	422	431	470	457	501	159	310	455	0
9	0	390	323	489	518	741	155	59	438	0

space {0, 1, . . ., 9999}. Positions 2, 4, 5, and 9 have the most uniform distribution of digits, so they are selected. For example, a key 1234567890 is transformed to the address 9542 by selecting digits in positions 2, 4, 5, and 9 and reversing their order. This hashing transformation technique has been used in conjunction with static key sets (that is, key sets that do not change over time).

THE LENGTH-DEPENDENT METHOD

Another hashing technique which has been commonly used in table-handling applications is called the *length-dependent method*. In this method the length of the key is used along with some portion of the key to produce either a table address directly or, more commonly, an intermediate key which is used, for example, with the division method to produce a final table address. One function which has produced good results sums the internal binary representation of the first and last characters and the length of the key shifted left four binary places (or, equivalently, the length multiplied by 16). As an example, the key PARTNO becomes $215 + 214 + (6 \times 16) = 525$ assuming EBCDIC representation. If we treat 525 as an intermediate key and apply the division method with a divisor of 49, then the resulting address is 36.

Thus far we have described how to perform key-to-address transformations using hashing functions. We have, however, ignored a very important aspect relevant to this process—the problem of colliding records. In general, a hashing function is a many-to-one mapping. That is, many keys can be transformed into the same address. In practice, such a phenomenon happens because there are many more keys than there are addresses. Clearly, two records cannot occupy the same location in a table. Consequently, such collisions among keys must be resolved. We now turn our attention to this problem.

10-10.2.3 Collision-resolution techniques

As mentioned earlier, a hashing function can, in general, map several keys into the same address. When this situation arises, the colliding records must be sorted and accessed as determined by a *collision-resolution technique*. There are two broad classes of such techniques: open addressing and chaining. In this subsection we formulate algorithms from both classes. Also, we examine certain variations of these basic techniques.

The general objective of a collision-resolution technique is to attempt to place colliding records elsewhere in the table. This requires the investigation of a series of table positions until an empty one is found to accommodate a colliding record. We require a mechanism to generate the series of table positions to be examined. The main criteria for this mechanism are speed (that it determine the positions quickly), coverage (that it will try *every* table position eventually), and reproducibility (that the series produced can be produced again, namely, when it comes time to find the placed record).

With *open addressing*, if a record with key x is mapped to an address location d and this location is already occupied, then other locations in the table are examined until a free location is found for the new record. If a record with key K_i is deleted, then K_i is set to a special value called DELETE, which is not equal to the value of any key. The sequence in which the locations of a table are examined can be formulated in many ways. One of the simplest techniques for resolving collisions is to use the following sequence of locations for a table of m entries:

$$d, d + 1, \ldots, m - 1, m, 1, 2, \ldots, d - 1$$

An unoccupied record location is always found if at least one is available; otherwise, the search halts unsuccessfully after scanning m locations. When retrieving a particular record, the same sequence of locations is examined until that record is found or until an unoccupied (or empty) record position is encountered. In this latter case the desired record is not in the table, so the search fails. This collision-resolution technique is called *linear probing*.

We now formulate an algorithm which performs the table-lookup and insertion operations for a hashed table using the linear probing technique just described. The record to be inserted or located is identified by the key argument X, any other information to be inserted is denoted by INFO, and the type of operation (lookup or insertion) to be performed is specified by the logical parameter INSERT. Values of *true* and *false* indicate the insert and lookup operations, respectively. Furthermore, it is assumed that if a record location has never contained a record, then the corresponding key field has a special value called EMPTY. The algorithm that we now give is in the form of a function.

Function OPENLP(X, INFOR, INSERT). Given the parameters X, INFO, and INSERT introduced earlier, this function performs the table lookup and insertion operations. The function returns the position of the record in question, if successful. Otherwise, an error condition is signaled by a negated position. The hashing function HASH is used to calculate an initial position.

1. [Calculate initial position]
   $d \leftarrow HASH(X)$
2. [Perform indicated operation if location is found]
   Repeat for $i = d, d + 1, \ldots, m - 1, m, 1, 2, \ldots, d - 1$
      If $X = K_i$
      then If not INSERT
         then Return(i)   (position of retrieved record)
         else Return(−i)   (error in insertion)
      If K_i = EMPTY or K_i = DELETE
      then If INSERT
         then (perform indicated insertion)
            $K_i \leftarrow X$
            $DATA_i \leftarrow INFO$

$$Return(i)$$

else If K_i = EMPTY

then Return($-i$)   (error in lookup)

3. [Table overflow]

Write('OVERFLOW OR LOOK-UP ERROR')

Return(0)   □

In step 1 an initial position in the table is calculated. Any of the hashing functions which were discussed in the previous subsection can be used in this step. The second step scans the table starting at the initial position d. If the key field and the search argument (X) match, then that table position is returned in the case of a lookup operation. For an insertion, the negated index is returned, indicating that an attempt has been made to insert a duplicate key. If position i of the table is either empty or contains a deleted record, then the new record is entered in this position during an insert operation. On the other hand, an empty location encountered during a lookup operation indicates that the given search key is not in the table. In this case a negated position index is returned. Step 3 indicates that either no record locations are available or that the lookup operation fails. Such a failure is indicated by returning a value of 0.

As an example of this open addressing technique, let us assume the following:

the names PAY and RATE are mapped into 1

the name TAX is mapped into 2

the name PENSION is mapped into 4

the names DEDUCT, STATUS, DEPENDENTS, SEX, and SALARIED are
mapped into 8

Assume that the insertions are performed in the order

PAY, RATE, TAX, PENSION, DEDUCT, STATUS, DEPENDENTS, SEX, and
SALARIED.

Figure 10-39 represents the resulting structure with m = 11. The first key is placed in a single probe. The second key (RATE), however, must go into position 2 instead of 1, which is already occupied. TAX, which is placed in position 3, also requires two probes for its placement, since position 2 is already occupied. PENSION is placed in position 4 in one probe. DEDUCT is also placed in position 8 in one probe, but STATUS, DEPENDENTS, and SEX take two, three, and four probes, respectively. Finally, SALARIED is placed in position 5 after nine probes since positions 8, 9, 10, 11, 1, 2, 3, and 4 are all occupied. A lookup operation is completed successfully when the desired record is found, or unsuccessfully if an empty record is encountered. Since all steps of the previous algorithm apply to both insertion and lookup opera-

Number of probes

R_1	PAY		1
R_2	RATE		2
R_3	TAX		2
R_4	PENSION		1
R_5	SALARIED		9
R_6	Empty		
R_7	Empty		
R_8	DEDUCT		1
R_9	STATUS		2
R_{10}	DEPENDENTS		3
R_{11}	SEX		4

FIGURE 10-39
Collision resolution with
open addressing.

tions, the number of probes required for lookups is the same as that for insertions.

For a uniformly distributed set of keys, the linear probing technique performs reasonably well when compared to the linear and binary search techniques, provided that the table is not too full. That is, the ratio of the number of records being entered (n) to the table size (m) must be less than approximately 0.8. This ratio is known as the *load factor*. For higher load factors, the linear probe method degenerates rapidly. This phenomenon is due to an increased number of collisions as more records are being stored in the table.

The linear probe method of collision resolution has a number of drawbacks. In particular, deletions from the table are difficult to perform. The strategy used here was to have a special table entry with a value of DELETE to denote the deletion of that entry. Such an approach enables the table to be searched in a proper way. As an example, let us assume that the record with key PAY in Fig. 10-39 is marked for deletion by assigning a value of DELETE to K_1. Now, if we desire to retrieve the record with a key value of RATE, the previously stated algorithm still behaves properly. The question which arises is: Why use a special value such as DELETE to denote deleted entries? If this approach were not used, duplicate entries could occur in the table. For example, in the previous case, our algorithm would find an empty location in position 1 and assume that RATE is not in the table. It would then proceed to insert a duplicate entry for RATE. It is possible to formulate a deletion algorithm which performs deletions immediately by moving records, if necessary. Such an algorithm eliminates the necessity for having records with a value of DELETE. In other words, a record position can be either occupied or empty. This approach is not taken, however, because better collision-resolution techniques (based on linked allocation) are available.

Another drawback of the linear probe method is caused by *clustering* effects, whose severity increases as the table becomes full. This phenomenon is observed by considering a trace of Fig. 10-39, which shows the state of the table after performing each insertion. This trace is given in Fig. 10-40.

After inserting record	Contents of table after current insertion										
PAY	PAY										
RATE	PAY	RATE									
TAX	PAY	RATE	TAX								
PENSION	PAY	RATE	TAX	PENSION							
DEDUCT	PAY	RATE	TAX	PENSION				DEDUCT			
STATUS	PAY	RATE	TAX	PENSION				DEDUCT	STATUS		
DEPENDENTS	PAY	RATE	TAX	PENSION				DEDUCT	STATUS	DEPENDENTS	
SEX	PAY	RATE	TAX	PENSION				DEDUCT	STATUS	DEPENDENTS	SEX
SALARIED	PAY	RATE	TAX	PENSION	SALARIED			DEDUCT	STATUS	DEPENDENTS	SEX

FIGURE 10-40
Trace of insertions using open addressing.

When the first insertion (PAY) is made, the probability of inserting this element in a particular position is $\frac{1}{11}$. For the second insertion, (RATE), however, the probability that position 2 will be selected is twice that of any other empty position. That is, the second entry will be placed in position 2 if the second key is hashed into either 1 or 2. Continuing in this manner, on the fourth insertion the probability that the new entry (PENSION) is placed in position 4 is four times as likely as it being placed in any other unoccupied position. Consequently, long sequences of occupied positions tend to become longer. This kind of clustering phenomenon is called *primary clustering*.

The detrimental effect of primary clustering can be reduced by selecting a different probing technique. Such a technique exists and is called *random probing*. This method generates a random sequence of positions rather than an ordered sequence as was the case in the linear probing method. The random sequence generated in this fashion must contain every position between 1 and m exactly once. A table is full when the first duplicate position is generated. An example of a random-number generator that produces such a random sequence of position numbers is the statement

$$y \leftarrow (y + c) \bmod m$$

where y is the initial position number of the random sequence and c and m are integers that are relatively prime to each other (that is, their greatest common divisor is 1). For example, assuming that m = 11 and c = 5, the previous statement, starting with an initial value of 2, generates the sequence 7, 1, 6, 0, 5, 10, 4, 9, 3, 8, and 2. Adding 1 to each number of this sequence transforms all numbers so that they belong to the interval [1, 11]. Algorithm OPENLP can easily be modified to incorporate the random probe method. However, this modification is left as an exercise.

The deletion problem becomes more severe with random probing than is the case with linear probing. Therefore, if a table is subjected to many deletions, random probing should not be used.

Although random probing has alleviated the problem of primary clustering, we have not eliminated all types of clustering. In particular, clustering occurs when two keys are hashed into the same value. In such an instance, the same sequence of positions is generated for both keys by the random probe method. This clustering phenomenon is called *secondary clustering*.

One approach to alleviating this secondary clustering problem is to have a second hashing function, independent of the first, select the parameter c in the random probing method. For example, let us assume that H_1 is the first hashing function, with $H_1(x_1) = H_1(x_2) = i$ for $x_1 \neq x_2$ where i is the hash value. Now, if we have a second hashing function, H_2, such that $H_2(x_1) \neq H_2(x_1)$ when $x_1 \neq x_2$ and $H_1(x_1) = H_1(x_2)$, we can use $H_2(x_1)$ or $H_2(x_2)$ as the value of parameter c in the random probe method. The two random sequences generated by this scheme are different when H_2 and H_1 are independent. The effects of secondary clustering can, therefore, be curtailed. This variation of open addressing is called *double hashing* (or *rehashing*).

As a concrete example of this method, let

$H_1(x) = x$ mod m and $H_2(x) = (x$ mod $(m - 2)) + 1$

for a key x and table size m. For a table size m = 11 and a key value x = 75, $H_1(75) = 9$ and $H_2(75) = 4$. Now, the repeated application of the statement

$y \leftarrow (y + c)$ mod 11

with y = 9 and c = 4 produces the sequence

9, 2, 6, 10, 3, 7, 0, 4, 8, 1, 5

Next, if we choose a key having a value of 42, then

$H_1(42) = 9$ [the same value as $H_1(75)$, but $H_2(42) = 7$]

The repeated application of the statement

$y \leftarrow (y + c)$ mod 11

with y = 9 and c = 7, yields the sequence

9, 5, 1, 8, 4, 0, 7, 3, 10, 6, 2

Note that even though $H_1(75) = H_1(42) = 9$, the two random sequences generated are different.

This double-hashing technique outperforms the linear probing method, especially as the table becomes full. For example, when a table is 95% full, the average number of probes for the linear probe method is 10.5, while for the double-hashing method it is 3.2—a significant improvement!

In summary, there are three main difficulties with the open addressing method of collision resolution. First, lists of colliding records for different hash values become intermixed. This phenomenon requires, on the average, more probes. Second, we are unable to handle a table overflow situation in a satisfactory manner. On detecting an overflow, the entire table must be reorganized. The overflow problem cannot be ignored in many table applications where table space requirements can vary drastically. Finally, the physical deletion of records is difficult. We now turn to linked-allocation techniques to resolve these problems.

One of the most popular methods of handling overflow records is called *separate chaining*. In this method the colliding records are chained into a special *overflow area* which is distinct from the *prime area*. This area contains that part of the table into which records are initially hashed. A separate linked list is maintained for each set of colliding records. Therefore, a pointer field is required for each record in the primary and overflow areas. Figure 10-41 shows a separate chaining representation of the sample keys used earlier in this subsection, with $m = 11$ and $n = 9$. The keys are assumed to be inserted in the order

PAY, RATE, TAX, PENSION, DEDUCT, STATUS, DEPENDENTS, SEX, and SALARIED.

Note that the colliding records in each linked list are not kept in alphabetical order. When a new colliding record is entered in the overflow area, it is placed at the front of those records in the appropriate linked list of the overflow area.

A useful and more efficient representation of the separate chaining technique involves the use of an intermediate table or hash table. Figure 10-42 illustrates how a hash table is used, assuming the same keys, hashing function, and collision-resolution technique as in Fig. 10-41. Note that in this representation all the records reside in the overflow area while the prime area contains only pointers. For a table having large records, this approach results in a densely packed overflow area. Furthermore, the hash table

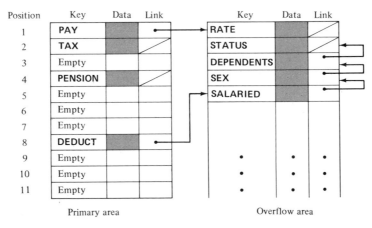

FIGURE 10-41
Collision resolution with separate chaining.

whose entries are pointers can be made large without wasting very much storage.

An algorithm for entering a key into an existing structure such as that of Fig. 10-42 is easily realized. Given a key X, which is to be entered if it is not already in the table, we first proceed to obtain its hash value. The hash-table value for this hash value can then be examined. A value of NULL in this position indicates that the corresponding linked list is empty. Therefore, we can insert the given key into an empty list. If, however, the particular linked list in question is not empty, we can search this linked list for the presence of the given key. If the search fails, X becomes the front entry of the updated linked list.

The following algorithm is a straightforward implementation of this process. We assume a node structure of the form

RECORD

KEY	DATA	LINK

where DATA represents some additional information for a table entry.

Algorithm ENTER. Given a pointer vector HASH_TABLE representing a hash table, each element of which contains a pointer to a linked list of colliding records, and a hashing function HASH, which maps a key into an integer (between 1 and m), this algorithm appends the given key X at the front of the appropriate linked list (if it is not already there). The node structure in the overflow area is of the form previously described. Assume that INFO represents the additional information which is to accompany the record to be inserted. P, NEW, and RANDOM represent pointer variables.

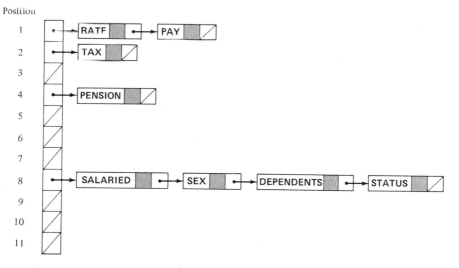

FIGURE 10-42
Separate chaining with a hash table.

1. [Compute the hash number]
     RANDOM ← HASH(X)
2. [Is linked list to which X belongs empty?]
     If HASH_TABLE[RANDOM] = NULL
     then NEW ⇐ RECORD
          HASH_TABLE[RANDOM] ← NEW
          KEY(NEW) ← X
          DATA(NEW) ← INFO
          LINK(NEW) ← NULL
          Exit
3. [Perform search for X]
     P ← HASH_TABLE[RANDOM]
     Repeat while P ≠ NULL
       If X = KEY(P)
       then Exit   (X is already present)
       else P ← LINK(P)
     (X is not in the table)
     NEW ⇐ RECORD
     KEY(NEW) ← X
     DATA(NEW) ← INFO
     LINK(NEW) ← HASH_TABLE[RANDOM]
     HASH_TABLE[RANDOM] ← NEW
     Exit   □

This algorithm can be modified to do either an insertion or a search for a particular key. Such a modification is left as an exercise. Similarly, a record can be easily deleted from its associated linked list. Also, for certain applications, such as symbol tables in compilers, it may be convenient to have all colliding records in a linked list ordered. In such applications an alphabetical list of all variable names is desirable.

In terminating this subsection we want to stress the importance of having a suitable hashing function. Ideally, each linked list of colliding records should have the same number of entries. The worst possible case occurs when all keys are mapped to the same hash number (that is, the same linked list), thereby causing the insertion and search operations to be no more efficient than those in a linear search method. In practice it is a nontrivial matter to obtain a good hashing function, since the size of the linked lists of colliding records which it induces depends on the keys being used.

Another important factor pertaining to the efficiency of the separate chaining method of collision resolution is the desirability of keeping the number of records in each linked list relatively small. For example, a hash table of 100 entries would nicely handle 125 records in the overflow area. The average number of comparisons (assuming a uniformly distributed hashing function) for accessing a particular entry in such a situation is slightly greater than 1. From this observation it is clear that the search time in such a table organization is independent of the number of entries in the table.

Finally, since our discussion has been concerned with internal tables (that is, tables that are stored in memory at one time), the hashing function which is used should be simple. The response time to insert or fetch a particular entry is the sum of the times taken to evaluate the hashing function and to perform the indicated operation. It may be more efficient to allow a greater number of comparisons for performing a table operation if a significant reduction in the complexity of the hashing function results.

The ideas introduced so far relate to searching, but the same ideas apply to sorting as well. We investigate this possibility in the next subsection.

10-10.2.4   Address Calculation Sorting

In applying a hashing function to the sorting process, a particular kind of hashing function is required. Let us assume that we have a hashing function H with the property

$$x_1 < x_2 \text{ implies that } H(x_1) \leqslant H(x_2)$$

A function which exhibits this property is called a *nondecreasing*, or *order-preserving*, *hashing function*. When such a function is used to hash a particular key into a particular number to which some previous keys have already been hashed (that is, a collision occurs), then the new key is placed in the set of colliding records so as to preserve the order of the keys. The result of hashing and inserting the sample key set

21, 51, 98, 87, 19, 25, 70, 83, 13, 65, 7, 34, 72, 43, 57

using a nondecreasing hashing function in which all the keys in the ranges 1–20, 21–40, 41–60, 61–80, and 81–100 are each hashed into a different set as sorting proceeds, is shown in Table 10-14. The five ordered sets obtained during the hash and insert phase can now be trivially merged to yield the desired sorted table.

The separate chaining method of collision resolution with a separate hash table can be used to represent this sorting process. Such a representation of the final sort is given in Fig. 10-43.

TABLE 10-14   Behavior of an Address-Calculation Sort

Set	3 Records Entered	6 Records Entered	9 Records Entered	12 Records Entered	15 Records Entered
1 (1–20)		19	13, 19	7, 13, 19	7, 13, 19
2 (21–40)	21	21, 25	21, 25	21, 25, 34	21, 25, 34
3 (41–60)	51	51	51	51	43, 51, 57
4 (61–80)			70	65, 70	65, 70, 72
5 (81–100)	98	87, 98	83, 87, 98	83, 87, 98	83, 87, 98

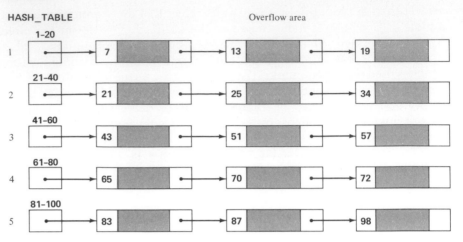

FIGURE 10-43
Representation of a table
in address-calculation
sorting.

A general algorithm for this sorting process follows:

1. Initialize hash table entries to null
2. Repeat thru step 4 while there are still input records
3. Input and hash a record
4. Insert record into appropriate linked list
5. Concatenate the nonempty linked lists into one

Let us examine some of the details of the general algorithm. We assume the following record structure in the overflow area:

RECORD

K	DATA	LINK

The important variables for this problem are:

M (integer)	Size of hash table
HEAD (pointer)	Address of the first record in the sorted table
HASH_TABLE (pointer)	Vector of pointers representing the hash table
HASH (integer)	Hashing function
RECORD (record)	Overflow area record
K (integer)	Key of a record
DATA (string)	Other relevant information in a record

LINK (pointer)	Location of the next record in the list
KEY (integer)	Key of the input record
INFO (string)	Other relevant information of the input record
RANDOM (integer)	Hash address of input record (between 1 and M)
NEW (pointer)	Address of newly created overflow record

The details of the first three steps of the general algorithm are obvious. The fourth step of the algorithm can be detailed by making minor modifications to the algorithm INS_ORDER (see Sec. 10-9.1). The general algorithm's final step is not difficult. The first task involves locating the first nonempty linked list. The following algorithm segment accomplishes this task.

```
I ← 1
Repeat while HASH_TABLE[I] = NULL and I < M
   I ← I + 1
HEAD ← HASH_TABLE[I]
J ← I + 1
```

The next task in obtaining a sorted table is to concatenate all nonempty linked lists. This task involves scanning each nonempty linked list for its last node. When this node is found, it is linked to the front node of the next list. The details of this task follow:

```
Repeat while J ≤ M
   If HASH_TABLE[J] ≠ NULL
   then (find tail of linked list)
        P ← HASH_TABLE[I]
        Repeat while LINK(P) ≠ NULL
           P ← LINK(P)
        (link tail of this linked list to the head of the next)
        LINK(P) ← HASH_TABLE[J]
        I ← J
        I ← I + 1
```

These details can be incorporated in a detailed algorithm.

Algorithm ADDR_CAL_SORT. Given input records of the form previously described, this algorithm sorts the records based on address calculation using a hash table with a separate overflow area. An order-preserving hashing function, HASH, is assumed. The sorted table generated is in the form of a linked list, where HEAD denotes the address of its first record. I and J are integer variables. P and S are pointer variables.

1. [Initialize hash table, that is, list heads]

```
                    Repeat for I = 1, 2, . . ., M
                        HASH_TABLE[I] ← NULL
    2. [Input and insert records into appropriate linked lists]
                    Repeat thru step 4 while there are still input records
    3. [Input and hash a record]
            Read(KEY, INFO)
            NEW ⇐ RECORD
            K(NEW) ← KEY
            DATA(NEW) ← INFO
            LINK(NEW) ← NULL
            RANDOM ← HASH(KEY)
    4. [Insert record into appropriate linked list]
            If HASH_TABLE[RANDOM] = NULL
            then (insert record into empty linked list)
                    LINK(NEW) ← HASH_TABLE[RANDOM]
                    HASH_TABLE[RANDOM] ← NEW
            else (insert record in middle or at end of linked list)
                    P ← HASH_TABLE[RANDOM]
                    S ← LINK(P)
                    Repeat while S ≠ NULL and K(S) < KEY
                      P ← S
                      S ← LINK(S)
                    LINK(P) ← NEW
                    LINK(NEW) ← S
    5. [Find first nonempty linked list]
            I ← 1
            Repeat while HASH_TABLE[I] = NULL and I < M
              I ← I + 1
            HEAD ← HASH_TABLE[I]
            J ← I + 1
    6. [Concatenate the nonempty linked lists]
            Repeat while J ≤ M
              If HASH_TABLE[J] ≠ NULL
              then (find tail of linked list)
                    P ← HASH_TABLE[I]
                    Repeat while link(P) ≠ NULL
                      P ← LINK(P)
                    (link tail of this linked list to the head of the next)
                    LINK(P) ← HASH_TABLE[J]
                    I ← J
              I ← I + 1
    7. [Finished]
            Exit                                              □
```

Assuming that the nondecreasing hashing function uniformly distributes the records of the table among the linked lists, this sort performs in a

linear manner; that is, the number of comparisons is $O(n)$. The worst case occurs when all keys are mapped into the same number. In this case the performance of the sorting method degenerates to $O(n^2)$.

EXERCISES 10-10.2

1. Using the division method of hashing with $m = 101$, obtain the hash values for the following set of keys:

    PAY
    AGE
    RATE
    NUMBER

    Assume an EBCDIC (see Sec. 5-1) representation of the keys.

2. Repeat exercise 1 for the ASCII representation of the keys.

3. Assuming an open addressing method of collision resolution with linear probing, obtain the hash table (as in Fig. 10-39) for the following set of keys:

    the name NODE is mapped into 1

    the name STORAGE is mapped into 2

    the names AN and ADD are mapped into 3

    the names FUNCTION, B, BRAND, and PARAMETER are mapped into 9

    Use the division method of hashing with $m = 11$. Also, assume that the insertions are performed in the following order:

    NODE, STORAGE, AN, ADD, FUNCTION, B, BRAND, and PARAMETER

4. Formulate an algorithm, based on the linear probe method, for deleting a record from a hash table. This algorithm is not to use a special value of DELETE. That is, each record position is to be either occupied or empty.

    One approach that can be used is first to mark the deleted record as empty. An ordered search is then made for the next empty position. If a record, say, y, is found whose hash value is not between the position of the record just marked for deletion and that of the present empty position, then record y can be moved to replace the deleted record. Then the position for record y is marked as empty and the entire process is repeated, starting at the position occupied by y.

5. Construct a searching algorithm based on double hashing.

6. Alter the algorithm ENTER in the text so that it can perform either an insertion or a search for a particular key.

7. Assuming a nondecreasing hashing function and the sample key set

42 23 74 11 65 57 94 36 99 87 70 81 61

in which the keys are hashed in the ranges 1–20, 21–40, 41–60, 61–80, and 81–100, describe (as in Table 10-13) the behavior of the algorithm ADDR_CAL_SORT for this key set.

10-10.3   RADIX SORTING

The notion of radix sorting was introduced in the exercises at the end of Chap. 4 (see exercise 2). The basic process consists of sorting on a particular digit position. All cards having the same digit value in that digit position are placed in the same pocket. Recall that there are 10 pockets, one for each digit value. An ascending-order sort is realized by performing several separate digit sorts in order. That is, each column (or digit position) is sorted in turn, starting with the lowest-order digit (that is, the rightmost column) first and processing through the other digits (or columns) from right to left.

As an example, let us consider the sorting of the following sequence of keys:

73, 65, 52, 77, 24, 83, 17, 35, 96, 62, 41, 87, 09, 11

After the first pass on the unit digit position of each number, we obtain:

							87		
	11	62	83		35		17		
	41	52	73	24	65	96	77		09
Pocket: 0	1	2	3	4	5	6	7	8	9

We now combine the contents of the 10 pockets in order so that the contents of the 0 pocket are on the bottom and the contents of the 9 pocket are on the top. The resulting sequence thus obtained is

41, 11, 52, 62, 73, 83, 24, 65, 35, 96, 77, 17, 87, 09

The result of performing the second pass of the sort on the higher-order digit yields the following arrangement:

							65	77	87	
	17									
	09	11	24	35	41	52	62	73	83	96
Pocket: 0	1	2	3	4	5	6	7	8	9

These 10 pockets are now combined in the same order as before, thus yielding the desired result.

The computerization of this mechanical process is straightforward. Two possible storage-allocation techniques for representing the data are possible. The sequential allocation method of storage cannot be applied readily to this application, since we cannot predict the size of each pocket. Actually, a pocket could contain all the keys. Using linked allocation, however, the unpredictability of the pocket sizes causes no problems. Each pocket can be represented by a linked FIFO queue. At the end of each pass, these queues can be easily combined into the proper order. Figure 10-44a represents the state of the sort after the first pass for the sample key set just given. Note that each linked queue has an associated top pointer (denoting the rear) and bottom pointer (denoting the front). The positions of the rear elements of the 10 queues (or pockets) are kept in a vector T. Similarly, vector B contains the positions of the bottom elements of the queues. Figure 10-44b represents the state of the queues after the second pass.

More generally, keys which contain m digits require m successive passes, from the unit digit to the highest-order digit, in order to complete a radix sort. We now formulate an algorithm for this sorting process. We assume that a key contains the m digits $b_m b_{m-1} \ldots b_1$ and that a selection mechanism is available for selecting each digit. The initial set of keys is assumed to be arranged as a singly linked list.

Algorithm RADIX_SORT. Given a table of n records arranged as a linked list, where each node in the list consists of a key field (K) and a pointer field (LINK), this algorithm performs a radix sort as previously described. The address of the first record in the linked table is given by the pointer variable FIRST. The vectors T and B are used to store the addresses of the rear and front records in each queue (pocket). In particular, the records T[i] and B[i] point to the top and bottom records in the ith pocket, respectively. The variable j is the pass index. The variable i is used as a pocket index, while r is a temporary index variable. The pointer variable R denotes the address of the current record being examined in the table and being directed to the appropriate pocket. Next is a pointer variable which denotes the address of the next record to be examined. PREV is a pointer variable which is used during the combining of the pockets at the end of each pass. The integer variable d denotes the current digit in a key being examined.

1. [Perform sort]
   Repeat thru step 4 for j = 1, 2, . . ., m
2. [Initialize the pass]
   Repeat for i = 0, 1, . . ., 9
       T[i] ← B[i] ← NULL
   R ← FIRST
3. [Distribute each record in the appropriate pocket]
   Repeat while R ≠ NULL
       d ← b_j (obtain jth digit of the key K(R))
       NEXT ← LINK(R)
       If T[d] = NULL
       then T[d] ← B[d] ← NULL
       else LINK(T[d]) ← R
           R ← NEXT

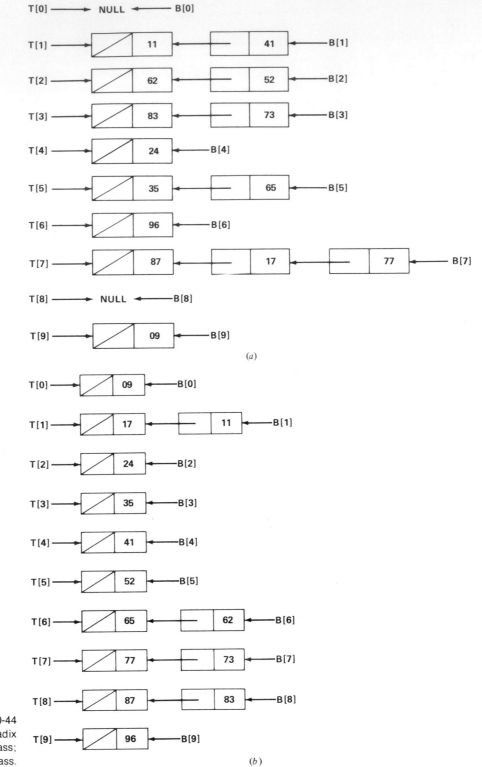

FIGURE 10-44
Representation of a radix
sort: (a) after first pass;
(b) after second pass.

```
        LINK(R) ← NULL
        R ← NEXT
4. [Combine pockets]
      r ← 0
      Repeat while B[r] = NULL
        r ← r + 1
      FIRST ← B[r]
      Repeat for i = r + 1, r + 2, . . ., 9
        PREV ← T[i − 1]
        If T[i] ≠ NULL
        then LINK(PREV) ← B[i]
        else  T[i] ← PREV
5. [Finished]
      Exit                                              □
```

Step 1 controls the number of passes required to perform the sort. The second step initializes the top and bottom pointers associated with each pocket. At the beginning of each pass the pockets are all empty. The variable R is assigned the address of the first record in the table. Step 3 directs each record to the appropriate pocket (depending on the value of d). The fourth step combines the 10 pockets into a new linked table. This revised linked list becomes the input to the next pass. The variable FIRST is initialized to the address of the bottom element of the first nonempty pocket.

This algorithm is efficient for a table which contains keys whose lengths are short. More specifically, for a key containing m digits, the radix sort requires m + n accesses.

EXERCISE 10-10.3

1. For the sample key set

   70   11   81   61   42   23   74   94   65   36   57   87   99

   describe the behavior (as in Fig. 10-44) of algorithm RADIX_SORT.

11

Trees

All data structures discussed thus far are linear. That is, the relationships that can be expressed by such data structures are essentially one-dimensional. In this chapter we introduce a nonlinear data structure called a tree. The tree is perhaps the most important structure in this class. A tree structure is capable of expressing more complex relationships than that of physical adjacency.

The chapter begins with a description of notation and concepts for trees. Section 11-2 contains a discussion of several operations and manipulations that are frequently performed on tree structures. The third section examines many storage structures for both binary trees and general trees. Particular attention is given to linked storage structures since, in general, they are suited to the representation of trees that are subjected to insertions and deletions. Finally, Sec. 11-4 presents several applications of trees. Among these are the applications of trees to searching and sorting.

11-1 NOTATION AND CONCEPTS

In this section we first consider the definition of a general tree and its associated terminology. A number of examples of trees are given. Several equivalent forms for representing trees are introduced. For reasons of simplicity and efficiency, it is convenient to define and manipulate binary trees instead of general trees. Each general tree is easily converted to an equivalent binary tree. An informal discussion of this conversion process is presented.

Trees are very useful in describing any structures which involve hierarchical relationships. Familiar examples of such structures are family trees, the decimal classification of books in a library, the hierarchy of positions in an organization, an algebraic expression involving operations for which certain rules of precedence are prescribed, and the structure of the contents of this chapter.

There are two types of charts which are used in the investigation of geneologies. The first type, exhibited in Fig. 11-1a, is called a *pedigree* chart. This chart specifies a person's ancestors. In the example chart given, André's

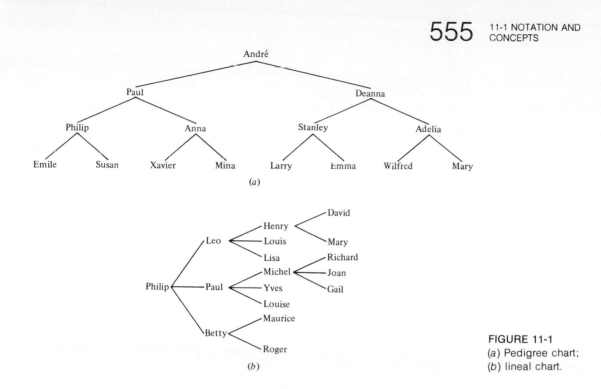

FIGURE 11-1
(a) Pedigree chart;
(b) lineal chart.

parents are Paul and Deanna. Paul's parents are Philip and Anna, who are André's grandparents on his father's side. The chart continues to the great-grandparent stage. Note that in this chart we have two-way branching. The second type of chart is called a *lineal* chart. Such a chart describes the descendants of an individual, as shown in Fig. 11-1b. In general, this chart can have many-way branching. The structure exhibited by a lineal chart is clearly different from the pedigree chart. Both are trees, however.

A partial organizational chart for a university is exemplified in Fig. 11-2. This chart exhibits the chain of command (or hierarchy) in the institution. The president is the highest level of command. The second level of command involves the academic and administrative vice presidents. The academic vice president controls a number of colleges, each of which has a dean. Each dean, in turn, controls several departments, each of which has a department head.

As a final example, consider the table of contents chart given for Chap. 11 (Fig. 11-3). This chapter contains four sections. The last two sections are each divided into four subsections.

All the tree structures introduced so far have a number of properties in common. First, in each tree there is a distinguished item or node called its *root*. For example, the root node in Fig. 11-1a is "André." A node which does not have any lines or branches emanating from it is called a *terminal node* or *leaf*; all other nodes are called *branch nodes*. Also, all trees have a finite number of nodes.

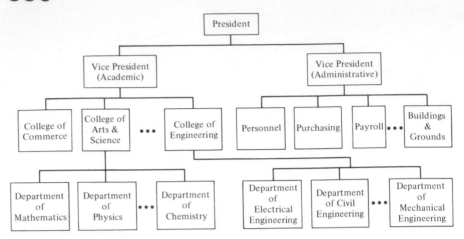

FIGURE 11-2
Partial organizational
chart of a university.

Let us now formalize the notion of a tree. Several definitions are possible. One popular definition involves defining a general graph (recall Sec.
6-5.3) and then restricting this graph to obtain a tree. Since we have not
dealt with graphs to any significant degree in this book, we do not use this
approach. As we shall see throughout this chapter, trees can be handled
easily within a recursive framework. As a first step in that direction, a tree
can be defined recursively as follows:

A *tree* is a finite set of one or more nodes such that:

1. There is a specially designated node called the root
2. The remaining nodes are partitioned into disjoint subsets $T_1, T_2, \ldots,$
   and T_n ($n \geq 0$), each of which is a tree. Each T_i ($1 \leq i \leq n$) is called a
   subtree of the root.

As an example, Fig. 11-4 represents a tree of 12 nodes. The root of the
tree is v_0. The three subtrees of this root node are the subsets $T_1 = \{v_1, v_2, v_3,$
$v_4, v_5, v_6\}$, $T_2 = \{v_7\}$, and $T_3 = \{v_8, v_9, v_{10}, v_{11}\}$. The root of T_1 is node v_1 and it has
three subtrees: $T_{11} = \{v_2\}$, $T_{12} = \{v_3\}$, and $T_{13} = \{v_4, v_5, v_6\}$. The tree T_{11} contains
a root node (that is, v_2), which is a leaf. The subtree T_{12} is also a leaf node.
T_{13} has a root node (v_4) and two subtrees: $\{v_5\}$ and $\{v_6\}$. The remaining parts
of the tree can be analyzed in a similar manner. Figure 11-4 is only one possible diagram for the example tree; many other diagrams are possible for the
same tree. These alternative diagrams are obtained by choosing different
relative positions of the nodes with respect to the root. The representation
chosen here is the one which is most commonly used in the literature.

An important notion in dealing with trees is that of the level of a node.
The *level* of any node is 1 plus the length of its path from the root. The level
of the root of a tree is 1, while the level of any other node is equal to 1 plus its
distance from the root. Another important property of a node is its degree.

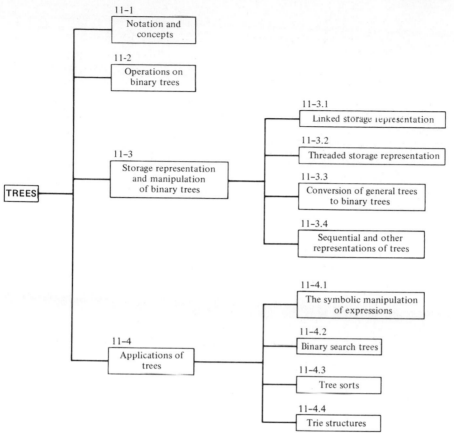

11-1 Notation and concepts

11-2 Operations on binary trees

11-3 Storage representation and manipulation of binary trees

11-3.1 Linked storage representation

11-3.2 Threaded storage representation

11-3.3 Conversion of general trees to binary trees

11-3.4 Sequential and other representations of trees

11-4 Applications of trees

11-4.1 The symbolic manipulation of expressions

11-4.2 Binary search trees

11-4.3 Tree sorts

11-4.4 Trie structures

TREES

FIGURE 11-3
Tree representation of the table of contents for Chap. 11.

The *degree* of a node is its number of subtrees. For example, the tree of Fig. 11-4 has one node of level 1, three nodes of level 2, five nodes of level 3, and three nodes of level 4. The root node has a degree of 3 and v_{10} is of degree 1. The degree of a leaf is 0.

In many applications the relative order of the children of a node at any particular level is important. In a storage representation of a tree, such an order, even if it is arbitrary, is automatically implied. It is easy to impose an order on the children of a node at each level by referring to a particular node as the first child, to another node as the second child, and so on. In the diagrams the ordering may be done from left to right. A tree whose nodes at each level are so ordered is called an *ordered tree*. For example, the trees in Fig. 11-5 are equivalent as far as trees are concerned, but they represent different ordered trees. Since we are interested primarily in ordered trees in this chapter, we use the term "tree" to mean "ordered tree" unless otherwise stated.

As mentioned earlier, family trees are perhaps one of the earliest applications of trees. They have left their mark as far as terminology is concerned.

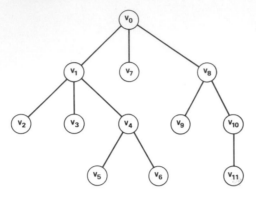

FIGURE 11-4

In fact, standard tree terminology is taken from the lineal chart. In particular, each root is called the parent, father or mother, of the roots of its subtrees which in turn are denoted as siblings, brothers or sisters. The latter nodes are called offspring, sons or daughters, of the parent node. For example, in Fig. 11-4 v_0 is the parent node, which has three offspring or direct descendants. Node v_1, in turn, also has three offspring, v_2, v_3, and v_4, which are siblings. v_0 is the grandparent of v_2 and the great grandparent of v_5. The node v_1 is an uncle or aunt of v_9.

If we delete the root and its associated branches which connect the nodes at level 1, we obtain a set of disjoint trees. Such a set of disjoint trees is called a *forest*. We have also seen that any node of a tree is the root of some subtree. Therefore, subtrees immediately below a certain node form a forest.

There are many other ways in which a tree can be represented graphically. These methods of representation for the tree of Fig. 11-4 are given in Fig. 11-6a–d. The first method uses a familiar technique known as Venn diagrams to show subtrees, the second involves the use of nested parentheses to show structure and adjacency, and the third method is the one used in the table of contents of a book. The last method, which is based on a level-number format, is similar to those techniques used in PL/I and COBOL for specifying hierarchical structures. Using this format, each node is assigned a number. The root of the tree has the smallest number. The number associated with a given node must be less than the numbers associated

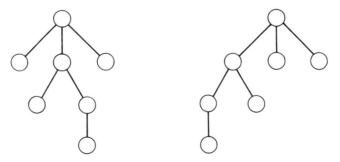

FIGURE 11-5
Equivalent trees with different ordered tree representations.

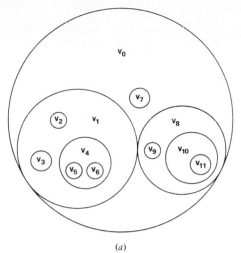

(a)

$$(v_0\,(v_1\,(v_2)\,(v_3)\,(v_4\,(v_5)\,(v_6)))\,(v_7)\,(v_8\,(v_9)\,(v_{10}\,(v_{11}))))$$

(b)

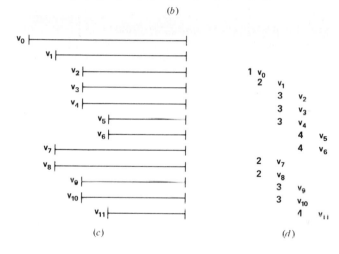

(c)   (d)

FIGURE 11-6
Different representations of trees: (a) Venn diagram; (b) nested parentheses; (c) bar chart; (d) level-number notation.

with the root nodes of its subtrees. Note that all the root nodes of the subtrees of a given node must have the same level number.

The method of representation given in Fig. 11-6b immediately indicates how any completely parenthesized algebraic expression can be represented by a tree structure. Naturally, it is not necessary to have a completely parenthesized expression if we prescribe a set of precedence rules, as discussed in Sec. 10-6. As an example, consider the expression

$$v_1 - v_2 * (v_3 * v_4 + v_5 \uparrow v_6)$$

The tree corresponding to this expression is given in Fig. 11-7.

So far we have not placed any restriction on the degree of a node (that

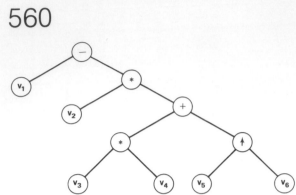

FIGURE 11-7
Tree representation of an
arithmetic expression.

is, the number of branches which emanate from a node). If the degree of every node is less than or equal to 2, then the tree is called a binary tree. Furthermore, in binary trees we distinguish between the left subtree and the right subtree of each node. No such distinction between subtrees was made in the case of an ordered tree. The following definition incorporates these restrictions:

> A binary tree *is a finite set of* m (m ⩾ 0) *nodes consisting of a root node which has two disjoint binary subtrees called the* left subtree *and the* right subtree. *A binary tree of zero nodes is said to be* empty.

A binary tree in which each node is of degree 0 or 2 is called a *full binary tree*.

Figure 11-8*a* exemplifies a binary tree, Fig. 11-8*b* shows a full binary tree, and Fig. 11-8*c* displays another binary tree. Note that, although the trees of Fig. 11-8*a* and *c* are not distinct ordered trees, these two trees are certainly two distinct binary trees. An application of full binary trees to sorting will be discussed in Sec. 11-4.3.

Since binary trees are easily represented and manipulated, it is convenient to convert any tree to an equivalent binary form, if possible. We shall now show that every tree can indeed be represented by an equivalent binary tree. This correspondence is a one-to-one relationship. This notation can be extended to represent a forest by a binary tree.

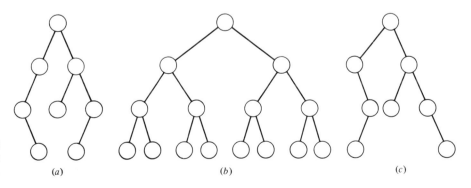

FIGURE 11-8
Examples of binary trees
and complete binary trees.

(*a*)   (*b*)   (*c*)

Figure 11-9 shows in two steps how to convert an ordered tree into its binary equivalent. In the first step, we delete all the branches originating in every node except the leftmost branch. Also, we connect all siblings at the same level by branches. That is, the first sibling is connected to the second, the second is connected to the third, and so on. The second step involves choosing the left and right offspring for each node. This task is accomplished in the following manner. The left offspring is the node immediately below the given node and the right offspring is the node to the immediate right of the given node on the same horizontal line. The resulting binary tree has an empty right subtree.

The preceding method of representing any ordered tree by a unique binary tree can be extended to an ordered forest, as shown in Figure 11-10.

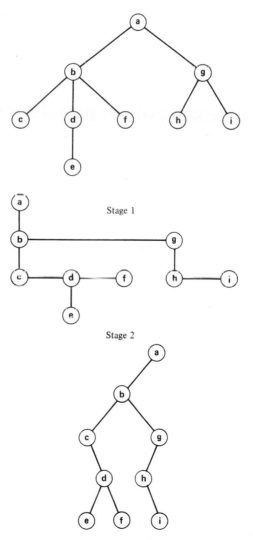

Stage 1

Stage 2

FIGURE 11-9
Binary tree representation of a tree.

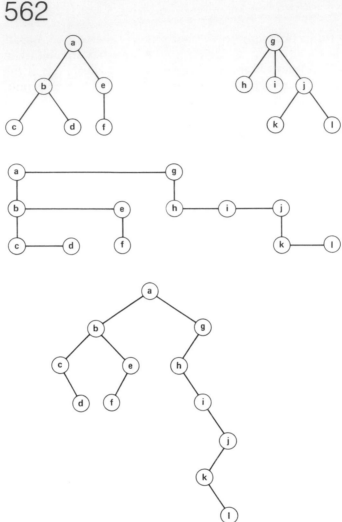

FIGURE 11-10
Binary tree representation
of a forest.

This correspondence is called the *natural correspondence* between ordered trees and binary trees, and also between ordered forests and binary trees. An algorithm to perform the indicated conversion will be given in Sec. 11-3.3.

In this section we have examined certain basic notions and concepts associated with tree structures. We have not, however, considered what operations might be performed on these structures. A number of such operations are the topic of the next section.

EXERCISES 11-1

1. How many different trees, ordered trees, and binary trees are there with three nodes?

2. Give an ordered tree representation of the formula

   $((a - b) * c) / (d + f \uparrow g)$

3. Give the subtrees of the tree given in Fig. 11-11. Also, give the level and degree of each node in this tree.

4. Obtain the binary trees that correspond to the tree and forest given in Figs. 11-11 and 11-12, respectively.

11-2   OPERATIONS ON BINARY TREES

The previous section introduced the notions of a binary tree structure. Little mention, however, was made of what types of operation are performed on such structures. In this section we introduce several of these operations, such as the traversal of trees, insertion, and deletion.

One of the most common operations performed on tree structures is that of traversal. This is a procedure by which each node in the tree is processed exactly once in a systematic manner. The meaning of "processed" depends on the nature of the application. For example, the tree of Fig. 11-13 represents an arithmetic expression. In this context the processing of a node which represents an arithmetic operation would probably mean performing or executing that operation. There are three main ways of traversing a binary tree: in preorder, in inorder, and in postorder. We now examine each traversal order. The easiest way to define each order is by using recursion.

The *preorder traversal* of a binary tree is defined as follows:

1. Process the root node
2. Traverse the left subtree in preorder
3. Traverse the right subtree in preorder

If a particular subtree is empty (that is, a node has no left or right descendant), the traversal is performed by doing nothing. In other words, a

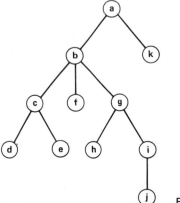

FIGURE 11-11

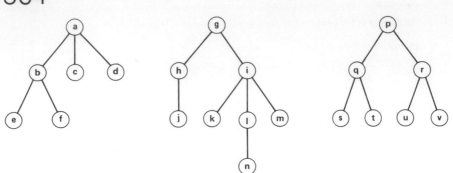

FIGURE 11-12

null subtree is considered to be fully traversed when it is encountered. The preorder traversal of the tree in Fig. 11-13 gives the following processing order:

$$+ - * A B \uparrow C D / E F$$

Note that this traversal order yields the same result as the prefix Polish equivalent of the expression discussed in Sec. 10-5.2.

The *inorder traversal* of a binary tree is given by the following steps:

1. Traverse the left subtree in inorder
2. Process the root node
3. Traverse the right subtree in inorder

The inorder traversal of the example tree given in Fig. 11-13 results in the following processing order:

$$A * B - C \uparrow D + E / F$$

This traversal order gives the infix form of the expression. A trace of the inorder traversal of the sample expression is given in Table 11-1.

Finally, we define the *postorder traversal* of a binary tree as follows:

1. Traverse the left subtree in postorder
2. Traverse the right subtree in postorder
3. Process the root node

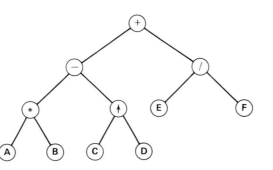

FIGURE 11-13
Binary tree representation
of an expression.

TABLE 11-1 Trace of Inorder Traversal of a Binary Tree

Invocation Level of Inorder Procedure	Information Content of Root of Tree	Action
0 (main)	'+'	
1	'−'	
2	'*'	
3	'A'	
4	Empty subtree	Process 'A'
4	Empty subtree	Process '*'
3	'B'	
4	Empty subtree	Process 'B'
4	Empty subtree	Process '−'
2	'↑'	
3	'C'	
4	Empty subtree	Process 'C'
4	Empty subtree	Process '↑'
3	'D'	
4	Empty subtree	Process 'D'
4	Empty subtree	Process '+'
1	'/'	
2	'E'	
3	Empty subtree	Process 'E'
3	Empty subtree	Process '/'
2	'F'	
3	Empty subtree	Process 'F'
3	Empty subtree	

The postorder traversal of the sample tree gives the following processing order:

A B * C D ↑ − E F / +

It should be noted by the reader that this processing order is equivalent to the expression's suffix Polish form.

If the words "left" and "right" are interchanged in the preceding definitions, we obtain three new traversal orders, which are called *converse preorder*, *converse inorder*, and *converse postorder*, respectively. The converse traversal orders for the example tree of Fig. 11-13 are

+ / F E − ↑ D C * B A   (converse preorder)

F / E + D ↑ C − B * A   (converse inorder)

F E / D C ↑ B A * − +   (converse postorder)

We will not, however, pursue the applicability of these converse traversals in this book.

So far, we have conveniently assumed that a tree somehow already

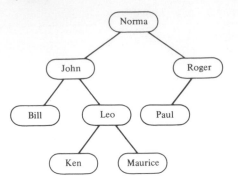

FIGURE 11-14
Lexically ordered binary
tree of eight names.

exists. Let us now examine the problem of constructing such a binary tree. The approach taken in constructing a tree is often application-dependent. For example, the binary tree representation of an expression, such as that given in Fig. 11-13, depends on the left and right operands associated with each binary operator. In this particular example, an algorithm, based on the Polish notation discussion of Sec. 10.5, could be formulated.

Rather than use the Polish notation approach here, however, we shall assume that the desired tree is to be kept in some kind of order. The order that we choose throughout the remainder of the discussion is based on the information content associated with each node. For example, assume that a list of names is to be kept in lexicographical order. That is, the left subtree of the tree (or subtree) is to contain nodes whose associated names are lexically less than the name associated with the root node of the tree (or subtree). Similarly, the right subtree of the tree (or subtree) is to contain nodes whose associated names are lexically greater than the name associated with the root node of the tree (or subtree). An example of such a tree for a set of first names is exemplified in Fig. 11-14. Note that the root node (Norma) lexically follows and precedes all names in its left subtree (Bill, John, Ken, Leo, and Maurice) and right subtree (Paul and Roger), respectively. The same relationships hold for the root node of each subtree.

The tree of Fig. 11-14 can be created by the repeated use of an algorithm which can insert one node into an existing binary tree such that the tree is still lexically ordered after insertion.

A general algorithm for performing such an insertion is as follows:

1. If the existing tree contains no nodes, then append the new node as the root node of the tree and Exit
2. Compare the new name with the name of the root node
    If the new name is lexically less than the root node name,
        then If the left subtree is not empty,
            then repeat step 2 on the left subtree;
            else append the new name as a left leaf to the present tree and Exit;

else If the right subtree is not empty
        then repeat step 2 on the right subtree;
        else append the new name as a right leaf to the present tree
              and Exit

This algorithm assumes that no attempt has been made to enter duplicate name entries. Assuming the insertion sequence of names

Norma, Roger, John, Bill, Leo, Paul, Ken, and Maurice

the behavior of the previous algorithm in the construction of the tree given in Fig. 11-14 is exhibited in Fig. 11-15.

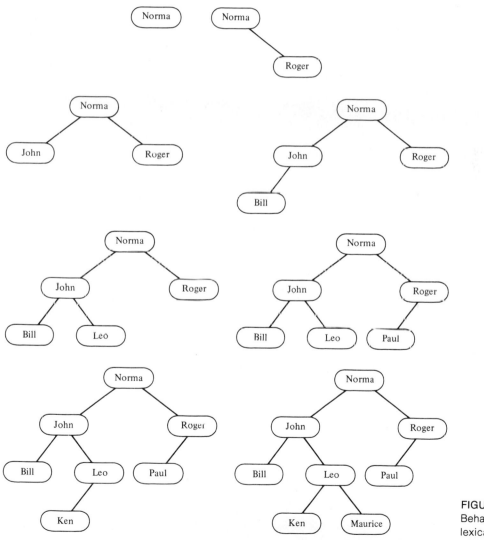

FIGURE 11-15
Behavior of creating a lexically ordered tree.

Another common operation which is performed on a lexically ordered binary tree is the deletion of an arbitrary node. By an arbitrary node, we mean that *any* node in the tree can be deleted, even its root. Consequently, a number of cases arise. A simple case occurs when the node marked for deletion contains an empty left and/or right subtree(s). Such a case is shown in Fig. 11-16*a*, where the node labeled Roger is deleted. If, however, it is required to delete a node whose left and right subtrees are nonempty, that node's inorder successor is first deleted and then used to replace the node initially marked for deletion. Note that the successor node in question always has an empty left subtree (by definition of inorder traversal). This second deletion possiblity is exemplified in Fig. 11-16*b*, where John is marked for deletion. In this case the inorder successor of John is Ken. This latter node replaces the former in the revised tree. In the deletion process the right subtree of Ken (that is, Kirk) becomes the left subtree of Ken's parent (Leo). Also, Ken becomes the new left offspring of John's parent (that is, Norma).

A general algorithm for deleting an arbitrary node from a lexically ordered tree follows:

1. Determine the parent node of the node marked for deletion, if it exists; note that it will not exist if we are deleting the root node
2. If the node being deleted has either a left or right empty subtree, then append this nonempty subtree to its grandparent node
   (that is, the node found in step 1) and Exit
3. Obtain the inorder successor of the node to be deleted
   Append the right subtree of this successor node to its grand parent
   Replace the node to be deleted by its inorder successor
   This is accomplished by appending the left and right subtrees (with the aforementioned successor node) of the node marked for deletion to the successor node
   Also, the successor node is appended to the parent of the node just deleted (that is, the node obtained in step 1)

Probably the most frequently performed operation on a binary tree is that of searching. Since searching is very often associated with a lexically ordered tree structure, we examine this case briefly here. Many more details of search trees are given in Sec. 11-4.2. Given a certain item which is to be searched for in a lexically ordered binary tree, the following general algorithm performs the desired task.

1. If the tree contains no nodes, then Write ('item not found') and Exit
2. Compare the given item with the item represented by the root node
   If the given item is lexically less than the root node item,
   then If the left subtree is not empty,
      then repeat step 2 on the left subtree;

else Write('item not found') and Exit;
  else If the given item is lexically greater than the root node item,
      then If the right subtree is not empty,
          then repeat step 2 on the right subtree;
          else Write('item not found') and Exit;
  else Write('item has been found') and Exit

Basically, this algorithm starts at the root of the tree and branches left or right repeatedly until either the given item is found or an empty subtree is encountered, in which case the item is not in the tree.

Another familiar operation associated with binary tree structures is that of sorting. For a lexically ordered binary tree, this sorting operation becomes equivalent to traversing the tree in inorder. For example, the inorder traversal of the binary tree given in Fig. 11-14 would result in the sequence

Bill, John, Ken, Leo, Maurice, Norma, Paul, and Roger

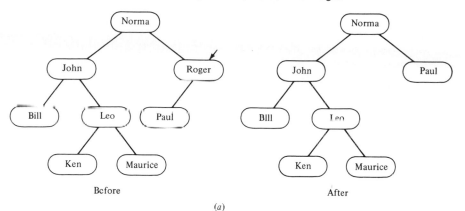

(a)

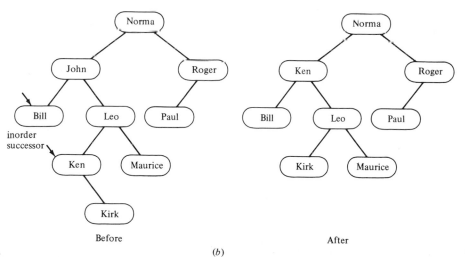

(b)

FIGURE 11-16
Deletion of a node from a lexically ordered binary tree.

Another application of trees to sorting is given in Sec. 11-4.3.

Many other operations on trees are possible. For example, the copying of a tree occurs in many applications. We will not, however, give an algorithm for this operation here. Rather, we leave the formulation of this algorithm to the next section, where several possible storage structures of binary trees are discussed.

EXERCISES 11-2

1. Traverse the tree of Fig. 11-17 in preorder, inorder, and postorder.

2. Give a trace (as in Table 11-1) of each traversal order obtained in exercise 1.

3. Assuming an insertion sequence of

   Ken, Roger, Bill, Leo, Paul, Norma, Maurice, and John

   construct (as in Fig. 11-15) a lexically ordered tree for this sequence of insertions.

4. Delete node D from the tree given in Fig. 11-18. Starting again from the original tree, delete node G.

11-3 STORAGE REPRESENTATION AND MANIPULATION OF BINARY TREES

In Chap. 10 the storage (computer) representation of certain elementary data structures such as linear lists and arrays was presented. In this section we extend these concepts to tree structures.

Linked and sequential allocation techniques will be used to represent these tree structures. The advantages and disadvantages of each allocation technique were also presented in Chap. 10. In this section we will give greater emphasis to linked storage structures. These linked structures are

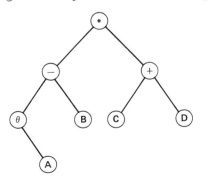

FIGURE 11-17

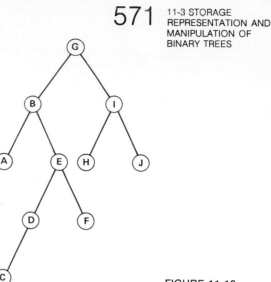

FIGURE 11-18

more popular than their corresponding sequential structures because, in performing insertions and deletions, the former structures are more easily altered than the latter. Furthermore, since the size of a tree structure is often unpredictable, linked allocation techniques are more appropriate.

The linked storage representation of binary trees is introduced first. Based on this representation, several algorithms, such as those for traversing and copying tree structures, are given. The concept of "threaded" binary trees is then introduced. The storage representation of a tree based on the threading concept is efficient from both time and space considerations. Next, an algorithm for the conversion of a general tree to an equivalent binary tree (see Sec. 11-1) is given. Finally, several sequential allocation techniques for binary trees are examined.

11-3.1 LINKED STORAGE REPRESENTATION

Since a binary tree consists of nodes which can have at most two offspring, an obvious linked representation of such a tree involves having storage nodes of the form

| LPTR | INFO | RPTR |

where LPTR and RPTR denote the addresses or locations of the left and right subtrees, respectively, of a particular root node. Empty subtrees are represented by a pointer value of NULL. INFO specifies the information contents of a node.

Figure 11-19a and b contains examples of a binary tree and its linked storage representation. The pointer variable T denotes the address of the root node. The two forms are remarkably similar. This similarity illustrates that the linked storage representation of a binary tree very closely reflects

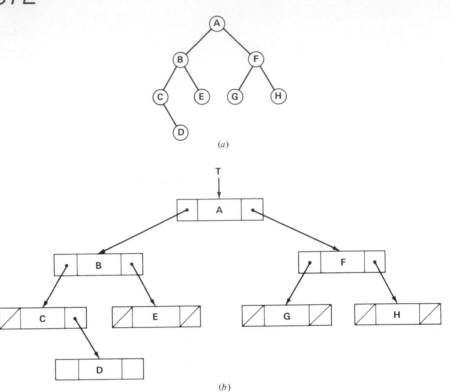

FIGURE 11-19
Binary tree and its linked
representation.

(a)

(b)

the logical structuring of the data involved. This property is very useful and desirable in designing algorithms which process binary tree structures.

We can now fill in the details of the general algorithms given in the previous section for the preorder, inorder, and postorder traversals of a binary tree. These algorithms are written as procedures of one parameter. The only parameter required is a pointer variable which contains the address of the root of the tree.

The algorithm for preorder traversal follows.

Procedure RPREORDER(T). Given a binary tree whose root node address is given by a pointer variable T and whose node structure is the same as previously described, this algorithm traverses the tree in preorder in a recursive manner. The tree is assumed to be nonempty.

1. [Process the root node]
    If T ≠ NULL then Write(INFO(T))
2. [Process the left subtree]
    If LPTR(T) ≠ NULL then Call RPREORDER(LPTR(T))
3. [Process the right subtree]
    If RPTR(T) ≠ NULL then Call RPREORDER(RPTR(T))
4. [Finished]
    Return □

A similar algorithm for the recursive inorder traversal of a binary tree is easily formulated.

Procedure RINORDER(T). Given a binary tree whose root node address is given by a pointer variable T and whose node structure is the same as previously described, this algorithm traverses the tree in preorder, again in a recursive manner. The tree is assumed to be nonempty.
1. [Process the left subtree]
      If LPTR(T) ≠ NULL then Call RINORDER(LPTR(T))
2. [Process the root node]
      Write(INFO(T))
3. [Process the right subtree]
      If RPTR(T) ≠ NULL then Call RINORDER(RPTR(T))
4. [Finished]
      Return   ☐

Since some programming languages do not allow recursion, it is sometimes necessary to formulate an equivalent iterative solution to each traversal order. Consequently, we now consider the traversal of binary trees by iteration. In traversing a tree, it is required to descend and subsequently to ascend parts of the tree. For this reason, information which will permit movement up the tree must be temporarily saved. Observe that downward movement in the tree is possible through the structural links of pointers. Because movement up the tree must be made in a reverse manner from that taken in descending the tree, temporary pointer values must be stacked as the tree is traversed. The following algorithm iteratively traverses a binary tree in preorder.

Algorithm IPREORDER. Given a binary tree whose root node address is given by a pointer variable T and whose node structure is the same as previously described, this algorithm traverses the tree in preorder, in an iterative manner. S and TOP denote the stack and its associated top index, respectively. The pointer variable P denotes the current node in the tree during processing.
1. [Initialize]
      If T = NULL
      then Write('EMPTY TREE')
            Exit
      else (initialize stack)
            TOP ← 1
            S[TOP] ← T
2. [Process each stacked branch address]
      Repeat step 3 while TOP > 0
3. [Get stored address and branch left]
      P ← S[TOP]

```
      TOP ← TOP − 1
      Repeat while P ≠ NULL
        Write(INFO(P))
        If RPTR(P) ≠ NULL
        then   (store address of nonempty right subtree)
                    TOP ← TOP + 1
                    S[TOP] ← RPTR(P)
            P ← LPTR(P)   (branch left)
  4. [Finished]
      Exit                                                         □
```

Step 1 checks for an empty tree and exits if T = NULL. Otherwise, it stacks the address of the root node. Step 2 controls the processing of the tree. The addresses of yet untraversed subtrees are kept on the stack. In the third step of the algorithm, we process each node. For our purposes this merely involves the writing out of the information content of the node in question. In general, of course, what is meant by "process" is application-dependent. The address of the right subtree, if it is not empty, is stacked and a left branch is taken. This left branching process (and the associated stacking of nonempty right subtree addresses) continues until we encounter an empty left link. At this point we unstack the address of the most recently encountered right subtree and continue processing it according to step 3. The behavior of this algorithm for the binary tree of Fig. 11-19*b* appears in Table 11-2, where the rightmost element in the stack is its top element. The notation NB, for example, denotes the address of node B. The processing of each node involves the output of the label associated with that node.

The next algorithm, which is similar to algorithm IPREORDER, traverses a tree in inorder.

TABLE 11-2   Behavior of Algorithm IPREORDER for Fig. 11-19*b*

Stack Contents	P	"Process" P	Output
NT			
	NT	A	A
NF	NB	B	AB
NF NE	NC	C	ABC
NF NE ND	NULL		
NF NE	ND	D	ABCD
NF NE	NULL		
NF	NE	E	ABCDE
NF	NULL		
	NF	F	ABCDEF
NH	NG	G	ABCDEFG
NH	NULL		
	NH	H	ABCDEFGH

Algorithm IINORDER. Given a binary tree with a root node address of T and a node structure as previously described, this algorithm traverses iteratively a binary tree in inorder. S and T denote the stack and its associated top index, respectively. P is a pointer variable which denotes the current node in the tree during processing.

1. [Initialize]
   If T − NULL
   then Write('EMPTY TREE')
      Exit
   else TOP ← 0
      P ← T
2. [Traverse the tree in inorder]
   Repeat thru step 4
3. [Stack addresses along a left chain]
   Repeat while P ≠ NULL
      TOP ← TOP + 1
      S[TOP] ← P
      P ← LPTR(P)
4. [Process node and right branch]
   If TOP > 0
   then P ← S[TOP]
      TOP ← TOP − 1
      Write(INFO(P))
      P ← RPTR(P)
   else Exit   □

Step 1 either exits, in the case of an empty tree, or initializes P to the root address of the tree. The second step controls the traversal of the tree. Observe that the control loop as it stands will never terminate. An exit, however, will eventually be made in step 4. Step 3 chains through a sequence of left branches until an empty left subtree is encountered.

We stack the address of each node on this chain. When P − NULL, we proceed to step 4. At this point the address of the most recently saved node is removed from the stack, the information content of this node is output, and the right branch of the node is followed. A return to step 3 then processes this right subtree. The behavior of this algorithm for the example tree of Fig. 11-19*b* is given in Table 11-3.

A similar algorithm can be formulated for the postorder traversal of a tree, but this task is left as an exercise.

It is often convenient in the manipulation of trees to have a tree with a head node as discussed in Sec. 10-9.2. Figure 11-20*a* represents a binary tree with a list head. An empty binary tree takes the form of a list head whose left link is NULL. Observe that we have chosen to append the tree as a left subtree to the list head node.

We now proceed to formulate an algorithm which performs an insertion

TABLE 11-3 Behavior of Algorithm IINORDER for Fig. 11-19*b*

Stack Contents	P	"Process" P	Output
	NA		
NA	NB		
NA NB	NC		
NA NB NC	NULL		
NA NB	NC	C	C
NA NB	ND		
NA NB ND	NULL		
NA NB	ND	D	CD
NA NB	NULL		
NA	NB	B	CDB
NA	NE		
NA NE	NULL		
NA	NE	E	CDBE
NA	NULL		
	NA	A	CDBEA
	NF		
NF	NG		
NF NG	NULL		
	NG	G	CDBEAG
NF	NULL		
	NF	F	CDBEAGF
	NH		
NH	NULL		
	NH	H	CDBEAGFH

into a lexically ordered binary tree. We assume that the type of key associated with each node is alphabetic. If the storage representation of such an ordered tree has a list head, care must be taken to preserve order between the list head and the rest of the tree. This is accomplished easily by making the unused information field in the head node lexically greater than the information of any other node in the tree. In this way the tree will always be appended as a left subtree of the list head. The list head representation of such a lexically ordered tree is given in Fig. 11-21.

The general algorithm for inserting a node in such a tree structure was discussed in Sec. 11-2. A new entry is appended as a left or right subtree (consisting of a leaf node) to some parent node. Since such an insertion algorithm can be used repeatedly to construct the desired storage tree, we formulate the following subalgorithm.

Procedure INSERT_NODE(HEAD, X). Given a binary tree with a list head (HEAD) and the information contents (X) of a node which is to be appended to the existing tree structure, this subalgorithm appends the indicated node as a leaf node, if it is not already there. The structure NODE, consisting of LPTR, INFO, and RPTR, is assumed to be global to this subalgorithm. The

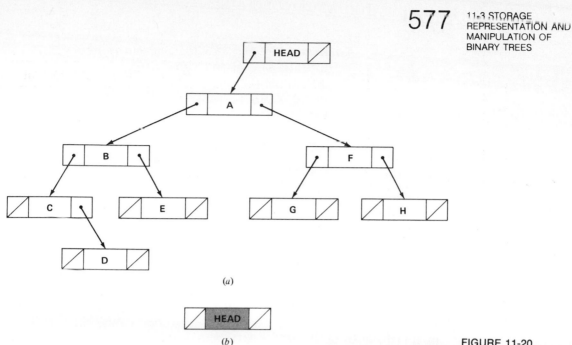

(a)

(b)

FIGURE 11-20

pointer variables NEW and PARENT represent the address of the new node
and its parent, respectively. T denotes the address of the current node being
examined.

1. [Initialize]
    PARENT ← T ← HEAD
2. [Perform indicated insertion, if required]
    Repeat step 3 while T ≠ NULL
3. [Find the location and append new node]
    If X < INFO(T)
    then (branch left)
        If LPTR(T) ≠ NULL
        then PARENT ← T
            T ← LPTR(T)
        else (append new node as a left subtree)
            NEW ⇐ NODE
            LPTR(NEW) ← RPTR(NEW) ← NULL
            INFO(NEW) ← X
            LPTR(PARENT) ← NEW
            Return
    else If X > INFO(T)
        then If RPTR(T) ≠ NULL
            then (branch right)
                PARENT ← T
                T ← RPTR(T)
            else (append new node as a right subtree)

```
            NEW ⇐ NODE
            LPTR(NEW) ← RPTR(NEW) ← NULL
            INFO(NEW) ← X
            RPTR(PARENT) ← NEW
            Return
    else    (node already there)
            Write('DUPLICATE NODE', X)
            Return                                    □
```

This algorithm is straightforward. Note that the repeat statement in step 2 never fails since we check for empty left and right subtrees in step 3. On encountering such an empty subtree, we append the new node to the existing tree structure. Also, duplicate names are ignored (with an appropriate message being generated). The following main algorithm repeatedly invokes the procedure INSERT_NODE in order to create the desired tree.

Algorithm CREATE_TREE. This algorithm creates a binary tree with a list head. The input consists of a series of information items, each of which is an alphabetic string (NAME). HEAD denotes the address of the head node. The previously described node structure is assumed.

1. [Create head node for the tree]
      HEAD ⇐ NODE
      LPTR(HEAD) ← RPTR(HEAD) ← NULL
      INFO(HEAD) ← 'ZZ...Z'
2. [Create desired tree]
      Repeat while there is no more input
        Read(NAME)
        Call INSERT_NODE(HEAD, NAME)
3. [Finished]
      Exit   □

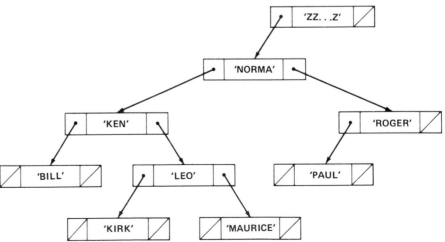

FIGURE 11-21
Lexically ordered binary
tree with a list head.

Another familiar operation that is required in manipulating binary trees involves making a duplicate copy of a tree. Frequently, the original tree may be destroyed during processing. Therefore, a copy of the tree is produced before such processing begins. The following recursive subalgorithm generates a copy of a tree.

Function COPY(T). Given a binary tree whose root node address is given by the pointer variable T and a node structure (NODE) as previously described, this subalgorithm generates a copy of the tree and returns the address of its root node. NEW is a temporary pointer variable.

1. [Null pointer?]
    If T = NULL
    then Return(NULL)
        Exit
2. [Create a new node]
    NEW ⇐ NODE
3. [Copy information field]
    INFO(NEW) ← INFO(T)
4. [Set the structural links]
    LPTR(NEW) ← COPY(LPTR(T))
    RPTR(NEW) ← COPY(RPTR(T))
5. [Return address of new node]
    Return(NEW)   □

Again, this subalgorithm is simple and requires no further comment. An iterative formulation of this operation is also possible. This version, however, is left as an exercise.

In terminating this subsection we formulate an algorithm for deleting a specified node from a binary tree. Recall from Sec. 11-2 that essentially two different cases arise. The first case, which is the simplest, involves deleting a node which has at least one empty subtree. The second (and more difficult) case concerns deleting a node whose two subtrees are nonempty. As mentioned in the previous section, we must find the inorder successor of the node to be deleted in order to perform the indicated deletion.

The following algorithm performs the required task.

Algorithm TREE_DELETE. Given a lexically ordered binary tree with the node structure previously described and the information value (X) of the node marked for deletion, this algorithm deletes the node whose information field is equal to X. PARENT is a pointer variable which denotes the address of the parent of the node marked for deletion. P denotes the address of the node to be deleted. PRED and SUC are pointer variables used to find the inorder successor of P. Q contains the address of the node to which either the left or right link of the parent of X must be assigned in order to complete the deletion. Finally, D contains the direction from the parent node to the node marked for deletion. Also, the tree is assumed to have a list head whose address is given by HEAD.

1. [Initialize]

    If LPTR(HEAD) ≠ NULL

    then P ← LPTR(HEAD)

        PARENT ← HEAD

        D ← 'L'

    else Write('NODE NOT FOUND')

        Exit

2. [Search for and delete the marked node]

    Repeat while P ≠ NULL

      If X < INFO(P)

      then   (branch left)

            PARENT ← P

            P ← LPTR(P)

            D ← 'L'

      else If X > INFO(P)

         then   (branch right)

               PARENT ← P

               P ← RPTR(P)

               D ← 'R'

         else   (indicated node has been found, so delete it)

             If LPTR(P) = NULL

             then   (empty left subtree)

                 Q ← RPTR(P)

             else If RPTR(P) = NULL

               then   (empty right subtree)

                   Q ← LPTR(P)

               else   (check right son)

                 PRED ← RPTR(P)

                 If LPTR(PRED) = NULL

                 then LPTR(PRED) ← LPTR(P)

                   Q ← PRED

                 else   (search for successor of P)

                   SUC ← LPTR(PRED)

                   Repeat while LPTR(SUC) ≠ NULL

                     PRED ← SUC

                     SUC ← LPTR(PRED)

                   (connect successor)

                   LPTR(PRED) ← RPTR(SUC)

                   LPTR(SUC) ← LPTR(P)

                   RPTR(SUC) ← RPTR(P)

                   Q ← SUC

           If D = 'L' then LPTR(PARENT) ← Q else RPTR(PARENT) ← Q

           Exit

3. [Search for indicated node has failed]
   Write('NODE NOT FOUND')
   Exit   □

The first step of the algorithm checks for an empty tree. If the tree is empty, then the required node cannot be found and the algorithm terminates; otherwise, a left branch is taken and the direction D is set to 'L' (indicating a left branch). Step 2, although lengthy, is not complex. The first part of this step searches for the node to be deleted. If it is not found, then the Repeat statement fails and control passes to step 3, where we print an appropriate message. When the desired node is found, it must be deleted from the tree. As mentioned earlier, two cases arise. These cases are exhibited in Fig. 11-16. The tracing of the algorithm is left as an exercise.

On examining the previously chosen storage representation for a binary tree, it is evident that there are many NULL links. In fact, it can be shown that there are exactly n + 1 such links for a tree of n nodes. This wasted memory space can be used in the reformulation of the previous representation of a binary tree. This representation is the topic of the next subsection.

11-3.2 THREADED STORAGE REPRESENTATION

The wasted NULL links in the storage representation of binary trees introduced in the previous subsection can be replaced by *threads*. A binary tree is threaded according to a particular traversal order. For example, the threads for the inorder traversal of a tree are pointers to its higher nodes. For this traversal order, if the left link of a node P is normally NULL, then this link is replaced by the address of the predecessor of P. Similarly, a normally NULL right link is replaced by the address of the successor of the node in question. Because the left or right link of a node can denote either a structural link or a thread, we must somehow be able to distinguish them. Assuming that valid pointer values are positive and nonzero, then structural links can be represented, as usual, by positive addresses. Threads, on the other hand, will be represented by negative addresses. Also, it is often desirable in processing a tree to have a list head associated with it. This head node is simply another node which serves as the predecessor and successor of the first and last tree nodes with respect to inorder traversal. Such an approach, in essence, imposes a circular structure on the tree in addition to its tree structure. Figure 11-22a gives the threaded version, for inorder traversal, of the tree in Fig. 11-20a. In this diagram a dashed arrow denotes a thread link. Note that the tree is attached to the left branch of the head node. Using this approach the address of the root node of the tree is LPTR(HEAD). Observe that this example tree of eight nodes contains nine threads. Figure 11-22b denotes an empty tree with the left link of the head (and only) node denoting a thread.

Another way of distinguishing a thread link from a structural link is to

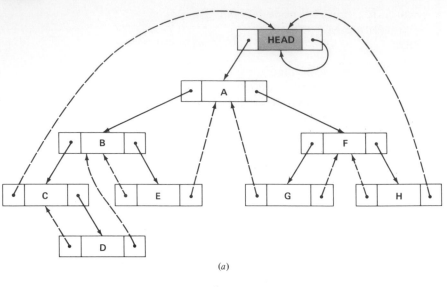

(a)

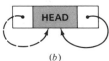

(b)

FIGURE 11-22
A threaded binary tree
for inorder traversal.

have a separate boolean flag for each of the left and right pointers. The node structure using this approach becomes

LPTR	LFLAG	INFO	RFLAG	RPTR

where **LFLAG** and **RFLAG** are boolean indicators associated with the left and right links, respectively. The following coding scheme is used to distinguish between a structural link and a thread.

LFLAG = true(1)   Denotes a left structural link

LFLAG = false(0)   Denotes a left thread link

RFLAG = true(1)   Denotes a right structural link

RFLAG = false(0)   Denotes a right thread link

Figure 11-23 shows an alternative storage representation of the tree in Fig. 11-22. This method of distinguishing between a thread and a structural link is preferred over the previous method if the programming language used does not allow negative-valued pointers. Note, however, that the negative-valued pointer approach is more efficient from the point of view of storage than is the flag method. In the remainder of our discussion we will use a negative pointer to denote a thread.

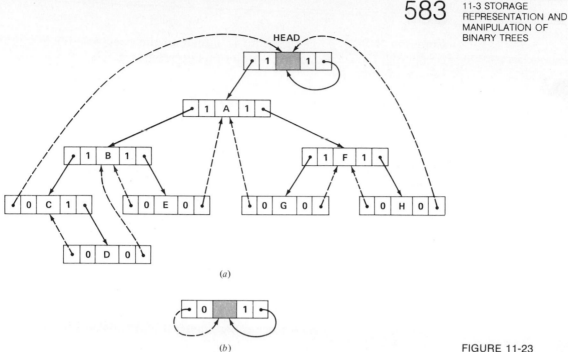

(a)

(b)

FIGURE 11-23

Given the threaded representation of a binary tree with respect to inorder traversal, it is a simple matter to formulate algorithms for obtaining the inorder predecessor and successor of a designated node. We now proceed to formulate these algorithms.

Function INS(X). Given X, the address of a particular node in a threaded binary tree of the form given in Fig. 11-22, this functional subalgorithm returns the address of the inorder successor of this node. P is a temporary pointer variable.

1. [A thread?]
    P ← | RPTR(X) |
    If RPTR(X) < 0 then Return(P)
2. [Branch left?]
    Repeat while LPTR(P) > 0
        P ← LPTR(P)
3. [Return address of successor]
    Return(P)   □

Step 1 first initializes P to the absolute value of the right link of X. If the original link value is negative (that is, denotes a thread), then the inorder successor of X has been found and the value of P is returned. If this test fails, however, control passes to step 2. In this step we repeatedly branch left until a left thread is encountered. At this point we enter step 3 where the required address is returned.

Function INP(X). This algorithm is similar to the previous subalgorithm except that the address of the inorder predecessor is returned. P is a temporary pointer variable.

1. [A thread?]

    P ← | LPTR(X) |

    If LPTR(X) < 0 then Return(P)

2. [Branch right?]

    Repeat while RPTR(P) > 0

    P ← RPTR(P)

3. [Return address of predecessor]

    Return(P)   ☐

This algorithm operates in a manner similar to that of the previous algorithm. The roles of LPTR and RPTR are simply interchanged.

The successor subalgorithm can be used repeatedly to traverse the threaded tree in inorder.

Algorithm TINORDER. Given the address of the list head of a binary tree which has been threaded for inorder traversal and subalgorithm INS previously discussed, this algorithm traverses the tree in inorder. P is a temporary pointer variable.

1. [Initialize]

    P ← HEAD

2. [Traverse threaded tree in inorder]

    Repeat while true

    P ← INS(P)

    If P = HEAD

    then Exit

    else Write(INFO(P))   ☐

The algorithm is simple. Observe that we have set up, in step 2, a repeat statement which seems to control an infinite loop. This, however, is not the case, since when the successor node becomes the list head, the entire tree has been traversed. At this point the algorithm terminates.

From this algorithm it can be seen that the threaded tree has certain advantages over its unthreaded counterpart. First, the inorder traversal of a threaded tree is somewhat faster than that of its unthreaded version, since no stack is required. Second, the threaded tree representation permits the efficient determination of the predecessor and successor of a particular node. For an unthreaded tree, however, this task is more difficult, since a stack is required to provide the upward-pointing information in the tree which threading provides.

Of course, a price must be paid for these advantages. First, threaded trees cannot share common subtrees, as can unthreaded trees. Second, if negative addressing is not permitted in the programming language being used, two additional fields are required to distinguish between thread and structural links. Finally, insertions into and deletions from a threaded tree are more time-consuming, since both thread and structural links must be maintained.

The following subalgorithm inserts a node into a threaded binary tree to the left of a designated node. There are two possible cases. The easiest case involves inserting the new node as a left subtree of the designated node, if that node has an empty left subtree. The remaining (and more difficult) case inserts the new node between the given node, say, X, and the LPTR(X). The two cases are exemplified in Fig. 11-24a and b, respectively.

Procedure LEFT(HEAD, X, DATA). Given the address of the head node of an inorder threaded binary tree (HEAD), the address of a designated node (X), and the information associated with a new node (DATA), this subalgorithm inserts a new node to the left of the designated node. NEW is a pointer variable which denotes the address of the node to be inserted.

1. [Create new node]
    NEW $\Leftarrow$ NODE
    INFO(NEW) $\leftarrow$ DATA
2. [Adjust pointer fields]
    LPTR(NEW) $\leftarrow$ LPTR(X)
    RPTR(NEW) $\leftarrow -$ X
    LPTR(X) $\leftarrow$ NEW
3. [Reset predecessor thread, if required]
    If LPTR(NEW) > 0
    then RPTR(INP(NEW)) $\leftarrow -$ NEW
        Return   ☐

The first step creates a new node and initializes its information field. Step 2 handles the case where the new node becomes the left subtree of node X. The last step handles the second insertion case where the right link of the inorder predecessor of X (before insertion) is set to a thread which points to the new node.

The notion of threading a binary tree can also be extended to preorder and postorder traversals. In these cases, however, thread pointers need not always point to higher nodes in the tree. Also, the algorithms associated with these traversals may sometimes be more complex than those obtained earlier for inorder traversals.

Thus far we have been concerned with the storage representation of binary trees. Clearly, there are applications in which the tree structures are not binary. As mentioned in Sec. 11-1, these more general tree structures can be converted easily to equivalent binary trees. This topic is described in detail in the next subsection.

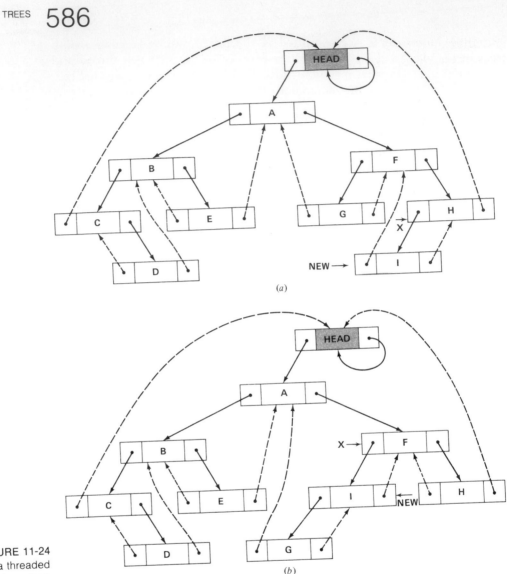

FIGURE 11-24
Insertion into a threaded
binary tree.

11-3.3 CONVERSION OF GENERAL TREES TO BINARY TREES

Recall from Sec. 11-2 that a general tree or, more generally, a forest of trees can be converted into an equivalent binary tree. This conversion process was called the natural correspondence between general and binary trees. Furthermore, this correspondence is a one-to-one relationship. In this subsection we formulate a detailed algorithm for converting general trees to binary trees. Before giving this algorithm, however, the specification of the input format to the algorithm is given.

Perhaps one of the most popular and convenient ways to specify a general tree (or forest) is to use a notation similar to that used in several pro-

gramming languages for writing record structures. As an example, the two
trees of Fig. 11-25a can be specified in the following manner:

```
1 A
     2 B
     2 C
          3 E
          3 F
     2 D
1 G
     2 H
          3 J
     2 I
```

In this notation the numbers associated with the nodes indicate the subtree
relationship. For example, those nodes with an associated number of 1
denote root nodes. Those with higher numbers indicate their lower position
in the tree structure. The equivalent binary tree for the previous forest of
two trees is given in Fig. 11-25b.

In the formulation of an algorithm for converting a forest to a binary
tree, we must connect a parent to its left offspring and connect from left
to right all the siblings at the same level within the same tree. This latter requirement implies that a stack must be maintained. We assume that the
input to the algorithm consists of a sequence of nodes for each tree in preorder. The representation of each node takes the form of a pair of elements;

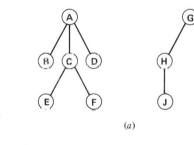

(a)

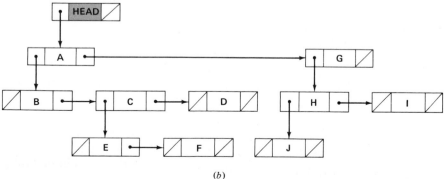

(b)

FIGURE 11-25
(a) Forest of two trees;
(b) binary tree representation of this forest.

the first and second elements represent the level number and name associated with that node, respectively. Each entry in the stack is made up of two items. The first item represents the level number associated with a node and the second item denotes the address of that node.

The following algorithm implements the notion of natural correspondence.

Algorithm CONVERT. Given a forest of trees whose input format is in the form just described, it is required to convert this forest into an equivalent binary tree with a list head (HEAD). Each element of the stack consists of two fields. The vectors NUMBER and LOC denote the level number and address associated with a node. TOP denotes the top of the stack. The variables LEVEL and NAME represent the level number and name of an input node, respectively. PRED_LEVEL and PRED_LOC give the level number and address, respectively, of a node which has been previously encountered. NEW is a temporary pointer variable.

1. [Initialize]
   HEAD ⇐ NODE
   LPTR(HEAD) ← RPTR(HEAD) ← NULL
   NUMBER[1] ← 0
   LOC[1] ← HEAD
   TOP ← 1
2. [Process the input]
   Repeat thru step 6 while there still remains a node
3. [Input a node]
   Read(LEVEL, NAME)
4. [Create a tree node]
   NEW ⇐ NODE
   LPTR(NEW) ← RPTR(NEW) ← NULL
   INFO(NEW) ← NAME
5. [Compare levels]
   PRED_LEVEL ← NUMBER[TOP]
   PRED_LOC ← LOC[TOP]
   If LEVEL > PRED_LEVEL
   then LPTR(PRED_LOC) ← NEW   (connect parent to its left offspring)
   else (remove nodes from stack)
        Repeat while PRED_LEVEL > LEVEL
           TOP ← TOP − 1
           PRED_LEVEL ← NUMBER[TOP]
           PRED_LOC ← LOC[TOP]
        If PRED_LEVEL < LEVEL
        then Write('MIXED LEVEL NUMBERS')
             Exit

```
        (connect siblings together)
          RPTR(PRED_LOC) ← NEW
          TOP ← TOP − 1
   6. [Push a new node onto the stack]
          TOP ← TOP + 1
          NUMBER[TOP] ← LEVEL
          LOC[TOP] ← NEW
   7. [Finished]
          Exit                                        □
```

The first step of the algorithm creates a list head for the required binary tree and places the level number and address of this node on the stack. Observe that a level number of zero is associated with the list head. This convenient choice will later cause the root of the first tree to be appended to the list head. Step 2 controls the input of the given forest and step 3 inputs a pair of values which represent a node in a tree. The fourth step creates a new node and initializes its links to NULL. The label of the new node is also copied. Step 5 first copies the level number and address of the top element in the stack into PRED_LEVEL and PRED_LOC, respectively. If the level number of the new node is greater than the level number of the topmost node on the stack, then the left link of the latter is set to the address of the former. This assignment connects a parent to its leftmost offspring. A transfer to step 6 then results. If, however, the level number of the new node is less than or equal to that of the stack top, then successive elements are removed from the stack until the level number of its topmost element is less than or equal to the level number of the new node. If the comparison gives a "less than" result, then an error exists in the numbering of the tree structures; otherwise, in the case of equality, the right link of the stack top node is set to NEW and removed from the stack. Step 6 stacks the level number and address of the new node.

The behavior of this algorithm for the forest of Fig. 11-25a is given in Table 11-4, where NA denotes the address of a node with label A, NB denotes the address of a node with label B, and so on. A stack entry written, for example, as 1NA means that a node with level number 1 and address NA having label A has been stacked. The stack top is the rightmost element in the string. Note that if a forest contains nodes with the same label, then each of these nodes will have a different address, even though our notation refers to each such address as NB. Finally, the table shows only changes which have occurred since the previous steps.

The preorder and inorder traversals of a binary tree which corresponds to a forest have a natural correspondence with these traversals on the forest. In particular, the preorder traversal of the associated binary tree is equivalent to visiting the nodes of the forest in *tree preorder*, which is defined as follows:

TABLE 11-4 Behavior of Algorithm **CONVERT** for the Forest of Fig. 11-25a

Current Input	Stack	LEVEL	NEW	PRED_ LEVEL	PRED_ LOC	LPTR(PRED_LOC)	RPTR(PRED_LOC)
	0HEAD					NULL	NULL
1, A	0HEAD 1NA	1	NA	0	HEAD	NA	NULL
2, B	0HEAD 1NA 2NB	2	NB	1	NA	NB	NULL
2, C	0HEAD 1NA 2NC	2	NC	2	NB	NULL	NC
3, E	0HEAD 1NA 2NC 3NE	3	NE	2	NC	NE	NULL
3, F	0HEAD 1NA 2NC 3NF	3	NF	3	NE	NULL	NF
2, D	0HEAD 1NA 2NC	2	ND	3	NF	NULL	NULL
	0HEAD 1NA 2ND			2	NC	NE	ND
1, G	0HEAD 1NA	1	NG	2	ND	NULL	NULL
	0HEAD 1NG			1	NA		NG
2, H	0HEAD 1NG 2NH	2	NH	1	NG	NH	NULL
3, J	0HEAD 1NG 2NG 3NJ	3	NJ	2	NH	NJ	
2, I	0HEAD 1NG 2NI	2	NI	3	NJ		
				2	NH		NI

1. Process the root of the first tree
2. Traverse the subtrees of the first tree in tree preorder
3. Traverse the remaining trees of the forest in tree preorder

The tree traversal of the forest of the two trees given in Fig. 11-25a gives the sequence

A B C E F D G H J I

This is exactly the same sequence obtained during the preorder traversal of the equivalent binary tree of Fig. 11-25b.

Similarly, the *tree inorder* traversal of a forest is defined as follows:

1. Traverse the subtrees of the first tree in tree inorder
2. Process the root of the first tree
3. Traverse the remaining trees of the forest in tree inorder.

The tree inorder traversal of the example forest of Fig. 11-25a yields the sequence of labels

B E F C D A J H I G

Again, this is the same sequence as that obtained by traversing the binary tree of Fig. 11-25b. No such direct correspondence exists for postorder traversal.

Now that we have discussed how to handle general trees in terms of binary trees, the topic of the next subsection is the sequential storage representation of general trees and binary trees.

11-3.4 SEQUENTIAL AND OTHER REPRESENTATIONS OF TREES

In this subsection we describe several representations of trees that are based on sequential-allocation techniques. These representations are efficient and convenient, provided that the tree structure does not change very much (as to insertions, deletions, etc.) during its existence. The particular representation chosen also depends on other types of application-dependent operations that are to be performed on the tree structure. Some of the representation methods that we examine are for binary trees, others are for general trees.

Perhaps one of the better known subclasses of binary trees is the set of complete binary trees. A *complete binary tree* is a binary tree in which every nonleaf node has exactly two children. An example of such a tree structure, together with its sequential representation, is shown in Fig. 11-26. In this representation the locations of the left and right sons of node i and 2i and 2i + 1, respectively. For example, in Fig. 11-26 the index of the left son of the node in position 3 (that is, E) is 6. Similarly, the index of the right son is 7. Conversely, the position of the parent of node j is the index TRUNC(j /2). For example, the parent of nodes 4 and 5 is node 2.

In the previous example we conveniently chose a tree with $2^3 - 1$ nodes. In general, a tree with $2^n - 1$ nodes for a particular value of n is easily represented by a vector of $2^n - 1$ elements. Binary trees which have more or less than $2^n - 1$ nodes for some n, however, can also be represented using the previous approach. An example of a tree which contains nine nodes is given in Fig. 11-27. Note that a substantial amount of memory is wasted in this case. Therefore, for large trees of this type, this method of representation may not be efficient in terms of storage.

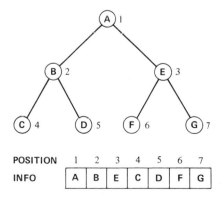

FIGURE 11-26
Sequential representation of a complete binary tree.

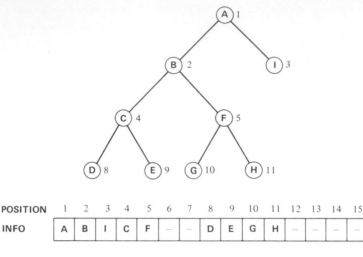

FIGURE 11-27
Sequential representation
of an incomplete binary
tree.

POSITION	1	2	3	4	5	6	7	8	9	10	11	12	13	14	15
INFO	A	B	I	C	F	–	–	D	E	G	H	–	–	–	–

A common method for the sequential representation of binary trees uses the physical adjacency relationship of the computer's memory to replace one of the link fields in the linked representation method introduced earlier. For example, consider an alternative representation of a tree structure, in which the left link (LPTR) from the usual doubly linked representation has been omitted. One possibility involves representing the tree sequentially, such that its nodes appear in preorder. Using this approach, the tree of Fig. 11-28a is represented by Fig. 11-28b, where RPTR, INFO, and TAG are vectors. Observe that in this representation we do not require the LPTR pointer, since for a nonnull link, it would point to the node to its immediate right. The bit (logical) vector TAG denotes, with a logical value of 1, a leaf node. This representation is wasteful of space because over one-half of the right links are null. This wasted space becomes useful by making the right link of each current node point to the node which immediately follows (with respect to preorder) the left subtree below this current node. The field RPTR is renamed RANGE in such a representation, as shown in Fig. 11-28c. Also observe that we do not need a TAG item for a leaf node, since a leaf node occurs when RANGE(P) = P + 1. This approach is applicable to general trees. A TAG bit of 1 refers to a leaf node in the original tree and the RPTR and RANGE fields are determined from the equivalent binary tree of the general tree.

Another popular method of representing a general tree sequentially is based on its postorder traversal. Such a representation takes the form of one vector, which represents the nodes of the tree in postorder, and a second vector, which denotes the number of children of the nodes. An example of the postorder representation is given in Fig. 11-29. Recall that this postorder representation of a tree is useful in evaluating functions that are defined on certain nodes of the tree. Section 10-5.2.3 contained an example of such a

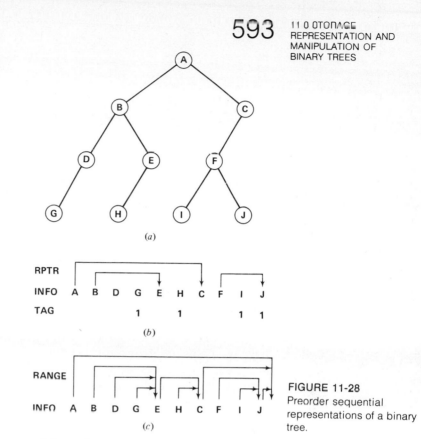

FIGURE 11-28
Preorder sequential
representations of a binary
tree.

case where object code was generated from the reverse Polish representation
of an expression.

As a final and straightforward sequential method of representing a tree,
consider a vector which contains the father of each node in the tree. As an
example, the tree of Fig. 11-30 is represented as follows:

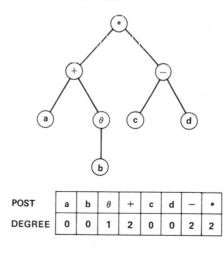

FIGURE 11-29
Postorder sequential
representation of a tree.

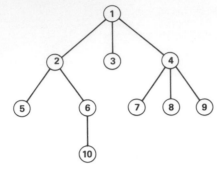

FIGURE 11-30

$$\begin{array}{lc} i & 1\ 2\ 3\ 4\ 5\ 6\ 7\ 8\ 9\ 10 \\ \text{FATHER}[i] & 0\ 1\ 1\ 1\ 2\ 2\ 4\ 4\ 4\ 6 \end{array}$$

where the branches in the tree are given by

$$\{(\text{FATHER}[i],\ i)\} \qquad \text{for } i = 2, 3, \ldots, 10$$

Observe that the root node (1) of the tree has no father; consequently, we have used a value of zero for its father. This method of representation can be extended to represent a forest. An obvious disadvantage of this method is that it fails to reflect certain orderings of the nodes. For example, if we interchange nodes 5 and 6, the representation of this new tree is the same as that of the previous tree.

This concludes our discussion of storage representations of tree structures. In the next section we explore several applications of these structures.

EXERCISES 11-3

1. Give unthreaded and threaded storage representations for the tree given in Fig. 11-31.

2. Formulate a recursive algorithm for traversing a binary tree in postorder.

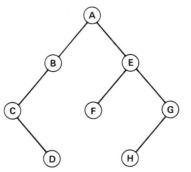

FIGURE 11-31

3. Trace the algorithm IPREORDER for the tree in Fig. 11-31.

4. Repeat exercise 3 for the algorithm IINORDER.

5. Formulate an iterative algorithm for traversing a binary tree in post-order.

6. Obtain an algorithm to obtain the "swapped" version of a binary tree. Figure 11-32 gives the swapped version of Fig. 11-31.

7. Trace the algorithm TREE_DELETE for the deletions of nodes D and then G from Fig. 11-18.

8. Given a threaded binary tree for inorder traversal, construct an algorithm (similar to algorithm LEFT) for inserting a node to the immediate right of a designated node.

9. Investigate the threading of a binary tree for preorder traversal. In particular, attempt to formulate algorithms for obtaining the preorder predecessor and successor of a designated node.

10. Repeat exercise 9 for postorder traversal.

11. Using the forest of Fig. 11-10, trace (as in Table 11-4) the algorithm CONVERT.

12. Formulate a recursive algorithm for converting a forest into an equivalent binary tree.

13. Based on the sequential representation of a complete binary tree, formulate an algorithm for its inorder traversal.

14. Based on the storage representation of a binary tree given in Fig. 11-28b, construct an algorithm for its postorder traversal.

15. Construct an algorithm for traversing, in inorder, a binary tree which is stored as in Fig. 11-28c.

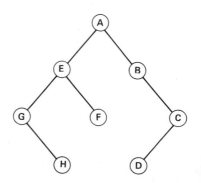

FIGURE 11-32

11-4 APPLICATIONS OF TREES

This section contains four applications of trees. The first application involves the symbolic manipulation of algebraic expressions. This was one of the earliest nonnumeric applications to which computers were applied. The second application is very broad and consists of applying tree structures to the process of searching. The version of a binary tree introduced earlier in this chapter is extended so that a greater efficiency in searching can be achieved. Trees are also used in sorting. We next describe the application of complete binary trees to this process. Finally, we discuss the applicability of general trees to searching.

11-4.1 THE SYMBOLIC MANIPULATION OF EXPRESSIONS

In Chap. 10 (see Sec. 10-10.1) we discussed the formal manipulation of polynomial expressions. These expressions were represented by linear linked lists. In this subsection we extend that discussion. This extended discussion, however, is based on the binary tree representation of expressions. In particular, we view certain operations on these expressions and their associated properties in terms of their tree representations. Since we have already introduced this topic in Chap. 10, it should be clear that it is the expressions themselves and not their values which must be processed. We may want

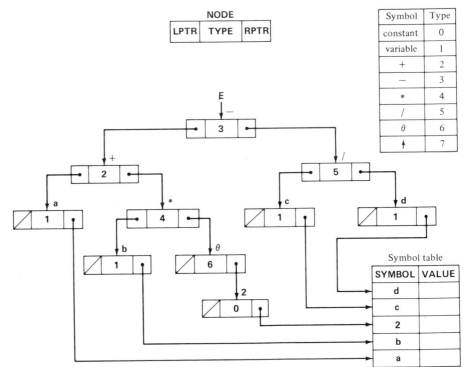

FIGURE 11-33
Binary tree representation
of an expression.

symbolically to add, subtract, multiply, divide, differentiate, compare for equivalence, and so on, such expressions.

The binary tree representation of expressions was introduced earlier in the chapter. Recall that a nonleaf node represents an operator and the left and right subtrees (for binary operators) are the left and right operands of that operator. The leaves of the tree are the variables and constants in the expression. Such a representation is given for the expression $a + b * \theta 2 - c / d$ in Fig. 11-33, where θ is used to denote the unary minus operator. Note that the operand of this unary operator is represented as a right subtree. In this figure E is a pointer variable which denotes the root of the tree. The node structure contains three fields: a left pointer field (LPTR), an information field (TYPE), and a right pointer field (RPTR). The values of TYPE associated with the operators a, $-$, $*$, $/$, θ, and $\uparrow$ are 2, 3, 4, 5, 6, and 7, respectively. Each leaf node represents a constant or a variable whose TYPE is given by 0 and 1, respectively. For each leaf node its right pointer contains the address in the symbol table which corresponds to that variable or constant. Also note that for each operator we have chosen to store the type of the operator rather than the operator itself. Such a choice simplifies the processing of these trees. The simple symbol-table representation chosen here contains the name of each variable or constant (SYMBOL) and its value (VALUE). Clearly, a more sophisticated representation, such as one of those given in the previous subsection, can also be adopted.

Let us first consider the evaluation of an expression which is represented by a binary tree. In other words, we require the value of the expression. The easiest way of obtaining the desired value is by formulating a recursive solution. Such a solution is given in the following recursive function.

Function EVAL(E). Given an expression which is represented by a binary tree with a root-node address of E, this algorithm returns the value of the given expression. F is a temporary pointer variable.

1. [Evaluate expression recursively]
   If TYPE = 0   (a constant)
   then F ← RPTR(E)
        Return(VALUE(F))
   else If TYPE = 1   (a variable)
        then F ← RPTR(E)
             Return(VALUE(F))
        else If TYPE = 2   (an addition operation)
             then Return(EVAL(LPTR(E)) + EVAL(RPTR(E)))
             else If TYPE = 3   (a subtraction operation)
                  then Return(EVAL(LPTR(E)) − EVAL(RPTR(E)))
                  else If TYPE = 4   (a multiplication operation)
                       then Return(EVAL(LPTR(E)) * EVAL(RPTR(E)))
                       else If TYPE = 5   (a division operation)
                            then Return(EVAL(LPTR(E)) / EVAL(RPTR(E)))

else If TYPE = 6   (a negation operation)
then Return(−EVAL(RPTR(E)))
else If TYPE = 7   (an exponentiation
operation)
then Return(EVAL(LPTR(E)) ↑
EVAL(RPTR(E)))

2. [Invalid expression]
Write('INVALID EXPRESSION')
Return(0)   □

In this subalgorithm we assume that the variables LPTR, RPTR, TYPE, and VALUE are global. The function is straightforward. If the node on entering the function is a leaf (that is, TYPE is 0 or 1), then the value of the constant or variable of this node is returned. This is achieved by using the right pointer of that node to reference the associated entry in the symbol table. For a nonleaf node, however, the recursive evaluation of the subtree(s) of this node which represents the operand(s) of the current operator is initiated. This evaluation is accomplished by invoking the function EVAL with the left and right pointers of that operator node as arguments in the case of a binary operator; otherwise, only the right pointer is used in the case of the negation operator. The process continues until a leaf node is encountered. When such a leaf node is detected, a value from its associated symbol-table entry is located.

Let us now consider the symbolic addition of two arithmetic expressions. Assume that E_1 and E_2 are pointer variables that denote the root nodes of the binary trees which represent the given expressions. The desired symbolic addition is easily represented by a new binary tree by first creating a root node for the required sum and then setting the left and right pointers of this node to E_1 and E_2, respectively. Finally, the type field of the new root node is set to 2. The algorithmic statements to accomplish this construction are

P ⇐ NODE
LPTR(P) ← E_1
RPTR(P) ← E_2
TYPE(P) ← 2

with P representing the address of the root node of the sum expression. This algorithm to add two expressions symbolically is trivial when we compare it with the algorithm developed in Sec. 10-10.1 for a very similar problem.

In practice, however, we would want to make certain obvious simplifications when adding two expressions symbolically. As an example, if the two expressions being summed are constants, then we should create a new leaf node which represents this constant sum. As another example, if one of the expressions being added is zero, then no new root node is required. Similar

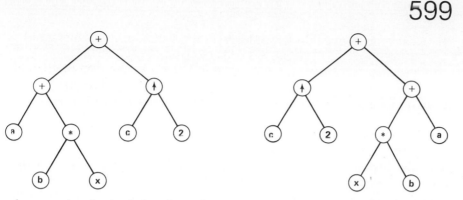

FIGURE 11-34
Two similar expression
trees.

rules can be devised for the other arithmetic operators and are left as
exercises.

As another example, let us consider the problem of deciding whether or
not two expressions which are represented as binary trees are similar. Two
binary trees which represent a pair of expressions are *similar* if they are
identical for all node types except for those which represent the commutative operators + and *. In the latter case we also consider two expression
trees to be equivalent if the right subtree of the first tree is identical to the
left subtree of the second tree and the left subtree of the first tree is identical
to the right subtree of the second tree. For example, the two expression
trees of Fig. 11-34 are equivalent. An algorithm for similarity can be formulated readily by dividing the operators into three categories: the binary commutative operators, the binary noncommutative operators, and the negation
operator. A recursive algorithm which uses this approach follows.

Function SIMILAR(A, B). Given two expressions which are represented by
binary trees with root nodes A and B, respectively, this algorithm determines
whether or not these two binary trees are similar.
1. [Check the root nodes]
       If TYPE(A) ≠ TYPE(B) then Return(false)
2. [Compare leaf nodes]
       If TYPE(A) = 0 and TYPE(B) = 0
       then If RPTR(A) ≠ RPTR(B)
             then Return(false)
             else Return(true)
       else If TYPE(A) = 1 and TYPE(B) = 1
             then If RPTR(A) ≠ RPTR(B)
                  then Return(false)
                  else Return(true)
3. [Check for commutativity of addition and multiplication operators]
       If TYPE(A) = 2 or TYPE(A) = 4
       then Return(SIMILAR(RPTR(A), RPTR(B)))
                   LPTR(B))) or (SIMILAR(LPTR(A), LPTR(B)) and
                   SIMILAR(RPTR(A), RPTR(B))))

4. [Check for identical binary subtrees]
    If TYPE(A) = 3 or TYPE(A) = 5 or TYPE(A) = 7
    then Return(SIMILAR(LPTR(A), LPTR(B)) and SIMILAR(RPTR(A),
          RPTR(B)))
5. [Check for identical unary subtrees]
    If TYPE = 6
    then Return(SIMILAR(RPTR(A), RPTR(B)))   □

The first step of the algorithm checks the types of the root nodes. If these differ, failure (that is, false) is reported. Step 2 compares leaf nodes. Identical variables (as to name and value) result in success (that is, true); otherwise, failure results. The same test is made for constants. The third step applies the commutativity test for addition and multiplication. Step 4 checks for identical binary subtrees. Finally, step 5 performs a test for identical unary subtrees.

There are several other classical applications involving the symbolic manipulation of expressions. Symbolic differentiation and integration are typical examples. The binary tree approach used in this subsection can also handle these applications. However, we leave these investigations as exercises.

11-4.2 BINARY SEARCH TREES

In Sec. 4-3.2 two searching techniques (that is, linear search and binary search) based on sequential allocation were examined. The better of the two techniques for searching was the binary search method. Its search time, for a set of n entries, was proportional to $\log_2 n$. Recall that, when using the binary search strategy, the set of entries being searched must be ordered. Such an ordering ensures that a binary search is efficient. For other operations, such as insertion and deletion, however, this sequential ordering presents severe drawbacks. In particular, an insertion into an existing ordered set of entries may require the movement of many entries in order to maintain order. More specifically, all entries at and below the position of insertion must be moved down one position. For a set of n entries, the average insert time is proportional to n.

Similarly, a deletion operation causes the same type of problem. In this instance, however, entries below the position of deletion must be moved up one position in order to close up the gap left by the entry being removed. Again, the time taken to perform a deletion is proportional to n.

The decisions made at each step in a binary search can be represented by a binary tree. Such a representation is obvious when we consider that the strategy at each point in the search is to split the set of remaining entries to be examined into two disjoint subsets. Figure 11-35 gives a tree representation of the search process for a sample table of seven entries. The root of the tree represents the middle entry in the original table. All entries which precede this entry are less than 33, while all items which follow it are greater.

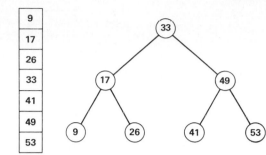

FIGURE 11-35
The decisions in a binary search viewed as a binary tree.

The next item to be examined in the first half of the table is 17. This item, in turn, can be preceded by 9 and followed by 26. The remainder of the tree can be obtained in a similar manner. Note that the ordering in the table is also present in the tree. For each nonleaf node in the tree, all entries in its left subtree are less than the value associated with that node, while all entries in its right subtree are greater than its associated value. As was mentioned earlier, a binary tree ordered in this manner is said to be *lexicographically* (or *lexically*) ordered.

The drawbacks of performing insertions and deletions on an ordered table whose entries are stored sequentially can be alleviated by using a linked-allocated binary tree. In this binary tree organization, a new entry is inserted as a leaf node into the tree. The insertion is performed so as to preserve the lexical ordering of the entries. For example, let us consider the representation of the following seven entries by a binary tree:

DOG, BEAR, HORSE, CAT, MONKEY, GIRAFFE, and BIRD

Assuming a linked representation of the tree with a node structure consisting of a left pointer (LPTR), a right pointer (RPTR), and an information item (INFO), Fig. 11-36 represents the given set of entries. If a new entry, INSECT, is inserted as a leaf in this tree, it is placed at the left of the entry MONKEY. This insertion is shown in Fig. 11-37.

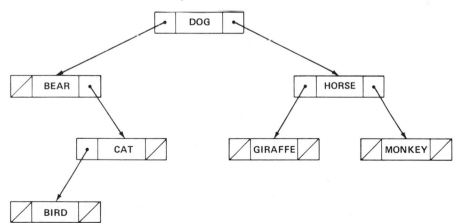

FIGURE 11-36
Binary tree representation of seven items.

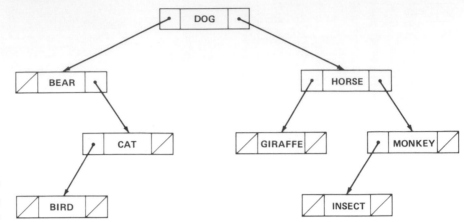

FIGURE 11-37
Performing an insertion
in a binary tree.

Note that because of this insertion strategy, the organization of the tree structure depends on the order in which entries are inserted. For example, the tree structure of Fig. 11-36 was obtained assuming that the entries were inserted in the order

DOG, BEAR, HORSE, CAT, MONKEY, GIRAFFE, and BIRD

However, if the insertion order is

CAT, BIRD, GIRAFFE, DOG, HORSE, MONKEY, and BEAR

a different tree is created. Figure 11-38 exhibits the behavior of the construction process for the latter insertion order.

We now proceed to formulate a combined algorithm which can perform either a search operation or an insertion. The solution to this problem takes the form of a function subalgorithm. The function is to have three parameters: the address of the head node in the tree (HEAD), a logical variable (INSERT) that denotes which operation is to be performed, and the entry which is to be either inserted or searched for (ITEM). A value of *true* or *false* denotes an insertion or a search operation, respectively. ITEM contains a string of characters. We assume that the information content of the list head of the tree is lexically greater than any other item that could be in the tree. For example, if each item is a set of alphabetic characters, then the list head information could be the string 'ZZ . . . Z'.

If the operation is an insertion and it is successful (that is, the item being entered is not already there), then the location of the new node is returned. If the operation is unsuccessful, then a negated pointer value, which indicates the position of a previously entered record with the same information, is returned.

For a search operation, a successful search returns the location of the entry with the same information as the entry being searched for. If the search fails to find the desired item, then a negated pointer value, which indicates the last node examined during the search, is returned. This loca-

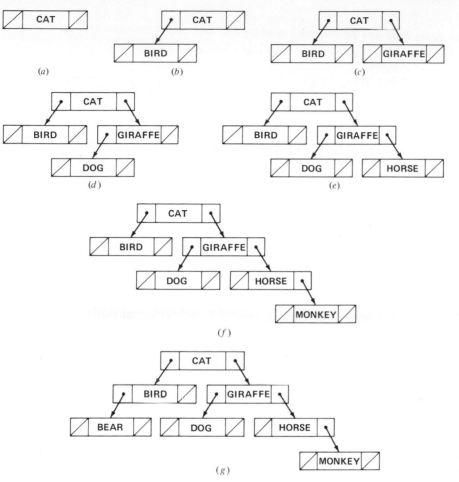

FIGURE 11-38
Construction of a tree-
structured table.

tion could be used to insert the missing item in an error-recovery procedure.
We will not, however, go into the details of such a procedure here. Rather,
we leave this problem as an exercise.

    Based on the previous discussion, a detailed algorithm can now be
formulated.

Function BINARY_TREE(HEAD, INSERT, ITEM). Given the parameters
HEAD, INSERT, and ITEM, as described previously, this algorithm performs
the requested operation on the tree structure. The node structure (NODE)
contains a left pointer (LPTR), an item description (INFO), and a right pointer
(RPTR). PARENT denotes the address of the parent node of the new item
which is to be inserted. T and P are temporary pointer variables.

1. [Initialize search variable]
      T ← HEAD

2. [Perform indicated operation]
   Repeat step 3 while T ≠ NULL
3. [Compare given item with the root entry of the subtree]
   If ITEM < INFO(T)
   then   (branch left)
           PARENT ← T
           T ← LPTR(T)
           If T = NULL
           then If not INSERT
                   then Return (−PARENT)   (search unsuccessful)
                   else   (create a new leaf and insert as a left subtree)
                           P ⇐ NODE
                           INFO(P) ← ITEM
                           LPTR(P) ← RPTR(P) ← NULL
                           LPTR(PARENT) ← P
                           Return(P)
   else If ITEM > INFO(T)
         then   (branch right)
                 PARENT ← T
                 T ← RPTR(T)
                 If T = NULL
                 then If not INSERT
                         then Return (−PARENT)   (search unsuccessful)
                         else   (create a new leaf and insert as a right subtree)
                                 P ⇐ NODE
                                 INFO(P) ← ITEM
                                 LPTR(P) ← RPTR(P) ← NULL
                                 RPTR(PARENT) ← P
                                 Return(P)
         else   (a match has occurred)
                 If INSERT then Return(−T) else Return(T)   □

The algorithm is easy to understand. The first step initializes T to contain the address of the list head node (recall that the left pointer of this node points to the root of the tree). Step 2 controls the descent through the tree. The test in this step will always succeed, since a NULL address is first detected in step 3 and an exit results. In step 3 we compare ITEM against a tree-table entry. If a match occurs, we have either found the desired entry or we have found a duplicate entry. If the match is unsuccessful, then depending on the value of ITEM, we branch either left or right and the entire process is repeated to make further comparisons. However, since a tree ordered in this way is such that every entry in a left subtree or right subtree lexicographically precedes or follows the root node, respectively, we know that whenever an attempt is made to descend an empty subtree, the matching process has failed. In such a situation, we know where the new entry should go for an insert operation. A new node is created, pertinent information is copied into it, and this node is inserted either to the left or to the right of the current node being examined in the existing tree. If we encounter a null link when performing a search operation, an error condition is reported.

The average number of comparisons for this algorithm is proportional to $\log_2 n$. From the discussion in Sec. 11-2, an algorithm for deleting a particular entry from a binary tree can be formulated. Such an algorithm was given in Sec. 11-3, and the average number of comparisons using the indicated approach is also proportional to $\log_2 n$. The worst case for performing an insertion and a search is of order n, a performance which is no better than the worst case for a sequential search. Figure 11-39 gives several examples of such cases. The first two cases involve trees which contain only non-null left or right links. The remaining two cases represent zigzag situations.

We now proceed to outline a method which prevents the occurrence of unbalanced trees such as that given in Fig. 11-39. The tree structure which results from this discussion has a worst case of $O(\log_2 n)$ for the operations of insertion, search, and deletion. The node structure is expanded to contain an extra item, which is called a *balance indicator*. This indicator specifies the "state of imbalance" of its associated node. There are three possible values for this indicator: left (L), right (R), or balance (B), according to the following definitions:

Left   A node is called *left heavy* if the longest path in its left subtree is exactly one longer than the longest path of its right subtree

Balance   A node is called *balanced* if the longest paths in its left and right subtrees are equal

Right   A node is called *right heavy* if the longest path in its right subtree is exactly one longer than the longest path in its left subtree

If each node in a binary tree is in one of these three states, then the tree is said to be *balanced*. A binary tree which is not balanced is said to be *unbal-*

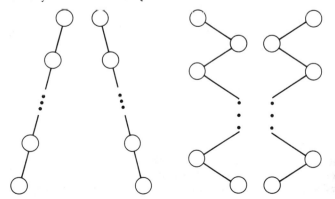

FIGURE 11-39
Worst-time cases for
binary trees.

anced. Figures 11-40 and 11-41 give examples of balanced trees and un-balanced trees, respectively.

Let us examine the problem of inserting a new node into a balanced tree. It is assumed that the new node is inserted at the leaf level. Note that the only nodes which can have their balance indicators changed by such an insertion are those nodes which lie on the path between the root node of the tree and the parent node of the newly inserted leaf. The three possibilities which can occur with respect to the state of a node on this path are as follows:

1. The node was either left- or right-heavy and has become balanced after the insertion
2. The node was balanced and has become left- or right-heavy
3. The node was left- or right-heavy and the new node has been appended in the heavy subtree, thus creating an unbalanced subtree; in such a case the node in question is called a *critical node*

If the first condition applies to a certain node, the balance indicator of each ancestor node of this node remains unchanged. In this case the longest path in the subtree (in which the node in question is its root) remains unchanged. When the second condition applies to a certain node, then the balance indicator of each ancestor of this node changes. The applicability of condition 3 to a certain node gives rise to an unbalanced tree and the node in question is *critical*. Figure 11-42 exhibits each of these three possibilities. The dotted branch and node notation denotes the new entry which is being inserted.

When a critical node is encountered, the tree must be rebalanced. There are essentially two general cases which can arise. Each of these cases can be further subdivided into two similar subcases. Figure 11-43 represents an occurrence of case 1, where the rectangles labeled T_1, T_2, and T_3 represent trees and the node labeled **NEW** denotes the new node being appended to the existing tree. The expression at the bottom of each rectangle denotes the

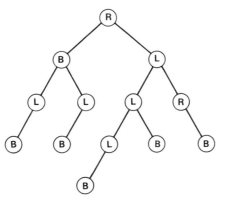
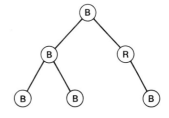

FIGURE 11-40
Examples of balanced trees.

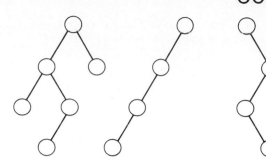

FIGURE 11-41
Examples of unbalanced trees.

maximum path length in that tree. The maximum path length in T_1 reflects the presence of the newly inserted node. For example, in Fig. 11-43a, X is a critical node and node Y must have been balanced before node NEW was appended to the tree. Note that in this case node Y has become left-heavy because node X was already left-heavy before insertion. A similar case arises for a right-heavy situation. In each subcase three pointers must be changed to achieve rebalancing (this includes changing the left or right link of the parent of X). An example of the first subcase is shown in Fig. 11-44, where node 5 is critical after the insertion of node 1. Note that only the balance indicators of those nodes which lie on the path between the list head of the tree and the new leaf can be affected by rebalancing. In the current example the balance indicators of nodes 2, 3, and 5 are affected.

The second general case which can occur in rebalancing a tree is somewhat more complicated. The two possible subcases of this case are shown in Fig. 11-45. The situation is similar to that encountered in case 1 except that node Y becomes heavy in the opposite direction to that in which X was heavy. Clearly, node Z must have been balanced prior to insertion. Again, note the similarity of the two subcases and the possibility of making the insertion into either the left or right subtree of Z. The diagram shows an insertion into the left subtree of Z. An example of the first subcase is illustrated in Fig. 11-46, where node 8 becomes critical after the insertion of node 4. From this discussion an insertion algorithm for balanced trees can be formulated. This task, however, is left as an exercise. This brief discussion concludes our treatment of balanced trees.

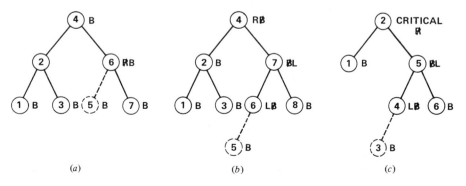

(a)   (b)   (c)

FIGURE 11-42
Examples of insertions into a balanced tree: (a) condition 1; (b) condition 2; (c) condition 3.

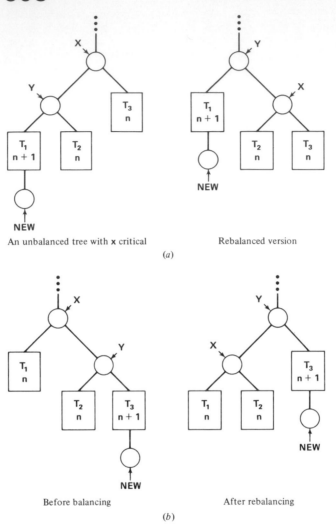

An unbalanced tree with **x** critical   Rebalanced version

(a)

Before balancing   After rebalancing

(b)

FIGURE 11-43
Case 1 for rebalancing
a tree.

Before ending this subsection we introduce the notion of a weighted tree. Function **BINARY_TREE** given at the beginning of the subsection does not take into consideration the probability of accessing a given node. We now examine the benefits of creating a binary tree which takes this probability into account. Intuitively, the most frequently accessed node in the tree should be its root. The next most frequently accessed node should also be high in the tree. The least frequently accessed node should be low in the tree.

Figure 11-47 is an example of a binary tree which does not take into consideration the frequency of access of a node. Each node in the tree contains the number of times that node has been accessed to this point. For example, nodes Jim and Maurice have been accessed three and seven times,

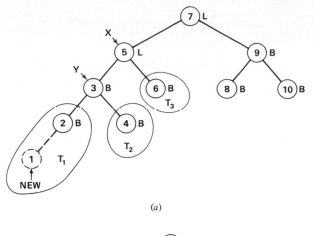

(a)

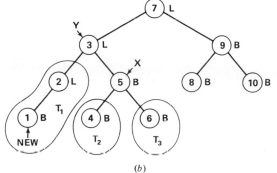

(b)

FIGURE 11-44
Example of case 1:
(a) before rebalancing;
(b) after rebalancing.

respectively. The Average Length Of Search (ALOS) of a tree T of n nodes can be calculated as follows:

$$\text{ALOS(T)} = \sum_{i=1}^{n} p_i d_i$$

where p_i is the probability that node i will be accessed based on the proportion of accesses for this node thus far and d_i is the depth of node i in the binary tree. The root node is assumed to have a depth of 1. For example, the binary tree of Fig. 11-47 has an ALOS of

$$\begin{aligned}
\text{ALOS(T}_1) &= \tfrac{1}{25} \times 1 + \tfrac{3}{25} \times 2 + \tfrac{4}{25} \times 2 + \tfrac{6}{25} \times 3 + \tfrac{7}{25} \times 3 + \tfrac{5}{25} \times 3 \\
&= .04 + .24 + .32 + .72 + .84 + .60 \\
&= 2.76
\end{aligned}$$

Figure 11-48 shows a different binary tree for the same set of entries. This structure, however, reflects the number of accesses of each entry. The

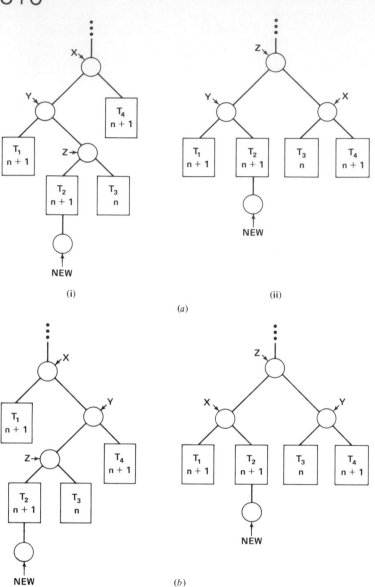

FIGURE 11-45
Case 2 for rebalancing
a tree.

greater the frequency of access of a particular node, the higher the position of this node in the tree. The ALOS for this tree is

$$\text{ALOS}(T_2) = \tfrac{7}{25} \times 1 + \tfrac{6}{25} \times 2 + \tfrac{4}{25} \times 2 + \tfrac{3}{25} \times 3 + \tfrac{4}{25} \times 3 + \tfrac{1}{25} \times 4$$
$$= .28 + .48 + .32 + .36 + .48 + .16$$
$$= 2.08$$

While this reduced average does not appear to be significant in this example, the savings can be considerable for large tables. The tree of Fig. 11-48 is called a *weighted binary tree*.

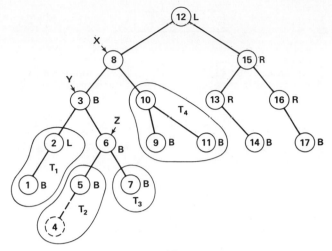

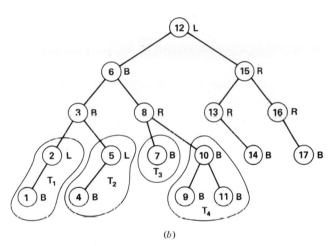
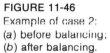

FIGURE 11-46
Example of case 2:
(a) before balancing;
(b) after balancing.

A general recursive algorithm for inserting an item into a weighted binary tree structure can be expressed as follows:

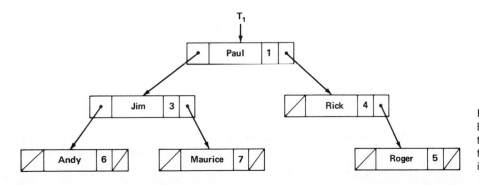

FIGURE 11-47
Binary tree which does not take into consideration the frequency of access of its nodes.

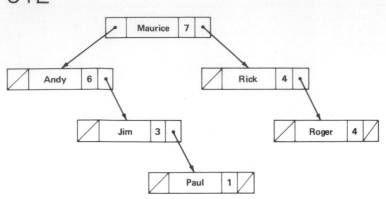

FIGURE 11-48
Weighted binary tree.

1. The root node of a tree (or subtree) is the node with the largest frequency of access count from the set of nodes constituting the tree (or subtree)
2. The left subtree of the tree (or subtree) are those nodes whose key values are lexically less than the root node
3. The right subtree of the tree (or subtree) are those nodes whose key values are greater than the root node

The implementation of this algorithm is left as an exercise.

In this subsection we have discussed the application of binary trees to the searching process. In Sec. 11-4.4 we will investigate the applicability of general trees to searching.

11-4.3 TREE SORTS

This subsection examines two sorting techniques which are based on the representation of a table by a tree. The first method, which is simple, is a binary tree sort. The second technique involves the use of a full binary tree to represent the given table. Although the second method is more complex than the first, it can also be significantly more efficient.

The basic concepts of a binary tree sort were introduced in the symbol-table discussion of Sec. 11-4.2, where the symbols were kept in a particular order. More specifically, the root symbol of each tree lexically followed and preceded all the symbols in its left and right subtrees, respectively. Once the binary tree for a given table has been constructed in this manner, it can then be traversed in order to yield the desired sorted table. For example, the key set

58, 86, 25, 44, 92, 61, 13, 39, 8, 75

yields the binary tree given in Fig. 11-49, assuming that the keys are inserted from left to right. The inorder traversal of this tree gives the desired results. The average number of comparisons for this sorting technique is $O(n \log_2 n)$ for a table of n entries. In the worst case, however, the number of

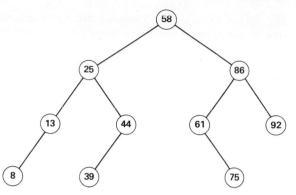

FIGURE 11-49
Binary tree representation
of a table of 10 keys.

comparisons required is O(n²), a case which arises when the binary tree for the table is severely unbalanced (recall Sec. 11-4.2).

We now turn to a completely different method of sorting, which also involves binary trees. Given n keys, it takes at least $n - 1$ comparisons to obtain either the smallest or largest key value. Once the smallest or largest key is removed, however, we can obtain the next smallest or largest element in much less than $n - 2$ comparisons. The way of achieving this result can be explained in terms of a conventional tennis tournament. Let us assume that this tournament has eight participants and is to be played according to the schedule given in Fig. 11-50. The diagram displays the outcome of the tournament with Jane beating Lorna, Janet beating Gail, and so on, and finally Jane beating Judy. Jane is therefore declared to be the winner of the tournament. Suppose that it were required to determine the second best player as well. This player can be Lorna or Janet or Judy. The second best player can be determined by having Lorna play Janet, and the winner of this match play Judy. Using this approach, only two additional games are required to determine the second best player. The important point to note is that the complete tournament need not be replayed with Jane absent.

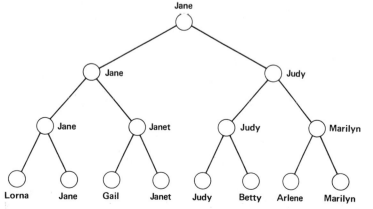

FIGURE 11-50
Tennis tournament.

We can use this tournament approach to sort a given table. First, the table is represented by a particular kind of binary tree called a *heap*. A heap is essentially a full binary tree (see Sec. 11-3.4). An example of a heap for 10 elements is given in Fig. 11-51. Note that the largest key is at the root of the tree. The next two largest keys in the table are the left and right off-spring of the root. Similar relationships exist for the remaining keys of the table. Also note that, except for the left subtree which contains the key whose value is **58**, the given tree is a full binary tree. This phenomenon occurs depending on the value of n, and its presence does not make much difference in terms of the final algorithm. In general, a heap which repre-sents a table of n records satisfies the property

$$K_j \leq K_i \quad \text{for } 2 \leq j \leq n \quad \text{and} \quad i = \text{TRUNC}(j/2)$$

This property clearly holds for the binary tree given in Fig. 11-51.

The heap can be allocated sequentially, as was done for full binary trees in Sec. 11-3.4. Recall that in such a representation the indices of the left and right children of record i (if they exist) are 2i and 2i + 1, respectively. Con-versely, the index of the parent of record j (if it is not the root) is **TRUNC(j/2)**. These simple relationships permit us to descend or to ascend the heap with relative ease. As a result, the sort algorithm based on a heap representation of a table tends to be relatively straightforward.

We now proceed to formulate an algorithm which constructs a heap from an unsorted sequentially allocated input table. In this construction process we start with a heap (for example, a one-record tree is a heap) and then insert a new record into the existing heap such that a new heap is formed after performing the indicated insertion. Such insertions are per-formed repeatedly until all records in the original table form a heap.

Let us examine the insertion process more carefully. As an example, assume that we wish to insert a new record with a key value of **92** into the existing heap given in Fig. 11-52. As indicated in the figure, the new key (**92**) is appended as a leaf to the existing tree in position **5**. Clearly, the new struc-ture is not a heap. The parent of the node is **TRUNC(5/2)** or **2**. The new key is

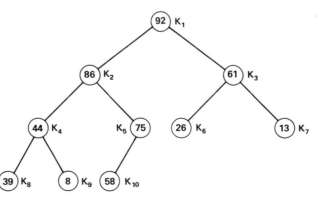

FIGURE 11-51
Heap representation of a sample key set of 10 numbers.

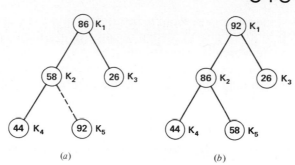

(a)   (b)

FIGURE 11-52
Insertion of a new record
into an existing heap.

larger than the key value of its parent, so we interchange the positions of keys 92 and 58. Since the new key is now in position 2, its new parent is in position 1. Again, the new key has a value greater than its parent (key 86). Consequently, keys 86 and 92 interchange positions. At this point the process terminates, with the revised tree structure again representing a heap.

We now formalize the previous notions in the following subalgorithm.

Procedure CREATE_HEAP(K). Given a vector K (type integer) containing the keys of the n records of a table, this algorithm creates a heap as previously described. The index variable q controls the number of insertions which are to be performed. The integer variable j denotes the index of the parent of key K_i. KEY (type integer) contains the key of the record being inserted into an existing heap.

1. [Build heap]
      Repeat thru step 7 for q = 2, 3, . . ., n
2. [Initialize construction phase]
      i ← q
      KEY ← K_q
3. [Obtain parent of new record]
      j ← TRUNC(i/2)
4. [Place new record in existing heap]
      Repeat thru step 6 while i > 1 and KEY > K_j
5. [Interchange record]
      K_i ← K_j
6. [Obtain next parent]
      i ← j
      j ← TRUNC(i/2)
      If j < 1 then j ← 1
7. [Copy new record into its proper place]
      K_i ← KEY
8. [Finished]
      Exit   □

Step 1 of the algorithm contains an iteration statement which controls the construction, by successive insertions, of the desired heap. The second step selects the record to be inserted into the existing heap and copies the key into KEY. The third step obtains the position of the parent of the new leaf node which is being inserted. Step 4 controls the placing of the new record into the existing heap. This record (initially a leaf node in the binary tree) potentially moves up the tree along the path between the new leaf and the top of the heap. This process continues (steps 5 and 6) until the new record reaches a position in the tree that satisfies the definition of a heap. The copying of the new key into its proper place in the tree occurs in step 7. The behavior of the construction of the heap of Fig. 11-51 for the sample key set

58, 86, 26, 44, 92, 61, 13, 39, 8, 75

is given in Fig. 11-53. Each tree in the diagram represents the state of construction after the insertion and reconstruction process is complete.

Now that we have a heap for the given table, we can use the notions of the tennis tournament to perform the sort. The record with the largest key is presently at the top of the heap and it can be written out directly. Since we want to perform the sort in place, we can interchange the records in positions 1 and n. When this is done, however, we must reconstruct a new heap consisting of $n - 1$ records. This reconstruction process is realized in a manner similar to that used in subalgorithm CREATE_HEAP. In particular, the heap of Fig. 11-51 is shown in Fig. 11-54a after interchanging keys 92 and 58 and also after the reconstruction of a new heap. Once the key having a value of 58 is placed at the root of the tree, it must be moved down since the present tree (Fig. 11-54a) is no longer a heap. This is accomplished by choosing the largest of its siblings (that is, its direct descendants), which is 86. In this instance 86 moves to the top of the heap. Now, we examine the greatest of the left and right sons of K_2 (that is, 44 and 75) and choose 75. In order to obtain a heap we must interchange 75 and 58. We then get the heap of Fig. 11-54b. This second record can now be exchanged with the ninth record. A new heap must then be constructed for $n - 2$ records. By repeating this exchange and reconstruction process, the initial table is sorted. The formalization of this technique yields the following algorithm.

Algorithm HEAP_SORT. Given a vector K (type integer) containing the keys of the n records of a table and the procedure CREATE_HEAP which has been previously described, this algorithm sorts the table into ascending order. The variable q represents the pass index. Index variables i and j are used, where the latter is the index of the left son of the former. KEY is an integer variable which contains the key of the record being swapped at each pass.

1. [Create the initial heap]
   Call CREATE_HEAP(K)
2. [Perform sort]
   Repeat thru step 10 for q = n, n − 1, . . ., 2
3. [Output and exchange record]
   Write(K_1)
   $K_1 \Leftrightarrow K_q$

4. [Initialize pass]
   $i \leftarrow 1$
   $KEY \leftarrow K_1$
   $j \leftarrow 2$
5. [Obtain index of largest son of new record]
   If $j + 1 < q$
   then If $K_{j+1} > K_j$
       then $j \leftarrow j + 1$
6. [Reconstruct the new heap]
   Repeat thru step 10 while $j \leqslant q - 1$ and $K_j > KEY$
7. [Interchange record]
   $K_i \leftarrow K_j$
8. [Obtain next left son]
   $i \leftarrow j$
   $j \leftarrow 2 * i$
9. [Obtain index of next largest son]
   If $j + 1 < q$
   then If $K_{j+1} > K_j$
       then $j \leftarrow j + 1$
   else if $j > n$
       then $j \leftarrow n$
10. [Copy record into its proper place]
    $K_i \leftarrow KEY$
11. [Finished]
    Write(K)
    Exit   ☐

The behavior of algorithm HEAP_SORT for the heap of Fig. 11-51 is given in terms of trees in Fig. 11-55. An alternative behavior of the algorithm for the same heap appears in Table 11-5.

TABLE 11-5   Behavior of the Algorithm HEAP_SORT

r	Initial Table K_r	Initial Heap	Pass Number (q)							Sorted	
			10	9	8	7	6	5	4	3	2
1	58	92	86	75	61	58	44	39	26	13	8
2	86	86	75	58	58	44	26	26	13	8	13
3	26	61	61	61	39	39	39	8	8	26	26
4	44	44	44	44	44	13	13	13	39	39	39
5	92	75	58	8	8	8	8	44	44	44	44
6	61	26	26	26	26	26	58	58	58	58	58
7	13	13	13	13	13	61	61	61	61	61	61
8	39	39	39	39	75	75	75	75	75	75	75
9	8	8	8	86	86	86	86	86	86	86	86
10	75	58	92	92	92	92	92	92	92	92	92

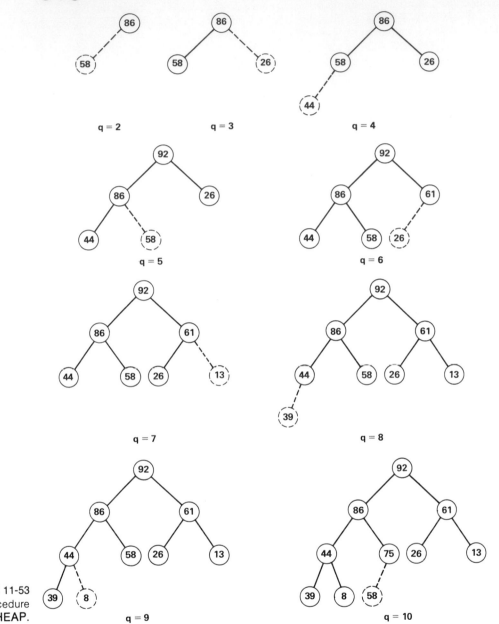

FIGURE 11-53
Behavior of the procedure
CREATE_HEAP.

11-4.4 TRIE STRUCTURES

Section 11-4.2 discussed the applicability of binary trees to searching. In that discussion the branching at any level in the tree was determined by the entire key value. In this subsection we examine briefly the feasibility of using

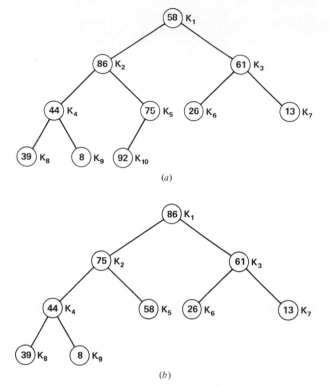

(a)

(b)

FIGURE 11-54

m-ary trees ($m \geq 2$) for searching. Also, the branching criterion at a particular level in such a tree will be based on a portion of the key value rather than its entire key value. A *trie* structure is a complete m-ary tree in which each node consists of m components. Typically, these components correspond to letters and digits. Trie structures occur frequently in the area of information organization and retrieval. The method of searching in tries is analogous to the notion of digital sorting (see Sec. 10-9.3). In particular, the branching at each node of level k depends on the kth character of a key. Table 11-6 contains an example of a trie structure for searching a set of 32 names. It consists of 12 nodes, each of which is a vector of 27 elements. Each element contains a dash, or the desired word, or a node number. A blank symbol ($\square$) is used to denote the end of a word during the scan of a key. Node 1 is the root of the trie.

As an example, we will trace through the search for the word **PEACOCK**. The letter **P** tells us that we should go from node 1 to node 6, since the entry which corresponds to this letter is 6. The second letter (**E**) selects the sixth component of node 6. The entry corresponding to **E** transfers us to node 12. At this node, the letter **A** is then used to find the desired word.

An algorithm for this search technique follows.

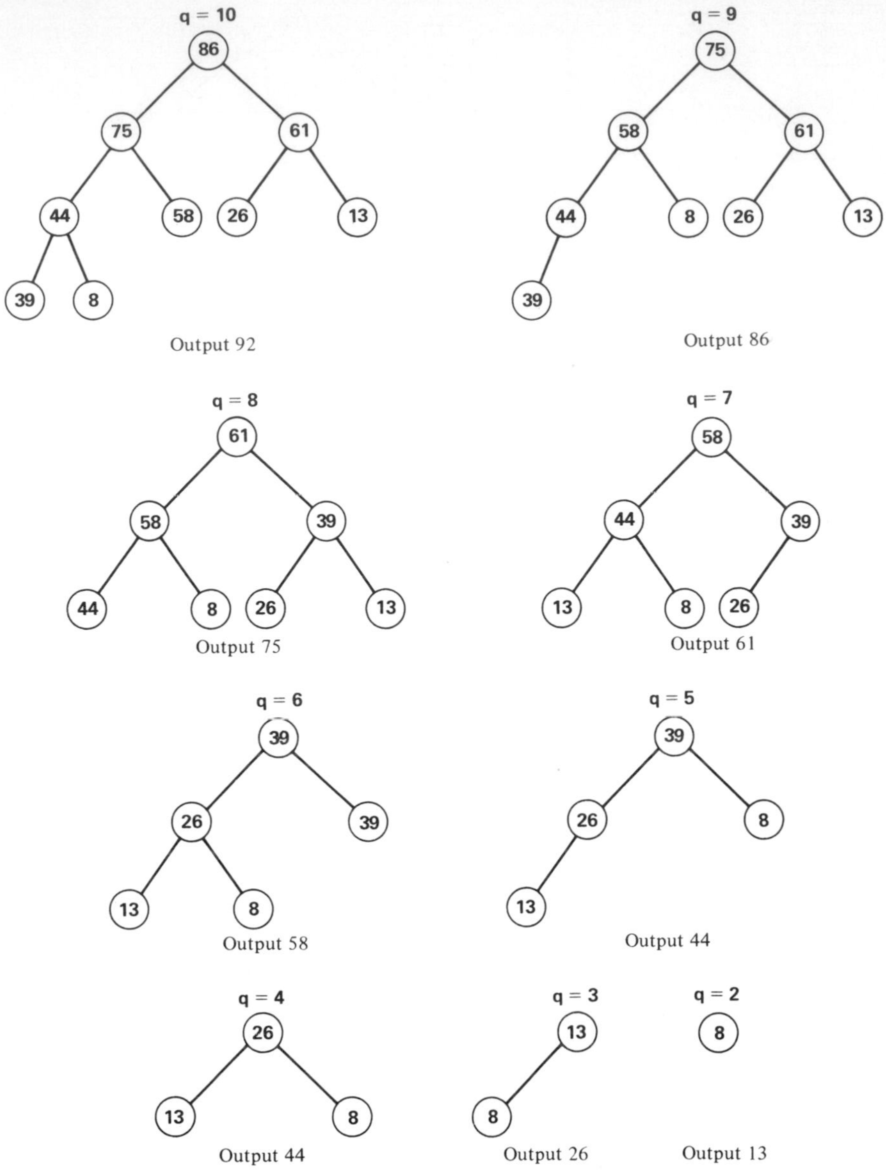

FIGURE 11-55
Behavior of the algorithm
HEAP_SORT.

Procedure TRIE_SEARCH(NAME, ROW, TRIE, COL). Given **NAME**, the name to be searched for, **TRIE**, an array of **27** by n elements representing a trie structure, where n represents the number of nodes required to represent the structure, this algorithm performs a search of the trie structure. The parameters **ROW** and **COL** hold the row and column indices of **NAME** within **TRIE**, respectively. **K** is an index which designates the position of the character currently being scanned. If the search fails, **ROW** and **COL** are assigned a value of zero. We assume that **NAME** contains only alphabetic characters.

TABLE 11-6 Trie Structure for a List

	Node Number											
	1	2	3	4	5	6	7	8	9	10	11	12
□	—	—	—	—	—	—	—	—	—	—		—
A	—	CANARY	EAGLE	—	—	—	—	SANDPIPER	—	CRANE	—	PEACOCK
B	—	—	—	—	—	—	—	—	—	—	—	—
C	2	—	—	—	—	—	—	—	—	—	—	—
D	—	—	—	—	—	—	—	—	—	—	—	—
E	3	—	—	—	—	12	REDHEAD	—	—	—	—	—
F	FALCON	—	—	—	—	—	—	—	—	—	—	—
G	4	—	—	—	—	—	—	—	—	—	—	—
H	HAWK	—	—	—	—	PHEASANT	—	SHOVELLER	—	—	—	—
I	—	—	—	—	—	PINTAIL	—	—	—	—	—	—
J	—	—	—	—	—	—	—	—	—	—	—	—
K	KIWI	—	—	—	—	—	—	—	—	—	—	—
L	LOON	—	—	—	—	—	—	—	—	—	GOLDFINCH	PELICAN
M	MALLARD	—	EMU	—	—	—	—	—	—	—	—	—
N	—	—	—	—	—	—	—	—	CONDOR	—	—	—
O	5	9	—	11	—	—	ROBIN	—	COOT	CROW	GOOSE	—
P	6	—	—	—	—	—	—	SPARROW	—	—	—	—
Q	—	—	—	—	—	—	—	—	—	—	—	—
R	7	10	—	GROUSE	ORIOLE	—	—	—	—	—	—	—
S	8	—	—	—	—	—	—	—	—	—	—	—
T	TEAL	—	—	—	—	—	—	—	—	—	—	—
U	—	—	—	GULL	—	PUFFIN	—	—	—	—	—	—
V	VULTURE	—	—	—	—	—	—	—	—	—	—	—
W	WREN	—	—	—	OWL	—	—	SWAN	—	—	—	—
X	—	—	—	—	—	—	—	—	—	—	—	—
Y	—	—	—	—	—	—	—	—	—	—	—	—
Z	—	—	—	—	—	—	—	—	—	—	—	—

1. [Initialize]
    COL ← 1
2. [Perform the search]
    Repeat for K = 1, 2, . . ., LENGTH(NAME)
      ROW ← INDEX('□ABC . . . XYZ', SUB(NAME, K, 1))
      Repeat while TRIE[ROW, COL] ≠ '—'
        If TRIE[ROW, COL] = NAME then Return
        If INDEX('0123456789', SUB(TRIE[ROW, COL], 1, 1)) = 0
        then Write('UNEXPECTED', NAME, 'FOUND')
            ROW ← COL ← 0
            Return
        else COL ← TRIE[ROW, COL]
3. [Missing name]
    Write('NAME NOT FOUND')
    ROW ← COL ← 0
    Return   □

The algorithm repeatedly scans the next character from **NAME** and branches accordingly. This scanning process stops either when the given name is found or when the search fails.

The example trie given is very wasteful of memory space. Memory can be saved at the expense of running time if each class of names is represented by a linked tree. Figure 11-56 represents the trie of Table 11-6. Note that this representation is a forest of trees.

The best compromise situation in terms of space and running time occurs when only a few levels of a trie are used for the first few characters of the key and then some other structure, such as a linked list or binary tree, is used in the remainder of the search.

CHAPTER EXERCISES

1. Based on the discussion of symbolic manipulation of expressions in the book, construct a binary tree (and associated symbol-table entries) as in Fig. 11-33 for the expression

   a * ln b + c / d

2. Using the binary tree obtained in exercise 1 and assuming values of $a = 2, b = 10, c = 8$, and $d = 4$, trace the operation of the function EVAL.

3. Trace the function SIMILAR for the pair of trees given in Fig. 11-34.

4. A classical example of symbol manipulation is finding the derivative of a formula with respect to a variable, say, x. This symbol-manipulation application was one of the first to be implemented on a digital computer. Such implementations have existed since the early 1950s. This application is of great importance in many scientific applications areas.

   The following rules define the derivative of a formula with respect to x where u and v denote functions of x:

   1. $D(x) = 1$
   2. $D(a) = 0$, if a is a constant or a variable other than x
   3. $D(\ln u) = D(u) / u$, where ln denotes the natural logarithm
   4. $D(-u) = -D(u)$
   5. $D(u + v) = D(u) + D(v)$
   6. $D(u - v) = D(u) - D(v)$
   7. $D(u * v) = D(u) * v + u * D(v)$
   8. $D(u / v) = D(u) / v - (u * D(v)) / v^2$
   9. $D(v \uparrow u) = (v \uparrow u) * (u * D(v) / v + D(u) * \ln v)$

   These rules permit evaluation of the derivative D(y) for any formula y composed of the preceding operators. Based on the binary tree representation of an expression given in the text, formulate an algorithm which differentiates a given expression according to the differentiation rules 1 through 9.

5. If we apply the differentiation rules of exercise 4 to the formula

$$y = \frac{2\ln(x+a)+b}{x^2}$$

we obtain

$$D(y) = \frac{\left[0 \cdot \ln(x+a) + 2\left(\frac{1+0}{x+a}\right) + 0\right]}{x^2}$$
$$- \frac{\left[2\ln(x+a)+b\right] \cdot x^2 \cdot \left(\frac{2 \cdot 1}{x}\right) + 0 \cdot \ln x}{(x^2)^2}$$

which is far from being satisfactory. Certain redundant operations can be avoided, however, by recognizing the special cases of adding or mul-

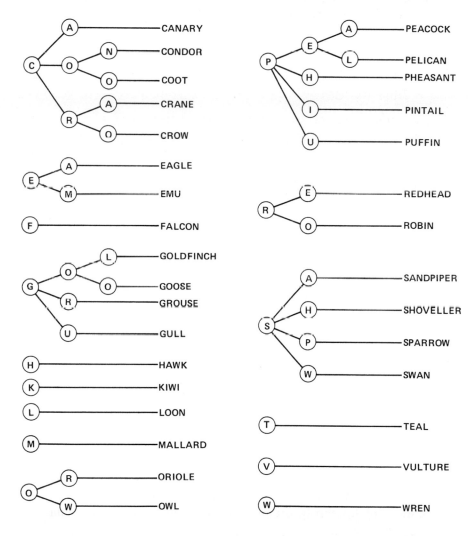

FIGURE 11-56
Forest representation of the trie given in Table 11-5.

tiplying by 0, multiplying by 1, or raising to the first power. These simplifications reduce the previous expression to

$$D(y) = 2 \frac{1}{x + a} + \frac{- \left[b(2x) \right]}{(x^2)^2}$$

which is somewhat more acceptable, yet not ideal. For our purposes, however, these simplifications will suffice. Modify the algorithm obtained in exercise 4 so that these simplifications are performed.

6. Construct an algorithm which uses the function BINARY_TREE repeatedly to perform a sequence of insertions and/or searches. Trace through your algorithm for the following sequence of inputs:

True	'THEN'
True	'DECLARE'
True	'ELSE'
True	'GET'
False	'ELSE'
True	'PUT'
True	'FREE'
False	'ALLOCATE'

7. Perform the indicated balance-tree insertions (shown with dashed lines) to the trees of Fig. 11-57. In each case change any balance indicators by the insertion.

8. Construct a detailed algorithm for performing an insertion into a balanced binary tree structure.

9. Obtain an algorithm for performing a deletion in a balanced binary tree.

10. Formulate a detailed algorithm for the construction of a weighted binary tree. The input to this algorithm is to contain a sequence of keys. Any of these keys can be repeated in the input.

11. One important task in the writing of a book is the construction of an index. In such an index, the major terms used in the book are presented in lexical order. Several subterms can be associated with a particular major term and must be written in lexical order immediately following that major term. Each major term and subterm is followed by a set of ascending numbers that identifies the pages where the corresponding term is discussed. It is required to formulate an algorithm to process a number of arbitrarily ordered major terms and subterms and their

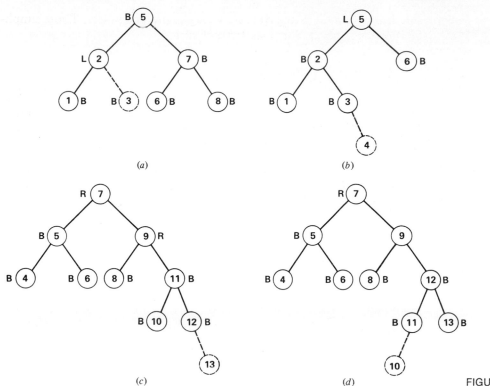

(a)

(b)

(c)

(d)

FIGURE 11-57

associated page numbers in a book, and subsequently print the required index.

The input data are to consist of a sequence of input items. Each item is to represent a major term string, a subterm string, or a major term and subterm string. A subterm string always corresponds to the major term most recently encountered in the input.

As an example, some input strings are

    DATA STRUCTURE@125@64@481
    #VECTOR@475
    DATA STRUCTURE#STACK@473@451@407

The symbols # and @ are delimiters for separating a major term from its subterm and page numbers, respectively. In the previous example subterms VECTOR and STACK correspond to the major term DATA STRUCTURE.

To print the required index, the index must first be represented in computer memory. One possible organization is given in Fig. 11-58, where each major term corresponds to a node with four fields, which we describe as MAJORNODE. The major term name is stored in the field TERM. The field MJLINK denotes a pointer to the node containing the

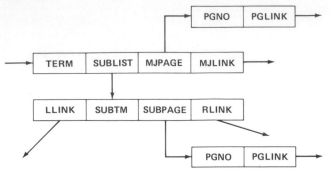

FIGURE 11-58
Possible structure for
an index.

next major term in a sequence of lexically ordered major terms. MJPAGE specifies a pointer to a linked linear list of page numbers where the major term is discussed. SUBLIST is a pointer to a binary tree structure consisting of nodes each of which is denoted as SUBNODE.

In each SUBNODE the field SUBTM contains a term which is subsidiary to the major term in the MAJORNODE predecessor. LINK and RLINK point to the previous and successor subnodes, respectively. This linked binary tree containing subterms is lexically ordered. SUBPAGE is a pointer to a linked linear list of page numbers where the subterm is discussed.

The lists of page numbers contain nodes, each of which is a PAGE_NODE. The field PGNO specifies a page number for the predecessor major term or subterm. PGLINK denotes the address of the next PAGE_NODE. The nodes are kept in increasing page number sequence. Using these node structures, the major term DATA STRUCTURE in the previous example is represented in Fig. 11-59.

12. Determine the behavior of the algorithm HEAP_SORT as in Table 11-5 for the following set of keys:

43, 23, 74, 11, 65, 58, 94, 36, 99, 87

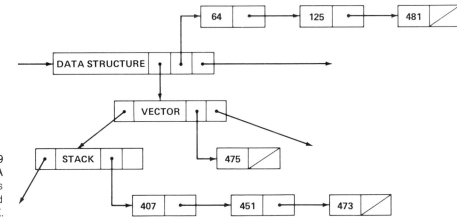

FIGURE 11-59
Representation of DATA
STRUCTURE and its
subterms VECTOR and
STACK.

13. Construct a trie structure (as in Table 11-6) for the following set of words:

> ALLOCATE, BEGIN, BY, CALL, CHECK, CLOSE, DCL, DO, ELSE, END, ENTRY, EXIT, FLOW, FORMAT, FREE, GET, GO, GOTO, IF, NO, OPEN, PROC, PUT, RETURN, STOP, THEN, TO, WHILE, and WRITE.

14. Construct an algorithm for the insertion of an element into the trie structure which is organized as in Table 11-6.

15. Repeat exercise 14 for the deletion of an element from a trie structure.

Index